Springer Monographs in Mathematics

Springer
Berlin
Heidelberg
New York
Barcelona
Budapest
Hong Kong
London
Milan
Paris
Santa Clara
Singapore
Tokyo

Thierry Aubin

Some Nonlinear Problems in Riemannian Geometry

Springer

Thierry Aubin
University of Paris VI
Mathematiques
4, Place Jussieu, Boite 172
F-75252 Paris
France

Library of Congress Cataloging-in-Publication Data

Aubin, Thierry.
Some nonlinear problems in Riemannian geometry / Thierry Aubin.
p. cm. -- (Springer monographs in mathematics)
Includes bibliographical references and index.
ISBN 3-540-60752-8 (hardcover)
1. Geometry, Riemannian. 2. Nonlinear theories. I. Title.
II. Series.
QA649.A833 1998
516.3'73--dc21 98-4150
CIP

Mathematics Subject Classification (1991): 35, 53, 58

ISBN 3-540-60752-8 Springer-Verlag Berlin Heidelberg New York

Typesetting: Data conversion by A. Leinz, Karlsruhe
SPIN 10518869 41/3143-5 4 3 2 1 0 – Printed on acid-free paper

Preface

This book is the union of two books: the new edition of the former one "Nonlinear Analysis on Manifolds. Monge-Ampere Equations" (Grundlehren 252 Springer 1982) mixed with a new one where one finds, among other things, up-to-date results on the problems studied in the earlier one, and new methods for solving nonlinear elliptic problems. We will give below successively the prefaces of the two books, and at the end of the volume, the two bibliographies (the references * are new).

A very interesting area of nonlinear partial differential equations lies in the study of special equations arising in Geometry and Physics. This book deals with some important geometric problems that are of interest to many mathematicians and scientists but have only recently been partially solved.

Each problem is explained, up-to-date results are given and proofs are presented. Thus the reader is given access, for each specific problem, to its present status of solution as well as to most up-to-date methods for approaching it.

The book deals with such important subjects as variational methods, the continuity method, parabolic equations on fiber bundles, ideas concerning points of concentration, blowing up technique, geometric and topological methods.

My book "Nonlinear Analysis on Manifolds. Monge–Ampère Equations" (Grundlehren 252) is self-contained, and is an introduction to research in nonlinear analysis on manifolds, a field that was almost unexplored when the book appeared. Ever since then, the field has undergone great development. This new book deals with concrete applications of the knowledge contained in the earlier one.

This book is adressed to researchers and advanced graduate students specializing in the field of partial differential equations, nonlinear analysis, Riemannian geometry, functional analysis and analytic geometry. Its objectives are to deal with some basic problems in Geometry and to provide a valuable tool for the researchers. It will allow readers to apprehend not only the latest results on most topics, but also the related questions, the open problems and the new techniques that have appeared recently. Some may find the pace of presentation rather fast, but ultimately, it represents an economy of time and effort for the reader. In the space of a few pages, for instance, the ideas and methods of proof of an important result may be sketched out completely here, whereas the full details are only to be found dispersed in several very long original articles.

Some problems studied here are not treated in any other book. For instance:

- Very few people know if the remaining cases of the Yamabe problem are really solved. The results were announced ten years ago, but parts of the proofs appeared only recently and in different articles, some not easily available.
- On prescribed scalar curvature. Between the author's first article on the topics in 1976, and the second one in 1991 which poses the problem again, only a few results appeared. Ever since, a lot of results have been proved. The same thing applies to the Nirenberg problem, the Kähler manifolds with $C_1(M) > 0$ and the problem of Einstein metrics. The last chapter of the book deals with a very broad topic, on which there are many books: it is discussed here so that the reader may obtain an idea of the subject.
- About the methods. There are books on the variational method or on topological methods, but is there any book where we can find so many methods together ? Of course it is of advantage, when we attack a problem, to have many methods at one's disposal, and in this book there are also new techniques.

The reader can find most of the backgroung knowledge needed in [*1]. Some additional material is given in Chapter 1.

Chapter 2 is devoted to the Yamabe Problem. Thirty years were necessary to solve it entirely. After a proof with all the details, we will find new proofs which do not use the method advocated by Yamabe (minimizing his functional). The study of the Yamabe functional is not completed. We know very little about $\bar{\mu} = \sup \mu_{[g]}$, where $\mu_{[g]}$ is the inf of the Yamabe functional in the conformal class $[g]$. This problem is related to Einstein metrics.

Chapter 3 is concerned with the problem of prescribing the scalar curvature by a conformal change of metrics. When the manifolds is the sphere (S_n, g_0) endowed with its canonical metric, the problem is very special: we study it in Chapter 4.

Chapter 5 deals with Einstein–Kähler metrics. Although there has been a great progress when $C_1(M) > 0$, not everything is clear yet.

Chapter 6 deals with Ricci curvature. A problem that remains open for the next few years is the existence (or the non-existence) of Einstein metrics on a given manifold.

Lastly, Chapter 7 studies harmonics maps. We present the pioneer article of Eells–Sampson on this topics, then we mention some new results. The subject is very large and is continually developing ; several books would be necessary to cover it!

There are many other interesting subjects, but it is not the ambition of this book to treat all the field of research ! To explain some methods and to apply them is our main aim.

It is my pleasure and privilege to express my deep thanks to my friends Melvyn Berger, Dennis DeTurck, Jerry Kazdan, Albert Milani and Joel Spruck who agreed to read one or two chapters. They suggested some mathematical improvements, and corrected many of my errors in English.

I am also extremely grateful, to Pascal Cherrier, Emmanuel Hebey and Michel Vaugon, who helped me in the preparation of the book.

February 1997 Thierry Aubin

Preface to "Grundlehren 252"

This volume is intended to allow mathematicians and physicists, especially analysts, to learn about nonlinear problems which arise in Riemannian Geometry.

Analysis on Riemannian manifolds is a field currently undergoing great development. More and more, analysis proves to be a very powerful means for solving geometrical problems. Conversely, geometry may help us to solve certain problems in analysis.

There are several reasons why the topic is difficult and interesting. It is very large and almost unexplored. On the other hand, geometric problems often lead to limiting cases of known problems in analysis, sometimes there is even more than one approach, and the already existing theoretical studies are inadequate to solve them. Each problem has its own particular difficulties.

Nevertheless there exist some standard methods which are useful and which we must know to apply them. One should not forget that our problems are motivated by geometry, and that a geometrical argument may simplify the problem under investigation. Examples of this kind are still too rare.

This work is neither a systematic study of a mathematical field nor the presentation of a lot of theoretical knowledge. On the contrary, I do my best to limit the text to the essential knowledge. I define as few concepts as possible and give only basic theorems which are useful for our topic. But I hope that the reader will find this sufficient to solve other geometrical problems by analysis.

The book is intended to be used as a reference and as an introduction to research. It can be divided into two parts, with each part containing four chapters. Part I is concerned with essential background knowledge. Part II develops methods which are applied in a concrete way to resolve specific problems.

Chapter 1 is devoted to Riemannian geometry. The specialists in analysis who do not know differential geometry will find, in the beginning of the chapter, the definitions and the results which are indispensable. Since it is useful to know how to compute both globally and in local coordinate charts, the proofs which we will present will be a good initiation. In particular, it is important to know Theorem 1.53, estimates on the components of the metric tensor in polar geodesic coordinates in terms of the curvature.

Chapter 2 studies Sobolev spaces on Riemannian manifolds. Successively, we will treat density problems, the Sobolev imbedding theorem, the Kondrakov theorem, and the study of the limiting case of the Sobolev imbedding theorem.

These theorems will be used constantly. Considering the importance of Sobolev's theorem and also the interest of the proofs, three proofs of the theorem are given, the original proof of Sobolev, that of Gagliardo and Nirenberg, and my own proof, which enables us to know the value of the norm of the imbedding, an introduction to the notion of best constants in Sobolev's inequalities. This new concept is crucial for solving limiting cases.

In Chapter 3 we will find, usually without proof, a substantial amount of analysis. The reader is assumed to know this background material. It is stated here as a reference and summary of the versions of results we will be using. There are as few results as possible. I choose only the most useful and applicable ones so that the reader does not drown in a host of results and lose the main point. For instance, it is possible to write a whole book on the regularity of weak solution for elliptic equations without discussing the existence of solutions. Here there are six theorems on this topic. Of course, sometimes other will be needed; in those cases there are precise references.

It is obvious that most of the more elementary topics in this Chapter 3 have already been needed in the earlier chapters. Although we do assume prior knowledge of these basic topics, we have included precise statements of the most important concepts and facts for reference. Of course, the elementary material in this chapter could have been collected as a separate "Chapter 0" but this would have been artificial, and probably less useful to the reader. And since we do not assume that the reader knows the material on elliptic equations in Sobolev spaces, the corresponding sections should follow the two first chapters.

Chapter 4 is concerned with the Green's function of the Laplacian on compact manifolds. This will be used to obtain both some regularity results and some inequalities that are not immediate consequences of the facts in Chapter 3.

Chapter 5 is devoted to the Yamabe problem concerning the scalar curvature. Here the concept of best constants in Sobolev's inequalities plays an essential role. We close the chapter with a summary of the status of related problems concerning scalar curvature such as Berger's problem, for which we also use the results from Chapter 2 concerning the limiting case of the Sobolev imbedding theorem.

In Chapter 6 we will study a problem posed by Nirenberg.

Chapter 7 is concerned with the complex Monge–Ampère equation on compact Kählerian manifolds. The existence of Einstein–Kähler metrics and the Calabi conjecture are problems which are equivalent to solving such equations.

Lastly, Chapter 8 studies the real Monge–Ampère equation on a bounded convex set of $\mathbb{R}^n$. There is also a short discussion of the complex Monge–Ampère equation on a bounded pseudoconvex set of $\mathbb{C}^n$.

Throughout the book I have restricted my attention to those problems whose solution involves typical application of the methods. Of course, there are many other very interesting problems. For example, we should at least mention that, curiously, the Yamabe equation appears in the study of Yang–Mills fields, while a corresponding complex version is very close to the existence of complex Einstein–Kähler metrics discussed in Chapter 7.

It is my pleasure and privilege to express my deep thanks to my friend Jerry Kazdan who agreed to read the manuscript from the beginning to end. He suggested many mathematical improvements, and, needless to say, corrected many blunders of mine in this English version. I also have to state in this place my appreciation for the efficient and friendly help of Jürgen Moser and Melvyn Berger for the publication of the manuscript. Pascal Cherrier and Philippe Delanoë deserve special mention for helping in the completion of the text.

May 1982 Thierry Aubin

Contents

Chapter 1

Riemannian Geometry

§1. Introduction to Differential Geometry . . . 1
1.1. Tangent Space . . . 2
1.2. Connection . . . 3
1.3. Curvature . . . 3
§2. Riemannian Manifold . . . 4
2.1. Metric Space . . . 5
2.2. Riemannian Connection . . . 6
2.3. Sectional Curvature. Ricci Tensor. Scalar Curvature . . . 6
2.4. Parallel Displacement. Geodesic . . . 8
§3. Exponential Mapping . . . 9
§4. The Hopf–Rinow Theorem . . . 13
§5. Second Variation of the Length Integral . . . 15
5.1. Existence of Tubular Neighborhoods . . . 15
5.2. Second Variation of the Length Integral . . . 15
5.3. Myers' Theorem . . . 16
§6. Jacobi Field . . . 17
§7. The Index Inequality . . . 18
§8. Estimates on the Components of the Metric Tensor . . . 20
§9. Integration over Riemannian Manifolds . . . 23
§10. Manifold with Boundary . . . 25
10.1. Stokes' Formula . . . 26
§11. Harmonic Forms . . . 26
11.1. Oriented Volume Element . . . 26
11.2. Laplacian . . . 27
11.3. Hodge Decomposition Theorem . . . 29
11.4. Spectrum . . . 31

Chapter 2

Sobolev Spaces

§1. First Definitions . . . 32
§2. Density Problems . . . 33
§3. Sobolev Imbedding Theorem . . . 35
§4. Sobolev's Proof . . . 37

§5. Proof by Gagliardo and Nirenberg . . . 38
§6. New Proof . . . 39
§7. Sobolev Imbedding Theorem for Riemannian Manifolds . . . 44
§8. Optimal Inequalities . . . 50
§9. Sobolev's Theorem for Compact Riemannian Manifolds with Boundary . . . 50
§10. The Kondrakov Theorem . . . 53
§11. Kondrakov's Theorem for Riemannian Manifolds . . . 55
§12. Examples . . . 56
§13. Improvement of the Best Constants . . . 57
§14. The Case of the Sphere . . . 61
§15. The Exceptional Case of the Sobolev Imbedding Theorem . . . 63
§16. Moser's Results . . . 65
§17. The Case of the Riemannian Manifolds . . . 67
§18. Problems of Traces . . . 69

Chapter 3

Background Material

§1. Differential Calculus . . . 70
1.1. The Mean Value Theorem . . . 71
1.2. Inverse Function Theorem . . . 72
1.3. Cauchy's Theorem . . . 72
§2. Four Basic Theorems of Functional Analysis . . . 73
2.1. Hahn–Banach Theorem . . . 73
2.2. Open Mapping Theorem . . . 73
2.3. The Banach–Steinhaus Theorem . . . 73
2.4. Ascoli's Theorem . . . 74
§3. Weak Convergence. Compact Operators . . . 74
3.1. Banach's Theorem . . . 74
3.2. The Leray–Schauder Theorem . . . 74
3.3. The Fredholm Theorem . . . 75
§4. The Lebesgue Integral . . . 75
4.1. Dominated Convergence Theorem . . . 76
4.2. Fatou's Theorem . . . 77
4.3. The Second Lebesgue Theorem . . . 77
4.4. Rademacher's Theorem . . . 77
4.5. Fubini's Theorem . . . 78
§5. The L_p Spaces . . . 78
5.1. Regularization . . . 80
5.2. Radon's Theorem . . . 81
§6. Elliptic Differential Operators . . . 83
6.1. Weak Solution . . . 84
6.2. Regularity Theorems . . . 85
6.3. The Schauder Interior Estimates . . . 88
§7. Inequalities . . . 88
7.1. Hölder's Inequality . . . 88
7.2. Clarkson's Inequalities . . . 89
7.3. Convolution Product . . . 89

7.4. The Calderon–Zygmund Inequality . . . 90
7.5. Korn–Lichtenstein Theorem . . . 90
7.6. Interpolation Inequalities . . . 93
§8. Maximum Principle . . . 96
8.1. Hopf's Maximum Principle . . . 96
8.2. Uniqueness Theorem . . . 96
8.3. Maximum Principle for Nonlinear Elliptic Operator of Order Two . . . 97
8.4. Generalized Maximum Principle . . . 98
§9. Best Constants . . . 99
9.1. Applications to Sobolev's Spaces . . . 100

Chapter 4

Complementary Material

§1. Linear Elliptic Equations . . . 101
1.1. First Nonzero Eigenvalue λ of Δ . . . 101
1.2. Existence Theorem for the Equation $\Delta\varphi = f$ . . . 104
§2. Green's Function of the Laplacian . . . 106
2.1. Parametrix . . . 106
2.2. Green's Formula . . . 107
2.3. Green's Function for Compact Manifolds . . . 108
2.4. Green's Function for Compact Manifolds with Boundary . . . 112
§3. Riemannian Geometry . . . 115
3.1. The First Eigenvalue . . . 115
3.2. Locally Conformally Flat Manifolds . . . 117
3.3. The Green Function of the Laplacian . . . 119
3.4. Some Theorems . . . 123
§4. Partial Differential Equations . . . 125
4.1. Elliptic Equations . . . 125
4.2. Parabolic Equations . . . 129
§5. The Methods . . . 134
§6. The Best Constants . . . 139

Chapter 5

The Yamabe Problem

§1. The Yamabe Problem . . . 145
1.1. Yamabe's Method . . . 146
1.2. Yamabe's Functional . . . 150
1.3. Yamabe's Theorem . . . 150
§2. The Positive Case . . . 152
§3. The First Results . . . 157
§4. The Remaining Cases . . . 160
4.1. The Compact Locally Conformally Flat Manifolds . . . 160
4.2. Schoen's Article . . . 161
4.3. The Dimension 3, 4 and 5 . . . 162
§5. The Positive Mass . . . 164
5.1. Positive Mass Theorem, the Low Dimensions . . . 166

5.2. Schoen and Yau's Article . . . 166
5.3. The Positive Energy . . . 169
§6. New Proofs for the Positive Case ($\mu > 0$) . . . 171
6.1. Lee and Parker's Article . . . 171
6.2. Hebey and Vaugon's Article . . . 171
6.3. Topological Methods . . . 172
6.4. Other Methods . . . 175
§7. On the Number of Solutions . . . 175
7.1. Some Cases of Uniqueness . . . 175
7.2. Particular Cases . . . 176
7.3. About Uniqueness . . . 178
7.4. Hebey–Vaugon's Approach . . . 178
7.5. The Structure of the Set of Minimizers of J . . . 179
§8. Other Problems . . . 179
8.1. Topological Meaning of the Scalar Curvature . . . 179
8.2. The Cherrier Problem . . . 180
8.3. The Yamabe Problem on CR Manifolds . . . 182
8.4. The Yamabe Problem on Non-compact Manifolds . . . 183
8.5. The Yamabe Problem on Domains in $\mathbb{R}^n$ . . . 185
8.6. The Equivariant Yamabe Problem . . . 187
8.7. An Hard Open Problem . . . 188
8.8. Berger's Problem . . . 191

Chapter 6

Prescribed Scalar Curvature

§1. The Problem . . . 194
1.1. The General Problem . . . 194
1.2. The Problem with Conformal Change of Metric . . . 196
§2. The Negative Case when M is Compact . . . 197
§3. The Zero Case when M is Compact . . . 204
§4. The Positive Case when M is Compact . . . 209
§5. The Method of Isometry-Concentration . . . 214
5.1. The Problem . . . 214
5.2. Study of the Sequence $\{\varphi_{q_j}\}$ . . . 216
5.3. The Points of Concentration . . . 218
5.4. Consequences . . . 221
5.5. Blow-up at a Point of Concentration . . . 224
§6. The Problem on Other Manifolds . . . 227
6.1. On Complete Non-compact Manifolds . . . 227
6.2. On Compact Manifolds with Boundary . . . 229
§7. The Nirenberg Problem . . . 230
§8. First Results . . . 231
8.1. Moser's Result . . . 232
8.2. Kazdan and Warner Obstructions . . . 233
8.3. A Nonlinear Fredholm Theorem . . . 235
§9. G-invariant Functions f . . . 238
§10. The General Case . . . 241
10.1. Functions f Close to a Constant . . . 241

10.2. Dimension Two . . . 243
10.3. Dimension $n \geq 3$ . . . 245
10.4. Rotationally Symmetric Functions . . . 247
§11. Related Problems . . . 247
11.1. Multiplicity . . . 247
11.2. Density . . . 248
11.3. The Problem on the Half Sphere . . . 249

Chapter 7

Einstein–Kähler Metrics

§1. Kähler Manifolds . . . 251
1.1. First Chern Class . . . 252
1.2. Change of Kähler Metrics. Admissible Functions . . . 253
§2. The Problems . . . 254
2.1. Einstein–Kähler Metric . . . 254
2.2. Calabi's Conjecture . . . 255
§3. The Method . . . 255
3.1. Reducing the Problem to Equations . . . 255
3.2. The First Results . . . 256
§4. Complex Monge–Ampère Equation . . . 257
4.1. About Regularity . . . 257
4.2. About Uniqueness . . . 258
§5. Theorem of Existence (the Negative Case) . . . 258
§6. Existence of Einstein–Kähler Metric . . . 259
§7. Theorem of Existence (the Null Case) . . . 260
§8. Proof of Calabi's Conjecture . . . 263
§9. The Positive Case . . . 263
§10. A Priori Estimate for $\Delta\varphi$ . . . 263
§11. A Priori Estimate for the Third Derivatives of Mixed Type . . . 266
§12. The Method of Lower and Upper Solutions . . . 267
§13. A Method for the Positive Case . . . 269
§14. The Obstructions When $C_1(M) > 0$ . . . 271
14.1. The First Obstruction . . . 271
14.2. Futaki's Obstruction . . . 272
14.3. A Further Obstruction . . . 272
§15. The C^0-Estimate . . . 273
15.1. Definition of the Functionals $I(\varphi)$ and $J(\varphi)$ . . . 273
15.2. Some Inequalities . . . 274
15.3. The C^0-Estimate . . . 276
15.4. Inequalities for the Dimension $m = 1$ . . . 277
15.5. Inequalities for the Exponential Function . . . 278
§16. Some Results . . . 281
§17. On Uniqueness . . . 285
§18. On Non-compact Kähler Manifolds . . . 288

Chapter 8

Monge–Ampère Equations

§1. Monge–Ampère Equations on Bounded Domains of $\mathbb{R}^n$ 289
1.1. The Fundamental Hypothesis 289
1.2. Extra Hypothesis 290
1.3. Theorem of Existence 291
§2. The Estimates 292
2.1. The First Estimates 292
2.2. C^2-Estimate 293
2.3. C^3-Estimate 296
§3. The Radon Measure $\mathscr{M}(\varphi)$ 301
§4. The Functional $\mathscr{I}(\varphi)$ 306
4.1. Properties of $\mathscr{I}(\varphi)$ 306
§5. Variational Problem 311
§6. The Complex Monge–Ampère Equation 314
6.1. Bedford's and Taylor's Results 314
6.2. The Measure $\mathfrak{M}(\varphi)$ 315
6.3. The Function $\mathfrak{I}(\varphi)$ 315
6.4. Some Properties of $\mathfrak{I}(\varphi)$ 315
§7. The Case of Radially Symmetric Fucntions 316
7.1. Variational Problem 317
7.2. An Open Problem 318
§8. A New Method 318

Chapter 9

The Ricci Curvature

§1. About the Different Types of Curvature 321
1.1. The Sectional Curvature 321
1.2. The Scalar Curvature 322
1.3. The Ricci Curvature 322
1.4. Two Dimension 323
§2. Prescribing the Ricci Curvature 323
2.1. DeTurck's Result 323
2.2. Some Computations 324
2.3. DeTurck's Equations 324
2.4. Global Results 325
§3. The Hamilton Evolution Equation 326
3.1. The Equation 326
3.2. Solution for a Short Time 327
3.3. Some Useful Results 330
3.4. Hamilton's Evolution Equations 333
§4. The Consequences of Hamilton's Work 343
4.1. Hamilton's Theorems 343
4.2. Pinched Theorems on the Concircular Curvature 344
§5. Recent Results 345
5.1. On the Ricci Curvature 345
5.2. On the Concircular Curvature 346

Chapter 10

Harmonic Maps

§1. Definitions and First Results . . . 348
§2. Existence Problems . . . 351
2.1. The Problem . . . 351
2.2. Some Basic Results . . . 352
2.3. Existence Results . . . 354
§3. Problems of Regularity . . . 356
3.1. Sobolev Spaces . . . 356
3.2. The Results . . . 357
§4. The Case of $\partial M \neq \emptyset$ . . . 359
4.1. General Results . . . 359
4.2. Relaxed Energies . . . 360
4.3. The Ginzburg–Landau Functional . . . 361

Bibliography . . . 365

Bibliography* . . . 375

Subject Index . . . 389

Notation . . . 393

Chapter 1

Riemannian Geometry

§1. Introduction to Differential Geometry

1.1 A *manifold* M_n, of dimension n, is a Haussdorff topological space such that each point of M_n has a neighborhood homeomorphic to $\mathbb{R}^n$. Thus a manifold is locally compact and locally connected. Hence a connected manifold is pathwise connected.

1.2 A *local chart* on M_n is a pair (Ω, φ), where Ω is an open set of M_n and φ a homeomorphism of Ω onto an open set of $\mathbb{R}^n$.

A collection $(\Omega_i, \varphi_i)_{i\in I}$ of local charts such that $\bigcup_{i\in I}\Omega_i = M_n$ is called an *atlas*. The *coordinates* of $P \in \Omega$, related to φ, are the coordinates of the point $\varphi(P)$ of $\mathbb{R}^n$.

1.3 An atlas of class C^k (respectively, C^∞, C^ω) on M_n is an atlas for which all changes of coordinates are C^k (respectively, C^∞, C^ω). That is to say, if $(\Omega_\alpha, \varphi_\alpha)$ and $(\Omega_\beta, \varphi_\beta)$ are two local charts with $\Omega_\alpha \cap \Omega_\beta \neq \emptyset$, then the map $\varphi_\alpha \circ \varphi_\beta^{-1}$ of $\varphi_\beta(\Omega_\alpha \cap \Omega_\beta)$ onto $\varphi_\alpha(\Omega_\alpha \cap \Omega_\beta)$ is a diffeomorphism of class C^k (respectively, C^∞, C^ω).

1.4 Two atlases of class C^k on M_n $(U_i, \varphi_i)_{i\in I}$ and $(W_\alpha, \psi_\alpha)_{\alpha\in\Lambda}$ are said to be equivalent if their union is an atlas of class C^k.
By definition, a *differentiable manifold* of class C^k (respectively, C^∞, C^ω) is a manifold together with an equivalence class of C^k atlases, (respectively, C^∞, C^ω).

1.5 A mapping f of a differentiable manifold C^k: W_p into another M_n, is called differentiable C^r $(r \leq k)$ at $P \in U \subset W_p$ if $\psi \circ f \circ \varphi^{-1}$ is differentiable C^r at $\varphi(P)$, and we define the *rank* of f at P to be the rank of $\psi \circ f \circ \varphi^{-1}$ at $\varphi(P)$. Here (U, φ) is a local chart of W_p and (Ω, ψ) a local chart of M_n with $f(P) \in \Omega$.

A C^r differentiable mapping f is an *immersion* if the rank of f is equal to p for every point P of W_p. It is an *imbedding* if f is an injective immersion such that f is a homeomorphism of W_p onto $f(W_p)$ with the topology induced from M_n.

1.1. Tangent Space

1.6 Let (Ω, φ) be a local chart and f a differentiable real-valued function defined on a neighborhood of $P \in \Omega$. We say that f is flat at P if $d(f \circ \varphi^{-1})$ is zero at $\varphi(P)$.

A *tangent vector* at $P \in M_n$ is a map $X: f \to X(f) \in \mathbb{R}$ defined on the set of functions differentiable in a neighborhood of P, where X satisfies:

(a) If $\lambda, \mu \in \mathbb{R}$, $X(\lambda f + \mu g) = \lambda X(f) + \mu X(g)$.
(b) $X(f) = 0$, if f is flat.
(c) $X(fg) = f(P)X(g) + g(P)X(f)$; this follows from (a) and (b).

1.7 The *tangent space* $T_P(M)$ at $P \in M_n$ is the set of tangent vectors at P. It has a natural vector space structure. In a coordinate system $\{x^i\}$ at P, the vectors $(\partial/\partial x^i)_P$ defined by $(\partial/\partial x^i)_P\,(f) = [\partial(f \circ \varphi^{-1})/\partial x^i]_{\varphi(P)}$, form a basis.

The tangent space $T(M)$ is $\bigcup_{P \in M} T_p(M)$. It has a natural vector fiber bundle structure. If $T_P^*(M)$ denotes the dual space of $T_P(M)$, the cotangent space is $T^*(M) = \bigcup_{P \in M} T_P^*(M)$. Likewise, the fiber bundle $T_s^r(M)$ of the tensor of type (r, s) is $\bigcup_{P \in M} \overset{r}{\otimes} T_P(M) \overset{s}{\otimes} T_P^*(M)$.

1.8 Let $P \in M_n$ and Φ be a differentiable map of M_n into W_p. Set $Q = \Phi(P)$. The map Φ induces a linear map Φ_* of the tangent space $T_P(M)$ into $T_Q(W)$ defined by $(\Phi_* X)(f) = X(f \circ \Phi)$, where $X \in T_p(M)$ and f is a differentiable function in a neighborhood of Q. We call Φ_* the *linear tangent mapping* of Φ.

By duality, we define the *linear cotangent mapping* Φ^* of $T^*(W)$ into $T^*(M)$ as follows: $T_Q^*(W) \ni \omega \to \Phi^*(\omega) \in T_P^*(M)$, which satisfies

$$\langle \Phi^*(\omega), X \rangle = \langle \omega, \Phi_*(X) \rangle, \text{ for all } X \in T_P(M).$$

One verifies easily that $\Phi^*(df) = d(f \circ \Phi)$.

1.9 A differentiable vector field is a section of $T(M)$. A section of vector fiber bundle (E, π, M) is a differentiable map ξ of M into E, such that $\pi \circ \xi =$ identity. If $E = T(M)$, π is the mapping of E onto M: $T_P(M) \ni X \to P$.

The *bracket* $[X, Y]$ of two vector fields X and Y is the vector field defined by

$$[X, Y](f) = X[Y(f)] - Y[X(f)].$$

A *differentiable tensor field* of type (r, s) is a section of $T_s^r(M)$.

1.10 An exterior differential *p-form* η is a section of $\Lambda^p T^*(M)$. In a local chart

$$\eta = \sum_{j_1 < j_2 < \cdots < j_p} a_{j_1 \cdots j_p}\, dx^{j_1} \wedge dx^{j_2} \wedge \cdots \wedge dx^{j_p},$$

and the exterior differentiation $d\eta$ of η is

$$d\eta = \sum_{j_1 < j_2 < \cdots < j_p} da_{j_1 \cdots j_p} \wedge dx^{j_1} \wedge \cdots \wedge dx^{j_p}.$$

Clearly $dd\eta = 0$.

Denote by $\Gamma_1(M)$ the space of differentiable vector fields and by $\Lambda^p(M)$ the space of exterior differential p-forms. For $\alpha \in \Lambda^p(M)$ and $\beta \in \Lambda^q(M)$, $d(\alpha \wedge \beta) = d\alpha \wedge \beta + (-1)^p \alpha \wedge d\beta$, as it is easy to verify.

1.2. Connection

1.11 A *connection* is a map D (called the covariant derivative) of $T(M) \times \Gamma_1(M)$ into $T(M)$ such that:

(a) $D(X_P, Y) = D_{X_P}(Y) \in T_P(M)$ when $X_P \in T_P(M)$.
(b) For any $P \in M$, the restriction of D to $T_P(M) \times \Gamma_1(M)$ is bilinear.
(c) If f is a differentiable function

$$D_{X_P}(fY) = X_P(f)Y + f(P)D_{X_P}(Y).$$

(d) If X and Y belong to $\Gamma_1(M)$, X of class C^r and Y of class C^{r+1}, then $D_X Y$ is of class C.

In a local chart (Ω, φ), denote $\nabla_i Y = D_{\partial/\partial x^i} Y$. Conversely, if we are given, for all pairs (i, j),

$$\nabla_i\left(\frac{\partial}{\partial x_j}\right) = \Gamma_{ij}^k \frac{\partial}{\partial x^k},$$

then a unique connection D is defined.

The functions Γ_{ij}^k are called the *Christoffel symbols* of the connection D with respect to the local coordinate system $x^1, \ldots, x^n$.

1.12 The *torsion* of the connection is the map of $\Gamma_1 \times \Gamma_1$ into Γ_1 defined by

$$T(X, Y) = D_X Y - D_Y X - [X, Y].$$

$T^k(\partial/\partial x^i, \partial/\partial x^j) = \Gamma_{ij}^k - \Gamma_{ji}^k$ are the components of a tensor.

1.3. Curvature

1.13 The *curvature* of the connection is the 2-form with values in $\mathrm{Hom}(\Gamma_1, \Gamma_1)$ defined by:

$$R(X, Y) = D_X D_Y - D_Y D_X - D_{[X, Y]}.$$

One verifies that $R(X, Y)Z$ at P depends only upon the values of X, Y, and Z at P.

In a local chart, denote by R^l_{kij} the lth component of $R(\partial/\partial x^i, \partial/\partial x^j)\partial/\partial x^k$. R^l_{kij} are the components of a tensor, called the curvature tensor, and

$$(1) \qquad R^l_{kij} Z^k = \nabla_i \nabla_j Z^l - \nabla_j \nabla_i Z^l.$$

It follows that

$$R^l_{kij} = \partial_i \Gamma^l_{jk} - \partial_j \Gamma^l_{ik} + \Gamma^l_{im} \Gamma^m_{jk} - \Gamma^l_{jm} \Gamma^m_{ik}.$$

1.14 The definition of *covariant derivative* extends to differentiable tensor fields as follows:

(a) For functions, $D_X f = X(f)$.
(b) D_X preserves the type of the tensor.
(c) D_X commutes with the contraction.
(d) $D_X(u \otimes v) = (D_X u) \otimes v + u \otimes (D_X v)$, where u and v are tensor fields.

For simplicity, we set $\nabla_{\alpha_1 \alpha_2 \cdots \alpha_l} u = \nabla_{\alpha_1} \nabla_{\alpha_2} \cdots \nabla_{\alpha_l} u$.

§2. Riemannian Manifold

1.15 A C^∞ *Riemannian manifold* is a pair (M_n, g), where M_n is a C^∞ differentiable manifold and g a C^∞ Riemannian metric. A Riemannian metric is a twice-covariant tensor field g (that is to say, a section of $T^*(M) \otimes T^*(M)$), such that at each point $P \in M$, g_P is a positive definite bilinear symmetric form:

$$g_P(X, Y) = g_P(Y, X) \quad \text{and} \quad g_P(X, X) > 0 \quad \text{if } X \neq 0.$$

Hereafter, unless otherwise stated, a Riemannian manifold M_n is a connected C^∞ Riemannian manifold of dimension n.

1.16 Theorem. *On a paracompact C^∞ differentiable manifold, there exists a C^∞ Riemannian metric g.*

Proof. Let $(\Omega_i, \varphi_i)_{i \in I}$ be an atlas and $\{\alpha_i\}$ a C^∞ partition of unity subordinate to the covering $\{\Omega_i\}$. Such $\{\alpha_i\}$ exists since the manifold M_n is paracompact. Set $\mathscr{E} = (\mathscr{E}_{jk})$ be the Euclidean metric on $\mathbb{R}^n$ (in an orthonormal basis $\mathscr{E}_{jk} = \delta^k_j$, Kronecker's symbol). Then $g = \sum_{i \in I} \alpha_i \varphi^*_i(\mathscr{E})$ is a Riemannian metric on M_n, as one can easily verify. ■

For an alternate proof of Theorem (1.16) one can also use *Whitney's theorem* and give M_n the imbedded metric. Whitney's theorem asserts that every differentiable manifold M_n has an immersion in $\mathbb{R}^{2n}$ and an imbedding in $\mathbb{R}^{2n+1}$. So let Φ be an imbedding of M_n in $\mathbb{R}^{2n+1}$. On M_n define the Riemannian metric g by $g(X, Y) = \mathscr{E}(\Phi_* X, \Phi_* Y)$. This is the metric induced by the imbedding.

Let $\{x_i\}$, $(i = 1, 2, \ldots, n)$, be a local coordinate system at a point $P \in M_n$ and $\{y^\alpha\}$ $(\alpha = 1, 2, \ldots, 2n + 1)$ the coordinates of $\Phi(P) \in \mathbb{R}^{2n+1}$. The components of g can be expressed as follows:

$$g_{ij} = \sum_{\alpha=1}^{2n+1} \frac{\partial y^\alpha}{\partial x^i} \frac{\partial y^\alpha}{\partial x^j}.$$

By definition, g^{ij} are the components of the inverse matrix of the metric matrix $((g_{ij}))$: $g_{ij} g^{ik} = \delta_j^k$.

2.1. Metric Space

1.17 Definition. Let C be a differentiable curve in a Riemannian manifold (M_n, g): $\mathbb{R} \supset [a, b] \ni t \to C(t) \in M_n$ with C differentiable (namely the restriction of a differentiable mapping of a neighborhood of $[a, b]$ into M_n).

Define the arc length of C by:

$$\text{(2)} \qquad L(C) = \int_a^b \sqrt{g_{C(t)}\left(\frac{dC}{dt}, \frac{dC}{dt}\right)}\, dt = \int_a^b \sqrt{g_{ij}[C(t)] \frac{dC^i}{dt} \frac{dC^j}{dt}}\, dt,$$

where $C^i(t)$ are the coordinates of $C(t)$ in a local chart, and dC^i/dt the components of the tangent vector at C: $dC/dt = C_*(\partial/\partial t)$, $\partial/\partial t$ being the unit vector of $\mathbb{R}$.

One verifies easily that the definition of $L(C)$ makes sense; the integral depends neither on the local chart, nor on a change of parametrization $s = s(t)$ with $ds/dt \neq 0$. Henceforth we suppose that the manifold is connected. This implies that it is pathwise connected. Two points P and Q of M_n are the endpoints of a differentiable curve. Indeed, a continuous curve from P to Q is covered by a finite number of open sets Ω_i homeomorphic to $\mathbb{R}^n$, and in each Ω_i one replaces the continuous curve by a differentiable one.

Set $d(P, Q) = \inf L(C)$ for all differentiable curves from P to Q.

1.18 Theorem. *$d(P, Q)$ defines a distance on M_n, and the topology determined by d is equivalent to the topology of M_n as a manifold.*

Proof. Clearly $d(P, Q) = d(Q, P)$ and $d(P, Q) \leq d(P, R) + d(R, Q)$. Since $d(P, P) = 0$, the only point remaining to be proved is that $d(P, Q) = 0 \Rightarrow P = Q$. Assume that $P \neq Q$ and let (Ω, φ) be a local chart with $\varphi(P) = 0$, $P \in \Omega$. There exists a ball of radius r, $B_r \subset \mathbb{R}^n$, with center 0, such that $\bar{B}_r \subset \varphi(\Omega)$ and $Q \notin \varphi^{-1}(\bar{B}_r)$. At a point M define $\lambda(M) = \inf_{\|\xi\|=1} g_M(\xi, \xi)$ and $\mu(M) = \sup_{\|\xi\|=1} g_M(\xi, \xi)$, where $\|\xi\|$ is the Euclidean norm of $\xi \in R^n$ and

$\tilde{\xi} = \varphi_*^{-1}\xi$. $\lambda(M)$ and $\mu(M)$ are strictly positive real numbers, because the sphere $\mathbb{S}_{n-1}(1)$ is compact. Clearly $\lambda = \inf \lambda(M)$ and $\mu = \sup \mu(M)$ for $M \in \varphi^{-1}(\bar{B}_r)$, satisfy $0 < \lambda \leq \mu < \infty$, because $\lambda(M)$ and $\mu(M)$ are strictly positive continuous functions on the compact set $\varphi^{-1}(\bar{B}_r)$.
Let Γ be the connected component in $\varphi^{-1}(\bar{B}_r)$ of P on the curve C from P to Q. Γ has P and R as extremities, $C(a) = P$, $C(b) = R$. We have

$$L(C) \geq L(\Gamma) = \int_a^b \sqrt{g_{ij}\frac{dC^i}{dt}\frac{dC^j}{dt}}\, dt \geq \lambda \int_a^b \left\| \frac{d(\varphi \circ C)}{dt} \right\| dt \geq \lambda r,$$

since the arc length of $\varphi(\Gamma)$ is at least r.

Therefore $d(P, Q) > 0$ if $P \neq Q$. Setting $S_P(r) = \{Q \in M, d(P, Q) \leq r\}$, we have $S_P(\lambda r) \subset \varphi^{-1}(\bar{B}_r)$, according to the above inequality. Likewise, it is possible to prove: $\varphi^{-1}(\bar{B}_r) \subset S_p(\mu r)$. Hence the topology defined by the distance d is the same as the manifold topology of M. ∎

2.2. Riemannian Connection

1.19 Definition. The Riemannian connection is the unique connection with vanishing torsion tensor, for which the covariant derivative of the metric tensor is zero.

Let us compute the expression of the Christoffel symbols in a local coordinate system. The computation gives a proof of the existence and uniqueness of the Riemannian connection.
The connection having no torsion, $\Gamma_{ij}^k = \Gamma_{ji}^k$. Moreover,

$$\nabla_k g_{ij} = \partial_k g_{ij} - \Gamma_{ki}^l g_{jl} - \Gamma_{kj}^l g_{il} = 0,$$
$$\nabla_i g_{jk} = \partial_i g_{jk} - \Gamma_{ik}^l g_{jl} - \Gamma_{ij}^l g_{kl} = 0,$$
$$\nabla_j g_{ik} = \partial_j g_{ik} - \Gamma_{jk}^l g_{il} - \Gamma_{ji}^l g_{kl} = 0.$$

Taking the sum of the last two equalities minus the first one, we obtain:

$$\Gamma_{ij}^l = \tfrac{1}{2}[\partial_i g_{kj} + \partial_j g_{ki} - \partial_k g_{ij}]g^{kl}. \tag{3}$$

2.3. Sectional Curvature. Ricci Tensor. Scalar Curvature

1.20 Consider the 4-covariant tensor $R(X, Y, Z, T) = g[X, R(Z, T)Y]$ with components $R_{lkij} = g_{lm} R^m_{kij}$. For the definition of the curvature tensor see 1.13. It has the following properties: $R_{ijkl} = -R_{ijlk}$ (by definition), $R_{ijkl} = R_{klij}$, and the *Bianchi identities*: $R_{ijkl} + R_{iklj} + R_{iljk} = 0$,

$$\nabla_m R_{ijkl} + \nabla_k R_{ijlm} + \nabla_l R_{ijmk} = 0. \tag{4}$$

1.21 Definition. $\sigma(X, Y) = R(X, Y, X, Y)$ is the *sectional curvature* of the 2-dimensional subspace of $T(M)$ defined by the vectors X and Y, which are chosen orthonormal (i.e., $g(X, X) = 1$, $g(Y, Y) = 1$, $g(X, Y) = 0$).

1.22 Definition. From the curvature tensor, only one nonzero tensor (or its negative) is obtained by contraction. It is called the *Ricci tensor.* Its components are $R_{ij} = R^k_{ikj}$. The Ricci tensor is symmetric and its contraction $R = R_{ij}g^{ij}$ is called the *scalar curvature.*
The Ricci curvature in the direction of the unit tangent vector $X = \{\xi^i\}$ is $R_{ij}\xi^i\xi^j$.

1.23 Definition. An *Einstein metric* is a metric for which the Ricci tensor and the metric tensor are proportional:

$$R_{ij}(P) = f(P)g_{ij}(P). \tag{5}$$

Contracting this equality, we obtain $f(P) = R(P)/n$, which is a constant when $n \geq 3$. Indeed, if we multiply the second Bianchi identity (4) by g^{jm}, we obtain:

$$\nabla^j R_{ijkl} + \nabla_k R_{il} - \nabla_l R_{ik} = 0,$$

which multiplied by g^{il} results in $\nabla_k R = 2\nabla^i R_{ik}$. But contracting the covariant derivative of (5) gives $\nabla_k R = n\nabla^i R_{ik}$. Hence when $n \neq 2$, the scalar curvature R must be constant.

1.24 Definition. A *normal coordinate system* at $P \in M_n$ is a local coordinate system $\{x^i\}$, for which the components of the metric tensor at P satisfy: $g_{ij}(P) = \delta^j_i$ and $\partial_k g_{ij}(P) = 0$, for all i, j, k (according to 1.19, $\partial_k g_{ij}(P) = 0$ is equivalent to $\Gamma^k_{ij}(P) = 0$).

1.25 Proposition. *At each point P, there exists a normal coordinate system.*

Proof. Let (Ω, φ) be a local chart with $\varphi(P) = 0$, and $\{x^i\}$ the corresponding coordinate system. At first we may choose in $\mathbb{R}^n$ an orthogonal frame, so that $g_{ij}(P) = \delta^j_i$. Then consider the change of coordinates defined by:

$$x^k - y^k = -\tfrac{1}{2}\Gamma^k_{ij}(P)y^i y^j.$$

In the coordinate system $\{y^k\}$, the components of the metric tensor are:

$$g'_{ij}(Q) = g_{kl}(Q)[\delta^k_i - \Gamma^k_{im}(P)y^m][\delta^l_j - \Gamma^l_{jm}(P)y^m],$$

since $\partial x^k/\partial y^i = \delta^k_i - \Gamma^k_{ij}(P)y^j$. Q is a point in the local charts corresponding to $\{x^i\}$ and $\{y^k\}$; $\{y^k\}$ is a coordinate system, according to the inverse function

theorem 3.10, the Jacobian matrix $((\partial x^k/\partial y^i))$ being equal to the unit matrix at P.
The first order in y^i of $g'_{ij}(Q) - g_{ij}(Q)$ is:

$$-[\Gamma^j_{ik}(P)y^k + \Gamma^i_{jk}(P)y^k] = -\left(\frac{\partial g_{ij}}{\partial x^k}\right)_P y^k = -\left(\frac{\partial g_{ij}}{\partial y^k}\right)_P y^k.$$

Hence $(\partial g'_{ij}/\partial y^k)_P = 0$ and all $\Gamma'^j_{ik}(P)$ are zero. ■

2.4. Parallel Displacement. Geodesic

1.26 Definition. Let $C(t)$ be a differentiable curve. A vector field X is said to be *parallel* along C if its covariant derivative in the direction of the tangent vector to C is zero. Letting $X(t) = X(C(t))$:

$$D_{dC(t)/dt}X(t) = \frac{dC^i(t)}{dt}\nabla_i X(t)$$

$$= \frac{dC^i(t)}{dt}[\partial_i X^j(t) + \Gamma^j_{ik}(C(t))X^k(t)]\frac{\partial}{\partial x^j} = 0.$$

Thus $X(t)$ is a parallel vector field along C if, in a local chart:

$$\frac{dX^j}{dt} + \Gamma^j_{ik}X^k\frac{dC^i}{dt} = 0. \tag{6}$$

1.27 Definition. Let P and Q be two points of M_n, $C(t)$ a differentiable curve from P to Q, $(C(a) = P, C(b) = Q)$, and X_0 a vector of $T_P(M)$.
According to Cauchy's theorem, 3.11, the initial value problem $X(a) = X_0$, of Equation (6), has a unique solution $X(t)$ defined for all $t \in [a, b]$ since (6) is linear. The vector $X(b)$ of this parallel vector field along C (with $X(a) = X_0$) is called the *parallel translate vector* of X_0 from P to Q along $C(t)$.

1.28 Definition. A differentiable curve $C(t)$ of class C^2 is a *geodesic* if its field of tangent vectors is parallel along $C(t)$. Thus $C(t)$ is a geodesic if and only if

$$\frac{d^2C^j(t)}{dt^2} + \Gamma^j_{ik}(C(t))\frac{dC^i(t)}{dt}\frac{dC^k(t)}{dt} = 0, \tag{7}$$

according to (6) with $X = dC/dt$.

Applying Cauchy's theorem, 3.11, to Equation (7) yields:

1.29 Proposition. *Given $P \in M_n$ and $X \in T_P(M)$, $X \neq 0$, there exists a unique geodesic, starting at P, such that X is its tangent vector at P. This geodesic depends differentiably on the initial conditions P and X.*

§3. Exponential Mapping

1.30 Let (Ω, φ) be a local chart related to a normal coordinate system $\{x^i\}$ at $P \in \Omega$, X a tangent vector of $T_P(M)$, $X = (\xi^1, \xi^2, \ldots, \xi^n) \neq 0$, and $C^i(t)$ the coordinates of the point $C(t)$, belonging to the geodesic defined by the initial conditions $C(0) = P$, $(dC/dt)_{t=0} = X$. $C(t)$ is defined for the values of t satisfying $0 \leq t < \beta$ (β given by the Cauchy theorem). Since

$$g_{ij}(C(t)) \frac{dC^i(t)}{dt} \frac{dC^j(t)}{dt}$$

is constant along C (the covariant derivative along C of each of the three terms is zero), s, the parameter of arc length, is proportional to t: $s = \|X\| t$. $C^i(t)$ are C^∞ functions not only of t, but also of the initial conditions. We may consider $C^i(t, x^1, x^2, \ldots, x^n, \xi^1, \xi^2, \ldots, \xi^n)$. According to the Cauchy theorem 3.11, β may be chosen valid for initial conditions in an entire open set, for instance for $P \in \varphi^{-1}(B_r)$ and $\|X\| < \alpha$, ($B_r \subset \varphi(\Omega)$ being a ball of radius $r > 0$, and $\alpha > 0$).
It is easy to verify that $C(t, \lambda X) = C(\lambda t, X)$ for all λ, when one of the two numbers exists. Thus in all cases, if α is small enough, we may assume $\beta > 1$, without loss of generality. By Taylor's formula:

$$C^i(t, \xi^1, \xi^2, \ldots, \xi^n) = x^i + t\xi^i + t^2\psi^i(t, \xi^1, \xi^2, \ldots, \xi^n).$$

1.31 Theorem. *The exponential mapping*: $\exp_P(X)$, *defined by*: $\mathbb{R}^n \supset \Theta \ni X \to C(1, P, \tilde{X}) \in M_n$ *is a diffeomorphism of Θ (a neighborhood of zero, where the mapping is defined) onto a neighborhood of P. By definition* $\exp_P(0) = P$, *and the identification of $\mathbb{R}^n$ with $T_P(M)$ is made by means of φ_*: $\tilde{X} = (\varphi_*^{-1})_P X$, ($\varphi$ is introduced in* 1.30).

Proof. $\exp_P(X)$ is a C^∞ map of a neighborhood of $0 \in \mathbb{R}^n$ into M_n. This follows from 1.30 (β may be chosen greater than 1). At P the Jacobian matrix of this map is the unit matrix; then, according to the inverse function theorem 3.10, the exponential mapping is locally a diffeomorphism: $\xi^1, \xi^2, \ldots, \xi^n$ can be expressed as functions of $C^1, C^2, \ldots, C^n$. ∎

1.32 Corollary. *There exists a neighborhood Ω of P, such that every point $Q \in \Omega$ can be joined to P by a unique geodesic entirely included in Ω. $(\Omega, \exp_P^{-1})$ is a local chart and the corresponding coordinate system is called a normal geodesic coordinate system.*

Proof. Let $\{\xi^i\}$ be the coordinates of a point $Q \in \Omega$, and $C(t) = \{C^i(t)\}$ the geodesic from P to Q lying in Ω. $C^i(t) = t\xi^i$, for $t \in [0, 1]$. Since the arc length $s = \|X\|t$,

$$(8) \qquad g_{ij}(Q)\xi^i\xi^j = \sum_{i=1}^{n} (\xi^i)^2 = \|X\|^2.$$

The length of the geodesic from P to Q is $\|X\|$.
Since $C(t)$ is a geodesic, by (1.28) we conclude that

$$\Gamma^i_{jk}[C(t)]\xi^j\xi^k = 0.$$

Letting $t \to 0$, we have $\Gamma^i_{jk}(P)\xi^j\xi^k = 0$ for all $\{\xi^i\}$. Thus $\Gamma^i_{jk}(P) = 0$; all Christoffel symbols are zero at P. ∎

1.33 Proposition. *Every geodesic through P is perpendicular to $\sum_P(r)$, the subset of the points $Q \in \Omega$ satisfying $\sum_{i=1}^n (\xi^i)^2 = r^2$, with r small enough (ξ^i are geodesic coordinates of Q).*

Proof. Let $Q \in \sum_P(r) \subset \Omega$. Choose an orthonormal frame of $\mathbb{R}^n$ such that the geodesic coordinates of Q are $\xi^1 = r$ and $\xi^2 = \xi^3 = \cdots = \xi^n = 0$.
We are going to prove that $g_{1i}(Q) = \delta^i_1$ for all i; thus the desired result will be established, because a vector in Q tangent to $\sum_P(r)$ has a zero first component; (if $\gamma(u)$ is a differentiable curve in $\sum_P(r)$ through Q, $\sum_{i=1}^n \gamma^i(u) \times (d\gamma^i(u)/du) = 0$, and that implies $d\gamma^1(u)/du = 0$ at Q).

Clearly, by (8), $g_{11}(Q) = 1$. Differentiation of (8) with respect to ξ^k yields:

$$\partial_k g_{ij}(Q)\xi^i\xi^j + 2g_{ik}(Q)\xi^i = 2\xi^k.$$

Hence, at Q, if $k \neq 1$:

$$r\partial_k g_{11}(r) + 2g_{1k}(r) = 0,$$

where $g_{ij}(r)$ are the components of g at the point with coordinates $\xi^1 = r$, $\xi^i = 0$ for $i > 1$. Moreover, $\Gamma^h_{ij}(r)\xi^i\xi^j = 0$ for all h leads to

$$2\partial_r g_{1k}(r) = \partial_k g_{11}(r).$$

Thus $g_{1k}(r) + r\partial_r g_{1k}(r) = 0$, $(\partial/\partial r)[rg_{1k}(r)] = 0$, and $rg_{1k}(r)$ is constant along the geodesic from P to Q, so

$$g_{1k}(Q) = 0, \quad \text{for } k \neq 1. \qquad ∎$$

1.34 Definition. C is called a *minimizing curve* from P to Q if $L(C) = d(P, Q)$. See 1.17 for the definition of $L(C)$ and $d(P, Q)$.

1.35 Proposition. *A minimizing differentiable curve C from P to Q is a geodesic.*

Proof. Consider C parametrized by arc length s ($[0, \tau] \ni s \to C(s) \in C$), and suppose that $C(s)$ is of class C^2 and lies in a chart (Ω, φ).
Let $\Gamma_\lambda(s)$ be a C^2 differentiable curve from P to Q close to C, defined by $\gamma^i(s) = C^i(s) + \lambda\xi^i(s)$, with $\xi^i(0) = \xi^i(\tau) = 0$ for all i, and λ small enough. The first variation of arc length $L(\Gamma_\lambda)$ at $\lambda = 0$, $(dL_\lambda(\Gamma)/d\lambda)_{\lambda=0}$ must be zero. Now

$$\left(\frac{dL_\lambda(\Gamma)}{d\lambda}\right)_{\lambda=0} = \int_0^\tau \left[\tfrac{1}{2}\partial_k g_{ij}(C(s))\xi^k \frac{dC^i}{ds}\frac{dC^j}{ds} + g_{ij}(C(s))\frac{d\xi^i}{ds}\frac{dC^j}{ds}\right] ds.$$

Integration by parts of the second term, using (3), leads to Euler's equation:

$$\text{(9)} \qquad \frac{d^2C^i}{ds^2} + \Gamma^i_{jk}(C(s))\frac{dC^j}{ds}\frac{dC^k}{ds} = 0.$$

Hence C is a geodesic. Moreover, let $S_P(r) = \{Q \in M_n, d(P, Q) \le r\}$. According to Corollary 1.32, there exists an r_0 such that every point of $S_P(r_0)$ can be joined to P by a unique geodesic lying in $S_P(r_0)$. If we suppose $Q \in S_P(r_0)$, this unique geodesic is the minimizing curve C, and its length is $d(P, Q)$. Indeed, any other curve from P to Q does not satisfy Euler's equation (9) if it is included in $S_P(r_0)$; if it is not its length is greater than r_0, and then it cannot be minimizing. Thus we have proved that a minimizing curve C is a geodesic, because, if at a point P the Euler equation (9) were not satisfied, the above result would imply C is not minimizing. Finally, a C^1 differentiable minimizing curve must satisfy (9) as a distribution and will be C^∞ (Theorem 3.54) by induction. ∎

1.36 Theorem. *There exists $\delta(P)$, a strictly positive continuous function on M, such that every point Q satisfying $d(P, Q) \le \delta(P)$ can be joined to P by a unique geodesic of length $d(P, Q)$. Moreover $\delta(P)$ can be chosen so that $S_P(\delta(P))$ is a convex neighborhood: every pair (Q, T) of points of $S_P(\delta(P))$ can be joined by a unique minimizing geodesic lying in $S_P(\delta(P))$.*

Proof. a) According to 1.30 and Corollary 1.32, for every $P \in M_n$ there exists $\alpha(P) > 0$, such that each $Q = \exp_P X$ with $\|X\| < \alpha(P)$ can be joined to P by a unique geodesic C included in Ω, where $(\Omega, \exp_P^{-1})$ is the chart related to the normal geodesic coordinate system. We have to prove that the length of this geodesic is $d(P, Q)$.

Let $\{\xi^i\}$ be the geodesic coordinates of Q. We suppose that $\xi^1 = r$ and $\xi^i = 0$ for $i > 1$. The equation of C is $[0, 1] \ni t \to C^i(t) = t\xi^i$, its length is r.

Consider $\gamma(t)$, $t \in [0, 1]$ a differentiable curve from P to Q lying in Ω. Its length is

$$\ell = \int_0^1 \sqrt{g_{ij}(\gamma(t)) \frac{d\gamma^i}{dt} \frac{d\gamma^j}{dt}}\, dt.$$

According to Proposition 1.33, if (ρ, θ) are geodesic polar coordinates, $ds^2 = (d\rho)^2 + \rho^2 g_{\theta_i \theta_j}\, d\theta^i\, d\theta^j$; therefore

$$g_{ij}[\gamma(t)] \frac{d\gamma^i}{dt} \frac{d\gamma^j}{dt} \geq \left(\frac{d\rho}{dt}\right)^2 = \left[\frac{d\|\gamma(t)\|}{dt}\right]^2.$$

Hence

$$\ell \geq \int_0^1 \left| \frac{d\|\gamma(t)\|}{dt} \right| dt \geq \int_0^1 \frac{d\|\gamma(t)\|}{dt}\, dt = \|\gamma(1)\| = r.$$

Consequently $r = d(P, Q)$.

Thus there exists an α such that $\exp_P X$ is a diffeomorphism of a ball with center 0 and radius α: $B_\alpha \subset \mathbb{R}^n$ onto $S_P(\alpha)$. Also, every geodesic through P is minimizing in $S_P(\alpha)$.

b) By 1.31, consider the following differentiable map ψ defined on a neighborhood of $(P, 0) \in T(M)$:

$$T(M) \ni (Q, \tilde{X}) \to (Q, \exp_Q \varphi_* \tilde{X}) \in M \times M.$$

The Jacobian matrix of ψ at $(P, 0)$ is invertible; thus, by the inverse function Theorem 3.10. The restriction of ψ to a neighborhood Θ of $(P, 0)$ in $T(M)$ is a diffeomorphism onto $\psi(\Theta)$. This result allows us to choose $\delta(P)$ to be continuous. Moreover, we choose Θ as $S_P(\beta) \times B_\beta$, with β small enough (in particular $\beta \leq \alpha/2$).
Pick λ small enough so that $S_P(\lambda) \times S_P(\lambda) \subset \psi(\Theta)$. Then $S_P(\lambda)$ is a neighborhood of P such that every pair (Q, T) of points belonging to $S_P(\lambda)$ can be joined by a geodesic.
Since $\lambda \leq \beta \leq \alpha/2$, the length of this geodesic is not greater than β. Thus it is included in $S_P(\alpha)$, and is minimizing and unique.

c) Let us prove that this geodesic γ is included in $S_P(\lambda)$ for λ small enough. Denote by R, the (or a) point of γ, whose distance to P is maximum. If R is not Q or T, $\|\gamma(t)\|^2 = \sum_{i=1}^n [\gamma^i(t)]^2$ has a maximum at R for $t = t_0$. Thus its second derivative at t_0 is less than or equal to zero:

$$\sum_{i=1}^n \gamma^i(t_0) \frac{d^2\gamma^i}{dt^2}(t_0) + \sum_{i=1}^n \left(\frac{d\gamma^i}{dt}(t_0)\right)^2 \leq 0.$$

Since γ is a geodesic,

$$\frac{d^2\gamma^i(t_0)}{dt^2} + \Gamma^i_{jk}(R)\frac{d\gamma^j(t_0)}{dt}\frac{d\gamma^k(t_0)}{dt} = 0.$$

Multiplying by $\gamma^i(t_0)$ and summing over i leads to:

$$\left[g_{jk}(P) - \sum_{i=1}^{n}\gamma^i(t_0)\Gamma^i_{jk}(R)\right]\frac{d\gamma^j}{dt}\frac{d\gamma^k}{dt} \leq 0,$$

since $\sum_{i=1}^n (d\gamma^i/dt)^2 = g_{jk}(d\gamma^j/dt)(d\gamma^k/dt)$.
But this inequality is impossible, if λ is small enough, because when $\lambda \to 0$, $R \to P$, $\gamma^i(t_0) \to 0$, and $\Gamma^i_{jk}(R) \to 0$. Hence, for λ small enough, R is Q or T, and $\gamma \subset S_P(\lambda)$. ∎

§4. The Hopf–Rinow Theorem

1.37. *The following four propositions are equivalent:*

(a) *The Riemannian manifold M is complete as a metric space.*
(b) *For some point $P \in M$, all geodesics from P are infinitely extendable.*
(c) *All geodesics are infinitely extendable.*
(d) *All bounded closed subsets of M are compact.*

Moreover, we also have the following:

1.38 Theorem. *If M is connected and complete, then any pair (P, Q) of points of M can be joined by a geodesic arc whose length is equal to $d(P, Q)$.*

Proof. a) ⇒ b) and c).
Let $P \in M$ and a geodesic $C(s)$ through P be defined for $0 \leq s < L$, where s is the canonical parameter of arc length. Consider s_p, an increasing sequence converging to L, and set $x_p = C(s_p)$. We have $d(x_p, x_q) \leq |s_p - s_q|$. Hence $\{x_p\}$ is a Cauchy sequence in M, and it converges to a point, say Q, which does not depend on the sequence $\{s_p\}$.
Applying Theorem 1.31 at Q, we prove that the geodesic can be extended for all values of s such that $L \leq s < L + \varepsilon$ for some $\varepsilon > 0$.

Proof. b) ⇒ d) and Theorem 1.38.
Denote by $E_P(r)$ the subset of the points $Q \in S_P(r)$, such that there exists a minimizing geodesic from P to Q. Recall $S_P(r) = \{Q \in M, d(P, Q) \leq r\}$.
We are going to prove that $E(r) = E_P(r)$ is compact and is the same as $S(r) = S_P(r)$.

Let $\{Q_i\}$ be a sequence of points in $E(r)$, $\tilde{X}_i$ (with $\|X_i\| = 1$ (recall $X_i = \varphi_* \tilde{X}_i$)) the corresponding tangent vectors at P to the minimizing geodesic (or one of them) from P to Q_i, and $s_i = d(P, Q_i)$. Since the sphere $\mathbb{S}_{n-1}(1)$ is compact and the sequence $\{s_i\}$ bounded, there exists a subsequence $\{Q_j\}$ of $\{Q_i\}$ such that $\{X_j\}$ converges to a unit vector $X_0 \in \mathbb{S}_{n-1}(1)$ and $s_j \to s_0$.
Assuming b), $Q_0 = \exp_P s_0 X_0$ exists. It follows that $Q_j \to Q_0$ and $d(P, Q_0) = s_0 \leq r$. Hence $E(r)$ is compact. Indeed, $\exp_P$ is continuous: We have only to consider a finite covering of the geodesic, from P to Q_0 by open balls, where we can apply Proposition 1.29.
According to Theorem 1.36, $E(r) = S(r)$ for $0 < r < \delta(P)$. Suppose $E(r) = S(r)$ for $0 < r < r_0$ and let us prove first, that equality occurs for $r = r_0$, then for $r > r_0$. Let $Q \in S(r_0)$ and $\{Q_i\}$ be a sequence, which converges to Q, such that $d(P, Q_i) < r_0$. Such a sequence exists because P and Q can be joined by a differentiable curve whose length is as close as one wants to r_0. $Q_i \in E(r_0)$, which is compact; hence $E(r_0) = S(r_0)$. By Theorem 1.36, $\delta(Q)$ is continuous. It follows that there exists a $\delta_0 > 0$ such that $\delta(Q) \geq \delta_0$ when $Q \in E(r_0)$, since $E(r_0)$ is compact.

Let us prove that $E(r_0 + \delta_0) = S(r_0 + \delta_0)$.
Pick $Q \in S(r_0 + \delta_0)$, $Q \notin S(r_0)$. For every $k \in \mathbb{N}$, there exists C_k, a differentiable curve from P to Q, whose length is smaller than $d(P, Q) + 1/k$. Denote by T_k the last point on C_k, which belongs to $E(r_0)$. After possibly passing to a subsequence, since $E(r_0)$ is compact, T_k converges to a point T. Clearly, $d(P, T) = r_0$, $d(T, Q) \leq \delta_0 \leq \delta(T)$, and

$$d(P, T) + d(T, Q) = d(P, Q), \quad \text{since } d(P, T_k) + d(T_k, Q) < d(P, Q) + 1/k.$$

There exists a minimizing geodesic from P to T and another from T to Q. The union of these two geodesics is a piecewise differentiable curve from P to Q, whose length is $d(P, Q)$. Hence it is a minimizing geodesic from P to Q.

This proves d) and Theorem 1.38, any bounded subset of M being included in $S(r)$ for r large enough, and $S(r) = E(r)$ being compact.

Finally, d) $\Rightarrow$ a), obviously. ∎

1.39 Definition. Cut-locus of a point P on a complete Riemannian manifold. According to Theorem 1.37, $\exp_P(rX)$ with $\|X\| = 1$ is defined for all $r \in \mathbb{R}$ and $X \in \mathbb{S}_{n-1}(1)$. Moreover the exponential mapping is differentiable.

Consider the following map $\mathbb{S}_{n-1}(1) \ni X \to \mu(X) \in]0, +\infty]$, $\mu(X)$ being the upper bound of the set of the r, such that the geodesic $[0, r] \ni s \to C(s) = \exp_P sX$ is minimizing. It is obvious that, for $0 < r \leq \mu(X)$, the geodesic $C(s)$ is minimizing.

The set of the points $\exp_P[\mu(X)\, X]$, when X varies over $\mathbb{S}_{n-1}(1)$, is called the *cut-locus* of P.

It is possible to show that $\mu(X)$ is a continuous function on $\mathbb{S}_{n-1}(1)$ with value in $]0, \infty]$ (Bishop and Crittenden [53]). Thus the cut-locus is a closed subset of M. So when M is complete, $\exp_P$, which is defined and differentiable on the whole $\mathbb{R}^n$, is a diffeomorphism of

$$\Theta = \{rX \in \mathbb{R}^n | 0 \leq r < \mu(X)\} \quad \text{onto } \Omega = \exp_P \Theta.$$

M is the union of the two disjoint sets: Ω and the cut-locus of P.

1.40 Definition. Let $\mu(X)$ be as above and $\delta_P = \inf \mu(X)$, $X \in \mathbb{S}_{n-1}(1)$. δ_P is called the *injectivity radius* at P. Clearly $\delta_P > 0$. The *injectivity radius* δ of a manifold M is the greatest real number such that $\delta \leq \delta_P$ for all $P \in M$. Clearly δ may be zero. But according to Theorem 1.36, δ is strictly positive if the manifold is compact.

§5. Second Variation of the Length Integral

5.1. Existence of Tubular Neighborhoods

1.41 Let $C(s)$ be an imbedded geodesic $[a, b] \ni s \to C(s) \in M$. At $P = C(a)$, fix an orthonormal frame of $T_P(M)$, $\{e_i\}$, $(i = 1, 2, \ldots, n)$ with $e_1 = (dC/ds)_{s=a}$, s being the parameter of arc length. Consider $e_i(s)$, the parallel translate vector of e_i from P to $C(s)$ (see Definition 1.27).
$\{e_i(s)\}$ forms an orthonormal frame of $T_{C(s)}(M)$ with $e_1(s) = dC(s)/ds$, since $g_{C(s)}(e_i(s), e_j(s))$ is constant along C.
Consider the following map Γ defined on an open subset of $\mathbb{R}^n$: $\mathbb{R} \times \mathbb{R}^{n-1} \ni (s, \tilde{\xi}) \to \exp_{C(s)} \xi$. To define Γ, associate to $\tilde{\xi} \in \mathbb{R}^{n-1}$ the vector $\xi \in \mathbb{R}^n$, whose first component ξ^1 is zero. According to Cauchy's theorem (see Proposition 1.29), Γ is differentiable. Moreover, by 1.30, the differential of Γ at each point $C(s)$ is the identity map of $\mathbb{R}^n$ if we identify the tangent space with $\mathbb{R}^n$; thus Γ is locally invertible in a neighborhood of C, by the inverse function theorem, 3.10.

For $\mu > 0$, define $T_\mu = \{$the set of the $\Gamma(s, \tilde{\xi})$ with $s \in [a, b]$ and $\|\tilde{\xi}\| < \mu\}$. T_μ is called a tubular neighborhood of C. The restriction Γ_μ of Γ to $[a, b] \times B_\mu \subset \mathbb{R}^n$ is a diffeomorphism onto T_μ, provided μ is small enough. Indeed, it is sufficient to show that for μ small enough Γ_μ is one-to-one. Suppose the contrary: there exists a sequence $\{Q_i\}$ of points belonging to $T_{1/i}$, such that $Q_i = \Gamma(s_i, X_i) = \Gamma(\sigma_i, Y_i)$ with $(s_i, X_i) \neq (\sigma_i, Y_i)$ and $\|X_i\| \leq \|Y_i\| < 1/i$. After possibly passing to a subsequence Q_j, when $j \to \infty$, Q_j converges to a point of C, say $C(s_0)$. Accordingly, $s_j \to s_0$ and $\sigma_j \to s_0$. This yields the desired contradiction, since Γ is locally invertible at $C(s_0)$, as proved above.

5.2. Second Variation of the Length Integral

1.42 Let C be a geodesic from P to Q, $[0, r] \ni s \to C(s) \in M$ being injective. Choose μ small enough so that Γ_μ is injective (for the definition of Γ_μ see 1.41).

On T_μ, the tubular neighborhood of C, (s, ξ) forms a coordinate system (called *Fermi coordinates*), which is normal at each point of C, as it is possible to show. We are going to compute the second variation of arc length in this chart (T_μ, Γ_μ^{-1}). Set $x^1 = s$ and $x^i = \xi^i$, for $i > 1$.
Let $\{C_\lambda\}$ be a family of curves close to C, defined by the C^2 differentiable mappings: $[0, r] \times]-\varepsilon, +\varepsilon[\ni (s, \lambda) \to x^i(s, \lambda)$, the coordinates of the point $Q(s, \lambda) \in C_\lambda$. In addition, suppose that $Q(s, 0) = C(s)$, $x^1(s, \lambda) = s$, and that $\varepsilon > 0$ is chosen small enough so that C_λ is included in T_μ for all $\lambda \in]-\varepsilon, +\varepsilon[$.
The first variation of the length integral

$$L(\lambda) = \int_0^r \sqrt{g_{ij}[Q(s, \lambda)] \frac{\partial x^i}{\partial s} \frac{\partial x^j}{\partial s}}\, ds$$

is zero at $\lambda = 0$, since $C_0 = C$ is a geodesic. A straightforward calculation leads to

$$(10) \qquad I = \left(\frac{\partial^2 L(\lambda)}{\partial \lambda^2}\right)_{\lambda=0} = \int_0^r \left[\sum_{i=2}^n \left(\frac{dy^i}{ds}\right)^2 - R_{1i1j}(C(s))y^i(s)y^j(s)\right] ds,$$

where $y^i(s) = [\partial x^i(s, \lambda)/\partial \lambda]_{\lambda=0}$. Indeed, by 1.13, $R_{1i1j} = -\frac{1}{2}\partial_{ij} g_{11}$ on C. Recall that on C, $g_{ij} = \delta_i^j$ and $\partial_k g_{ij} = 0$.

5.3. Myers' Theorem

1.43 *A connected complete Riemannian manifold M_n with Ricci curvature $\geq (n-1)k^2 > 0$ is compact and its diameter is $\leq \pi/k$.*

Proof. Let P and Q be two points of M_n and let C be the (or a) minimizing geodesic from P to Q, r its length.
Consider the second variation I_j $(j \geq 2)$ related to the family C_λ defined by $x^j(s, \lambda) = \lambda \sin(\pi s/r)$ and $x^i(s, \lambda) = 0$ for all $i > 1$, $i \neq j$. According to (10):

$$I_j = \int_0^r \left[\frac{\pi^2}{r^2} \cos^2 \frac{\pi s}{r} - R_{1j1j}(s) \sin^2 \frac{\pi s}{r}\right] ds.$$

Adding these equations and using the hypothesis $R_{11} \geq (n-1)k^2$, it follows that

$$\sum_{j=2}^n I_j = \int_0^r \left[(n-1)\frac{\pi^2}{r^2} \cos^2 \frac{\pi s}{r} - R_{11}(s) \sin^2 \frac{\pi s}{r}\right] ds \leq (n-1)\frac{r}{2}\left(\frac{\pi^2}{r^2} - k^2\right).$$

If $r > \pi/k$, this expression will be negative and at least one of the I_j must be negative. It follows that C is not minimizing, since there exists a curve from P to Q with length smaller than r. Hence $d(P, Q) \leq \pi/k$ for all pair of points P and Q. By Theorem 1.37, M is compact. ∎

§6. Jacobi Field

1.44 Definition. A vector field $Z(s)$, along a geodesic C, is a Jacobi field if its components $\xi^i(s)$ satisfy the equations:

$$(\xi^i)''(s) = -R^i_{1j1}(s)\xi^j(s) \tag{11}$$

in a Fermi coordinate system (see 1.42).

The set of the Jacobi fields along C forms a vector space of dimension $2n$, because by Cauchy's Theorem, 3.11, there is a unique Jacobi field which satisfies $Z(s_0) = Z_0$ and $Z'(s_0) = Y_0$, $s_0 \in [0, r]$, when Z_0 and Y_0 belong to $T_{C(s_0)}(M)$. The subset of the Jacobi fields which vanish at a fixed s_0 forms a vector subspace of dimension n. Those, which are in addition, orthogonal to C, form a vector subspace of dimension $(n-1)$. Indeed, if $\xi^1(s_0) = 0$ and $(\xi^1)'(s_0) = 0$, $\xi^1(s) = 0$ for all $s \in [0, r]$, since $(\xi^1)''(s) = 0$, for all s (by definition 1.44).

1.45 Definition. If there exists a non-identically-zero Jacobi field which vanishes at P and Q, two points of C, then Q is called a *conjugate point* to P.

1.46 Theorem. $\exp_P X$ *is singular at* X_0 *if and only if* $Q = \exp_P X_0$ *is a conjugate point to* P.

Proof. $\exp_P X$ is singular at X_0 if and only if there exists a vector $Y \neq 0$ orthogonal to X_0 such that

$$\left(\frac{\partial \exp_P(X_0 + \lambda Y)}{\partial \lambda}\right)_{\lambda=0} = 0. \tag{12}$$

Consider the family $\{C_\lambda\}$ of geodesics through P, defined by $[0, r] \ni s \to Q_\lambda(s) = \exp_P[(s/r)(X_0 + \lambda Y)] \in C_\lambda$, with $r = \|X_0\|$.
In a Fermi coordinate system (see 1.41) on a tubular neighborhood of C_0, the coordinates $x^i(s, \lambda)$ of $Q_\lambda(s)$ satisfy:

$$\frac{\partial^2 x^i(s, \lambda)}{\partial s^2} = -\Gamma^i_{jk}(Q_\lambda(s))\frac{\partial x^j}{\partial s}\frac{\partial x^k}{\partial s}; \tag{13}$$

for λ small enough, $\lambda \in]-\varepsilon, +\varepsilon[$, by (7), since C_λ is a geodesic.
The first order term in λ of (13) leads to

$$\frac{d^2 y^i(s)}{ds^2} = -\partial_j \Gamma^i_{11}(Q_0(s))y^j(s) = -R_{1i1j}(Q_0(s))y^j(s),$$

where $y^i(s) = (\partial x^i(s, \lambda)/\partial \lambda)_{\lambda=0}$ (recall that Christoffel's symbols are zero on C_0).

Hence $\{y^i(s)\}$ are the components of a Jacobi field $Z(s)$ along C_0, orthogonal to C_0.
If (12) holds, the preceding Jacobi field $Z(s)$ vanishes at P and Q, and it is not identically zero, since $Z'(0) = Y/r$. Conversely, if there exists a Jacobi field $Z(s) \not\equiv 0$, which vanishes at P and Q, then (12) holds with $Y = rZ'(0) \neq 0$. $\mathbb{R}^n$ and $T_P(M)$ are identified by $(\Gamma_\mu)_*$ (for the definition of $(\Gamma_\mu)_*$ see 1.8 and 1.41). ∎

1.47 Theorem. *If Q belongs to the cut-locus of P, then one at least of the following two situations occurs:*

(a) *Q is a conjugate point to P;*
(b) *There exist at least two minimizing geodesics from P to Q.*

For the proof see Kobayashi and Nomizu [167].

1.48 Theorem. *On a complete Riemannian manifold with nonpositive curvature, two points are never conjugate.*

Proof. Let $\{y^i(s)\}$ be the components of $Z(s) \not\equiv 0$, a Jacobi field which vanishes at P, as above. Then

$$\frac{1}{2}\left[\sum_{i=2}^{n}(y^i)^2\right]'' = \sum_{i=2}^{n}[(y^i)']^2 + \sum_{i=2}^{n}(y^i)(y^i)''$$

$$= \sum_{i=2}^{n}[(y^i)']^2 - R_{1i1j}(s)y^i(s)y^j(s) \geq \sum_{i=2}^{n}[(y^i)']^2.$$

Now $f(s) = \|Z(s)\|^2 = \sum_{i=2}^n [y^i(s)]^2$ cannot be zero for $s > 0$, since $f(0) = f'(0) = 0$ and $f''(0) > 0$, with $f''(s) \geq 0$ for all $s > 0$. ∎

§7. The Index Inequality

1.49 Proposition. *Let Y and Z be two Jacobi fields along (C), as in 1.44. Then $g(Y, Z') - g(Y', Z)$ is constant along (C). In particular, if Y and Z vanish at P, then $g(Y, Z') = g(Y', Z)$.*

Indeed, $[\sum_{i=1}^n (y^i z'^i - y'^i z^i)]' = 0$.

1.50 Definition (The Index Form). Let Z be a differentiable (or piecewise differentiable) vector field along a geodesic (C): $[0, r] \ni t \to C(t) \in M$. For Z orthogonal to dC/dt, the *index form* is

$$I(Z) = \int_0^r \left\{g(Z'(t), Z'(t)) + g\left[R\left(\frac{dC}{dt}, Z\right)\frac{dC}{dt}, Z\right]\right\} dt. \tag{14}$$

1.51 Theorem (The Index Inequality). *Let P and Q be two points of M_n, and let (C) be a geodesic from P to Q: $[0, r] \ni s \to C(s) \in M$ such that P admits no conjugate point along (C). Given a differentiable (or piecewise differentiable) vector field Z along (C), orthogonal to dC/dt and vanishing at P, consider the Jacobi field Y along (C) such that $Y(0) = 0$ and $Y(r) = Z(r)$. Then $I(Y) \leq I(Z)$. Equality occurs if and only if $Z = Y$.*

Proof. First of all, such a Jacobi field exists. Indeed, by 1.44, the Jacobi fields V, vanishing at P and orthogonal to dC/dt, form a vector space $\mathscr{V}$ of dimension $n - 1$.
Since P has no conjugate point on (C), the map $V'(0) \to V(r)$ is one-to-one, from the orthogonal complement of dC/dt in $T_P(M)$ to that of dC/dt in $T_Q(M)$. Thus this map is onto. And given $Z(r)$, Y exists.
Let $\{V_i\}$ $(i = 2, 3, \ldots, n)$ be a basis of $\mathscr{V}$. For the same reason as above, $\{V_i(s)\}$ $(2 \leq i \leq n)$ and dC/ds form a basis of $T_{C(s)}(M)$. Hence there exist differentiable (or piecewise differentiable) functions $f_i(s)$, such that $Z(s) = \sum_{i=2}^n f_i(s) V_i(s)$.
Furthermore, set $W(s) = \sum_{i=2}^n f'_i(s) V_i(s)$ and $e_1 = dC/ds$. Then by (11), $g[R(e_1, Z)e_1, Z] = \sum_{i=2}^n f_i g[R(e_1, V_i)e_1, Z] = \sum_{i=2}^n f_i g(V''_i, Z)$. Thus:

$$I(Z) = \int_0^r \Bigg[g(W, W) + \sum_{i,j} g(f_i V'_i, f_j V'_j) + \sum_{i,j} g(f_i V'_i, f'_j V_j)$$

$$+ \sum_{i,j} g(f'_i V_i, f_j V'_j) + \sum_{i,j} g(f_i V''_i, f_j V_j) \Bigg] dS.$$

By virtue of Proposition 1.49, $g(V_i, V'_j) = g(V'_i, V_j)$. Thus, integrating the last term of $I(Z)$ by parts gives

$$I(Z) = \int_0^r g(W, W)\, ds + g[Y'(r), Y(r)],$$

because $Y(s) = \sum_{i=2}^n f_i(r) V_i(s)$ and $Y'(s) = \sum_{i=2}^n f_i(r) V'_i(s)$.
If f_i are constant for all i, we find:

$$I(Y) = g[Y'(r), Y(r)]. \tag{15}$$

Hence $I(Z) \geq I(Y)$ and equality occurs if and only if $W = 0$, which is equivalent to $f'_i = 0$ for all i, that is to say, if $Y = Z$. ■

1.52 Proposition. *Let b^2 be an upper bound for the sectional curvature of M and δ its injectivity radius. Then the ball $S_P(r)$ is convex, if r satisfies $r < \delta/2$ and $r \leq \pi/4b$.*

Proof. Let $Q \in S_P(r)$ with $d(P, Q) = r$, and (C) the minimizing geodesic from P to Q. In a tubular neighborhood of (C), we consider a Fermi coordinate system, (see 1.42).
Given a geodesic γ through Q orthogonal to (C) at Q, so that $]-\varepsilon, +\varepsilon[\ni\lambda \to \gamma(\lambda) \in M$, with $\gamma(0) = Q$, set $Y_0 = (d\gamma/d\lambda)_{\lambda=0}$. The first coordinate of $\gamma(\lambda)$ is equal to r, for all λ.
By (10), the second variation of $d(P, \gamma(\lambda))$ at $\lambda = 0$ is $I(Y)$, where Y is the Jacobi field along (C) satisfying $Y(P) = 0$, $Y(Q) = Y_0$. But

$$I(Y) \geq \int_0^r [g(Y', Y') - b^2 g(Y, Y)]\, ds = I_b(Y);$$

$I_b(Y)$ is the index form (14) on a manifold with constant sectional curvature b^2.
On such a manifold, the solutions of (11) vanishing at $s = 0$ are of the type $\xi^i = \beta^i \sin bs$, for $i \geq 2$, where β^i are some constants. If $br < \pi$, a solution does not vanish for some $s \in]0, r]$, without being identically zero. In that case, according to Theorem 1.51, and by (15):

$$I_b(Y) \geq I_b\left(\frac{\sin bs}{\sin br} Y_0\right) = b \cot br\, g(Y_0, Y_0).$$

If $r < \pi/2b$, then $I(Y) > 0$ and for ε small enough, the points of γ, except Q, lie outside $S_P(r)$. Henceforth suppose $r < \delta/2$ and $r \leq \pi/4b$.

Consider Q_1 and Q_2, two points of $S_P(r)$, and γ a minimizing geodesic from Q_1 to Q_2 (see Theorem 1.38). Since $d(Q_1, Q_2) \leq 2r < \delta$, γ is unique and included in $S_P(2r)$. Let T be the (or a) point of γ, whose distance to P is maximum. Since $d(P, T) < 2r \leq \pi/2b$, T is one end point of γ. Indeed, if T is not Q_1 or Q_2, γ is orthogonal at T to the geodesic from P to T and by virtue of the above result, γ is not included in $S_P(d(P, T))$ and that contradicts the definition of T. ∎

§8. Estimates on the Components of the Metric Tensor

1.53 Theorem. *Let M_n be a Riemannian manifold whose sectional curvature K satisfies the bounds $-a^2 \leq K \leq b^2$, the Ricci curvature being greater than $a' = (n-1)\alpha^2$. Let $S_P(r_0)$ be a ball of M with center P and radius $r_0 < \delta_P$ the injectivity radius at P. Consider $(S_P(r_0), \exp_P^{-1})$, a normal geodesic coordinate system. Denote the coordinates of a point $Q = (r, \theta) \in [0, r_0] \times \mathbb{S}_{n-1}(1)$, locally by $\theta = \{\theta^i\}$, $(i = 1, 2, \ldots, n-1)$. The metric tensor g can be expressed by*

$$ds^2 = (dr)^2 + r^2 g_{\theta^i\theta^j}(r, \theta)\, d\theta^i\, d\theta^j.$$

For convenience let $g_{\theta\theta}$ be one of the components $g_{\theta^i\theta^i}$ and $|g| = \det((g_{\theta^i\theta^j}))$. Then $g_{\theta\theta}$ and $|g|$ satisfy the following inequalities:

(α) $\partial/\partial r \log \sqrt{g_{\theta\theta}(r,\theta)} \geq \partial/\partial r \log[\sin(br)/r]$, $g_{\theta\theta}(r,\theta) \geq [\sin(br)/br]^2$ *when* $br < \pi$;

(β) $\partial/\partial r \log \sqrt{g_{\theta\theta}(r,\theta)} \leq \partial/\partial r \log[\sinh(ar)/r]$, $g_{\theta\theta}(r,\theta) \leq [\sinh(ar)/ar]^2$;

(γ) $\partial/\partial r \log \sqrt{|g(r,\theta)|} \leq (n-1)(\partial/\partial r)\log[\sin(\alpha r)/r] \leq -a'r/3$,

(16)

$$\sqrt{|g(r,\theta)|} \leq \left[\frac{\sin(\alpha r)}{\alpha r}\right]^{n-1};$$

(δ) $\partial/\partial r \log \sqrt{|g(r,\theta)|} \geq (n-1)(\partial/\partial r)\log[\sin(br)/r]$,

$$\sqrt{|g(r,\theta)|} \geq \left[\frac{\sin(br)}{br}\right]^{n-1} \qquad \textit{when } br < \pi.$$

As usual, if $a' = \alpha = 0$, we set $\sin(\alpha r)/\alpha = r$, while if $(n-1)\alpha^2 = a' < 0$, $\sinh i\alpha r = i \sin \alpha r$ *and* $\cosh i\alpha r = \cos \alpha r$.

Proof. Let Y be a Jacobi field along (C), the minimizing geodesic from P to Q, $[0, r] \ni s \to C(s) \in M$, Y satisfying $Y(0) = 0$ and $Y \not\equiv 0$. When $br < \pi$, according to the proof of Proposition 1.52, and using (15):

$$g[Y'(r), Y(r)] = I(Y) \geq I_b(Y) \geq b \cot br\, g[Y(r), Y(r)],$$

where $I_b(Y)$ is the index form (14) on a manifold with constant sectional curvature b^2.
Moreover, according to the proof of Theorem 1.46,

$$Y(r) = \left(\frac{\partial \exp_P(X_0 + \lambda Y'(0))}{\partial \lambda}\right)_{\lambda=0},$$

where $X_0 = \exp_P^{-1} Q$ and we identify $T_P(M)$ with $\mathbb{R}^n$.
Thus $g[Y(r), Y(r)] = r^2 g_{\theta\theta}(r,\theta)\|Y'(0)\|^2$, θ being in the direction defined by $Y(r)$. Differentiating this equality, we obtain:

$$(\partial/\partial r)\log \sqrt{g_{\theta\theta}(r,\theta)} \geq g[Y'(r), Y(r)]/g[Y(r), Y(r)] - 1/r \geq b \cot br - 1/r.$$

The inequality α) follows, since $g_{\theta\theta}$ is equal to 1 at P and $\lim_{r\to 0}[\sin(br)/br] = 1$. To establish β), let us use the index inequality, Theorem 1.51:

$$g(Y'(r), Y(r)) = I(Y) \le I\left[\frac{\sinh as}{\sinh ar}\,Y(r)\right]$$

$$\le \left[a^2\int_0^r\left(\frac{\cosh as}{\sinh ar}\right)^2 ds + a^2\int_0^r\left(\frac{\sinh as}{\sinh ar}\right)^2 ds\right] g[Y(r), Y(r)]$$

$$= a\coth ar\; g[Y(r), Y(r)].$$

Thus $(\partial/\partial r)\log\sqrt{g_{\theta\theta}(r,\theta)} \le a\coth ar - 1/r$.

Let us now prove γ). Consider $\{e_i(s)\}$, an orthonormal frame on $T_{C(s)}(M)$, as in 1.41. Denote by $Y_2, Y_3, \ldots, Y_n$, the Jacobi fields along (C), such that $Y_i(0) = 0$ and $Y_i(r) = e_i(r)$, for $2 \le i \le n$. Using the index inequality, Theorem 1.51, yields:

$$g(Y_i'(r), Y_i(r)) = I(Y_i) \le I\left(\frac{\sin \alpha s}{\sin \alpha r}\,e_i(r)\right).$$

The possibility that $\sin\alpha r = 0$ for some $r > 0$ does not occur, even if $\alpha^2 > 0$, since $r < \delta_P \le \pi/\alpha$ (Myers' Theorem, 1.43).
Adding these inequalities leads to:

$$\sum_{i=2}^n g(Y_i'(r), Y_i(r)) \le (n-1)\alpha^2\int_0^r\left(\frac{\cos\alpha s}{\sin\alpha r}\right)^2 ds - \sum_{i=2}^n\int_0^r R_{1i1i}(s)\left(\frac{\sin\alpha s}{\sin\alpha r}\right)^2 ds.$$

Since

$$\sum_{i=2}^n R_{1i1i}(s) = R_{11}(s) \ge (n-1)\alpha^2,$$

$$\sum_{i=2}^n g(Y_i'(r), Y_i(r)) \le (n-1)\alpha\cot\alpha r.$$

Therefore $(\partial/\partial r)\log\sqrt{|g(r,\theta)|} = \sum_{i=2}^n g(Y_i'(r), Y_i(r)) - (n-1)/r \le (n-1) \times (\partial/\partial r)\log[\sin(\alpha r)/r]$ and the properties of $\cot u$ give the second inequality for γ). The third inequality follows by integrating the first, since $|g| = 1$ at P. To prove δ), we have only to add $n-1$ inequalities α) in the $n-1$ directions $e_i(r)$, $i = 2, \ldots, n$. ■

§9. Integration over Riemannian Manifolds

1.54 Definition. A differentiable manifold is said to be orientable if there exists an atlas all of whose changes of coordinate charts have positive Jacobian.

Given two charts of the atlas, (Ω, φ) and (Θ, ψ), with $\Omega \cap \Theta \neq \emptyset$, denote by $\{x^i\}$ the coordinates corresponding to (Ω, φ) and by $\{y^\alpha\}$ those corresponding to (Θ, ψ). In $\Omega \cap \Theta$, let $A_i^\alpha = \partial y^\alpha/\partial x^i$ and $B_\beta^j = \partial x^j/\partial y^\beta$, the Jacobian matrix $A = ((A_i^\alpha)) \in GL(\mathbb{R}^n)^+$, the subgroup of $GL(\mathbb{R}^n)$ consisting of those matrices A for which $\det A = |A| > 0$.

1.55 Theorem. *A differentiable manifold M_n is orientable if and only if there exists an exterior differential n-form, everywhere nonvanishing.*

Proof. Suppose M_n orientable. Let $(\Omega_i, \varphi_i)_{i\in I}$ be an atlas, all of whose changes of charts have positive Jacobian, and $\{\alpha_i\}$ a partition of unity subordinated to the covering $\{\Omega_i\}$.
Consider the differential n-forms $\omega_i = \alpha_i\, dx^1 \wedge dx^2 \wedge \cdots \wedge dx^n$ ($x^1, x^2, \ldots, x^n$ being the coordinates on Ω_i). It is easy to verify that the differential n-form $\omega = \sum_{i\in I} \omega_i$ is nowhere zero.

Conversely, let ω be a nonvanishing differentiable n-form, and $\mathscr{A} = (\Omega_i, \varphi_i)_{i\in I}$ an atlas such that all Ω_i are connected. On Ω_i there exists f_i, a nonvanishing function, such that $\omega = f_i\, dx^1 \wedge dx^2 \wedge \cdots \wedge dx^n$. Since Ω_i is connected, f_i has a fixed sign. If f_i is positive, we keep the chart (Ω_i, φ_i). In that case set $\tilde{\varphi}_i = \varphi_i$. Otherwise, whenever f_j is negative, we consider $\tilde{\varphi}_j$, the composition of φ_j with the transformation $(x^1, x^2, \ldots, x^n) \to (-x^1, x^2, \ldots, x^n)$ of $\mathbb{R}^n$. So from $\mathscr{A}$, we construct an atlas $\mathscr{A}_0$.
The charts of $\mathscr{A}_0$ are (Ω_i, φ_i) or $(\Omega_j, \tilde{\varphi}_j)$, depending on whether $f_i > 0$ or $f_j < 0$. Set $\tilde{f}_j = -f_j \circ \varphi_j \circ \tilde{\varphi}_j^{-1}$. All changes of charts of $\mathscr{A}_0$ have positive Jacobian. Indeed, at $x \in \Omega_i \cap \Omega_j$, denoting by $|A|$ the determinant of the Jacobian of $\tilde{\varphi}_j \circ \tilde{\varphi}_i^{-1}$, we have $\tilde{f}_j|A| = \tilde{f}_i$. Since $\tilde{f}_j$ and $\tilde{f}_i$ are positive, $|A| > 0$.

1.56 Definition. Let M be a connected orientable manifold. On the set of nonvanishing differentiable n-forms, consider the equivalence relation: $\omega_1 \sim \omega_2$ if there exists $f > 0$ such that $\omega_1 = f\omega_2$. There are two equivalence classes. Choosing one of them defines an *orientation* of M; then M is called *oriented.* There are two possible orientations of an orientable connected manifold.

Some examples of nonorientable manifolds: Möbius' band, Klein's bottle, the real projective space $\mathbb{P}_{2m}$ of even dimension $2m$.

Some examples of orientable manifolds: the sphere $\mathbb{S}_n$, the tangent space of any manifold, the complex manifolds.

1.57 Definition. Let M_n be a differentiable oriented manifold. We define *the integral of* ω, a differentiable n-form with compact support, as follows: Let $(\Omega_i, \varphi_i)_{i \in I}$ be an atlas compatible with the orientation chosen, and $\{\alpha_i\}_{i \in I}$ a partition of unity subordinate to the covering $\{\Omega_i\}_{i \in I}$. On Ω_i, ω is equal to $f_i(x)\, dx^1 \wedge \cdots \wedge dx^n$. By definition

$$\int_M \omega = \sum_{i \in I} \int_{\varphi_i(\Omega_i)} [\alpha_i(x) f_i(x)] \circ \varphi_i^{-1}\, dx^1 \wedge dx^2 \wedge \cdots \wedge dx^n.$$

One may verify that the definition makes sense. The integral does not depend on the partition of unity (see 1.73) and the sum is finite.

1.58 Theorem. *If M_n is nonorientable, there exists a covering manifold $\tilde{M}$ of M with two sheets, such that $\tilde{M}$ is orientable.*

For the proof see Narasimhan [212].

1.59 Definition. $\tilde{M}$ is called a *covering manifold* of M, if there exists $\pi: \tilde{M} \to M$, a differentiable map, such that for every $P \in M$:

α) $\pi^{-1}(P)$ is a discrete space, F;

β) there exists a neighborhood Ω of P, such that $\pi^{-1}(\Omega)$ is diffeomorphic to $\Omega \times F$. Each point $P' \in \pi^{-1}(P)$ has a neighborhood $\Omega' \subset \tilde{M}$, such that the restriction π' of π to Ω' is a diffeomorphism of Ω' onto Ω.
The map π is a 2-sheeted covering, if F consists of two points.

If (M, g) is a Riemannian manifold, on a covering manifold $\tilde{M}$ of M, we can consider the Riemannian metric $\tilde{g} = \pi^* g$. We call $(\tilde{M}, \tilde{g})$ a Riemannian covering of M.

1.60 Theorem. *If M is simply connected, then M is orientable.*

For the proof see Narasimhan [212].

1.61 Definition. Let E be the half-space of $\mathbb{R}^n$ ($x^1 < 0$), x^1 the first coordinate of $\mathbb{R}^n$. Consider $\bar{E} \subset \mathbb{R}^n$ with the induced topology. We identify the hyperplane of $\mathbb{R}^n$, $x^1 = 0$, with $\mathbb{R}^{n-1}$.
Letting Ω and θ be two open sets of $\bar{E}$, and $\varphi: \Omega \to \Theta$ a homeomorphism, it is possible to prove that the restriction of φ to $\Omega \cap \mathbb{R}^{n-1}$ is a homeomorphism of $\Omega \cap \mathbb{R}^{n-1}$ onto $\Theta \cap \mathbb{R}^{n-1}$. B will denote B_1, the unit ball with center 0 in $\mathbb{R}^n$, and we set $D = B \cap \bar{E}$.

§10. Manifold with Boundary

1.62 Definition. M_n is a manifold with boundary if each point of M_n has a neighborhood homeomorphic to an open set of $\bar{E}$.

The points of M_n which have a neighborhood homeomorphic to $\mathbb{R}^n$ are called interior points. They form the inside of M_n. The other points are called boundary points. We denote the set of boundary points by ∂M.

As in 1.4, we define a C^k-differentiable manifold with boundary. By definition, a function is C^k-differentiable on $\bar{E}$, if it is the restriction to $\bar{E}$, of a C^k-differentiable function on $\mathbb{R}^n$.

1.63 Theorem. *Let M_n be a (C^k-differentiable) manifold with boundary. If ∂M is not empty, then ∂M is a (C^k-differentiable) manifold of dimension $(n-1)$, without boundary*: $\partial(\partial M) = \varnothing$.

Proof. If $Q \in \partial M$, there exists a neighborhood Ω of Q homeomorphic by φ, to an open set $\Theta \subset \bar{E}$. The restriction $\tilde{\varphi}$ of φ to $\tilde{\Omega} = \Omega \cap \partial M$ is a homeomorphism of a neighborhood $\tilde{\Omega}$ of $Q \in \partial M$ onto an open set $\tilde{\Theta} \subset \mathbb{R}^{n-1}$. Thus ∂M is a manifold (without boundary) of dimension $(n-1)$ (Definition 1.1). If M_n is C^k-differentiable, let $(\Omega_i, \varphi_i)_{i \in I}$ be a C^k-atlas. Clearly, $(\tilde{\Omega}_i, \tilde{\varphi}_i)_{i \in I}$ form a C^k-atlas for ∂M. ■

1.64 Definition. By $\overline{W}_n$ a *compact Riemannian manifold with boundary* of *class* C^k, we understand the following: $\overline{W}_n$ is a C^k-differentiable manifold with boundary and $\overline{W}_n$ is a compact subset of M_n, a C^∞ Riemannian manifold. We set $W = \mathring{\overline{W}}$. We always suppose that the boundary is C^1, or at least Lipschitzian (Remark 2.35).

1.65 Theorem. *If M_n is a C^k-differentiable oriented manifold with boundary, ∂M is orientable. An orientation of M_n induces a natural orientation of ∂M.*

Proof. Let $(\Omega_j, \varphi_j)_{j \in I}$ be an allowable atlas with the orientation of M_n, and $(\tilde{\Omega}_j, \tilde{\varphi}_j)_{j \in I}$ the corresponding atlas of ∂M, as above. Set $\mathrm{i}: \partial M \to M$, the canonical imbedding of ∂M into M. We identify Q with $\mathrm{i}(Q)$, and $X \in T_Q(\partial M)$ with $\mathrm{i}_*(X) \in T_Q(M)$. Given $Q \in \partial M$, pick $e_1 \in T_Q(M)$, $e_1 \notin T_Q(\partial M)$, e_1 being oriented to the outside, namely, $e_1(f) \geq 0$ for all functions differentiable on a neighborhood of Q, which satisfy $f \leq 0$ in M_n, $f(Q) = 0$. We choose a basis of $T_Q(\partial M) = \{e_2, e_3, \ldots, e_n\}$, such that the basis of $T_Q(M)$: $\{e_1, e_2, \ldots, e_n\}$, belongs to the positive orientation given on M_n. ■

This procedure defines a canonical orientation on ∂M, as one can see.

10.1. Stokes' Formula

1.66 *Let M_n be a C^k-differentiable oriented compact manifold with boundary, and ω a differentiable $(n-1)$-form on M_n; then*

$$\int_M d\omega = \int_{\partial M} \omega, \tag{17}$$

where ∂M is oriented according to the preceding theorem. For convenience we have written $\int_{\partial M} \omega$ instead of $\int_{\partial M} \mathrm{i}^*\omega$, (for the definition of i^* see (1.8)).

Proof Let $(\Omega_i, \varphi_i)_{i\in I}$ be a finite atlas compatible with the orientation of M_n; such an atlas exists, because M_n is compact. Set $\Theta_i = \varphi_i(\Omega_i)$. Consider $\{\alpha_i\}$, a C^k-partition of unity subordinate to $\{\Omega_i\}$. By definition $\int_M d\omega = \sum_{i\in I} \int_{\Theta_i} d(\alpha_i\omega)$. Thus we have only to prove that $\int_{\Theta_i} d(\alpha_i\omega) = \int_{\tilde{\Theta}_i} \alpha_i\omega$, where we recall that $\tilde{\Omega}_i = \Omega_i \cap \partial V$ and have set $\tilde{\Theta}_i = \varphi_i(\tilde{\Omega}_i) = \Theta_i \cap \mathbb{R}^{n-1}$. In (Ω_i, φ_i), $\alpha_i\omega = \sum_{j=1}^n f_j(x)\, dx^1 \wedge \cdots \wedge \widehat{dx_j} \wedge \cdots \wedge dx^n$, $f_j(x)$ are C^k-differentiable functions with compact support included in $\varphi_i(\Omega_i)$; $\widehat{dx_j}$ means: this term is missing. Now,

$$d(\alpha_i\omega) = \left[\sum_{j=1}^n (-1)^{j-1} \frac{\partial f_j(x)}{\partial x^j}\right] dx^1 \wedge dx^2 \wedge \cdots \wedge dx^n,$$

by Definition 1.10. According to Fubini's theorem:

$$\int_{\Theta_i} d(\alpha_i\omega) = \int_{\tilde{\Theta}_i} f_1(x)\, dx^2 \wedge dx^3 \wedge \cdots \wedge dx^n = \int_{\tilde{\Theta}_i} \alpha_i\omega. \qquad \blacksquare$$

§11. Harmonic Forms

11.1. Oriented Volume Element

1.67 Definition. Let M_n be an oriented Riemannian manifold, and $\mathscr{A}$ an atlas compatible with the orientation. In the coordinate system $\{x^i\}$ corresponding to $(\Omega, \varphi) \in \mathscr{A}$, define the differential n-form η by:

$$\eta = \sqrt{|g|}\, dx^1 \wedge dx^2 \wedge \cdots \wedge dx^n, \tag{18}$$

where $|g|$ is the determinant of the metric matrix $((g_{ij}))$. η is a global differentiable n-form, called *oriented volume element*, and is nowhere zero.

Indeed, in another chart $(\Theta, \psi) \in \mathscr{A}$, such that $\Theta \cap \Omega \neq \emptyset$, consider the differentiable n-form: $\eta' = \sqrt{|g'|}\, dy^1 \wedge dy^2 \wedge \cdots \wedge dy^n$. But $g'_{\alpha\beta} = B^i_\alpha B^j_\beta g_{ij}$, hence $\sqrt{|g'|} = \sqrt{|B|^2}\sqrt{|g|}$ (for the definition of the matrices A and B see 1.54). Thus on $\Theta \cap \Omega$:

$$\eta' = \sqrt{|B|^2}\sqrt{|g|}|A|\, dx^1 \wedge dx^2 \wedge \cdots \wedge dx^n = \eta,$$

since $|A| > 0$ and $|A||B| = 1$. Moreover, η does not vanish.

1.68 Definition (Adjoint operator $*$). Let M_n be a Riemannian oriented manifold and η its oriented volume element. We associate to a p-form α, a $(n-p)$-form $*\alpha$, called the *adjoint* of α, defined as follows:
In a chart $(\Omega, \varphi) \in \mathscr{A}$, the components of $*\alpha$ are

(19) $$(*\alpha)_{\lambda_{p+1}, \lambda_{p+2}, \ldots, \lambda_n} = \frac{1}{p!}\eta_{\lambda_1, \lambda_2, \ldots, \lambda_n}\alpha^{\lambda_1, \lambda_2, \ldots, \lambda_p}.$$

We can verify that:

(20) $$*1 = \eta, \qquad **\alpha = (-1)^{p(n-p)}\alpha, \qquad \alpha \wedge (*\beta) = (\alpha, \beta)\eta,$$

where β is a p-form, and (α, β) denotes the scalar product of α and β:

$$(\alpha, \beta) = \frac{1}{p!}\alpha_{\lambda_1\lambda_2,\ldots,\lambda_p}\beta^{\lambda_1\lambda_2,\ldots,\lambda_p}.$$

Note that the adjoint operator is an isomorphism between the spaces $\Lambda^p(M)$ and $\Lambda^{n-p}(M)$.

11.2. Laplacian

1.69 Definition. (Co-differential δ, Laplacian Δ). Let $\alpha \in \Lambda^p(M)$. We define $\delta\alpha$, by its components in a chart $(\Omega, \varphi) \in \mathscr{A}$, as follows:

(21) $$(\delta\alpha)_{\lambda_1, \ldots, \lambda_{p-1}} = -\nabla^\nu \alpha_{\nu\lambda_1, \ldots, \lambda_{p-1}}.$$

The differentiable $(p-1)$-form $\delta\alpha$ is called the *co-differential* of α and has the properties:

(22) $$\delta = (-1)^p *^{-1} d\, *, \quad \delta\delta = -*^{-1} dd\, *, \quad \text{hence } \delta\delta = 0.$$

The *Laplacian operator* Δ is defined by:

(23) $$\Delta = d\delta + \delta d.$$

If $\alpha \in \Lambda^p(M)$, $\Delta\alpha \in \Lambda^p(M)$.

The Laplacian commutes with the adjoint operator:

(24) $$*\Delta = \Delta *.$$

For a function φ, $\delta\varphi = 0$, and

(25) $$\Delta\varphi = \delta d\varphi = -\nabla^\nu\nabla_\nu\varphi.$$

α is said to be *closed* if $d\alpha = 0$, co-closed if $\delta\alpha = 0$, harmonic if $\Delta\alpha = 0$.
α is said to be *exact* if there exists a differential form β, such that $\alpha = d\beta$.
α is said to be *co-exact* if there exists a differential form γ, such that $\alpha = \delta\gamma$.

Two p-forms are homologous if their difference is exact.

1.70 Definition (Global scalar product). On a compact oriented Riemannian manifold, we define the global scalar product $\langle\alpha, \beta\rangle$ of two p-forms α and β, as follows:

$$\langle\alpha, \beta\rangle = \int_M (\alpha, \beta)\eta.$$

Recall that $(\alpha, \beta) = \frac{1}{p!}\alpha_{\lambda_1,\lambda_2,\ldots,\lambda_p}\beta^{\lambda_1,\lambda_2,\ldots,\lambda_p}$.

The name of the operator δ comes from the formula:

(26) $$\langle d\alpha, \gamma\rangle = \langle\alpha, \delta\gamma\rangle \quad \text{for all } \gamma \in \Lambda^{p+1}(M) \quad \text{and } \alpha \in \Lambda^p(M).$$

Let us verify this. Using (1.10) we have:

(27) $$d(\alpha \wedge *\gamma) = d\alpha \wedge (*\gamma) + (-1)^p\alpha \wedge d(*\gamma),$$

while by (22), $*\delta\gamma = (-1)^{p+1}d(*\gamma)$. According to Stokes' formula (17), integrating (27) over M leads to:

$$0 = \int_M d\alpha \wedge (*\gamma) + (-1)^p \int_M \alpha \wedge d(*\gamma).$$

That is the equality (26) (see (20)). By (26), obviously, α and β being any p-forms:

(28) $$\langle\Delta\alpha, \beta\rangle = \langle\delta\alpha, \delta\beta\rangle + \langle d\alpha, d\beta\rangle = \langle\alpha, \Delta\beta\rangle.$$

Δ is an elliptic selfadjoint differential operator (for the definition see 3.51). If $\varphi \in C^2(M)$:

(29) $$\langle\Delta\varphi, \varphi\rangle = \int_M \nabla^\nu\varphi\nabla_\nu\varphi\, dV.$$

1.71 Theorem. *On a compact oriented Riemannian manifold, any harmonic form is closed and co-closed. A harmonic function is necessarily a constant.*

Proof. By (28), if α is harmonic:

$$0 = \langle \Delta\alpha, \alpha \rangle = \langle \delta\alpha, \delta\alpha \rangle + \langle d\alpha, d\alpha \rangle.$$

Thus $\delta\alpha = 0$ and $d\alpha = 0$. For a harmonic function φ, this implies $d\varphi = 0$, $\varphi = \text{const}$. ∎

11.3. Hodge Decomposition Theorem

1.72 *Let M_n be a compact and orientable Riemannian manifold. A p-form α may be uniquely decomposed into the sum of three p-forms:*

$$\alpha = d\lambda + \delta\mu + H\alpha,$$

where $H\alpha$ is a harmonic p-form.
Uniqueness comes from the orthogonality of the three spaces for the global scalar product

$$\langle \alpha, \alpha \rangle = \langle d\lambda, d\lambda \rangle + \langle \delta\mu, \delta\mu \rangle + \langle H\alpha, H\alpha \rangle.$$

For the proof see De Rham [106].

The dimension of $H_p(M_n)$, the space of harmonic p-forms, is called the pth Betti number of M. It is finite. By (24), $\Delta * = * \Delta$, $*$ defines an isomorphism between the spaces $H_p(M_n)$ and $H_{n-p}(M_n)$. Hence $b_p(M) = b_{n-p}(M)$. Clearly, $b_0(M_n) = b_n(M_n) = 1$ (Theorem 1.71). We set $\chi(M_n) = \sum_{p=0}^{n} (-1)^p b_p$.

1.73 Definition (The *Lebesgue Integral*). Let M_n be a Riemannian manifold and (Ω, φ) a local chart, with $\{x^i\}$ the associated coordinate system. We set:

$$\int_M f\,dV = \int_{\varphi(\Omega)} (\sqrt{|g|}\, f) \circ \varphi^{-1}\, dx^1\, dx^2 \cdots dx^n \tag{30}$$

for the continuous functions f on M_n with compact support lying in Ω.

Let (Θ, ψ) be another chart, $\{y^\alpha\}$ the associated coordinate system. We check that the definition (30) makes sense. Suppose $\operatorname{supp} f \subset \Omega \cap \Theta$; set $\partial y^\alpha / \partial x^i = A_i^\alpha$ and $\partial x^j / \partial y^\beta = B_\beta^j$ (see [1.54]). Then,

$$\int_{\varphi(\Omega\cap\Theta)} (\sqrt{|g|}\, f) \circ \varphi^{-1}\, dx^1\, dx^2 \cdots dx^n$$

$$= \int_{\psi(\Omega\cap\Theta)} (\sqrt{|g'|}\, f) \circ \psi^{-1}\, dy^1\, dy^2 \cdots dy^n.$$

Indeed $|g'| = |B|^2 |g|$ and $dy^1\, dy^2 \cdots dy^n = ||A||\, dx^1\, dx^2 \cdots dx^n$.

Consider $\{\Omega_i, \varphi_i\}_{i\in I}$ and $\{\Theta_j, \psi_j\}_{j\in J}$, two atlases, and $\{\alpha_i\}_{i\in I}$ (respectively, $\{\beta_j\}_{j\in J}$), a partition of unity subordinate to the covering $\{\Omega_i\}_{i\in I}$, (respectively, $\{\Theta_j\}_{j\in J}$). Since only a finite number of terms are nonzero, we have

$$\sum_{i\in I}\int_M \alpha_i f\,dV = \sum_{i\in I}\sum_{j\in J}\int_M (\alpha_i\beta_j) f\,dV = \sum_{j\in J}\int_M \beta_j f\,dV.$$

Thus $f \to \int_M f\,dV = \sum_{i\in I}\int_M \alpha_i f\,dV$ defines a positive Radon measure and the theory of the Lebesgue integral can be applied.

1.74 Definition. $dV = \sqrt{|g|}\,dx^1\,dx^2\cdots dx^n$ is called the *Riemannian volume element*.

1.75 Proposition. *Let M_n be a compact Riemannian manifold, and ω a 1-form. Then $\int_M \delta\omega\,dV = 0$. In particular if $f \in C^2(M)$, $\int_M \Delta f\,dV = 0$.*

Proof. Consider $\tilde{M}$ an orientable Riemannian covering manifold of M with two sheets, Theorem (1.58) and Definition 1.59. Let π be the covering map: $\tilde{M} \to M$ and let $\tilde{\omega} = \pi^*\omega$. Since $\tilde{M}$ is orientable, let $\tilde{\eta}$ be one of its two oriented volume elements.
According to (26):

$$\int_M \delta\tilde{\omega}\tilde{\eta} = \int_{\tilde{M}} (\delta\tilde{\omega}, 1)\tilde{\eta} = \int_{\tilde{M}} (\tilde{\omega}, d1)\tilde{\eta} = 0.$$

Moreover, from Definition 1.59 it follows that

$$\left|\int_{\tilde{M}} \delta\tilde{\omega}\tilde{\eta}\right| = 2\left|\int_M \delta\omega\,dV\right|.$$

If $f \in C^2(M)$, $\Delta f = \delta df$ and the preceding result applied to $\omega = df$ gives $\int_M \Delta f\,dV = 0$. Or else, by using the Stokes' formula (17), (20), and (24):

$$d\tilde{\delta}(*\tilde{f}) = \tilde{\Delta}(*\tilde{f}) = *\tilde{\Delta}\tilde{f} = (\tilde{\Delta}\tilde{f})\tilde{\eta},$$

and $\int_{\tilde{M}} (\tilde{\Delta}\tilde{f})\tilde{\eta} = 0$, where $\tilde{f} = f \circ \pi$. ■

1.76 Theorem. *Let M be a compact Riemannian manifold with strictly positive Ricci curvature. Then $b_1(M)$, the first Betti number of M, is zero.*

Proof. Let α be a harmonic 1-form. Then, by Theorem 1.71, $d\alpha = 0$ and $\delta\alpha = 0$. That is to say, in a local coordinate system with $\alpha = \alpha_i\,dx^i$ we have $\nabla_j\alpha_i = \nabla_i\alpha_j$ and $\nabla^i\alpha_i = 0$.

Contracting (1), ($i = l$), with $Z^i = \alpha^i$, gives

$$R_{ij}\alpha^i = \nabla_i(\nabla_j\alpha^i) - \nabla_j(\nabla_i\alpha^i). \tag{31}$$

Multiplying (31) by α^j and integrating over M lead to:

$$\int_M R_{ij}\alpha^i\alpha^j\,dV = \int_M \nabla_i[\alpha^j(\nabla_j\alpha^i)]\,dV - \int_M (\nabla_i\alpha^j)(\nabla_j\alpha^i)\,dV. \tag{32}$$

According to Proposition 1.75,

$$\int_M \nabla_i[\alpha^j(\nabla_j\alpha^i)]\,dV = -\int_M \delta[\alpha^j(\nabla_j\alpha^i)]\,dV = 0.$$

Hence, if α does not vanish everywhere, the first member of (32) will be strictly positive, while the second member is ≤ 0, since $\nabla_j\alpha_i = \nabla_i\alpha_j$. ∎

11.4. Spectrum

1.77 Definition. Let M be a compact Riemannian manifold. $Sp(M) = \{\lambda \in \mathbb{R}$, such that there exists $f \in C^2(M)$, $f \not\equiv 0$, satisfying $\Delta f = \lambda f\}$ is called the *spectrum* of M. λ is called an *eigenvalue* of the Laplacian and f an *eigenfunction.*

If $\lambda \in Sp(M)$, $\lambda \geq 0$, because

$$\lambda \int_M f^2\,dV = \int_M f\Delta f\,dV = \int_M \nabla^\nu f\nabla_\nu f\,dV \geq 0.$$

The eigenvalues of the Laplacian form an infinite sequence $0 = \lambda_0 < \lambda_1 < \lambda_2, \ldots$ going to $+\infty$. And for each eigenvalue λ_i, the set of the corresponding eigenfunctions forms a vector space of finite dimension (Fredholm's theorem (3.24). For λ_0 the vector space has one dimension.

1.78 Lichnerowicz's theorem. *If the Ricci tensor of M_n, a compact Riemannian manifold, is such that the 2-tensor $R_{ij} - kg_{ij}$ is non-negative for some $k > 0$, then $\lambda_1 \geq nk/(n-1)$.*

Proof. Let f be an eigenfunction: $\Delta f = \lambda f$ with $\lambda > 0$. Multiplying formula (31), with $\alpha = df$, by $\nabla^j f$, and integrating over M_n lead to:

$$\lambda \int_M \nabla^i f\nabla_i f\,dV - \int_M \nabla_i\nabla_j f\nabla^i\nabla^j f\,dV = \int_M R_{ij}\nabla^i f\nabla^j f\,dV.$$

As $(\nabla_i\nabla_j f + (1/n)\Delta f g_{ij})(\nabla^i\nabla^j f + (1/n)\Delta f g^{ij}) \geq 0$, it follows that $\nabla_i\nabla_j f\nabla^i\nabla^j f \geq (1/n)(\Delta f)^2$, hence $\lambda(1 - 1/n) \geq k$. ∎

Chapter 2

Sobolev Spaces

§1. First Definitions

2.1 We are going to define *Sobolev spaces* of integer order on a Riemannian manifold. First we shall be concerned with density problems. Then we shall prove the Sobolev imbedding theorem and the Kondrakov theorem. After that we shall introduce the notion of best constant in the Sobolev imbedding theorem. Finally, we shall study the exceptional case of this theorem (i.e., H_1^n on n-dimensional manifolds).

For Sobolev spaces on the open sets in n-dimensional, real Euclidean space $\mathbb{R}^n$, we recommend the very complete book of Adams [1].

2.2 Definitions. Let (M_n, g) be a smooth Riemannian manifold of dimension n (smooth means C^∞). For a real function φ belonging to $C^k(M_n)$ ($k \geq 0$ an integer), we define:

$$|\nabla^k \varphi|^2 = \nabla^{\alpha_1}\nabla^{\alpha_2}\cdots\nabla^{\alpha_k}\varphi\nabla_{\alpha_1}\nabla_{\alpha_2}\cdots\nabla_{\alpha_k}\varphi$$

In particular, $|\nabla^0\varphi| = |\varphi|$, $|\nabla^1\varphi|^2 = |\nabla\varphi|^2 = \nabla^\nu\varphi\nabla_\nu\varphi$. $\nabla^k\varphi$ will mean any kth covariant derivative of φ.

Let us consider the vector space $\mathfrak{C}_k^p$ of C^∞ functions φ, such that $|\nabla^\ell\varphi| \in L_p(M_n)$, for all ℓ with $0 \leq \ell \leq k$, where k and ℓ are integers and $p \geq 1$ is a real number.

2.3 Definitions. The *Sobolev space* $H_k^p(M_n)$ is the completion of $\mathfrak{C}_k^p$ with respect to the norm

$$\|\varphi\|_{H_k^p} = \sum_{\ell=0}^{k} \|\nabla^\ell\varphi\|_p.$$

$\overset{\circ}{H}{}_k^p(M_n)$ is the closure of $\mathcal{D}(M_n)$ in $H_k^p(M_n)$. $\mathcal{D}(M_n)$ is the space of C^∞ functions with compact support in M_n and $H_0^p = L_p$.

It is possible to consider some other norms which are equivalent; for instance, we could use

$$\left[\sum_{\ell=0}^{k} \|\nabla^\ell \varphi\|_p^p\right]^{1/p}.$$

When $p = 2$, H_k^2 is a Hilbert space, and this norm comes from the inner product. For simplicity we will write H_k for the Hilbert space H_k^2.

§2. Density Problems

2.4 Theorem. *$\mathscr{D}(\mathbb{R}^n)$ is dense in $H_k^p(\mathbb{R}^n)$.*

Proof. Let $f(t)$ be a C^∞ decreasing function on $\mathbb{R}$, such that $f(t) = 1$ for $t \le 0$ and $f(t) = 0$ for $t \ge 1$.
It is sufficient to prove that a function $\varphi \in C^\infty(\mathbb{R}^n) \cap H_k^p(\mathbb{R}^n)$ can be approximated in $H_k^p(\mathbb{R}^n)$ by functions of $\mathscr{D}(\mathbb{R}^n)$. We claim that the sequence of functions $\varphi_j(x) = \varphi(x) f(\|x\| - j)$, of $\mathscr{D}(\mathbb{R}^n)$, converges to $\varphi(x)$ in $H_k^p(\mathbb{R}^n)$.

Let us verify this for the functions and the first derivatives, that is, in the case of $H_1^p(\mathbb{R}^n)$. When $j \to \infty$, $\varphi_j(x) \to \varphi(x)$ everywhere and $|\varphi_j(x)| \le |\varphi(x)|$, which belongs to L_p. So by the Lebesgue dominated convergence theorem $\|\varphi_j - \varphi\|_p \to 0$. Moreover, when $j \to \infty$, $|\nabla\varphi_j(x)| \to |\nabla\varphi(x)|$ everywhere, and $|\nabla\varphi_j(x)| \le |\nabla\varphi(x)| + |\varphi(x)| \sup_{t\in[0,1]} |f'(t)|$ which belongs to L_p. Thus $\|\nabla(\varphi_j - \varphi)\|_p \to 0$.
This proves the density assertion for $H_1^p(\mathbb{R}^n)$. For $k > 1$, we have to use Leibnitz's formula. ■

2.5 Remark. The preceding theorem is not true for a bounded open set Ω in Euclidean space. Indeed, let us verify that $\mathring{H}_1^2(\Omega)$ is strictly included in $H_1^2(\Omega)$. For this purpose consider the inner product

$$\langle \varphi, \psi \rangle = \int_\Omega \varphi\psi \, dx + \sum_{i=1}^{n} \int_\Omega \partial_i\varphi \partial_i\psi \, dx.$$

For $\psi \in C^\infty(\Omega) \cap H_1^2(\Omega)$ and $\varphi \in \mathscr{D}(\Omega)$,

$$\langle \varphi, \psi \rangle = \int_\Omega \left(\psi - \sum_{i=1}^{n} \partial_{ii}\psi\right)\varphi \, dx.$$

If $\psi \not\equiv 0$ satisfies $\psi = \sum_{i=1}^n \partial_{ii}\psi$, then for all $\varphi \in \mathscr{D}(\Omega)$, $\langle \varphi, \psi \rangle = 0$, so that $\psi \notin \mathring{H}_1^2(\Omega)$.
Such a function ψ exists on a bounded open set Ω; for instance, $\psi = \sinh x_1$ (x_1 the first coordinate of x), $\int_\Omega |\sinh x_1|^p \, dx$, and $\int_\Omega |\cosh x_1|^p \, dx$ are finite.

For this reason we only try to prove the following theorem for complete Riemannian manifolds.

2.6 Theorem. *For a complete Riemannian manifold* $\mathring{H}_1^p(M_n) = H_1^p(M_n)$.

Proof. It is not useful to consider a function $f \in C^\infty$ on $\mathbb{R}$, as in the proof of the preceding theorem, because for a Riemannian manifold $[d(P, Q)]^2$ is only a Lipschitz function in $Q \in M_n$, P being a fixed point of M_n. So let us consider the function $f(t)$ on $\mathbb{R}$, defined by $f(t) = 1$ for $t \leq 0, f(t) = 1 - t$ for $0 < t < 1$, and $f(t) = 0$ for $t > 1$.

Let $\varphi(Q)$ be a C^∞ function belonging to $H_1^p(M_n)$, and P a fixed point of M_n. The sequence of functions $\varphi_j(Q) = \varphi(Q) f[d(P, Q) - j]$ belongs to $\mathring{H}_1^p(M_n)$, because the gradient exists almost everywhere, is bounded, and equals zero outside a compact set. One proves that the sequence $\varphi_j(Q)$ converges to $\varphi(Q)$ in $H_1^p(M_n)$, as in the proof of Theorem 2.4. Now we consider a regularization of $\varphi_j(Q)$.

Let K be the support of φ_j, and $\{\Theta_i\}$ be a finite covering of K such that Θ_i is homeomorphic to the open unit ball B of $\mathbb{R}^n$, (Θ_i, ψ_i) being the corresponding chart. Let $\{\alpha_i\}$ be a partition of unity of K subordinate to the covering $\{\Theta_i\}$. We approximate each function $\alpha_i \varphi_j$.
A function like $h = (\alpha_i \varphi_j) \circ \psi_i^{-1}$ has its support in B and is a uniformly Lipschitz function. Thus there exists a sequence $h_k \in \mathscr{D}(B)$, such that $h_k \to h$ in $H_1^p(B)$ (the usual regularization; for instance, see Adams [1]). It is now easy to show that $h_k \circ \psi_i$ converges, when $k \to \infty$, to $\alpha_i \varphi_j$ in $H_1^p(M_n)$.

2.7 Remark. It is possible to prove that if the manifold has an injectivity radius $\delta_0 > 0$, and if the curvature is bounded, then $\mathring{H}_2^p(M_n) = H_2^p(M_n)$. But for the proof of $\mathring{H}_k^p(M_n) = H_k^p(M_n)$, $(k > 2)$, besides these assumptions we need some hypothesis on the covariant derivatives of the components of the curvature tensor. (See Aubin [17] p. 154).

If there exists an injectivity radius $\delta_0 > 0$, it is simpler to consider, instead of the preceding functions $\varphi_j(Q)$, the following C^∞ functions $\tilde{\varphi}_j(Q)$, that tend to $\varphi(Q)$ in H_1^p or in H_k^p the space considered, under some conditions. Let T be a point of M_n, $B_T(q)$ the set of the points $Q \in M_n$ such that $d(T, Q) < q$ with $q \in \mathbb{N}$, and $\chi_q(Q)$ the characteristic function of $B_T(q)$. Let us consider $\gamma(t)$, a C^∞ decreasing function, which is equal to 1 for $t \leq 0$ and to zero for $t \geq \delta$, $(0 < \delta < \delta_0)$ and $\psi(P, Q) = \gamma[d(P, Q)]$. Now define the functions

$$h_q(P) = \int_{M_n} \psi(P, Q)\chi_q(Q)\, dV(Q) \Big/ \int_{M_n} \psi(P, Q)\, dV(Q).$$

These functions are C^∞, equal to 1 when $d(T, P) < q - \delta$, and equal to zero when $d(T, P) > q + \delta$. The functions $\tilde{\varphi}_j(P) = \varphi(P) h_j(P)$ have the desired property.

2.8 Theorem. *$C^\infty(\bar{E})$ is dense in $H_k^p(E)$, where E is the half-space:*

$$E = \{x \in \mathbb{R}^n / x_1 < 0\}.$$

By definition $C^\infty(\bar{E})$ is the set of functions that are restrictions to $\bar{E}$ of C^∞ functions on $\mathbb{R}^n$.

Proof. Let f belong to $C^\infty(E) \cap H_k^p(E)$. Consider the sequence of functions f_m, which are the restrictions to $\bar{E}$ of the functions $f(x_1 - 1/m, x_2, \ldots, x_n)$. It is obvious that $f_m \in C^\infty(\bar{E})$ and it is well known that, if $f \in L_p(E)$, $f_m \to f$ in $L_p(E)$. The same result holds for the derivatives of order $\leq k$.

For manifolds, we have the following theorem:

2.9 Theorem. *Let $\overline{W}_n$ be a compact Riemannian manifold with boundary of class C^r. Then $C^r(\overline{W})$ is dense in $H_k^p(W)$ for $k \leq r$.*

Proof. Let (Ω_i, φ_i) be a finite C^r atlas of $\overline{W}$, each Ω_i being homeomorphic either to a ball B of $\mathbb{R}^n$, or to a half ball $D \subset \bar{E}(D = B \cap \bar{E})$. $C^r(\overline{W})$ is the set of functions belonging to $C^r(W) \cap C^0(\overline{W})$, whose derivatives of order $\leq r$, in each Ω_i, can be extended to continuous functions on $\overline{W} \cap \Omega_i$.

Consider a C^∞ partition of unity $\{\alpha_i\}$ subordinate to the covering $\{\Omega_i\}$ of $\overline{W}$. Let $f \in H_k^p(W) \cap C^\infty(W)$. We have to prove that each function $\alpha_i f$ can be approximated in $H_k^p(W)$ by functions of $C^r(\overline{W})$. There is only a problem for the Ω_i homeomorphic to D. Let Ω_i be one of them.
The sequence of functions h_m defined, for m sufficiently large, as the restriction to D of $[(\alpha_i f) \circ \varphi_i^{-1}](x_1 - 1/m, x_2, \ldots, x_n)$ converges to $(\alpha_i f) \circ \varphi_i^{-1}$ in $H_k^p(D)$, where D has the Euclidean metric. Since the metric tensor, and all its derivatives are bounded on Ω_i (by a proper choice of the Ω_i, without loss of generality), $h_m \circ \varphi_i \in C^r(\overline{W})$ and converges to $\alpha_i f$ in $H_k^p(W)$ for $k \leq r$, when $m \to \infty$. ■

§3. Sobolev Imbedding Theorem

2.10 First part of the theorem.
Let k and ℓ be two integers ($k > \ell \geq 0$), p and q two real numbers ($1 \leq q < p$) satisfying $1/p = 1/q - (k - \ell)/n$. The Sobolev imbedding theorem asserts that for $\mathbb{R}^n$, $H_k^q \subset H_\ell^p$ and that the identity operator is continuous.

Second part.
If $(k - r)/n > 1/q$, $H_k^q \subset C_B^r$ and the identity operator is continuous. Here $r \geq 0$ is an integer and C_B^r is the space of C^r functions which are bounded as well as their derivatives of order $\leq r$, ($\|u\|_{C^r} \equiv \max_{0 \leq \ell \leq r} \sup |\nabla^\ell u|$).
If $(k - r - \alpha)/n \geq 1/q$, $H_k^q \subset C^{r+\alpha}$, where α is a real number satisfying $0 < \alpha < 1$ and $C^{r+\alpha}$ the space of the C^r functions, the rth derivatives of

which satisfy a Hölder condition of exponent α. Furthermore, the identity operator $H_k^q \subset C^\alpha$ is continuous. The norm of C^α is:

$$\|u\|_{C^\alpha} = \sup|u| + \sup_{P \neq Q}\{|u(P) - u(Q)|[d(P, Q)]^{-\alpha}\}.$$

We shall also denote the space $C^{r+\alpha}$, by $C^{r,\alpha}$ when $0 < \alpha \leq 1$. $C^{r,0} = C^r$. We will mainly discuss the first part of the theorem, because the other part concerns local properties (except the continuity of the imbedding) and so there is no difference in the case of manifolds. One will find the complete proof in Theorem 2.21.

But first of all, let us prove that the first part of the Sobolev imbedding theorem holds for all k, assuming it is true for $k = 1$.

2.11 Proposition. *Let M_n be a C^∞ Riemannian manifold. If $H_1^{q_0}(M_n)$ is imbedded in $L_{p_0}(M_n)$, with $1/p_0 = 1/q_0 - 1/n$ $(1 \leq q_0 < n)$, then $H_k^q(M_n)$ is imbedded in $H_\ell^{p_\ell}(M_n)$, with $1/p_\ell = 1/q - (k - \ell)/n > 0$.*

Proof. Let r be an integer and let $\psi \in C^{r+1}$. Then

$$|\nabla|\nabla^r\psi|| \leq |\nabla^{r+1}\psi|. \tag{1}$$

To establish this inequality, it is sufficient to develop

$$\begin{aligned}&(\nabla_\nu\nabla_{\alpha_1}\cdots\nabla_{\alpha_r}\psi\nabla_{\beta_1}\cdots\nabla_{\beta_r}\psi - \nabla_\nu\nabla_{\beta_1}\cdots\nabla_{\beta_r}\psi\nabla_{\alpha_1}\cdots\nabla_{\alpha_r}\psi)\\ &\quad\times g^{\nu\mu}g^{\alpha_1\lambda_1}g^{\alpha_2\lambda_2}\cdots g^{\alpha_r\lambda_r}g^{\beta_1\gamma_1}\cdots g^{\beta_r\gamma_r}(\nabla_\mu\nabla_{\lambda_1}\cdots\nabla_{\lambda_r}\psi\nabla_{\gamma_1}\cdots\nabla_{\gamma_r}\psi\\ &\quad - \nabla_\mu\nabla_{\gamma_1}\cdots\nabla_{\gamma_r}\psi\nabla_{\lambda_1}\cdots\nabla_{\lambda_r}\psi) \geq 0.\end{aligned}$$

We find $4|\nabla^{r+1}\psi|^2|\nabla^r\psi|^2 - |\nabla|\nabla^r\psi|^2|^2 \leq 0$.
Since $H_1^{q_0}(M_n)$ is imbedded in $L_{p_0}(M_n)$, there exists a constant A, such that for all $\varphi \in H_1^{q_0}(M_n)$:

$$\|\varphi\|_{p_0} \leq A(\|\nabla\varphi\|_{q_0} + \|\varphi\|_{q_0}).$$

Let us apply this inequality with $\varphi = |\nabla^r\psi|$, assuming φ belongs to $H_1^{q_0}$:

$$\begin{aligned}\|\nabla^r\psi\|_{p_0} &\leq A(\|\nabla|\nabla^r\psi|\|_{q_0} + \|\nabla^r\psi\|_{q_0}).\\ &\leq A(\|\nabla^{r+1}\psi\|_{q_0} + \|\nabla^r\psi\|_{q_0}).\end{aligned} \tag{2}$$

Now let $\psi \in H_k^q(M_n) \cap C^\infty(M_n)$. Applying inequalities (1) and (2) with $q = q_0$ and $r = k - 1, k - 2, \ldots$, we find:

$$\|\nabla^{k-1}\psi\|_{p_{k-1}} \leq A(\|\nabla^k\psi\|_q + \|\nabla^{k-1}\psi\|_q),$$

$$\|\nabla^{k-2}\psi\|_{p_{k-1}} \leq A(\|\nabla^{k-1}\psi\|_q + \|\nabla^{k-2}\psi\|_q),$$

$$\|\psi\|_{p_{k-1}} \leq A(\|\nabla\psi\|_q + \|\psi\|_q);$$

thus

$$\|\psi\|_{H^{p_{k-1}}_{k-1}} \leq 2A\|\psi\|_{H^q_k}.$$

Therefore a Cauchy sequence in H^q_k of C^∞ functions is a Cauchy sequence in $H^{p_{k-1}}_{k-1}$, and the preceding inequality holds for all $\psi \in H^q_k$.

Similarly, one proves the following imbeddings: $H^q_k \subset H^{p_{k-1}}_{k-1} \subset H^{p_{k-2}}_{k-2} \subset \cdots \subset H^{p_\ell}_\ell$.

§4. Sobolev's Proof

2.12 Sobolev's lemma. *Let $p' > 1$ and $q' > 1$ two real numbers. Define λ by $1/p' + 1/q' + \lambda/n = 2$. If λ satisfies $0 < \lambda < n$, there exists a constant $K(p', q', n)$, such that for all $f \in L_{q'}(\mathbb{R}^n)$ and $g \in L_{p'}(\mathbb{R}^n)$:*

$$\int_{\mathbb{R}^n}\int_{\mathbb{R}^n} \frac{f(x)g(y)}{\|x-y\|^\lambda}\,dx\,dy \leq K(p', q', n)\|f\|_{q'}\|g\|_{p'} \tag{3}$$

$\|x\|$ being the Euclidean norm.

The proof of this lemma is difficult (Sobolev [255]), we assume it.

Corollary. *Let λ be a real number, $0 < \lambda < n$, and $q' > 1$. If r, defined by $1/r = \lambda/n + 1/q' - 1$, satisfies $r > 1$, then*

$$h(y) = \int_{\mathbb{R}^n} \frac{f(x)}{\|x-y\|^\lambda}\,dx \quad \text{belongs to } L_r, \text{ when } f \in L_{q'}(\mathbb{R}^n).$$

Moreover, there exists a constant $C(\lambda, q', n)$ such that for all $f \in L_{q'}(\mathbb{R}^n)$

$$\|h\|_r \leq C(\lambda, q', n)\|f\|_{q'}.$$

Proof. For all $g \in L_{p'}(\mathbb{R}^n)$, with $1/r + 1/p' = 1$:

$$\int_{\mathbb{R}^n} h(y)g(y)\,dy \leq K(p', q', n)\|f\|_{q'}\|g\|_{p'};$$

therefore

$$h \in L_r \simeq L^*_{p'} \quad \text{and} \quad \|h\|_r \leq K(p', q', n)\|f\|_{q'}. \qquad \blacksquare$$

Now we will prove the existence of a constant $C(n, q)$ such that all $\varphi \in \mathcal{D}(\mathbb{R}^n)$ satisfy:

$$\|\varphi\|_p \leq C(n, q)\|\nabla\varphi\|_q, \tag{4}$$

with $1/p = 1/q - 1/n$ and $1 < q < n$.

Since $\mathscr{D}(\mathbb{R}^n)$ is dense in $H_1^q(\mathbb{R}^n)$ Theorem 2.4 the first part of the Sobolev imbedding theorem will be proved, according to Proposition 2.11.

Let x and y be points in $\mathbb{R}^n$, and write $r = \|x - y\|$. Let $\theta \in \mathbb{S}_{n-1}(1)$, the sphere of dimension $n - 1$ and radius 1. Introduce spherical polar coordinates (r, θ), with origin at x. Obviously, because $\varphi \in \mathscr{D}(\mathbb{R}^n)$:

$$\varphi(x) = -\int_0^\infty \frac{\partial\varphi(r, \theta)}{\partial r}\, dr = -\int_0^\infty \|x - y\|^{1-n} \frac{\partial\varphi(r, \theta)}{\partial r} r^{n-1}\, dr$$

and

$$|\varphi(x)| \le \int_0^\infty \|x - y\|^{1-n} |\nabla\varphi(r, \theta)| r^{n-1}\, dr.$$

Integrating over $\mathbb{S}_{n-1}(1)$, we obtain:

$$|\varphi(x)| \le \frac{1}{\omega_{n-1}} \int_{\mathbb{R}^n} \frac{|\nabla\varphi(y)|}{\|x - y\|^{n-1}}\, dy,$$

where ω_{n-1} is the volume of $\mathbb{S}_{n-1}(1)$.
According to Corollary 2.12 with $\lambda = n - 1$, inequality (4) holds. ∎

§5. Proof by Gagliardo and Nirenberg (1958)

2.13 Gagliardo [118] and Nirenberg [220] proved that for all $\varphi \in \mathscr{D}(\mathbb{R}^n)$:

(5) $$\|\varphi\|_{n/(n-1)} \le \frac{1}{2} \prod_{i=1}^n \left\| \frac{\partial\varphi}{\partial x^i} \right\|_1^{1/n}.$$

It is easy to see that the Sobolev imbedding theorem follows from this inequality. First $|\partial\varphi/\partial x^i| \le |\nabla\varphi|$; therefore $\|\varphi\|_{n/(n-1)} \le \frac{1}{2}\|\nabla\varphi\|_1$. Then setting $|\varphi| = u^{p(n-1)/n}$ and applying Hölder's inequality, we obtain:

$$\|u\|_p^{p(n-1)/n} = \|\varphi\|_{n/(n-1)} \le \tfrac{1}{2} p \frac{n-1}{n} \|u^{p'} |\nabla u|\|_1$$

$$\le p \frac{n-1}{2n} \|\nabla u\|_q \|u^{p'}\|_{q'},$$

where $1/q + 1/q' = 1$ and $p' = p(n-1)/n - 1$. But $p'q' = p$ since $1/p = 1/q - 1/n$; hence:

$$\|u\|_p \le p \frac{n-1}{2n} \|\nabla u\|_q.$$

We now prove inequality (5). For simplicity we treat only the case $n = 3$; but the proof for $n \neq 3$ is similar.

Let P be a point of $\mathbb{R}^3$, (x, y, z) the coordinates in $\mathbb{R}^3$, (x_0, y_0, z_0) those of P, and D_x (respectively, D_y, D_z) the straight line through P parallel to the x-axis, (respectively, y-, z-axis). Since $\varphi \in \mathscr{D}(\mathbb{R}^n)$,

$$\varphi(P) = \int_{-\infty}^{x_0} \frac{\partial \varphi}{\partial x}(x, y_0, z_0)\,dx = -\int_{x_0}^{+\infty} \frac{\partial \varphi}{\partial x}(x, y_0, z_0)\,dx.$$

Thus $|\varphi(P)| \leq \frac{1}{2}\int_{D_x} |\partial_x \varphi|\,dx$. Likewise for D_y and D_z:

$$|\varphi(P)|^{3/2} \leq \left(\frac{1}{2}\right)^{3/2} \left[\int_{D_x} |\partial_x \varphi|\,dx \int_{D_y} |\partial_y \varphi|\,dy \int_{D_z} |\partial_z \varphi|\,dz\right]^{1/2}.$$

Integration of x_0 over R yields, by Holder's inequality,

$$\begin{aligned}\int_{D_x} &|\varphi(x, y_0, z_0)|^{3/2}\,dx \\ &\leq \left(\frac{1}{2}\right)^{3/2} \left[\int_{D_x} |\partial_x \varphi(x, y_0, z_0)|\,dx \int_{D_{xy}} |\partial_y \varphi(x, y, z_0)|\,dx\,dy \right.\\ &\qquad \left.\times \int_{D_{xz}} |\partial_z \varphi(x, y_0, z)|\,dx\,dz\right]^{1/2},\end{aligned}$$

where D_{xy} means the plane through P parallel to the x- and y-axes.

Integration of y_0 over R gives, by Hölder's inequality,

$$\begin{aligned}\int_{D_{xy}} |\varphi(x, y, z_0)|^{3/2}\,dx\,dz \leq \left(\frac{1}{2}\right)^{3/2} &\left[\int_{D_{xy}} |\partial_x \varphi(x, y, z_0)|\,dx\,dy\right.\\ &\left.\times \int_{D_{xy}} |\partial_y \varphi(x, y, z_0)|\,dx\,dy \int_{R^3} |\partial_z \varphi|\,dx\,dy\,dz\right]^{1/2}.\end{aligned}$$

Finally, integrating z_0 over R, we obtain inequality (5). ■

§6. New Proof

2.14 Next we give a new proof of the Sobolev imbedding theorem (Aubin (1974)), which yields the explicit value of the norm of the imbedding.

Theorem (Aubin [13] or [17], see also Talenti (257)).
If $1 \leq q < n$, all $\varphi \in H_1^q(\mathbb{R}^n)$ satisfy:

(6) $$\|\varphi\|_p \leq K(n, q)\|\nabla \varphi\|_q,$$

with $1/p = 1/q - 1/n$ *and*

$$K(n, q) = \frac{q-1}{n-q}\left[\frac{n-q}{n(q-1)}\right]^{1/q}\left[\frac{\Gamma(n+1)}{\Gamma(n/q)\Gamma(n+1-n/q)\omega_{n-1}}\right]^{1/n}$$

for $1 < q < n$, *and*

$$K(n, 1) = \frac{1}{n}\left[\frac{n}{\omega_{n-1}}\right]^{1/n}.$$

$K(n, q)$ *is the norm of the imbedding* $H_1^q \subset L_p$, *and it is attained by the functions*

$$\varphi(x) = (\lambda + \|x\|^{q/(q-1)})^{1-n/q},$$

where λ *is any positive real number.*

When $q = 1$, this gives the usual isoperimetric inequality, Federer [113]; the extremum functions are then the characteristic functions of the balls of $\mathbb{R}^n$.

The proof is carried out in three steps.

First step, Proposition 2.16: Since $\mathcal{D}(\mathbb{R}^n)$ is dense in $H_1^q(\mathbb{R}^n)$, we have only to prove inequality (6) for the functions in question in Proposition (2.16).

Second step, Proposition 2.17: it is sufficient to establish inequality (6) for functions of the kind $\varphi(x) = f(\|x\|)$, f being a positive Lipschitzian function decreasing on $[0, \infty]$ and equal to zero at infinity.

Third step, Proposition 2.18: the proof of inequality (6) for these functions.

2.15 Proposition (Milnor [200] p. 37. This is actually due to Morse). *Let M be a Riemannian manifold. Any bounded smooth function $f : M \to \mathbb{R}$ can be uniformly approximated by a smooth function g which has no degenerate critical points. Furthermore, g can be chosen so that the* ith *derivatives of g on the compact set K uniformly approximate the corresponding derivatives of f for $i \le k$.*

We recall that a point $P \in M$ is called a critical point of f if $|\nabla f(P)| = 0$, the real number $f(P)$ is a critical value of f. A critical point P is called nondegenerate if and only if the matrix $\nabla_i \nabla_j f(P)$ is nonsingular. Nondegenerate critical points are isolated.

2.16 Proposition. *Let $f \not\equiv 0$ be a C^∞ function on M_n with compact support K. f can be approximated in $H_1^q(M_n)$ by a sequence of continuous functions f_p with compact support $K_p \subset K$ (the boundary of K_p being a sub-manifold of dimension $n - 1$); moreover $f_p \in C^\infty(K_p)$ and has only nondegenerate critical points on K_p.*

Since they are isolated, the number of critical points of f_p on K_p is finite.

Proof. According to Proposition 2.15 there is a sequence of C^∞ functions g_p which have no degenerate critical points and which satisfy $|f - g_p| < 1/p$ on M_n and $|\nabla(f - g_p)| < 1/p$ on K. Choose a real number α_p satisfying $1/p < \alpha_p < 2/p$, such that neither α_p nor $-\alpha_p$ is a critical value of g_p. Then $g_p^{-1}(\alpha_p)$ and $g_p^{-1}(-\alpha_p)$ are sub-manifolds of dimension $n - 1$, unless they are empty. Let $A_p = \{x \in M_n | g_p(x) \geq \alpha_p\}$ and $A_{-p} = \{x \in M_n | g_p(x) \leq -\alpha_p\}$. Define f_p by:

$$f_p(x) = [g_p(x) - \alpha_p]\chi_{A_p}(x) + [g_p(x) + \alpha_p]\chi_{A_{-p}}(x),$$

where χ_E is the characteristic function of the set E.
The support $K_p = A_p \cup A_{-p}$ of f_p is included in K because for $x \in K_p$, $|g_p(x)| > 1/p$; thus $|f(x)| > 0$. $f_p \in C^\infty(K_p)$ and f_p is Lipschitzian, hence $f_p \in H_1^q(M_n)$.
Since $|f(x) - f_p(x)| \leq (3/p)\chi_K(x)$, $\|f - f_p\|_q \to 0$ when $p \to \infty$. Moreover, at a point x where $f(x) \neq 0$, we have $|\nabla[f(x) - f_p(x)]| \to 0$, because $x \in \bigcup_{p=1}^\infty K_p$. But the set of the points where, simultaneously, $f(x) = 0$ and $|\nabla f(x)| \neq 0$, has zero measure; consequently $|\nabla[f(x) - f_p(x)]| \to 0$ almost everywhere. Therefore $\|\nabla(f - f_p)\|_q \to 0$, according to Lebesgue's theorem, since $|\nabla(f - f_p)| \leq (\sup|\nabla f| + 1/p)\chi_K$. ■

2.17 Proposition. *Let $f \geq 0$ be a continuous function on Σ (Σ denoting the sphere $\mathbb{S}_n$, the euclidean or hyperbolic space), which is C^∞ on its compact support K, whose boundary (if it is nonempty) is a submanifold of dimension $n - 1$, and assume f has only nondegenerate critical points. Pick P a point of Σ, and define $g(r)$, a decreasing function on $[0, \infty[$, by*

$$\mu\{Q | g[d(P, Q)] \geq a\} = \mu\{Q | f(Q) \geq a\} = \psi(a).$$

Then

$$\|\nabla g\|_q \leq \|\nabla f\|_q \quad \text{for} \quad 1 \leq q < \infty.$$

Proof. Let $d(P, Q)$ be the distance between P and Q on Σ, and let $a > 0$ be a real number; μ denote the measure defined by the metric, and write $g(Q) = g[d(P, Q)]$.
Let $Q_i (i = 1, \dots, k)$ be the critical points of f in K. Consider the set $\Sigma_a = f^{-1}(a)$ and note that, if $Q \in \Sigma_a$ is not one of the points Q_i, then $|\nabla f(Q)| \neq 0$. If $d\sigma(Q)$ denotes the area element on Σ_a, then we may write

$$\int_\Sigma |\nabla f|^q \, dV = \int_0^\infty \left(\int_{\Sigma_a} |\nabla f|^{q-1} \, d\sigma \right) da$$

Furthermore, when a is not a critical value of f, $\varphi(a) = \int_{\Sigma_a} |\nabla f|^{-1}\, d\sigma$ exists. $\varphi(a)$ is continuous and locally admits $-\psi(a)$ as primitive.

We consider $\varphi(a) = -\psi'(a)$ as given. Therefore $\int_{\Sigma_a} |\nabla f|^{q-1}\, d\sigma$ has a minimum, in the case $q > 1$, when $|\nabla f|$ is constant on Σ_a, according to Hölder's inequality:

$$\int_{\Sigma_a} d\sigma \le \left(\int_{\Sigma_a} |\nabla f|^{-1}\, d\sigma\right)^{(q-1)/q} \left(\int_{\Sigma_a} |\nabla f|^{q-1}\, d\sigma\right)^{1/q}.$$

But Σ_a is the boundary of a set, whose measure $\psi(a)$ is given. Hence $\int_{\Sigma_a} d\sigma$ is greater than or equal to the area of the boundary of the ball of volume $\psi(a)$ (by A. Dinghas [110]). This completes the proof. ∎

Furthermore, one verifies that $g(r)$ is absolutely continuous and even Lipschitzian on $[0, \infty[$.

2.18 Proposition. *Let $g(r)$ be a decreasing function absolutely continuous on $[0, \infty[$, and equal to zero at infinity. Then:*

$$(7) \qquad (\omega_{n-1})^{-1/n}\left(\int_0^\infty |g(r)|^p r^{n-1}\, dr\right)^{1/p} \le \mathsf{K}(n, q)\left(\int_0^\infty |g'(r)|^q r^{n-1}\, dr\right)^{1/q},$$

where $\mathsf{K}(n, q)$ is from Theorem 2.14.

Proof. Let us consider the following variational problem, when $q > 1$:

Maximize $I(g) = \int_0^\infty |g(r)|^p r^{n-1}\, dr$, when $J(g) = \int_0^\infty |g'(r)|^q r^{n-1}\, dr$ is a given positive constant.

The Euler equation is

$$(8) \qquad (|g'|^{q-1} r^{n-1})' = k g^{p-1} r^{n-1},$$

where k is a constant.

It is obvious that we have only to consider decreasing functions. One verifies that the functions $y = (\lambda + r^{q/(q-1)})^{1-n/q}$ are solutions of (8), $\lambda > 0$ being a real number:

$$|y'|^{q-1} r^{n-1} = \left(\frac{n-q}{q-1}\right)^{q-1} r^n(\lambda + r^{q/(q-1)})^{-n(q-1)/q},$$

$$(|y'|^{q-1} r^{n-1})' = n\lambda\left(\frac{n-q}{q-1}\right)^{q-1} r^{n-1} y^{p-1}$$

According to Bliss, Lemma 2.19, the corresponding value of the integral $I(y)$ is an absolute maximum.

The value of $\mathsf{K}(n, q)$, the best constant, is

$$\mathsf{K}(n, q) = (\omega_{n-1})^{-1/n}[I(y)]^{1/p}[J(y)]^{-1/q}.$$

Letting $q \to 1$, we establish the inequality (7) for $q = 1$:

$$\mathsf{K}(n, 1) = \lim_{q \to 1} \mathsf{K}(n, q).$$

Let us compute $\mathsf{K}(n, q)$.

$$\int_0^\infty |y'|^q r^{n-1}\, dr = \left(\frac{n-q}{q-1}\right)^q \int_0^\infty [\lambda + r^{q/(q-1)}]^{-n} r^{n+1/(q-1)}\, dr.$$

Setting $\lambda = 1$ and $r = t^{(q-1)/q}$, we obtain:

$$\int_0^\infty |y'|^q r^{n-1}\, dr = \left(\frac{n-q}{q-1}\right)^q \frac{q-1}{q} \int_0^\infty (1+t)^{-n} t^{n-n/q}\, dt = \left(\frac{n-q}{q-1}\right)^q \frac{q-1}{q} A,$$

$$\int_0^\infty y^p r^{n-1}\, dr = \frac{q-1}{q} \int_0^\infty (1+t)^{-n} t^{n-1-n/q}\, dt = \frac{q-1}{q} B.$$

Furthermore, $B/A = (n-q)/n(q-1)$, because

$$A = \int_0^\infty (1+t)^{-n} t^{n-n/q}\, dt = \frac{n}{n-1} \frac{q-1}{q} \int_0^\infty (1+t)^{1-n} t^{n-1-n/q}\, dt$$
$$= \frac{n}{n-1} \frac{q-1}{q} (A + B).$$

Hence:

$$\mathsf{K}(n, q) = (\omega_{n-1})^{-1/n} \left(\frac{q-1}{q} B\right)^{1/q-1/n} A^{-1/q} \frac{q-1}{n-q} \left(\frac{q}{q-1}\right)^{1/q},$$

$$\mathsf{K}(n, q) = \frac{q-1}{n-q} \left(\frac{B}{A}\right)^{1/q} \left(\frac{q-1}{q} B\omega_{n-1}\right)^{-1/n}, \quad \text{with} \quad B = \frac{\Gamma(n/q)\Gamma(n-n/q)}{\Gamma(n)}.$$

■

2.19 Lemma. *Let $h(x) \geq 0$ a measurable, real-valued function defined on $\mathbb{R}$, such that $J = \int_0^\infty h^q(x)\, dx$ is finite and given. Set $g(x) = \int_0^x h(t)\, dt$. Then $I = \int_0^\infty g^p(x) x^{\alpha-p}\, dx$ attains its maximum value for the functions $h(x) = (\lambda x^\alpha + 1)^{-(\alpha+1)/\alpha}$, with p and q two constants satisfying $p > q > 1$, $\alpha = (p/q) - 1$ and $\lambda > 0$ a real number.*

This is proved in Bliss [55]. The change of variable $x = r^{(q-n)/(q-1)}$ now yields the result used in the Proposition 2.18, above. Recall that here $1/p = (1/q) - (1/n)$ and so we have $\alpha = p/n$, $(\partial x/\partial r)^{1-q} = r^{n-1}$ and $x^{1+\alpha-p} = r^n$.

§7. Sobolev Imbedding Theorem for Riemannian Manifolds

2.20 Theorem. *For compact manifolds the Sobolev imbedding theorem holds. Moreover H_k^q does not depend on the Riemannian metric.*

Proof. We are going to give the usual proof of the first part of the theorem, because it is easy for compact manifolds. But for a more precise result and a more complete proof see Theorem 2.21. Let $\{\Omega_i\}$ be a finite covering of M, $(i = 1, 2, \dots, N)$, and (Ω_i, φ_i) the corresponding charts. Consider $\{\alpha_i\}$ a C^∞ partition of unity subordinate to the covering $\{\Omega_i\}$. We have only to prove there exist constants C_i such that every C^∞ function f on M satisfies:

$$\|\alpha_i f\|_p \le C_i \|\alpha_i f\|_{H_1^q}. \tag{9}$$

Indeed, since $|\nabla(\alpha_i f)| \le |\nabla f| + |f||\nabla\alpha_i|$,

$$\|f\|_p \le \sum_{i=1}^N \|\alpha_i f\|_p \le \sup_{1\le i\le N} C_i N\left[\|\nabla f\|_q + \left(1 + \sup_{1\le i\le N} |\nabla\alpha_i|\right)\|f\|_q\right],$$

and by density the theorem holds for $k = 1$.
In view of Proposition 2.11, this establishes the first part of the theorem.

On the compact set $K_i = \operatorname{supp} \alpha_i \subset \Omega_i$, the metric tensor and its derivatives of all orders are bounded in the system of coordinates corresponding to the chart (Ω_i, φ_i). Hence:

$$\begin{aligned} f \in H_k^q(M_n, g) &\Leftrightarrow [\alpha_i f \in H_k^q(M_n, g), \text{ for all } i] \\ &\Leftrightarrow [\alpha_i f \circ \varphi_i^{-1} \in H_k^q(\mathbb{R}^n), \text{ for all } i]. \end{aligned}$$

We define the functions $\alpha_i f \circ \varphi_i^{-1}$ to be zero outside $\varphi_i(K_i)$.
In particular, there exist two real numbers $\mu \ge \lambda > 0$, such that for all vectors $\xi \in \mathbb{R}^n$ and every $x \in K_i$, g_x being the metric tensor at x:

$$\lambda\|\xi\|^2 \le g_x[(\varphi_i^{-1})_*(\xi), (\varphi_i^{-1})_*(\xi)] \le \mu\|\xi\|^2.$$

And now according to Theorem 2.14, for any $f \in C^\infty$:

$$\left(\int_{\mathbb{R}^n} |\alpha_i f \circ \varphi_i^{-1}|^p\, dE\right)^{1/p} \le \mathrm{K}(n, q)\left(\int_{\mathbb{R}^n} |\nabla(\alpha_i f \circ \varphi_i^{-1})|^q\, dE\right)^{1/q}.$$

Thus we obtain inequality (9):

$$\|\alpha_i f\|_p \leq \mu^{n/2p}\left(\int_{\mathbb{R}^n} |\alpha_i f \circ \varphi_i^{-1}|^p \, dE\right)^{1/p}$$
$$\leq \mu^{n/2p} \mathsf{K}(n, q)\mu^{1/2}\lambda^{-n/2q}\|\nabla(\alpha_i f)\|_q. \qquad \blacksquare$$

2.21 Theorem. *The Sobolev imbedding theorem holds for M_n a complete manifold with bounded curvature and injectivity radius $\delta > 0$.*
Moreover, for any $\varepsilon > 0$, there exists a constant $A_q(\varepsilon)$ such that every $\varphi \in H_1^q(M_n)$ satisfies:

(10) $\|\varphi\|_p \leq [\mathsf{K}(n, q) + \varepsilon]\|\nabla\varphi\|_q + A_q(\varepsilon)\|\varphi\|_q$, with $1/p = 1/q - 1/n > 0$,

where $\mathsf{K}(n, q)$ is the smallest constant having this property.

According to Proposition 2.11 and Theorem 2.6, to prove the first part of the Sobolev imbedding theorem, it is sufficient to establish inequality (10) for the functions of $\mathscr{D}(M_n)$. The proof will be given at the end of 2.27 using Lemmas 2.24 and 2.25. First we will establish the second part of the Sobolev imbedding theorem.

2.22 Lemma. *Let M_n be a complete Riemannian manifold with injectivity radius $\delta_0 > 0$ and sectional curvature K, satisfying the bound $K \leq b^2$. There exists a constant $C(q)$ such that for all $\varphi \in \mathscr{D}(M_n)$:*

(11) $$\sup|\varphi| \leq C(q)\|\varphi\|_{H_1^q} \quad \text{if } q > n.$$

Proof. Let $f(t)$ be a C^∞ decreasing function on $\mathbb{R}$, which is equal to 1 in a neighbourhood of zero, and to zero for $t \geq \delta$ ($\delta \leq \delta_0$ satisfying $2b\delta \leq \pi$). Let P be a given point of M_n, then

$$\varphi(P) = -\int_0^\delta \partial_r[\varphi(r, \theta)f(r)]\, dr,$$

where (r, θ) is a system of geodesic polar coordinates with center P. Thus we have the estimate:

$$|\varphi(P)| \leq \int_0^\delta |\nabla[\varphi(r, \theta)f(r)]|(r^{1-n})r^{n-1}\, dr.$$

Integration with respect to θ over $\mathbb{S}_{n-1}(1)$ leads, by Hölder's inequality, to

$$|\varphi(P)| \leq (\omega_{n-1})^{-1}\left(\int_{B_P(\delta)} |\nabla[\varphi(r, \theta)f(r)]|^q r^{n-1}\, dr\, d\theta\right)^{1/q}$$
$$\times \left(\omega_{n-1}\int_0^\delta r^{(n-1)(1-q')}\, dr\right)^{1/q'},$$

with $1/q' = 1 - 1/q$. According to Theorem 1.53, for $r \le \delta$, $r^{n-1}\,dr\,d\theta \le (\pi/2)^{n-1}\,dV$. Thus:

$$|\varphi(P)| \le \left(\frac{\pi}{2}\right)^{(n-1)/q} (\omega_{n-1})^{-1/q}\left(\|\nabla\varphi\|_q + \sup_{0<r<\delta} |f'(r)|\,\|\varphi\|_q\right)$$
$$\times \left[\frac{q-1}{q-n}\,\delta^{(q-n)/(q-1)}\right]^{1-1/q} \quad \blacksquare$$

2.23 *Proof of the Sobolev imbedding theorem* 2.21 (Second part). $\mathscr{D}(M_n)$ is dense in $H_1^q(M_n)$ by Theorem 2.6. So let $f \in H_1^q$ and $\{\varphi_i\}$ be a sequence of functions of $\mathscr{D}(M_n)$ such that $\|f - \varphi_i\|_{H_1^q} \to 0$ when $i \to \infty$. Clearly $\{\varphi_i\}$ is a Cauchy sequence in H_1^q. By (11), $\sup|\varphi_i - \varphi_j| \le C(q)\|\varphi_i - \varphi_j\|_{H_1^q}$, so φ_i is a Cauchy sequence in C_B^0. Therefore $\varphi_i \to f$ in C_B^0 and $f \in C_B^0$.
Letting $i \to \infty$ in $\sup|\varphi_i| \le C(q)\|\varphi_i\|_{H_1^q}$, we establish, for all $f \in H_1^q$ when $q > n$, the inequality:

$$\|f\|_{C^0} \le C(q)\|f\|_{H_1^q}.$$

Let $f \in H_k^q \cap C^\infty$. Then (1) implies that $|\nabla^r f| \in H_{k-r}^q$. If $(k-r)/n > 1/q$, according to the first part of Theorem 2.21, that we are going to prove below,

$$H_{k-r}^q \subset H_1^{\tilde q} \quad \text{with} \quad \frac{1}{q} - \frac{k-r-1}{n} = \frac{1}{\tilde q} < \frac{1}{n}.$$

Therefore

$$\|\nabla^r f\|_{C^0} \le C(\tilde q)\|\nabla^r f\|_{H_1^{\tilde q}} \le \text{Const} \times \|f\|_{H_k^q} \text{ and } \|f\|_{C^r} \le \text{Const} \times \|f\|_{H_k^q}.$$

This inequality holds for all $f \in H_k^q$ since $H_k^q \cap C^\infty$ is dense in H_k^q. Let us now prove that if $q = n/(1-\alpha)$, where α satisfies $0 < \alpha < 1$, then $H_1^q \subset C^\alpha$ and the identity operator is continuous. First of all, $f \in H_1^q$ implies f is continuous.
Choose δ as in Lemma 2.22 ($\delta \le \delta_0$ and $2b\delta \le \pi$). When $d(P, Q) \ge \delta$, we may write:

$$|f(P) - f(Q)|[d(P,Q)]^{-\alpha} \le 2\delta^{-\alpha}C(q)\|f\|_{H_1^q}.$$

When $d = d(P, Q) < \delta$, consider a ball $\mathfrak{B}$ of radius $d/2$ and center O, with P and Q in $\mathfrak{B}$; note $y = \exp_O^{-1} P$ and $z = \exp_O^{-1} Q$; y and z belong to a ball $\tilde B \subset \mathbb{R}^n$ of radius $d/2$. Consider the function $h(x) = f(\exp_O x)$ defined in $\tilde B$ and let (r, θ) be polar coordinates with center y.
For $\tilde B \ni x = (r, \theta), 0 \le r \le \rho(\theta), (\rho(\theta), \theta)$ belonging to $\partial\tilde B$, the boundary of $\tilde B$:

$$h(x) - h(y) = \int_0^r \partial_\rho h(\rho, \theta)\,d\rho = r\int_0^1 \partial_\rho h(rt, \theta)\,dt.$$

Integration with respect to x over $\tilde{B}$ leads to

$$\int_{\tilde{B}} h(x)\,dx - \frac{1}{n}\,\omega_{n-1}(d/2)^n h(y) = \int_{\mathbb{S}_{n-1}} \int_0^{\rho(\theta)} r^n\,dr\,d\theta \int_0^1 \partial_\rho h(rt, \theta)\,dt.$$

Hence, putting $u = rt$ and using the inequality $r \leq \rho(\theta) < d$, we obtain:

$$\begin{aligned}\left|\int_{\tilde{B}} h(x)\,dx - \frac{1}{n}\,\omega_{n-1}\left(\frac{d}{2}\right)^n h(y)\right| &\leq d\int_0^1 t^{-n}\,dt \int_{\mathbb{S}_{n-1}} \int_0^{t\rho(\theta)} |\partial_\rho h(u, \theta)|\,u^{n-1}\,du\,d\theta\\ &\leq d\left(\int_{\tilde{B}} |\nabla_E h(x)|^q\,dE\right)^{1/q} \int_0^1 (\operatorname{vol} \tilde{B}_t)^{1-1/q} t^{-n}\,dt,\end{aligned}$$

where we have applied Hölder's inequality. Let $\tilde{B}_t$ be the ball homothetic to $\tilde{B}$, with ratio t. Then $\operatorname{vol} \tilde{B}_t = (1/n)\omega_{n-1}(d/2)^n t^n$. Thus, since $q > n$, the second integral converges:

$$\begin{aligned}\left|h(y) - \frac{n}{\omega_{n-1}}\left(\frac{2}{d}\right)^n \int_{\tilde{B}} h(x)\,dx\right| \leq \left(\frac{2^n n}{\omega_{n-1}}\right)^{1/q} \frac{q}{q-n}\, d^{1-n/q} \left(\int_{\tilde{B}} |\nabla_E h(x)|^q\,dE\right)^{1/q}.\end{aligned}$$

A similar inequality holds with z in place of y, and so:

$$|h(y) - h(z)| \leq \frac{2q}{q-n}\left(\frac{2^n n}{\omega_{n-1}}\right)^{1/q} d^\alpha \left(\int_{\tilde{B}} |\nabla_E h(x)|^q\,dE\right)^{1/q}.$$

According to Theorem 1.53, since $d < \delta$:

$$|f(P) - f(Q)|[d(P, Q)]^{-\alpha} \leq \frac{2q}{q-n}\left(\frac{2^n n}{\omega_{n-1}}\right)^{1/q} \frac{\sinh(a\delta)}{a\delta}\left(\frac{\pi}{2}\right)^{(n-1)/q} \|\nabla f\|_q.$$

The other results are local and do not differ from the case of $\mathbb{R}^n$. ■

Proof of the Sobolev imbedding theorem 2.21 (First part). According to Proposition 2.11, if we can prove that inequality (10) holds then Theorem 2.21 will be proved. First two lemmas.

2.24 Lemma. *Let $\delta \in \mathbb{R}$ satisfy $0 < \delta < \delta_0$ and $2b\delta \leq \pi$. If $B_P(\delta)$ is a ball of M_n with center P and radius δ, where the sectional curvature σ satisfies*

$-a^2 \leq \sigma \leq b^2$, *there exists a constant* $\mathsf{K}_\delta(n, q)$, *such that for all functions* $f \in H_1^q(M_n)$ *with compact support included in* $B_P(\delta)$:

$$(12) \qquad \|f\|_p \leq \mathsf{K}_\delta(n, q)\|\nabla f\|_q$$

For δ *sufficiently small we can make* $\mathsf{K}_\delta(n, q)$ *is as close as we like to* $\mathsf{K}(n, q)$. $\mathsf{K}_\delta(n, q)$ *depends on a and b, but does not depend on P.*

Proof. Let (r, θ_i), $(i = 1, 2, \ldots, n-1)$ be a system of geodesic polar coordinates with center P. According to Theorem 1.53, the components g_{ii} of the metric tensor satisfy:

$$\frac{\sin br}{br} \leq \sqrt{g_{ii}(r, \theta)} \leq \frac{\sinh(ar)}{ar}, \ g_{rr} = 1.$$

Let $\varepsilon > 0$ be given; one may choose δ small enough so that, when $r \leq \delta$:

$$\sinh(ar)/ar \leq 1 + \varepsilon \quad \text{and} \quad \sin(br)/br \geq 1 - \varepsilon.$$

Setting $\tilde{f}(x) = f(\exp_p x)$, when $\|x\| \leq \delta$, then according to Theorem 2.14 we have

$$\left(\int |\tilde{f}|^p \, dE\right)^{1/p} \leq \mathsf{K}(n, q)\left(\int |\nabla_E \tilde{f}|^q \, dE\right)^{1/q}.$$

Since $(1 - \varepsilon)^{n-1} \, dE \leq dV \leq (1 + \varepsilon)^{n-1} \, dE$ and $|\nabla_E \tilde{f}| \leq (1 + \varepsilon)|\nabla f|$; this establishes inequality (12) with

$$\mathsf{K}_\delta(n, q) = (1 - \varepsilon)^{(1-n)/q}(1 + \varepsilon)^{1+(n-1)/p}\mathsf{K}(n, q). \qquad \blacksquare$$

2.25 Lemma (Calabi). *Let* M_n *be a Riemannian manifold with injectivity radius* $\delta_0 > 0$; *then for all* $\delta > 0$, *there exist two real numbers* γ *and* β $(0 < \gamma < \beta < \delta)$, *a sequence of points* $P_i \in M_n$, *and* $\{\Omega_i\}$ *a partition of* M_n, *by sets, satisfying* $B_{P_i}(\gamma) \subset \Omega_i \subset B_{P_i}(\beta)$ *for all i*, $B_P(\rho)$ *being the ball with center P and radius* ρ.

According to this lemma, we are able to prove:

2.26 Lemma. *Let* M_n *be a Riemannian manifold with injectivity radius* $\delta_0 > 0$ *and bounded curvature; then there exists, if* δ *is small enough, a uniformly locally finite covering of* M_n *by a family of open balls* $B_{P_i}(\delta)$.

Uniformly locally finite means: there exists a constant k, which may depend on δ, such that each point $P \in M_n$ has a neighborhood whose intersection with each $B_{P_i}(\delta)$, at most except k, is empty.

Let us prove that the balls $B_{P_i}(\beta)$ form a uniformly locally finite covering of M_n. Let $B_{P_j}(\beta)$ be given, and suppose that k balls $B_{P_i}(\beta)$ have a nonempty intersection with $B_{P_j}(\beta)$, $i \neq j$. Since the curvature is bounded, the set of the points Q satisfying $d(P_j, Q) < 2\beta + \gamma$ has a measure less than a constant w independent of j. By theorem (1.53), if the Ricci curvature is greater than $-(n-1)\alpha^2$; therefore

$$w \leq \omega_{n-1} \int_0^{2\beta+\gamma} [\sinh(\alpha r)/\alpha]^{n-1}\, dr.$$

Also, if the sectional curvatures are less than b^2, then the measure of $B_{P_i}(\gamma)$ is greater than y, with $y = \omega_{n-1}\int_0^\gamma [\sin(br)/b]^{n-1}\, dr$. Therefore $(k+1)\, y \leq w$, since the balls $B_{P_i}(\gamma)$ are disjoint. ■

2.27 *Proof of Theorem* 2.21 (Continued). Consider a partition of unity $h_i \in C^\infty$ subordinate to the covering $\{B_{P_i}(\delta)\}$, such that $|\nabla(h_i^{1/q})|$ is uniformly bounded, ($|\nabla(h_i^{1/q})| \leq H$, for all i, where H is a constant). Such functions h_i exist, because the covering is uniformly locally finite.
Let $\{\tilde{h}_i\} \in C^\infty$ be a partition of unity subordinate to the covering $\{B_{P_i}(\delta)\}$ such that $|\nabla \tilde{h}_i| <$ Const. We may set $h_i = \tilde{h}_i^m/(\Sigma \tilde{h}_i^m)$, with m an integer greater than q. Let I be a finite subset of $\mathbb{N}$. By Lemma 2.24,

$$\begin{aligned}\sum_{i\in I} \|\varphi^q h_i\|_{p/q} &= \sum_{i\in I} \|\varphi h_i^{1/q}\|_p^q \leq \mathsf{K}_\delta^q(n,q) \sum_{i\in I} \|\nabla(\varphi h_i^{1/q})\|_q^q \\ &\leq \mathsf{K}_\delta^q(n,q) \sum_{i\in I} \int (|\nabla\varphi|\, h_i^{1/q} + \varphi|\nabla h_i^{1/q}|)^q\, dV \\ &\leq \mathsf{K}_\delta^q(n,q) \int \sum_{i\in I} \Big[|\nabla\varphi|^q h_i + \mu |\nabla\varphi|^{q-1} h_i^{(q-1)/q} |\varphi|\, |\nabla h_i^{1/q}| \\ &\qquad + \nu |\varphi|^q |\nabla h_i^{1/q}|^q \Big], \end{aligned}$$

$$\leq \mathsf{K}_\delta^q(n,q)[\|\nabla\varphi\|_q^q + \mu k H \|\nabla\varphi\|_q^{q-1}\|\varphi\|_q + \nu k H^q \|\varphi\|_q^q], \tag{13}$$

using Hölder's inequality. μ and ν are two constants such that for $t \geq 0$

$$(1+t)^q \leq 1 + \mu t + \nu t^q;$$

for instance, $\mu = \nu q$ and $\nu = \sup(1, 2^{q-2})$, if $q > 1$. Recall that $\beta \leq \delta$. If we choose δ small enough, then $K_\delta(n,q) < \mathsf{K}_\delta(n,q) < \mathsf{K}(n,q) + \varepsilon/2$. Since the last expression of (13) is independent of I, we establish the inequality

$$\begin{aligned}\|\varphi\|_p^q = \|\varphi^q\|_{p/q} &= \Big\| \sum_{i\in N} \varphi^q h_i \Big\|_{p/q} \leq \sum_{i\in N} \|\varphi^q h_i\|_{p/q} \\ &\leq [\mathsf{K}(n,q) + \varepsilon/2]^q [(1+\varepsilon_0)\|\nabla\varphi\|_q^q + A(\varepsilon_0)\|\varphi\|_q^q]\end{aligned}$$

by virtue of the inequality:

(14) $$qx^{q-1}y \leq \lambda(q-1)x^q + \lambda^{1-q}y^q,$$

valid with any x, y, and λ, three positive real numbers. To complete the proof of inequality (10), we have only to set $\lambda = q\varepsilon_0/\mu k H(q-1)$, $x = \|\nabla\varphi\|_q$, and $y = \|\varphi\|_q$, where ε_0 is small enough so that $[\mathrm{K}(n,q)+\varepsilon/2][1+\varepsilon_0]^{1/q} \leq \mathrm{K}(n,q)+\varepsilon$, and $A_q(\varepsilon) = [\mathrm{K}(n,q)+\varepsilon/2][A(\varepsilon_0)]^{1/q}$ ∎

§8. Optimal Inequalities

2.28 Theorem. *Let M_n be a C^∞ Riemannian manifold with injectivity radius $\delta_0 > 0$. If the curvature is constant or if the dimension is two and the curvature bounded, then $A_q(0)$ exists and every $\varphi \in H_1^q(M_n)$ satisfies*

$$\|\varphi\|_p \leq \mathrm{K}(n,q)\|\nabla\varphi\|_q + A_q(0)\|\varphi\|_q.$$

For $\mathbb{R}^n$ and $\mathbb{H}_n$ the hyperbolic space, the inequality holds with $A_q(0) = 0$.

For the proof, see Aubin [13] pp. 595 and 597.

2.29 Theorem. *There exists a constant $A(q)$ such that every $\varphi \in H_1^q(\mathbb{S}_n)$ satisfies:*

$$\|\varphi\|_p^q \leq \mathrm{K}^q(n,q)\|\nabla\varphi\|_q^q + A(q)\|\varphi\|_q^q \quad \text{if} \quad 1 \leq q \leq 2,$$

$$\|\varphi\|_p^{q/(q-1)} \leq \mathrm{K}^{q/(q-1)}(n,q)\|\nabla\varphi\|_q^{q/(q-1)} + A(q)\|\varphi\|_q^{q/(q-1)} \quad \text{if} \quad 2 \leq q < n.$$

Let $M_n (n \geq 3)$ be a Riemannian manifold, with constant curvature and injectivity radius $\delta_0 > 0$. There exists a constant A, such that every $\varphi \in H_1^2(M_n)$ satisfies:

$$\|\varphi\|_{2n/(n-2)}^2 \leq \mathrm{K}^2(n,2)\|\nabla\varphi\|_2^2 + A\|\varphi\|_2^2.$$

For the sphere of volume 1, the inequality holds with $A = 1$.

See Aubin [13] pp. 588 and 598. For the proof of the last part of the theorem see Aubin [14] p. 293.

§9. Sobolev's Theorem for Compact Riemannian Manifolds with Boundary

2.30 Theorem. *For the compact manifolds $\overline{W}_n$ with C^r-boundary, $(r \geq 1)$, the Sobolev imbedding theorem holds. More precisely:*
First part. The imbedding $H_k^q(W) \subset H_\ell^p(W)$ is continuous with $1/p = 1/q - (k-\ell)/n > 0$. Moreover, for any $\varepsilon > 0$, there exists a constant $A_q(\varepsilon)$ such

that every $\varphi \in \mathring{H}_1^q(W_n)$ *satisfies inequality* (10) *and such that every* $\varphi \in H_1^q(W_n)$ *satisfies*:

$$\|\varphi\|_p \leq [2^{1/n}K(n, q) + \varepsilon]\|\nabla\varphi\|_q + A_q(\varepsilon)\|\varphi\|_q. \tag{15}$$

Second part. The following imbeddings are continuous:

(a) $H_k^q(W) \subset C_B^s(W)$, *if* $k - n/q > s \geq 0$, s *being an integer*,
(b) $H_k^q(W) \subset C^s(\overline{W})$, *if in addition* $s < r$;
(c) $H_k^q(W) \subset C^\alpha(\overline{W})$, *if* α *satisfies* $0 < \alpha < 1$ *and* $\alpha \leq k - n/q$.

Proof of the first part. Let (Ω_i, φ_i) be a finite C^r-atlas of $\overline{W}_n$, each Ω_i being homeomorphic either to a ball of $\mathbb{R}^n$ or to a half ball $D \subset \bar{E}$. As in the proof of Theorem 2.20, we have only to prove inequality (9) for all $f \in H_1^q(W) \cap C^\infty(W)$, α_i being a C^r partition of unity subordinate to the covering Ω_i. When Ω_i is homeomorphic to a ball, the proof is that of theorem (2.20). When Ω_i is homeomorphic to a half ball, the proof is similar. But one applies the following lemma:

2.31 Lemma. *Let* ψ *be* C^1-*function on* $\bar{E}$, *whose support belongs to* D, *then* ψ *satisfies*:

$$\|\psi\|_p \leq 2^{1/n}K(n, q)\|\nabla\psi\|_q, \quad \text{with } 1/p = 1/q - 1/n > 0.$$

Proof. Recall that E is the half-space of $\mathbb{R}^n$ and $D = B \cap \bar{E}$, where B is the open ball with center 0 and radius 1.
Consider $\tilde{\psi}$ defined, for $x \in \bar{E}$, by $\tilde{\psi}(x) = \psi(x)$ and $\tilde{\psi}(\tilde{x}) = \psi(x)$, when $\tilde{x} = (-x_1, x_2, \ldots, x_n)$, $(x_1, x_2, \ldots, x_n)$ being the coordinates of x. $\tilde{\psi}$ is a Lipschitzian function with compact support, thus $\tilde{\psi} \in H_1^q(\mathbb{R}^n)$ and according to Theorem 2.14:

$$\|\tilde{\psi}\|_p \leq K(n, q)\|\nabla\tilde{\psi}\|_q.$$

The lemma follows, since

$$2\int_{\bar{E}} |\psi|^p \, dE = \int_{\mathbb{R}^n} |\tilde{\psi}|^p \, dE \quad \text{and} \quad 2\int_{\bar{E}} |\nabla\psi|^q \, dE = \int_{\mathbb{R}^n} |\nabla\tilde{\psi}|^q \, dE. \qquad \blacksquare$$

2.32 The proof that every $\varphi \in \mathring{H}_1^q(W_n)$ satisfies inequality (10) is similar to that in 2.27. But here it is easier because the covering is finite.

Using Lemma 2.31, we can prove that all $\varphi \in H_1^q(W_n)$ satisfy (15); for a complete proof see Cherrier [97].

Proof of the second part of Theorem 2.30. a) There exist constants $C_i(\tilde{q})$ such that for all $f \in H_1^{\tilde{q}}(W) \cap C^\infty(W)$

$$\sup|\alpha_i f| \leq C_i(\tilde{q})\|f\|_{H_1^{\tilde{q}}}, \text{ if } \tilde{q} > n. \tag{16}$$

Set $h(x) = 0$ for $x \notin D$, and $h(x) = (\alpha_i f) \circ \varphi_i^{-1}(x)$ for $x \in D$. Consider a half straight line through x, defined by $\theta \in \mathbb{S}_{n-1}(1)$, entirely included in E. We have

$$|h(x)| \leq \int_0^1 |\nabla h(r, \theta)|\, dr.$$

Now we proceed as in the proof of Lemma 2.22, but integration with respect to θ is only over half of $\mathbb{S}_{n-1}(1)$:

$$|h(x)| \leq (\omega_{n-1}/2)^{-1}\left(\int_E |\nabla h|^{\tilde{q}}\, dx\right)^{1/\tilde{q}} \left(\frac{\omega_{n-1}}{2}\int_0^1 r^{(n-1)(1-q')}\, dr\right)^{1/q'},$$

with $1/q' = 1 - 1/\tilde{q}$.
Since the metric tensor is bounded on Ω_i (by proper choice of the (Ω_i, φ_i), without loss of generality), for some constant $\tilde{C}_i(\tilde{q})$ we obtain:

$$\sup |\alpha_i f| \leq \tilde{C}_i(\tilde{q}) \|\nabla(\alpha_i f)\|_{\tilde{q}}.$$

Inequality (16) follows; thus, recalling that I is finite, we have

$$\sup |f| \leq C(\tilde{q}) \|f\|_{H_1^{\tilde{q}}}, \quad \text{with } C(\tilde{q}) = \sum_{i \in I} C_i(\tilde{q}).$$

Since $k > n/q$, there exists $\tilde{q} > n$, such that the imbedding $H_k^q(W) \subset H_1^{\tilde{q}}(W)$ is continuous; we have only to choose $1/\tilde{q} \geq 1/q - (k-1)/n$. So there exists a constant C, such that every $f \in H_k^q(W) \cap C^\infty(W)$ satisfies:

$$\sup |f| \leq C \|f\|_{H_k^q}.$$

Thus a Cauchy sequence of C^∞ functions in $H_k^q(W)$ is a Cauchy sequence in $C^0(W)$ and the preceding inequality holds for all $f \in H_k^q(W)$.
For $s > 0$, apply the preceding result to $|\nabla^\ell f|$, $0 \leq \ell \leq s$, and the continuous imbedding $H_k^q(W) \subset C_B^s(W)$ is established (the proof is similar to that of 2.23).

b) Instead of taking $f \in C^\infty(W)$, we may establish an inequality of the type:

$$\|f\|_{C^s} \leq A \|f\|_{H_{s+1}^{\tilde{q}}}, \quad \text{when } 0 \leq s < r, \tag{17}$$

for the functions $f \in C^r(\overline{W})$, with A a constant and $\tilde{q} > n$.
According to Theorem 2.9, $C^r(\overline{W})$ is dense in $H_{s+1}^{\tilde{q}}(W)$. Thus $H_{s+1}^{\tilde{q}}(W) \subset C^s(\overline{W})$ and inequality (17) holds for all $f \in H_{s+1}^{\tilde{q}}(W)$. When $k - n/q > s$, we may choose $\tilde{q} > n$, such that $1/\tilde{q} \geq 1/q - (k-s-1)/n$. In this case the imbedding $H_k^q(W) \subset H_{s+1}^{\tilde{q}}(W)$ is continuous and so $H_k^q(W) \subset C^s(\overline{W})$.

c) And now for the last part of Theorem 2.30:
Let $f \in H_k^q(W)$; according to the preceding result $f \in C^0(\overline{W})$ because $k - n/q > 0$. Consider the function on D, defined by $h(x) = (\alpha_i f) \circ \varphi_i^{-1}(x)$, for a given $i \in I$. By a proof similar to that in 2.23, we establish the existence of a constant B such that every $f \in H_1^{\tilde q}(W)$ satisfies

$$|h(x) - h(y)| \|x - y\|^{-\alpha} \le B\left(\int_D |\nabla h|^{\tilde q}\, dx\right)^{1/\tilde q},$$

where $\tilde q = n/(1 - \alpha)$. Instead of considering a ball of radius $\|x - y\|/2$, we must integrate over a cube K with edge $\|x - y\|$, included in $\bar E$, with x and y belonging to K (see Adams [1] p. 109).
Then, since the metric tensor is bounded on Ω_i, there exists a constant B_i such that, for every pair (P, Q) of points of $\overline{W}$, any $f \in H_1^{\tilde q}$ satisfies:

$$|\alpha_i(P) f(P) - \alpha_i(Q) f(Q)| [d(P, Q)]^{-\alpha} \le B_i \|f\|_{H_1^{\tilde q}}.$$

Thus we establish the desired inequality:

$$|f(P) - f(Q)| |d(P, Q)|^{-\alpha} \le \left(\sum_{i \in I} B_i\right) \|f\|_{H_1^{\tilde q}} \le \text{Const} \times \|f\|_{H_k^q},$$

where the last inequality follows from the first part of the Sobolev imbedding theorem, since $\tilde q = n/(1 - \alpha)$ satisfies $1/\tilde q \ge 1/q - (k - 1)/n$. ■

§10. The Kondrakov Theorem

2.33 *Let $k \ge 0$ be an integer, p and q two real numbers satisfying $1 \ge 1/p > 1/q - k/n > 0$. The Kondrakov Theorem asserts that, if Ω, a bounded open set of $\mathbb{R}^n$, has a sufficiently regular boundary $\partial\Omega$ ($\partial\Omega$ of class C^1, or only Lipschitzian):*

(a) *the imbedding $H_k^q(\Omega) \subset L_p(\Omega)$ is compact.*
(b) *With the same assumptions for Ω, the imbedding $H_k^q(\Omega) \subset C^\alpha(\bar\Omega)$ is compact, if $k - \alpha > n/q$, with $0 \le \alpha < 1$.*
(c) *For Ω a bounded open set of $\mathbb{R}^n$, the following imbeddings are compact:*

$$\mathring{H}_k^q(\Omega) \subset L_p(\Omega),\ \mathring{H}_k^q(\Omega) \subset C^\alpha(\bar\Omega).$$

Proof. Roughly, the proof consists in proving that if the Sobolev imbedding theorem holds for a bounded domain Ω, then the Kondrakov theorem is true for Ω.

a) According to the Sobolev imbedding theorem 2.30, the imbedding $H_k^q \subset H_1^{\tilde q}$ is continuous with $1/\tilde q = 1/q - (k - 1)/n$. Thus we have only to prove that the imbedding of $H_1^{\tilde q} \subset L_p$ is compact when $1 \geq 1/p > 1/\tilde q - 1/n > 0$, since the composition of two continuous imbeddings is compact if one of them is compact.

Let $\mathscr{A}$ be a bounded subset of $H_1^{\tilde q}(\Omega)$, so if $f \in \mathscr{A}$,

$$\|f\|_{H_1^{\tilde q}} \leq C, \text{ a constant.}$$

By hypothesis $H_1^{\tilde q}(\Omega) \subset L_r$, with $1/r = 1/\tilde q - 1/n$, and there exists a constant A such that for $f \in H_1^{\tilde q}(\Omega)$,

$$\|f\|_r \leq A\|f\|_{H_1^{\tilde q}}.$$

Set $K_j = \{x \in \Omega/\text{dist}(x, \partial\Omega) \geq 2/j\}, j \in \mathbb{N}$. For $f \in \mathscr{A}$, by Hölder's inequality:

$$\int_{\Omega - K_j} |f|\,dx \leq \left(\int_{\Omega - K_j} |f|^r\,dx\right)^{1/r}\left(\int_{\Omega - K_j} dx\right)^{1-1/r} \leq AC\left(\int_{\Omega - K_j} dx\right)^{1-1/r},$$

which goes to zero, when $j \to \infty$. Thus, given $\varepsilon > 0$, there exists $j_0 \in \mathbb{N}$, such that $[\text{vol}(\Omega - K_{j_0})]^{1-1/r} \leq \varepsilon/AC$. Now, by Fubini's theorem:

$$\int_{K_{j_0}} |f(x+y) - f(x)|\,dx \leq \int_{K_{j_0}} dx \int_0^1 \left|\frac{d}{dt} f(x+ty)\right| dt$$

$$\leq \|y\| \int_{K_{2j_0}} |\nabla f|\,dx \leq \|y\| \|\nabla f\|_1$$

for $\|y\| < 1/j_0$ since $x + y \in K_{2j_0}$, if $x \in K_{j_0}$. Since $C^\infty(\Omega)$ is dense in $H_1^{\tilde q}(\Omega)$, the preceding inequality holds for any $f \in H_1^{\tilde q}(\Omega)$. Moreover, by Hölder's inequality, $\|\nabla f\|_1 \leq \|\nabla f\|_r(\text{vol}\,\Omega)^{1-1/r} \leq B$, a constant.
Theorem 3.44 with $\delta = \varepsilon/B$ then implies that $\mathscr{A}$ is precompact in $L_1(\Omega)$. Hence $\mathscr{A}$ is precompact in $L_p(\Omega)$, because if $f_m \in \mathscr{A}$ is a Cauchy sequence in L_1, it is a Cauchy sequence in L_p:

$$\|f_m - f_\ell\|_p \leq \|f_m - f_\ell\|_1^\mu \|f_m - f_\ell\|_r^{1-\mu} \leq (2AC)^{1-\mu}\|f_m - f_\ell\|_1^\mu,$$

by Hölder's inequality, with $\mu = [(r/p) - 1]/(r - 1)$.

b) *Proof of the second part of Theorem* 2.33. Let λ satisfy $\alpha < \lambda < \inf(1, k - n/q)$. Then by the Sobolev imbedding theorem 2.30, $H_k^q(\Omega)$ is included in $C^\lambda(\bar\Omega)$, and there exists a constant A such that $\|f\|_{C^\lambda} \leq A\|f\|_{H_k^q}$.

Let $\mathscr{A}$ be a bounded subset of $H_k^q(\Omega)$; if $f \in \mathscr{A}$, $\|f\|_{H_k^q} \leq C$, a constant, and $\|f\|_{C^\lambda} \leq AC$.
Thus we can apply Ascoli's Theorem, 3.15. $\mathscr{A}$ is a bounded subset of equicontinuous functions of $C^0(\bar\Omega)$, and $\bar\Omega$ is compact. So $\mathscr{A}$ is precompact in $C^0(\bar\Omega)$.

Then, since

$$|f(x) - f(y)| \|x - y\|^{-\alpha} = (|f(x) - f(y)| \|x - y\|^{-\lambda})^{\alpha/\lambda} |f(x) - f(y)|^{1-\alpha/\lambda}$$

if a sequence $f_m \in \mathscr{A}$ converges to f in $C^0(\bar{\Omega})$, $\|f\|_{C^\lambda} \leq AC$ and

$$\|f - f_m\|_{C^\alpha} \leq (2AC)^{\alpha/\lambda}(\|f - f_m\|_{C^0})^{1-\alpha/\lambda} + \|f - f_m\|_{C^0}.$$

Thus $\mathscr{A}$ is precompact in $C^\alpha(\bar{\Omega})$.

c) $\mathscr{D}(\Omega)$ is included in $\mathscr{D}(\mathbb{R}^n)$, so we can apply the theorem of Sobolev, 2.10, to the space $\mathring{H}_k^q(\Omega)$. A proof similar to those of a) and b) gives the desired result.

§11. Kondrakov's Theorem for Riemannian Manifolds

2.34 Theorem. *The Kondrakov theorem, 2.33, holds for the compact Riemannian manifolds M_n, and the compact Riemannian manifolds $\overline{W}_n$ with C^1-boundary. Namely, the following imbeddings are compact:*

(a) $H_k^q(M_n) \subset L_p(M_n)$ *and* $H_k^q(W_n) \subset L_p(W_n)$, *with* $1 \geq 1/p > 1/q - k/n > 0$.

(b) $H_k^q(M_n) \subset C^\alpha(M_n)$ *and* $H_k^q(W_n) \subset C^\alpha(\overline{W}_n)$, *if* $k - \alpha > n/q$, *with* $0 \leq \alpha < 1$.

Proof. Let (Ω_i, φ_i), $(i = 1, 2, \ldots, N)$ be a finite atlas of M_n (respectively, C^1-atlas of $\overline{W}_n$), each Ω_i being homeomorphic either to a ball of $\mathbb{R}^n$ or to a half ball $D \subset \bar{E}$. We choose the atlas so that in each chart the metric tensor is bounded. Consider a C^∞ partition of unity $\{\alpha_i\}$ subordinate to the covering $\{\Omega_i\}$. It is sufficient to prove the theorem in the special case $k = 1$ for the same reason as in the preceding proof, 2.33.

a) Let $\{f_m\}$ be a bounded sequence in H_1^q. Consider the functions defined on B (or on D), i being given:

$$h_m(x) = (\alpha_i f_m) \circ \varphi_i^{-1}(x).$$

Since the metric tensor is bounded on Ω_i, the set $\mathscr{A}_i$ of these functions is bounded in $H_1^q(\Omega)$ with $\Omega = B$ or D. (The boundary of D is only Lipschitzian, but, since $\operatorname{supp}(\alpha_i \circ \varphi_i^{-1})$ is included in B, we may consider a bounded open set Ω with smooth boundary which satisfies $D \subset \Omega \subset E$).
According to Theorem 2.33, $\mathscr{A}_i$ is precompact. Thus there exists a subsequence which is a Cauchy sequence in L_p. Repeating this operation successively for $i = 1, 2, \ldots, N$, we may select a subsequence $\{\tilde{f}_m\}$ of the sequence $\{f_m\}$, such that $\alpha_i \tilde{f}_m$ is a Cauchy sequence in L_p for each i.

Thus $\{\tilde{f}_m\}$ is a Cauchy sequence in L_p, since

$$|\tilde{f}_m - \tilde{f}_\ell| \leq \sum_{i=1}^{N} |\alpha_i f_m - \alpha_i f_\ell|.$$

b) Let λ satisfy $\alpha < \lambda < \inf(1, k - n/q)$. According to Theorems 2.21 and 2.30, the imbeddings $H_k^q(M_n) \subset C^\lambda(M_n)$ and $H_k^q(W_n) \subset C^\lambda(\overline{W}_n)$ are continuous. Thus the same proof used to show 2.33, b) establishes the result.

2.35 Remark. We have given only the main results concerning the theorems of Sobolev and Kondrakov. These theorems are proved for the compact manifolds with Lipschitzian boundary in Aubin [17]. To obtain complete results for domains of $\mathbb{R}^n$ see Adams [1].

2.36 Remark. Instead of the spaces $H_k^p(M_n)$, it is possible to introduce the spaces $H_k'^p(M_n)$, which are the completion of $\mathscr{E}_k'^p$ with respect to the norm

$$\|\varphi\|_{H_k'^p} = \sum_{0 = \ell \leq k/2} \|\Delta^\ell \varphi\|_p + \sum_{0 = \ell \leq (k-1)/2} \|\nabla \Delta^\ell \varphi\|_p,$$

with $\mathscr{E}_k'^p$ the vector space of the functions $\varphi \in C^\infty(M_n)$, such that $\Delta^\ell \varphi \in L_p(M_n)$ for $0 \leq \ell \leq k/2$ and such that $|\nabla \Delta^\ell \varphi| \in L_p(M_n)$ for $0 \leq \ell \leq (k-1)/2$. For these spaces, the Kondrakov theorem holds, as well as the Sobolev imbedding theorem when $p > 1$ (see Aubin [17]).

For $k \leq 1$, $H_k^p = H_k'^p$. But for $k > 1$, the spaces $H_k'^p$ may be more convenient for the study of differential equations.

§12. Examples

2.37 The exceptional case of the Sobolev imbedding theorem (i.e., $H_k^{n/k}$ on n-dimensional manifolds).

Consider the function f on $\mathbb{R}^2$ defined by:

$$\mathbb{R}^2 \ni x \to f(x) = \log|\log\|x\||, \quad \text{for } 0 < \|x\| < 1/e, \text{ and } f(x) = 0,$$

for the other points x of $\mathbb{R}^2$.

$$\|\nabla f\|_2^2 = 2\pi \int_0^{1/e} \frac{dr}{r|\log r|^2} = 2\pi$$

and f^2 is integrable; thus $f \in H_1^2(\mathbb{R}^2)$.

As $1/q - 1/n = 0$ here, we could hope that the function f would be bounded, but it is not: $H_k^{n/k} \not\subset L_\infty$. On the other hand, e^f is integrable:

$$\|e^f\|_1 = 2\pi \int_0^{1/e} r|\log r|\,dr, \quad \text{see (2.46).}$$

2.38 The Sobolev imbedding theorem $H_k^q \subset L_p$ holds when $1/p = 1/q - k/n > 0$, but the Kondrakov theorem does not.

Consider the sequence of functions f_k defined on $\mathbb{R}^n$ ($n > 4$) by:

$$f_k(x) = \left(\frac{1}{k}\right)^{(n-2)/4} \left(\frac{1}{k} + \|x\|^2\right)^{1-n/2}.$$

Let us verify $f_k \in H_1^2(\mathbb{R}^n)$. Now

$$\|f_k\|_2^2 = \omega_{n-1} k^{1-n/2} \int_0^\infty \left(\frac{1}{k} + r^2\right)^{2-n} r^{n-1}\, dr = \frac{\omega_{n-1}}{k} \int_0^\infty (1+t^2)^{2-n} t^{n-1}\, dt$$

is finite and so is

$$\|\nabla f_k\|_2^2 = \omega_{n-1} k^{1-n/2} \int_0^\infty (n-2)^2 \left(\frac{1}{k} + r^2\right)^{-n} r^{n+1}\, dr$$

$$= \omega_{n-1}(n-2)^2 \int_0^\infty (1+t^2)^{-n} t^{n+1}\, dt = A.$$

Also, f_k belongs to L_N, with $N = 2n/(n-2)$, because

$$\int_0^\infty f_k^N r^{n-1}\, dr = \left(\frac{1}{k}\right)^{n/2} \int_0^\infty \left(\frac{1}{k} + r^2\right)^{-n} r^{n-1}\, dr$$

$$= \int_0^\infty (1+t^2)^{-n} t^{n-1}\, dt = C < \infty.$$

Let $h_k(x) = f_k(x) - (\sqrt{k} + 1/\sqrt{k})^{1-n/2}$ for $\|x\| \leq 1$.
Then $h_k \in \mathring{H}_1^2(B)$, where B is the unit ball of $\mathbb{R}^n$ with center 0. Clearly, $\|h_k\|_{H_1^2(B)} \to \sqrt{A}$ when $k \to \infty$, the sequence h_k is bounded in $\mathring{H}_1^2(B)$, and $\|h_k\|_{L_N(B)} \to C^{1/N} \neq 0$.
Now $h_k(x) \to 0$ when $\|x\| \neq 0$. Thus a subsequence of $\{h_k\}$ cannot converge in L_N, without the limit being zero in L_N. But this contradicts the above result ($C \neq 0$). The imbedding $H_1^2(B)$ in $L_N(B)$ is not compact.

§13. Improvement of the Best Constants

2.39 Let M_n be a complete Riemannian manifold with bounded curvature and injectivity radius $\delta > 0$. According to Theorem 2.21, if $1/p = 1/q - 1/n > 0$ then every $\varphi \in H_1^q(M_n)$ satisfies:

$$(18) \qquad \|\varphi\|_p \leq B\|\nabla\varphi\|_q + A\|\varphi\|_q,$$

where A and B are constants. It is proved in 3.78 and 3.79 that $\mathsf{K} = \{\inf B$ such that all $\varphi \in H_1^q(M_n)$ satisfy inequality (18) for a certain value, $A(B)\}$ is strictly positive. By Theorem 2.21, K depends only on the dimension n and q: $\mathsf{K} = \mathsf{K}(n, q)$. For the value of $\mathsf{K}(n, q)$, see Theorem 2.14. We are going to show that the best constants $\mathsf{K}(n, q)$ can be lowered if the functions φ satisfy some additional natural orthogonality conditions.

2.40 Theorem. *Let M_n be a complete Riemannian manifold with bounded curvature and injectivity radius $\delta > 0$, and $f_i (i \in I)$ functions of class C^1, with the following properties: they change sign, their gradients are uniformly bounded, and the family $\{|f_i|^q\}$ forms a partition of unity (subordinate to a uniformly locally finite cover by bounded open sets).*
Then the functions $\varphi \in H_1^q(M_n)$, which satisfy the conditions

$$(19) \qquad \int_{M_n} |\varphi|^p f_i |f_i|^{p-1}\, dV = 0 \quad \textit{for all} \quad i \in I,$$

satisfy inequality (18) *with pairs $(B, A(B))$, with B as close as one wants to $2^{-1/n}\mathsf{K}(n, q)$. Thus the best constant B of inequality* (18) *is $2^{1/n}$ times smaller for functions φ satisfying* (19).

Proof. Set $\check{f} = \sup(f, 0)$ and $\tilde{f} = \sup(-f, 0)$; thus $f = \check{f} - \tilde{f}$. Unless otherwise stated, integration is over M_n. Since $|\nabla|\varphi|| = |\nabla\varphi|$ almost everywhere, Proposition 3.49, we can suppose, without loss of generality, $\varphi \geq 0$. On the other hand, the functions f_i may be chosen more generally, for instance, uniformly Lipschitzian.

By hypothesis, $\int \varphi^p \check{f}_i^p\, dV = \int \varphi^p \tilde{f}_i^p\, dV$, and $\varphi\tilde{f}_i$, as well as $\varphi\check{f}_i$, belong to $H_1^q(M_n)$.

Let $K > \mathsf{K}(n, q)$ and $A_0 = A(K)$, the corresponding constant in (18).

$$\|\varphi\check{f}_i\|_p^q \leq K^q\|\nabla(\varphi\check{f}_i)\|_q^q + A_0\|\varphi\check{f}_i\|_q^q,$$

$$\|\varphi\tilde{f}_i\|_p^q \leq K^q\|\nabla(\varphi\tilde{f}_i)\|_q^q + A_0\|\varphi\tilde{f}_i\|_q^q.$$

Suppose, for instance, that $\|\nabla(\varphi\check{f}_i)\|_q \geq \|\nabla(\varphi\tilde{f}_i)\|_q$. Then write:

$$\begin{aligned}\|\varphi f_i\|_p^q = 2^{q/p}\|\varphi\tilde{f}_i\|_p^q &\leq 2^{q/p}(K^q\|\nabla(\varphi\tilde{f}_i)\|_q^q + A_0\|\varphi\tilde{f}_i\|_q^q)\\ &\leq 2^{-q/n}K^q(\|\nabla(\varphi\tilde{f}_i)\|_q^q + \|\nabla(\varphi\check{f}_i)\|_q^q) + 2^{q/p}A_0\|\varphi\tilde{f}_i\|_q^q.\end{aligned}$$

If $\|\nabla(\varphi\check{f}_i)\|_q < \|\nabla(\varphi\tilde{f}_i)\|_q$, we obtain a similar inequality, using $\varphi\check{f}_i$ instead of $\varphi\tilde{f}_i$. Thus in all cases:

$$\|\varphi f_i\|_p^q \leq 2^{-q/n}K^q\|\nabla(\varphi f_i)\|_q^q + 2^{q/p}A_0\|\varphi f_i\|_q^q.$$

Consider again the computation in (2.27). Set $H = \sup_{i \in I} \sup_M |\nabla f_i|$ and pick k an integer such that, at every point of M, at most k of the functions f_i are nonzero. Let J be a finite subset of I. Then there exist constants μ and ν such that

$$\sum_{i \in J} \|\varphi f_i\|_p^q \leq 2^{-q/n} K^q [\|\nabla\varphi\|_q^q + \mu k H \|\nabla\varphi\|_q^{q-1} \|\varphi\|_q + \nu k H^q \|\varphi\|_q^q]$$
$$+ 2^{q/p} A_0 \|\varphi\|_q^q.$$

The second member does not depend on J, hence the corresponding series is convergent. There exist two constants β and γ, such that

$$\|\varphi\|_p^q = \|\varphi^q\|_{p/q} = \left\| \sum_{i \in I} \varphi^q |f_i|^q \right\|_{p/q} \leq \sum_{i \in I} \|\varphi^q |f_i|^q\|_{p/q}$$
$$\leq 2^{-q/n} K^q \|\nabla\varphi\|_q^q + \beta \|\nabla\varphi\|_q^{q-1} \|\varphi\|_q + \gamma \|\varphi\|_q^q.$$

Using inequality (14), for any $\varepsilon_1 > 0$, there exists an M_0 such that

(20) $$\|\nabla\varphi\|_q^{q-1} \|\varphi\|_q \leq \varepsilon_1 \|\nabla\varphi\|_q^q + M_0 \|\varphi\|_q^q.$$

Thus the stated result is proved:

$$\|\varphi\|_p^q \leq (2^{-q/n} K^q + \beta\varepsilon_1) \|\nabla\varphi\|_q^q + (\gamma + M_0 \beta) \|\varphi\|_q^q.$$

To establish that $2^{-1/n} \mathrm{K}(n, q)$ is the best constant, when the functions φ satisfy (19), we have only to use the functions which are defined in Theorem 2.14. Indeed, for convenience, suppose $\tilde{f}_1$ is equal to 1 over a ball $\tilde{B}$ with center $\tilde{P}$ and radius $\rho < \delta$, and $\check{f}_1$ is equal to 1 over a ball $\check{B}$ with center $\check{P}$ and radius ρ.

Consider the sequence of functions $\psi_m (m \in N)$, which vanish outside $\tilde{B} \cup \check{B}$ and which are defined by

$$\psi_m(Q) = (r^{q/(q-1)} + 1/m)^{1-n/q} - (\rho^{q/(q-1)} + 1/m)^{1-n/q}$$

for $Q \in \tilde{B}$ with $r = d(\tilde{P}, Q)$, and for $Q \in \check{B}$ with $r = d(\check{P}, Q)$. This sequence satisfies:

$$\lim_{m \to \infty} \|\psi_m\|_p \|\nabla\psi_m\|_q^{-1} = 2^{-1/n} \mathrm{K}(n, q),$$

while

$$\lim_{m \to \infty} \|\psi_m\|_q \|\psi_m\|_p^{-1} = 0. \qquad \blacksquare$$

2.41 For applications to differential equations, the useful result is that concerning H_1. Since the best constant $\mathrm{K}(n, 2)$ is attained for the manifolds with constant curvature, Theorem 2.29, $2^{-1/n}\mathrm{K}(n, 2)$ is attained in that case.

Corollary. *Let $M_n (n \geq 3)$ be a Riemannian manifold with constant curvature and injectivity radius $\delta > 0$, and $f_i (i \in I)$ C^2-functions, having the properties required in Theorem (2.40), and satisfying $\Delta f_i^2 \leq$ Const., for all $i \in I$. Then there exists a constant A_2, such that all functions $\varphi \in \mathrm{H}_1$ satisfy:*

$$\|\varphi\|_N^2 \leq 2^{-2/n}\mathrm{K}^2(n, 2)\|\nabla\varphi\|_2^2 + A_2\|\varphi\|_2^2, \tag{21}$$

when they satisfy the conditions (19) *with* $p = N = 2n/(n-2)$.

Proof. This is similar to the preceding one. Use (14) instead of (18) and write:

$$\begin{aligned}\int |\nabla(\varphi f_i)^2|\, dV &= \int f_i^2 |\nabla\varphi|^2\, dV + \tfrac{1}{2}\int \nabla^\nu\varphi^2\nabla_\nu f_i^2\, dV + \int \varphi^2|\nabla f_i|^2\, dV\\ &= \int f_i^2|\nabla\varphi|^2\, dV + \int (|\nabla f_i|^2 + \tfrac{1}{2}\Delta f_i^2)\varphi^2\, dV.\end{aligned}$$

Since $|\nabla f_i|^2 + \frac{1}{2}\Delta f_i^2 \leq$ Const, (21) follows. ∎

2.42 The preceding theorem can be generalized. A similar proof establishes the following

Theorem. *Let M_n, $(n \geq 3)$, be a Riemannian manifold with bounded curvature and injectivity radius $\delta > 0$, and let $f_{i,j}$ $(i \in I, j = 1, 2, \ldots, m, m \geq 2$ an integer) be uniformly Lipschitzian and non-negative functions, with compact support $\Omega_{i,j}$, having the following properties: $\mathring{\Omega}_{i,j} \cap \mathring{\Omega}_{i,\ell} = \emptyset$ (for $1 \leq j < \ell \leq m$ and all $i \in I$), at each point $P \in M$, only k of all the functions f_{ij} can be nonzero, and*

$$\sum_{i \in I}\sum_{j=1}^{m} |f_{i,j}|^q = 1.$$

Then the functions $\varphi \in \mathrm{H}_1$, which satisfy the conditions

$$\int |\varphi|^p f_{i,j}^p\, dV = \int |\varphi|^p f_{i,1}^p\, dV \quad (\textit{for all } i \in I \textit{ and each } j = 2, \ldots, m),$$

satisfy inequality (18), *where B can be chosen equal to $m^{-1/n}\mathrm{K}(n, q) + \varepsilon$, with $\varepsilon > 0$, as small as one wants. (The best constant B of inequality* (18) *is now $m^{1/n}$ times smaller than $\mathrm{K}(n, q)$).*

2.43 Remark. On a compact manifold, if we consider the Rayleigh quotient $\inf \|\nabla\varphi\|_2 \|\varphi\|_2^{-1}$, when φ satisfies some well-known orthogonality conditions, we obtain, successively, the eigenvalues of the Laplacian $\lambda_0 = 0, \lambda_1, \lambda_2, \ldots$. Even if we know some properties of the sequence λ_i, we cannot compute λ_i from λ_1. It is therefore somewhat surprising that in the nonlinear case, the sequence is entirely known, the mth term being $m^{1/n}\mathsf{K}^{-1}(n, q)$.

§14. The Case of the Sphere

2.44 Definition. On the sphere $\mathbb{S}_n$, Λ will denote the vector space of functions ψ, which satisfy $\Delta\psi = \lambda_1\psi$, where λ_1 is the first nonzero eigenvalue of the Laplacian.
Recall that Λ is of dimension $n + 1$. One verifies that the eigenfunctions are $\psi(Q) = \gamma \cos[\alpha d(P, Q)]$, for any constant γ and any point $P \in \mathbb{S}_n$, with $\alpha^2 = R/n(n-1)$, R being the scalar curvature of the sphere.
There exists a family ξ_i $(i = 1, 2, \ldots, n+1)$ of functions in Λ, orthogonal in L_2 and satisfying $\sum_{i=1}^{n+1} \xi_i^2 = 1$ (see Berger (37)). In fact, if $x = (x_1, x_2, \ldots, x_{n+1})$ are the standard coordinates on $\mathbb{R}^{n+1}$, ξ_i is the restriction of x_i to $\mathbb{S}_n$.

Thus, we can apply Theorem 2.40 with $f_i = \xi_i$ when $n > 2$, $q = 2$, and $p = N = 2n/(n-2)$. But to solve the problem 5.11, we need the somewhat different conditions:

$$\int_{\mathbb{S}_n} \xi_i |\varphi|^N \, dV = 0, \quad \text{instead of} \int_{\mathbb{S}_n} \xi_i |\xi_i|^{N-1} |\varphi|^N \, dV = 0.$$

If we want to use Theorem 2.40, we must choose as functions f_i, the functions $\xi_i |\xi_i|^{1/N - 1}$. But this is impossible for two good reasons. On the one hand, $\{|\xi_i|^{2/N}\}$ does not form a partition of unity; on the other, the functions $|\xi_i|^{1/N}$ do not belong to $\mathsf{H}_1(\mathbb{S}_n)$. Nevertheless, these difficulties can be overcome. We are going to establish the following.

2.45 Theorem. *The functions ξ_i are a basis of Λ; then all $\varphi \in H_1^q(\mathbb{S}_n)$, $1 \le q < n$, satisfying $\int \xi_i |\varphi|^p \, dV = 0$ (for $i = 1, 2, \ldots, n+1$) satisfy:*

$$\|\varphi\|_p^q \le [2^{-1/n}\mathsf{K}(n, q) + \varepsilon]^q \|\nabla\varphi\|_q^q + A(\varepsilon)\|\varphi\|_q^q, \tag{22}$$

where $A(\varepsilon)$ is a constant which depends on $\varepsilon > 0$, ε as small as one wants.

Proof. Let $0 < \eta < 1/2$ be a real number, which we are going to choose very small. There exists a finite family of functions $\xi_i \in \Lambda$ $(i = 1, 2, \ldots, k)$, such that:

$$1 + \eta < \sum_{i=1}^{k} |\xi_i|^{q/p} < 1 + 2\eta \quad \text{with} \quad |\xi_i| < 2^{-p}.$$

Indeed, let $Q \in \mathbb{S}_n$ and $\xi_Q(P)$ the eigenfunction of Λ, such that ξ_Q attains its maximum 1 at $P = Q$; since $\xi_Q(P) = \xi_P(Q)$, we have

$$\int |\xi_Q(P)|^{q/p}\, dV(Q) = \int |\xi_P(Q)|^{q/p}\, dV(Q) = \text{Const.}$$

From this property of the family $\{\xi_Q\}_{Q \in \mathbb{S}_n}$, clearly the above family $\{\xi_i\}$ exists.

Consider h_i, C^1 functions, such that everywhere $h_i \xi_i \geq 0$ and such that $\|\, |h_i|^q - |\xi_i|^{q/p}| < (\eta/k)^p$. Then

$$1 < \sum_{i=1}^{k} |h_i|^q < 1 + 3\eta \tag{23}$$

and

$$\left| |h_i|^p - |\xi_i| \right| \leq \frac{p}{q} \left[|\xi_i|^{q/p} + \left(\frac{\eta}{k}\right)^p \right]^{p/q - 1} \left| |h_i|^q - |\xi_i|^{q/p} \right| \leq \frac{p}{q} \left(\frac{\eta}{k}\right)^p = \varepsilon_0^p,$$

according to Theorem 3.6, and since $|\xi_i|^{q/p} + (\eta/k)^p < 1$. As in the proof of Theorem 2.40, we suppose that $\varphi \geq 0$:

$$\|\varphi\|_p^q = \|\varphi^q\|_{p/q} \leq \left\| \sum_{i=1}^{k} \varphi^q |h_i|^q \right\|_{p/q} \leq \sum_{i=1}^{k} \|\varphi h_i\|_p^q.$$

If $\|\check{h}_i |\nabla\varphi|\|_q \geq \|\tilde{h}_i |\nabla\varphi|\|_q$, using (23) and the hypothesis, from Theorem 2.29 we obtain

$$\begin{aligned}
\|\varphi h_i\|_p^q &\leq \left[\int (\check{\xi}_i + \varepsilon_0^p)\varphi^p\, dV + \int \varphi^p \tilde{h}_i^p\, dV \right]^{q/p} \\
&= \left[\int (\tilde{\xi}_i + \varepsilon_0^p)\varphi^p\, dV + \int \varphi^p \tilde{h}_i^p\, dV \right]^{q/p} \\
&\leq 2^{q/p} \left[\int \varphi^p (\tilde{h}_i^p + \varepsilon_0^p)\, dV \right]^{q/p} \leq 2^{q/p} \|\varphi(\tilde{h}_i + \varepsilon_0)\|_p^q \\
&\leq 2^{q/p} [\mathrm{K}^q(n,q) \|\nabla[\varphi(\tilde{h}_i + \varepsilon_0)]\|_q^q + A(q) \|\varphi(\tilde{h}_i + \varepsilon_0)\|_q^q].
\end{aligned}$$

Set $H = \sup_{1 \leq i \leq k} \sup_{\mathbb{S}_n} |\nabla h_i|$. Then there exist constants μ and ν such that:

$$\begin{aligned}
\|\nabla[\varphi(\tilde{h}_i + \varepsilon_0)]\|_q^q \leq \int (\tilde{h}_i + \varepsilon_0)^q |\nabla\varphi|^q\, dV + \mu H \|\nabla\varphi\|_q^{q-1} \|\varphi\|_q \\
+ \nu H^q \|\varphi\|_q^q.
\end{aligned}$$

Since $(\tilde{h}_i + \varepsilon_0)^q \leq \tilde{h}_i^q + q(\tilde{h}_i + \varepsilon_0)^{q-1}\varepsilon_0 < \tilde{h}_i^q + q\varepsilon_0$, then

$$\|\varphi\|_p^q \leq 2^{q/p}\mathrm{K}^q(n, q)\left[\frac{1}{2}\sum_{i=1}^{k}\int |h_i|^q |\nabla\varphi|^q \, dV + qk\varepsilon_0\|\nabla\varphi\|_q^q + k\mu H\|\nabla\varphi\|_q^{q-1}\|\varphi\|_q + \nu H^q k\|\varphi\|_q^q\right] + 2^{q/p}kA(q)\|\varphi\|_q^q.$$

Using (20) and (23), this leads to:

$$\|\varphi\|_p^q \leq \mathrm{K}^q(n, q)[2^{-q/n}(1 + 3\eta) + 2^{q/p}qk\varepsilon_0 + k\mu H\varepsilon_1]\|\nabla\varphi\|_q^q + 2^{q/p}k[\mathrm{K}^q(n, q)(\mu H M_0 + \nu H^q) + A(q)]\|\varphi\|_q^q.$$

Since $\varepsilon_0 k = (p/q)^{1/p}\eta$, setting $\varepsilon_1 = \eta/Hk$ and taking η small enough we obtain the stated result.

Thus B in (18) is as close as desired to $2^{-1/n}\mathrm{K}(n, q)$.
We applied Theorem 2.29 for $q \leq 2$. In the case $q > 2$ the proof is not different, because ε is nonzero in (22). ∎

§15. The Exceptional Case of the Sobolev Imbedding Theorem

2.46 We will expound the topic chronologically. The exceptional case of the Sobolev imbedding theorem concerns the Sobolev space $H_1^n(M)$, where n is the dimension of the manifold M, or more generally, the spaces $H_k^{n/k}(M)$.

When $\varphi \in H_1^n$ we might hope that $\varphi \in L_\infty$. Unfortunately, this is not the case. Recall Example 2.37; the function $x \to \log|\log\|x\||$ defined on the ball $B_{1/e} \subset \mathbb{R}^2$ is not bounded but belongs, however, to $\mathring{H}_1^2(B_{1/e})$. But when $\varphi \in H_1^n$ it is possible to show that e^φ, and even $\exp[\alpha|\varphi|^{n/(n-1)}]$, are locally integrable, if α is small ehough (Trüdinger [261], Aubin [10]). More precisely,

Theorem 2.46. *Let M_n be a compact Riemannian manifold with or without boundary. If $\varphi \in H_1^n(M_n)$, then e^φ and $\exp[\alpha(|\varphi|\,\|\varphi\|_{H_1^n}^{-1})^{n/(n-1)}]$ are integrable for α a sufficiently small real number which does not depend on φ. Moreover, there exist constants C, μ and ν such that all $\varphi \in H_1^n$ satisfy:*

$$\int_M e^\varphi \, dV \leq C\exp[\mu\|\nabla\varphi\|_n^n + \nu\|\varphi\|_n^n] \tag{24}$$

and the mapping $H_1^n \ni \varphi \to e^\varphi \in L_1$ is compact.

Proof. α) Using a finite partition of unity we see that we have only to prove Theorem 2.46 for functions belonging to $\mathring{H}_1^n(B)$, where B is the unit ball of $\mathbb{R}^n$. Indeed, if the ball carries a Riemannian metric we use the inequalities of Theorem 1.53, and if the function obtained has its support included in the half ball we consider by reflection the x_1-even function which belongs to $\overset{\circ}{H}{}_1^n(B)$, as in 2.31.

β) Now $\varphi \in \overset{\circ}{H}{}_1^n(B)$. For almost all $P \in B$

$$|\varphi(P)| \le \omega_{n-1}^{-1} \int_B |\nabla\varphi(Q)| [d(P,Q)]^{1-n}\, dV(Q), \tag{25}$$

(see the end of 2.12). Then by Proposition 3.64, we obtain for all real $p \ge \eta$

$$\|\varphi\|_p \le \omega_{n-1}^{-1} \|\nabla\varphi\|_n \sup_{P\in B} \left[\int_B [d(P,Q)]^{k(1-n)}\, dV(Q)\right]^{1/k},$$

with $1/k = 1/p - 1/n + 1$. This yields:

$$\begin{aligned}\|\varphi\|_p &\le \omega_{n-1}^{-1+1/k} \|\nabla\varphi\|_n \left[\int_0^1 r^{(k-1)(1-n)}\, dr\right]^{1/k}\\ &= \omega_{n-1}^{-1+1/k} \|\nabla\varphi\|_n \left[\frac{p+1-p/n}{n}\right]^{1/k}.\end{aligned}$$

Thus there is a constant K such that for any $p \ge 1$:

$$\|\varphi\|_p \le K \|\nabla\varphi\|_n p^{(n-1)/n} \tag{26}$$

and we obtain

$$\int_B e^{\varphi}\, dV \le \sum_{p=0}^{\infty} K^p \|\nabla\varphi\|_n^p (p!)^{-1} p^{(n-1)p/n}.$$

But according to Stirling's formula, when $p \to \infty$

$$(p/n)!(p!)^{-1} p^{(n-1)p/n} \sim (p/en)^{p/n} n^{-1/2} (e/p)^p p^{(n-1)p/n} \sim e^{(n-1)p/n} n^{-p/n-1/2},$$

so that

$$\int_B e^{\varphi}\, dV \le C \exp(\mu \|\nabla\varphi\|_n^n),$$

where C and μ are two constants.

Remark. When M is compact, using Equation (15) of 4.13 after integrating by parts, and the properties of the Green's function yield immediately (without step α) an inequality of the kind (25). Thereby we can obtain a similar result when $\Delta^k\varphi \in L_{n/2k}$ instead of $\varphi \in H_{2k}^{n/2k}$ (see Aubin [10]). When

$\varphi \in H_m^s$ with $ms = n$, $\exp[\alpha|\varphi|^{s/(s-1)}]$ is locally integrable for $\alpha > 0$ small enough (see Cherrier [96]). Brezis and Wainger [67] obtained fine results in this field by using Lorentz spaces.

γ) Using (26) leads to

$$(27) \quad \int_B \exp[\nu|\varphi|^{n/(n-1)}]\, dV \leq \sum_{p=0}^{\infty} \nu^p (p!)^{-1} [K\|\nabla\varphi\|_n]^{pn/(n-1)} [pn/(n-1)]^p.$$

According to Stirling's formula $(p!)^{-1}(p/e)^p p^{1/2} \leq$ Const, thus the series in (27) converges for $(K\|\nabla\varphi\|_n)^{n/(n-1)}\nu\, en/(n-1) < 1$.

δ) To prove the last statement of Theorem 2.46 we will use the Kondrakov theorem 2.34: the imbedding $H_1^1(M) \subset L_1(M)$ is compact. Let $\mathscr{A}$ be a bounded set in $H_1^n(M)$. Rewriting (24) with $q\varphi$ instead of φ $(q \geq 1)$ implies that the set $\{e^\varphi\}_{\varphi \in \mathscr{A}}$ is bounded in L_q for all q. Then $\|\nabla e^\varphi\|_1 \leq \|\nabla\varphi\|_n \|e^\varphi\|_{n/(n-1)}$ shows that the set $\{e^\varphi\}_{\varphi \in \mathscr{A}}$ is bounded in H_1^1. Thus the result follows. ■

§16. Moser's Results

2.47 For applications, the best values of α and μ in Theorem (2.46) are essential. On this question the following result of Moser [209] was the first.

Theorem 2.47. *Let Ω be a bounded open set in $\mathbb{R}^n$ and set $\alpha_n = n\omega_{n-1}^{1/(n-1)}$. Then all $\varphi \in \mathring{H}_1^n(\Omega)$ such that $\|\nabla\varphi\|_n \leq 1$ satisfy for $\alpha \leq \alpha_n$:*

$$(28) \quad \int_\Omega \exp[\alpha|\varphi|^{n/(n-1)}]\, dV \leq C \int_\Omega dV.$$

Here the constant C depends only on n.
α_n is the best constant: if $\alpha > \alpha_n$ the integral on the left in (28) is finite but it can be made arbitrarily large by an appropriate choice of φ.

Proof. Making use of symmetrization as in 2.17 reduces the problem to a one-dimensional one. We have only to consider radially symmetric functions in B_ρ, the ball in $\mathbb{R}^n$ of radius ρ which has the same volume as Ω. Set $g(\|x\|) = \varphi(x)$. On the other hand, we can suppose $\varphi \in C^\infty(\bar{B}_\rho)$. Indeed, if $\varphi \in \mathring{H}_1^n(B_\rho)$ there exists $\{\psi_i\}_{i\in N}$, a sequence of C^∞ functions on $\bar{B}_\rho$ vanishing on the boundary, such that $\psi_i \to \varphi$ in H_1^n when $i \to \infty$ and such that $\psi_i \to \varphi$ almost everywhere (Proposition 3.43). Thus if we prove (28) for smooth functions, (28) holds for all $\varphi \in \mathring{H}_1^n(B_\rho)$ since

$$\int_{B(\rho)} \exp(\alpha|\varphi|^{n/(n-1)})\, dV \leq \liminf_{i\to\infty} \int_{B(\rho)} \exp(\alpha|\psi_i|^{n/(n-1)})\, dV.$$

So the problem is now:
For which $\alpha \in \mathbb{R}$, do all functions $g \in C^\infty([0, \rho])$ which vanish at ρ satisfy the inequality $\int_0^\rho \exp(\alpha |g|^{n/(n-1)})\, dV \le C\rho^n/n$ when $\int_0^\rho |g'|^n r^{n-1}\, dr \le \omega_{n-1}^{-1}$? Applying Proposition 2.48 below with $q = n$ and $k = \omega_{n-1}^{-1}$, C exists if and only if $\alpha \le \alpha_n = n\omega_{n-1}^{1/(n-1)}$. ∎

The following proposition discusses the existence of the integral for all α.

2.48 Proposition. *Let g be a Lipschitzian function on $[0, \rho]$ which vanishes at ρ. If $\int_0^\rho |g'|^q r^{q-1}\, dr \le k$ for some $q > 1$ and some $k > 0$, then $\int_0^\rho \exp(\beta |g|^{q/(q-1)}) r^{n-1}\, dr \le C\rho^n/n$, with C a constant which depends only on q and n, if and only if $\beta \le \beta_q = nk^{1/(1-q)}$. Moreover, the integral exists for any β, although the inequality holds only for $\beta \le \beta_q$.*

Proof. Moser stated the result only for $q = n$, but he gave the proof for arbitrary $q > 1$. We will follow his proof.
Set $e^{-t} = (r/\rho)^n$ and $f(t) = g(\rho e^{-t/n})$. Then $f(0) = 0$ and we have $(1/q)\, d(r^q) = -(1/n)\rho^q e^{-qt/n}\, dt$ and $f'(t) = -(\rho/n) g'(r) e^{-t/n}$.
Thus $n^{q-1} \int_0^\infty |f'|^q\, dt \le k$ and we want to have:

$$\int_0^\infty \exp(\beta |f|^{q/(q-1)} - t)\, dt \le C. \tag{29}$$

By Hölder's inequality:

$$|f(t)| \le \int_0^t |f'|\, dt \le \left(\int_0^t |f'|^q\, dt\right)^{1/q} t^{(q-1)/q} \le k^{1/q}(t/n)^{(q-1)/q} = (t/\beta_q)^{(q-1)/q}.$$

Hence for $\beta < \beta_q$ the result follows at once:

$$\int_0^\infty \exp(\beta |f|^{q/(q-1)} - t)\, dt \le \int_0^\infty \exp[(\beta/\beta_q - 1)t]\, dt = (1 - \beta/\beta_q)^{-1}.$$

If $\beta = \beta_q$ it is not easy to establish the result (see Moser [209] for the proof).
When $\beta > \beta_q$ the integral we are studying exists, but it can be made arbitrarily large. Consider for $\tau > 0$ the function f_τ defined as follows: $f_\tau(t) = (\tau/\beta_q)^{(q-1)/q} t/\tau$ for $t \le \tau$ and $f_\tau(t) = (\tau/\beta_q)^{(q-1)/q}$ for $t \ge \tau$. Clearly these functions satisfy the hypotheses and

$$\int_0^\infty \exp(\beta |f_\tau|^{q/(q-1)} - t)\, dt \ge \int_\tau^\infty \exp(\beta\tau/\beta_q - t)\, dt = \exp[(\beta/\beta_q - 1)\tau]$$

tends to infinity as $\tau \to \infty$.

It remains to prove the convergence of the integral. Applying Hölder's inequality we find that for $t > \tau$

$$|f(t) - f(\tau)| \leq \int_\tau^t |f'|\,dt \leq \left(\int_\tau^\infty |f'|^q\,dt\right)^{1/q} (t - \tau)^{1-1/q}.$$

Since we can choose τ so that $\int_\tau^\infty |f'|^q\,dt$ is as small as one wants, $t^{1/q-1}f(t) \to 0$ as $t \to \infty$. Hence $\beta|f(t)|^{q/(q-1)}t^{-1} \to 0$ as $t \to \infty$ and the integral in (29) exists. ■

2.49 Proposition. *Let g be as in Proposition 2.48. Then there are constants C and λ such that*

$$\int_0^\rho e^g r^{n-1}\,dr \leq C \exp\left[\lambda \int_0^\rho |g'|^q r^{q-1}\,dr\right]; \tag{30}$$

the inf of λ such that C exists is equal to $\lambda_q = ((q-1)/n)^{q-1}q^{-q}$.

Proof. It is easy to verify that all real numbers u satisfy

$$u \leq k\lambda_q + \beta_q|u|^{q/(q-1)};$$

thus according to Proposition 2.48,

$$\int_0^\rho e^g r^{n-1}\,dr \leq \frac{C}{n}\rho^n e^{k\lambda_q}, \quad \text{where we pick } k = \int_0^\rho |g'|^q r^{q-1}\,dr. \quad ■$$

Corollary 2.49. *Let Ω be a bounded open set of $\mathbb{R}^n$ and set $\mu_n = (n-1)^{n-1} n^{1-2n}\omega_{n-1}^{-1}$. Then all $\varphi \in \overset{\circ}{H}{}_1^n(\Omega)$ satisfy*

$$\int_\Omega e^\varphi\,dV \leq C \int_\Omega dV \exp(\mu_n \|\nabla\varphi\|_n^n), \tag{31}$$

where C depends only on n.

Proof. After symmetrization we use Proposition 2.49 with $q = n$ and we get $\mu_n = \lambda_n \omega_{n-1}^{-1}$. This result may also be obtained from (28) by using the inequality: $uv \leq \alpha_n|u|^{n/(n-1)} + \mu_n|v|^n$ with $v = \|\nabla\varphi\|_n$ and $u = \varphi\|\nabla\varphi\|_n^{-1}$. ■

§17. The Case of the Riemannian Manifolds

2.50 Return to Theorem 2.46. Set $\tilde{\alpha}_n$, the sup of α, such that $\exp[\alpha(|\varphi|\,\|\varphi\|_{H_1^n}^{-1})^{n/(n-1)}]$ is integrable and $\tilde{\mu}_n$, the inf of μ, such that C and ν exist in inequality (24). Two questions arise. Does $\tilde{\mu}_n$ depend on the manifold?

Is $\tilde{\mu}_n$ attained? (I.e., is $\mu = \tilde{\mu}_n$ allowed.) The answers were first found in Cherrier [95]. In Cherrier [96] there are similar results when $\varphi \in H_k^{n/k}$ or $\Delta^k \varphi \in L_{n/2k}$. He proved the following:

Theorem 2.50. *For Riemannian manifolds M_n with bounded curvature and global injectivity radius (in particular this is true if M is compact), the best constants $\tilde{\mu}_n$ and $\tilde{\alpha}_n$ in Theorem (2.46) depend only on n. They are equal to*

$$\mu_n = (n-1)^{n-1} n^{1-2n} \omega_{n-1}^{-1} \quad \text{and} \quad \alpha_n = n\omega_{n-1}^{1/(n-1)}.$$

For compact Riemannian manifolds with C^1 boundary the best constants are equal, respectively, to $2\mu_n$ and $2^{-1/(n-1)}\alpha_n$.
μ_n is attained for the sphere $\mathbb{S}_n$ and μ_2 is attained for compact Riemannian manifolds of dimension 2.

2.51 The case of the sphere. We have seen that we have the best possible inequality (31) for $\varphi \in \overset{\circ}{H}{}_1^n(\Omega)$ when Ω is a bounded open set of $\mathbb{R}^n$: $\mu = \mu_n$ and $\nu = 0$ in (24).
This is also the case for the sphere $\mathbb{S}_n$. The following was proved by Moser [209] when $n = 2$, and by Aubin [21], p. 156.

Theorem 2.51. *All $\varphi \in H_1^n(\mathbb{S}_n)$ with integral equal zero ($\int_{\mathbb{S}_n} \varphi \, dV = 0$) satisfy*

$$\int_{\mathbb{S}_n} e^{\varphi} \, dV \le C \exp(\mu_n \|\nabla\varphi\|_n^n), \tag{32}$$

where C depends only on n and $\mu_n = (n-1)^{n-1} n^{1-2n} \omega_{n-1}^{-1}$; in particular $\mu_2 = 1/16\pi$.

2.52 As in other inequalities concerning Sobolev spaces, the best constants can be lowered when the functions φ also satisfy some natural orthogonality conditions. Theorems similar to those in 2.40 and 2.42 are proved in Aubin [21], p. 157. The sequence of best constants is $\{\mu_n/m\}_{m\in N}$. For the sphere $\mathbb{S}_n$ the following is proved.

Theorem 2.52. *Let Λ be the eigenspace corresponding to the first nonzero eigenvalue. The functions $\varphi \in H_1^n(\mathbb{S}_n)$ satisfying $\int_{\mathbb{S}_n} \xi e^{\varphi} \, dV = 0$ for all $\xi \in \Lambda$ and $\int_{\mathbb{S}_n} \varphi \, dV = 0$, satisfy the inequality*

$$\int_{\mathbb{S}_n} e^{\varphi} \, dV \le C(\mu) \exp(\mu \|\nabla\varphi\|_n^n), \tag{33}$$

where it is possible to choose $\mu > \mu_n/2$ as close to $\mu_n/2$ as one wants. $C(\mu)$ is a constant which depends on μ and n.

2.53 The case of the real projective space $\mathbb{P}_n$.

Theorem. *For any $\varepsilon > 0$ there is a constant $C(\varepsilon)$ which depends only on n such that all $\psi \in H_1^n(\mathbb{P}_n)$ with integral zero ($\int_{\mathbb{P}_n} \psi \, dV = 0$) satisfy*

$$\int_{\mathbb{P}_n} e^{\psi} \, dV \leq C(\varepsilon) \exp[(\mu_n + \varepsilon)\|\nabla\psi\|_n^n]. \tag{34}$$

Proof. $\mathfrak{p}: \mathbb{S}_n \to \mathbb{P}_n$, the universal covering of $\mathbb{P}_n$ has two sheets. We associate to $\psi \in H_1^n(\mathbb{P}_n)$ the function φ on $\mathbb{S}_n$ defined by $\varphi(Q) = \psi(\mathfrak{p}(Q))$ for $Q \in \mathbb{S}_n$.

The function φ so obtained satisfies the hypotheses of Theorem 2.52. $\int_{\mathbb{S}_n} \varphi \, dV = 2\int_{\mathbb{P}_n} \psi \, dV = 0$ and e^{φ} is orthogonal to Λ. Indeed, if Q and $\tilde{Q}$ are antipodally symmetric on $\mathbb{S}_n$, $\xi(Q)e^{\varphi(Q)} = -\xi(\tilde{Q})e^{\varphi(\tilde{Q})}$ for $\xi \in \Lambda$. Thus $\int_{\mathbb{S}_n} \xi(Q)e^{\varphi(Q)} \, dV(Q) = -\int_{\mathbb{S}_n} \xi(\tilde{Q})e^{\varphi(\tilde{Q})} \, dV(\tilde{Q})$, and so vanishes.
By Theorem 2.52, for any $\varepsilon > 0$ there is $\tilde{C}(\varepsilon)$ which depends only on n, such that all $\psi \in H_1^n(\mathbb{P}_n)$ satisfy

$$2\int_{\mathbb{P}_n} e^{\psi} \, dV = \int_{\mathbb{S}_n} e^{\varphi} \, dV \leq \tilde{C}(\varepsilon) \exp\left[(\mu_n/2 + \varepsilon) \int_{\mathbb{S}n} |\nabla\varphi|^n \, dV\right]$$

$$= \tilde{C}(\varepsilon) \exp\left[(\mu_n + 2\varepsilon) \int_{\mathbb{P}_n} |\nabla\psi|^n \, dV\right].$$

Thus we get (34) with $C(\varepsilon) = \tilde{C}(\varepsilon/2)/2$. ■

§18. Problems of Traces

2.54 Let M be a Riemannian manifold and let $V \subset M$ be a Riemannian sub-manifold. If f is a C^k function on M, we can consider $\tilde{f}$ the restriction of f to V, $\tilde{f} \in C^k(V)$.

Now if $f \in H_k^p(M)$, it is often possible to define the trace $\tilde{f}$ of f on V by a density argument and there are imbedding theorems similar to those of Sobolev. Adams [1] discusses the case of Euclidean space. In Cherrier [97] the problem of traces is studied for Riemannian manifolds; he also considers the exceptional case.

The same problems arise for a Riemannian manifold W with boundary ∂W. We can try to define the trace on ∂W of a function belonging to $H_k^p(W)$. The results are useful for problems with prescribed boundary conditions.

Chapter 3

Background Material

§1. Differential Calculus

3.1 Definition. A *normed space* is a vector space $\mathfrak{F}$(over $\mathbb{C}$ or $\mathbb{R}$), which is provided with a norm. A *norm*, denoted by $\| \ \|$, is a real-valued functional on $\mathfrak{F}$, which satisfies:

(a) $\mathfrak{F} \ni x \to \|x\| \geq 0$, with equality if and only if $x = 0$,
(b) $\|\lambda x\| = |\lambda| \|x\|$ for every $x \in \mathfrak{F}$ and $\lambda \in \mathbb{C}$,
(c) $\|x + y\| \leq \|x\| + \|y\|$ for every $x, y \in \mathfrak{F}$.

A *Banach space* $\mathfrak{B}$ is a complete normed space: every Cauchy sequence in $\mathfrak{B}$ converges to a limit in $\mathfrak{B}$.

A *Hilbert space* $\mathfrak{H}$ is a Banach space where the norm comes from an *inner product*:

$$\mathfrak{H}^2 \ni (x, y) \to \langle x, y\rangle \in \mathbb{C}, \quad \text{so} \quad \|x\|^2 = \langle x, x\rangle.$$

$\mathfrak{P}$ is an inner product provided that $\mathfrak{P}$ is linear in x, that it satisfies $\langle x, y\rangle = \langle y, x\rangle$, and that $\langle x, x\rangle = 0$ if and only if $x = 0$.

3.2 Definition. Let $\mathfrak{F}$ and $\mathfrak{G}$ be two normed spaces. We denote by $\mathscr{L}(\mathfrak{F}, \mathfrak{G})$ the space of the continuous linear mappings u from $\mathfrak{F}$ to $\mathfrak{G}$. $\mathscr{L}(\mathfrak{F}, \mathfrak{G})$ has the natural structure of a normed space. Its norm is

$$\|\mathrm{u}\| = \sup\|\mathrm{u}(x)\| \quad \text{for all } x \in \mathfrak{F} \quad \text{with } \|x\| \leq 1.$$

$\mathfrak{F}^*$, the *dual space* of $\mathfrak{F}$, is $\mathscr{L}(\mathfrak{F}, \mathbb{C})$ or $\mathscr{L}(\mathfrak{F}, \mathbb{R})$, according to whether $\mathfrak{F}$ is a vector space on $\mathbb{C}$ or $\mathbb{R}$. $\mathfrak{F}$ is said to be reflexive if the natural imbedding $\mathfrak{F} \ni x \to \tilde{x} \in \mathfrak{F}^{**}$, defined by $\tilde{x}(\mathrm{u}) = \mathrm{u}(x)$ for $\mathrm{u} \in F^*$, is surjective.

3.3 Proposition. *A linear mapping* u *from* $\mathfrak{F}$ *to* $\mathfrak{G}$ *(where* $\mathfrak{F}$ *and* $\mathfrak{G}$ *are two normed spaces) is continuous if and only if there exists a real number* M *such that* $\|\mathrm{u}(x)\| \leq M\|x\|$ *for all* $x \in \mathfrak{F}$.

If $\mathfrak{B}$ *is a Banach space, then* $\mathscr{L}(\mathfrak{F}, \mathfrak{B})$ *is a Banach space.*

3.4 Definition. If Ω is an open subset of $\mathfrak{F}$, ($\mathfrak{F}$ and $\mathfrak{G}$ being two normed spaces), then $f:\Omega \to \mathfrak{G}$ is called *differentiable* at $x \in \Omega$ if there exists a $\mathrm{u} \in \mathscr{L}(\mathfrak{F}, \mathfrak{G})$ such that:

$$(1) \qquad f(x + y) - f(x) = \mathrm{u}(y) + \|y\|\omega(x, y),$$

where $\omega(x, y) \to 0$ when $y \to 0$, for all y such that $x + y \in \Omega$. u, called the differential of f at x, is denoted by $f'(x)$ or $Df(x)$.

3.5 Definition. Let f be as above. f is called differentiable on Ω if f is differentiable at each point $x \in \Omega$.
f is continuously differentiable on Ω, written $f \in C^1(\Omega, \mathfrak{G})$, if the map $\psi: \Omega \ni x \to f'(x) \in \mathscr{L}(\mathfrak{F}, \mathfrak{G})$ is continuous. $f \in C^1(\Omega, \mathfrak{G})$ is twice differentiable at x if ψ is differentiable at x. We write $D^2 f(x) = \psi'(x) \in \mathscr{L}(\mathfrak{F}, \mathscr{L}(\mathfrak{F}, \mathfrak{G})) = \mathscr{L}(\mathfrak{F} \times \mathfrak{F}, \mathfrak{G})$.

f is C^2, written $f \in C^2(\Omega, \mathfrak{G})$, if $\psi \in C^1(\Omega, \mathscr{L}(\mathfrak{F}, \mathfrak{G}))$. In this case $D^2 f(x)$ is symmetric. $D^2 f(x) \in \mathscr{L}_2(\mathfrak{F}, \mathfrak{G})$, the space of continuous bilinear maps from $\mathfrak{F} \times \mathfrak{F}$ to $\mathfrak{G}$. Continuing by induction, we can define the pth differential of f at x (if it exists): $D^p f(x) = D[D^{p-1} f(x)]$.
If $x \to D^p f(x)$ is continuous on Ω, f is said to be C^p, $f \in C^p(\Omega, \mathfrak{G})$, $D^p f(x) \in \mathscr{L}_p(\mathfrak{F}, \mathfrak{G}) = \mathscr{L}(\mathfrak{F}^p, \mathfrak{G})$.

1.1. The Mean Value Theorem

3.6 Theorem. *Let $\mathfrak{F}$ and $\mathfrak{G}$ be two normed vector spaces and $f \in C^1(\Omega, \mathfrak{G})$, with $\Omega \subset \mathfrak{F}$. If a and b are two points of Ω, set*

$$[a, b] = \{x \in \mathfrak{F} \quad \textit{such that } x = a + t(b - a) \quad \textit{for some } t \in [0, 1]\}.$$

If $[a, b] \subset \Omega$ and if $\|f'(x)\| \leq M$ for all $x \in [a, b]$, then

$$(2) \qquad \|f(b) - f(a)\| \leq M\|b - a\|.$$

3.7 Definitions. When $\mathfrak{F}$ and $\mathfrak{G}$ have finite dimension: $\mathfrak{F} = \mathbb{R}^n$, $\mathfrak{G} = \mathbb{R}^p$, a mapping f is defined on $\Omega \subset \mathfrak{F}$ by p real-valued functions $f^\alpha(x^1, x^2, \ldots, x^n)$, $(\alpha = 1, 2, \ldots, p)$. Then $f \in C^1(\Omega, \mathfrak{G})$ if and only if each function f^α has continuous partial derivatives.
The matrix $(n \times p)$, whose general entry is $\partial f^\alpha(x)/\partial x^i$, is called the *Jacobian matrix* of f at $x \in \Omega$.
The *rank* of f at $x \in \Omega$, is the rank of $f'(x)$, that is to say the dimension of the range of $f'(x)$.

3.8 Taylor's formula. *Let $f \in C^n(\Omega, \mathfrak{G})$, $\Omega \subset \mathfrak{F}$, ($\mathfrak{F}$ and $\mathfrak{G}$ two normed vector spaces), and $[x, x+h] \subset \Omega$. Then*

$$f(x+h) = f(x) + f'(x)h + \tfrac{1}{2}D^2 f(x)h^2 + \cdots$$

$$+ \frac{1}{n!} D^n f(x)h^n + \|h\|^n \omega_n(x, h) \tag{3}$$

where $\omega_n(x, h) \to 0$, when $h \to 0$.
$D^k f(x)h^k$ means $D^k f(x)(h, h, \ldots, h)$, the h repeated k times. If $f \in C^{n+1}(\Omega, \mathfrak{G})$ and if $\mathfrak{G}$ is complete, then

$$\|h\|^n \omega_n(x, h) = \frac{1}{n!} \int_0^1 (1-t)^n D^{n+1} f(x+th)h^{n+1}\, dt.$$

3.9 Definition. A *homeomorphism* of a topological space into another is a continuous one to one map, such that the inverse function is also continuous. A C^k-diffeomorphism of an open set $\Omega \subset \mathfrak{F}$ onto an open set in $\mathfrak{G}$ is a C^k-differentiable homeomorphism, whose inverse map is C^k, where $\mathfrak{F}$ and $\mathfrak{G}$ are two normed spaces.

1.2. Inverse Function Theorem

3.10 Theorem. *Let $\mathfrak{B}$ and $\mathfrak{G}$ be two Banach spaces and $f \in C^1(\Omega, \mathfrak{G})$, $\Omega \subset \mathfrak{B}$. If at $x_0 \in \Omega$, $f'(x_0)$ is a homeomorphism of $\mathfrak{B}$ onto $\mathfrak{G}$, then there exists a neighborhood Θ of x_0, such that Φ, the restriction of f to Θ, is a homeomorphism of Θ onto $f(\Theta)$.*
If f is of class C^k, Φ is a C^k-diffeomorphism.

Implicit function theorem. *Let $\mathfrak{E}$, $\mathfrak{F}$ and $\mathfrak{B}$ be Banach spaces and let $\mathfrak{U}$ be an open set of $\mathfrak{E} \times \mathfrak{F}$. Suppose $f \in C^p(\mathfrak{U}, \mathfrak{B})$ and let $D_y f(x_0, y_0) \in \mathscr{L}(\mathfrak{F}, \mathfrak{B})$ be the differential at y_0 of the mapping $y \to f(x_0, y)$. If at $(x_0, y_0) \in \mathfrak{U}$, $D_y f(x_0, y_0)$ is invertible, then the map $(x, y) \to (x, f(x, y))$ is a C^p diffeomorphism of a neighborhood $\Omega \subset \mathfrak{U}$ of (x_0, y_0) onto an open set of $\mathfrak{E} \times \mathfrak{B}$.*

1.3. Cauchy's Theorem

3.11 *Let $\mathfrak{B}$ be a Banach space and $f(t, x)$ a continuous function on an open subset $\mathfrak{U} \subset \mathbb{R} \times \mathfrak{B}$ with range in $\mathfrak{B}$.*
Consider the initial value problem, for functions: $t \to x(t) \in \mathfrak{B}$:

$$\frac{dx}{dt} = f(t, x), \; x(t_0) = x_0 \quad \text{with } (t_0, x_0) \in \mathfrak{U}. \tag{4}$$

If, on a neighborhood of (t_0, x_0), $f(t, x)$ *is a uniformly Lipschitzian in* x, *then there exists one and only one continuous solution of* (4), *which is defined on a neighborhood of* t_0.
If f *is* C^p, *the solution is* C^{p+1}. *Moreover the solution depends on the initial conditions* (t_0, x_0); *set* $x(t, t_0, x_0)$ *the unique solution of* (4). *The map* ψ: $(t, t_0, x_0) \to x(t, t_0, x_0) \in \mathfrak{B}$ *is continuous on an open subset of* $\mathbb{R} \times \mathbb{R} \times \mathfrak{B}$. *If* f *is* C^p, ψ *is* C^p.

Recall that we say $f(t, x)$ is uniformly Lipschitzian in x on $\Theta \subset \mathbb{R} \times \mathfrak{B}$ if there exists a k such that for any $(t, x_1) \in \Theta$ and $(t, x_2) \in \Theta$,

$$\|f(t, x_1) - f(t, x_2)\| \leq k\|x_1 - x_2\|. \tag{5}$$

It is possible to have a more precise result on the interval of existence of the solution. By continuity of f, there exist M, α, and ρ, three positive numbers, such that $\|f(t, x)\| \leq M$, for any $(t, x) \in I \times \bar{B}_{x_0}(\rho) \subset \mathfrak{U}$, with $I = [t_0 - \alpha, t_0 + \alpha]$ and $B_{x_0}(\rho) = \{x \in \mathfrak{B} \mid \|x - x_0\| < \rho\}$. If $M\alpha \leq \rho$, the solution of (4) exists on I.

§2. Four Basic Theorems of Functional Analysis

2.1. Hahn–Banach Theorem

3.12 *Let* $p(x)$ *be a seminorm defined on a normed space* $\mathfrak{G}$, $\mathfrak{F}$ *a linear subspace of* $\mathfrak{G}$, *and* $f(x)$ *a linear functional defined on* $\mathfrak{F}$, *with* $|f(x)| \leq p(x)$ *for* $x \in \mathfrak{F}$. *Then* f *can be extended to a continuous linear function* $\tilde{f}$ *on* $\mathfrak{G}$ *with* $|\tilde{f}(x)| \leq p(x)$ *for all* $x \subset \mathfrak{G}$.

A seminorm is a positive real-valued functional on $\mathfrak{G}$ which satisfies b) and c) of 3.1.

2.2. Open Mapping Theorem

3.13 *Under a continuous linear map* u *of one Banach space onto all of another, the image of every open set is open. If* u *is one-to-one,* u *has a continuous linear inverse.*

2.3. The Banach–Steinhaus Theorem

3.14 *Let* $\mathfrak{B}$ *and* $\mathfrak{F}$ *be Banach spaces and a family of* $\mathrm{u}_\alpha \in \mathscr{L}(\mathfrak{B}, \mathfrak{F})$, ($\alpha \in A$ *a given set*). *If for each* $x \in \mathfrak{B}$, *the set* $\{\mathrm{u}_\alpha(x)\}_{\alpha \in A}$ *is bounded, then there exists* M, *such that* $\|\mathrm{u}_\alpha\| \leq M$ *for all* $\alpha \in A$. *In particular, if* $\mathrm{u}_i \in \mathscr{L}(\mathfrak{B}, \mathfrak{F})$ *and if* $\lim_{i\to\infty} \mathrm{u}_i(x)$ *exists for each* $x \in \mathfrak{B}$, *then there exists an* M *such that* $\|\mathrm{u}_i\| \leq M$ *for all* $i \in \mathbb{N}$, *and there exists a* $\mathrm{u} \in \mathscr{L}(\mathfrak{B}, \mathfrak{F})$ *such that* $\mathrm{u}_i(x) \to \mathrm{u}(x)$ *for all* $x \in \mathfrak{B}$.
But u_i does not necessarily converge to u in $\mathscr{L}(\mathfrak{B}, \mathfrak{F})$.

2.4. Ascoli's Theorem

3.15 *Let $\mathfrak{K}$ be a compact Haussdorff space and $C(\mathfrak{K})$ the Banach space of the continuous functions on $\mathfrak{K}$ with the norm of uniform convergence.*
A subset $A \subset C(\mathfrak{K})$ is precompact ($\overline{A}$ is compact), if and only if it is bounded and equicontinuous.
(Recall A is said to be equicontinuous if, to every $\varepsilon > 0$ and every $x \in \mathfrak{K}$, there corresponds a neighborhood $\mathfrak{U}$ of x such that $|f(x) - f(y)| < \varepsilon$ for all $y \in \mathfrak{U}$ and all $f \in A$).

§3. Weak Convergence. Compact Operators

3.16 Definition. $\{x_i\}$, a sequence in $\mathfrak{F}$ a normed space, is said to *converge weakly* to $x \in \mathfrak{F}$ if $u(x_i) \to u(x)$ for every $u \in \mathfrak{F}^*$, the dual space of $\mathfrak{F}$ (see definition (3.2)). A subset A is said to be *weakly sequentially compact*, if every sequence in A contains a subsequence which converges weakly to a point in A.

3.17 Theorem. *A weakly convergent sequence $\{x_i\}$ in a normed space $\mathfrak{F}$ has a unique limit x, is bounded, and*

(6) $$\|x\| \leq \liminf_{i\to\infty} \|x_i\|$$

3.1. Banach's Theorem

3.18 Theorem. *A Banach space $\mathfrak{B}$ is reflexive, if and only if its closed unit ball $\overline{B}_1(0)$ is weakly sequentially compact.*

Particular case. In a Hilbert space, a bounded subset is weakly sequentially compact.

3.19 Definition. Let $\mathfrak{F}$ and $\mathfrak{G}$ be normed spaces and $\Omega \subset \mathfrak{F}$. A map $f : \Omega \to \mathfrak{G}$ (not necessarily linear) is said to be compact if f is continuous and maps bounded subsets of $\mathfrak{U}$ into precompact subsets of $\mathfrak{G}$.

3.20 Schauder fixed point theorem. *A compact mapping, f, of a closed bounded convex set Ω in a Banach space $\mathfrak{B}$ into itself has a fixed point.*

3.2. The Leray–Schauder Theorem

3.21 *Let T be a compact mapping of a Banach space $\mathfrak{B}$ into itself, and suppose there exists a constant M such that $\|x\| \leq M$ for all $x \in \mathfrak{B}$ and $\sigma \in [0, 1]$ satisfying $x = \sigma T x$. Then T has a fixed point.*

3.22 Definition. Let $\mathfrak{F}$ be a normed space on $\mathbb{C}$ and $T \in \mathscr{L}(\mathfrak{F}, \mathfrak{F})$. A number $\lambda \in \mathbb{C}$ is called an *eigenvalue* of T if there exists a non-zero element x in $\mathfrak{F}$ (called an *eigenvector*) satisfying $Tx = \lambda x$. The dimension of the null space of the operator $\lambda I - T$ is called the multiplicity of λ.

3.23 Theorem. *The eigenvalues of a compact linear mapping T of a normed space $\mathfrak{F}$ into itself form either a finite set, or a countable sequence converging to 0. Each non-zero eigenvalue has finite multiplicity. If $\lambda \neq 0$ is not an eigenvalue, then for each $f \in \mathfrak{F}$ the equation $\lambda x - Tx = f$ has a uniquely determined solution $x \in \mathfrak{F}$ and the operator $(\lambda I - T)^{-1}$ is continuous.*

3.3. The Fredholm Theorem

3.24 *Let T be a compact linear operator in a Hilbert space $\mathfrak{H}$ and consider the equations:*

(7) $$x - Tx = f,$$

(8) $$y - T^*y = g,$$

where T^* is the adjoint operator of T, ($\langle Tx, y\rangle = \langle x, T^*y\rangle$ for any x and y in $\mathfrak{H}$). Then the following alternative holds:

(α) *either there exists a unique solution of* (7) *and* (8) *for any f and g in $\mathfrak{H}$, or*
(β) *the homogeneous equation $x - Tx = 0$ has non trivial solutions. In that case the dimension of the null space of $I - T$ is finite and equals the dimension of the null space $\mathcal{N}^*$ of $I - T^*$. Furthermore* (7) *has a solution (not unique of course) if and only if $\langle f, y\rangle = 0$ for every $y \in \mathcal{N}^*$.*

§4. The Lebesgue Integral

3.25 Definition. Let $\mathscr{H}$ be a locally compact Haussdorff space and $C_0(\mathscr{H})$ the space of real-valued continuous functions on $\mathscr{H}$ with compact support. A *positive Radon measure* μ is a linear functionnal on $C_0(\mathscr{H})$, μ: $f \to \mu(f) \in \mathbb{R}$, such that $\mu(f) \geq 0$ for any $f \geq 0$.

3.26 Definition. Let μ be a positive Radon measure as above. We define an *upper integral* for the non-negative functions as follows. If $g \geq 0$ is lower semicontinuous:

$$\mu^*(g) = \sup \mu(f) \quad \text{for all } f \in C_0(\mathscr{H}) \quad \text{satisfying } f \leq g,$$

and for any functions $h \geq 0$: $\mu^*(h) = \inf \mu^*(g)$ for all lower semicontinuous functions g satisfying $h \leq g$.

3.27 Definition. Let μ be a positive Radon measure as above. A function f is said to be *μ-integrable*, if there exists a sequence $f_n \in C_0(\mathscr{H})$ such that $\mu^*(|f - f_n|) \to 0$. Then $\{\mu(f_n)\}$ converges, and we set $\int f\, d\mu = \lim_{n\to\infty} \mu(f_n)$.

3.28 Definition. A set $A \subset \mathscr{H}$ is *measurable* and with finite measure $\mu(A)$, if its characteristic function χ_A is integrable (by definition $\chi_A(x) = 1$ for $x \in A$, $\chi_A(x) = 0$ for $x \notin A$). We set $\mu(A) = \int \chi_A\, d\mu$. We say that a property holds almost everywhere on $\mathscr{H}$ if it holds for all $x \in \mathscr{H}$ except on a set A of measure zero. *Almost all points* means all points, except possibly those of a set with measure zero.

3.29 Remark. By Definitions 3.26 and 3.27, a non-negative lower semi-continuous function g, with $\mu^*(g)$ finite, is integrable and $\mu^*(g) = \int g\, d\mu$. One can prove that an integrable function f is equal almost everywhere to $g_1 - g_2$, with g_1 and g_2 non-negative lower semicontinuous integrable functions.
A compact subset $\mathfrak{K} \subset \mathscr{H}$ is measurable and with finite measure. An open subset Ω is measurable (χ_Ω is lower semicontinuous).

3.30 Definition. f is said *μ-measurable* (or measurable, when there is no ambiguity) if for all compact sets $\mathfrak{K}$ and all $\varepsilon > 0$, there exists a compact set $\mathfrak{K}_\varepsilon \subset \mathfrak{K}$, such that $\mu(\mathfrak{K} - \mathfrak{K}_\varepsilon) < \varepsilon$ and such that the restriction $f/\mathfrak{K}_\varepsilon$ is continuous on $\mathfrak{K}_\varepsilon$.

If a sequence of μ-measurable functions converges almost everywhere, then the limit function is μ-measurable.

3.31 Remark. For convenience, we develop the theory for real-valued functions; however, the theory of the Lebesgue integral is similar for the functions $\vec{f}$ on $\mathscr{H}$ with values in $\mathfrak{B}$, a Banach space.

Given μ, first we define $\int \vec{f}\, d\mu$ for continuous functions $\vec{f}$ with compact support. Let $\{\vec{f}_k\}$ be a sequence of simple functions ($\vec{f}_k = \sum_{j=1}^{k} \vec{a}_{kj}\chi_{A_j}$, with $\vec{a}_{kj} \in \mathfrak{B}$ and A_j measurable sets), which converges uniformly to $\vec{f}$. By definition $\int \vec{f}\, d\mu = \lim_{k\to\infty} \sum_{j=1}^{k} \mu(A_j)\vec{a}_{kj}$.
Then we define the integrable functions $\vec{g}$, as in definition 3.27. $\vec{g}$ is μ-integrable if there exists a sequence of continuous functions $\vec{f}_n$ with compact support, such that $\mu^*(\|\vec{g} - \vec{f}_n\|) \to 0$.

4.1. Dominated Convergence Theorem

3.32 The first Lebesgue theorem. *Given μ a positive Radon measure on $\mathscr{H}$ a locally compact Haussdorff space, let $\{\vec{f}_n\}$ be a sequence of μ-integrable functions, ($\vec{f}_n : \mathscr{H} \to \mathfrak{B}$ a Banach space) which converges almost everywhere to $\vec{f}$. Then $\vec{f}$ is integrable, $\int \vec{f}_n\, d\mu \to \int \vec{f}\, d\mu$ and $\int \|\vec{f} - \vec{f}_n\|\, d\mu \to 0$, if there exists h,* *a non-negative function, satisfying* $\mu^*(h) < \infty$ *and* $\|\vec{f}_n(x)\| \leq h(x)$ *for all n and* $x \in \mathscr{H}$.

Recall $\mu^*(h)$ is finite, in particular, if h is integrable.

4.2. Fatou's Theorem

3.33 *Given* μ *as above. Let* $\{h_i\}$ *be an increasing sequence of non-negative functions on* $\mathscr{H}$, $0 \leq h_1 \leq \cdots \leq h_i \leq h_{i+1} \leq \cdots$. *Then*

$$\lim_{i\to\infty} \mu^*(h_i) = \mu^*\left(\lim_{i\to\infty} h_i\right).$$

3.34 Theorem. *From now on, we suppose that* $\mathscr{H}$, *the locally compact Haussdorff space, is a denumerable union of compact sets, and that the Banach space* $\mathfrak{B}$ *is separable. A measurable function* $\vec{f}\colon \mathscr{H} \to \mathfrak{B}$ *is* μ*-integrable if and only if* $\mu^*(\|\vec{f}\|)$ *is finite.*

4.3. The Second Lebesgue Theorem

3.35 *Let f be a real-valued function defined on* $[a, b] \subset \mathbb{R}$. *If f is integrable on* $[a, b]$ *with the Lebesgue measure, then* $F(x) = \int_a^x f(t)\,dt$ *has a derivative almost everywhere, and almost everywhere* $F'(x) = f(x)$, *for* $a \leq x \leq b$.

The Lebesgue measure on $\mathbb{R}^n$ corresponds to the positive Radon measure, defined by the Jordan integral of the continuous functions with compact support.

3.36 Theorem. *If a function* $F(x)$ *is absolutely continuous on* $[a, b]$, *then there exists* $f(x)$, *an integrable function on* $[a, b]$, *such that* $F(x) - F(a) = \int_a^x f(t)\,dt$, *and conversely. Also* $F(x)$ *has a derivative almost everywhere, which is* $f(x)$.

Recall that $F(x)$ is said to be absolutely continuous on an interval I if, for each $\varepsilon > 0$, there exists a $\delta > 0$, such that $\sum_{i=1}^k |f(y_i) - f(x_i)| < \varepsilon$, whenever $]x_i, y_i[$, $i = 1, 2, \ldots, k$, are nonoverlapping subintervals of I, satisfying $\sum_{i=1}^k |y_i - x_i| < \delta$. In particular, a Lipschitzian function is absolutely continuous.

4.4. Rademacher's Theorem

3.37 *A Lipschitzian function from an open set of* $\mathbb{R}^n$ *to* $\mathbb{R}^p$ *is differentiable almost everywhere.*

4.5. Fubini's Theorem

3.38 *Let $\mathscr{H}$ and $\mathscr{F}$ be locally compact separable metric spaces. Given two positive Radon measures μ on $\mathscr{H}$ and ν on $\mathscr{F}$, if $f(x, y)\colon \mathscr{H} \times \mathscr{F} \to \mathbb{R}$ is $\mu \otimes \nu$-integrable, then for ν-almost all y, $f_y(x) = f(x, y)$ is μ-integrable and for μ-almost all x, $f_x(y) = f(x, y)$ is ν-integrable. Moreover*

$$\iint f(x, y)\, d\mu(x)\, d\nu(y) = \int \left[\int f(x, y)\, d\nu(y) \right] d\mu(x) \tag{9}$$
$$= \int \left[\int f(x, y)\, d\mu(x) \right] d\nu(y).$$

Fubini's Theorem is very useful, but for most applications, we don't know that $f(x, y)$ is $\mu \otimes \nu$-integrable.
We overcome this difficulty as follows: More often than not, it is obvious that $f(x, y)$ is $\mu \otimes \nu$-measurable. (Is it not a recognized fact, that function is measurable, when it is defined without using the axiom of the choice!)

Then by using Theorems (3.34) and (3.39), we shall know if $f(x, y)$ is $\mu \otimes \nu$-integrable or not. Recall that a locally compact metric space is separable if and only if it is a denumerable union of compact sets.

3.39 Theorem. *Let $(\mathscr{H}, \mu)$ and $(\mathscr{F}, \nu)$ be as above, and $f(x, y)$ a $\mu \otimes \nu$-measurable function. Then*

$$(\mu \otimes \nu)^*(|f|) = \mu^*[\nu^*(|f_x|)] = \nu^*[\mu^*(|f_y|)].$$

§5. The L_p Spaces

3.40 Definition. Let $\mathscr{H}$ be a locally compact separable metric space and μ a positive Radon measure. Given $p \geq 1$ a real number, we denote by $L_p(\mathscr{H})$ the class of all measurable functions f on $\mathscr{H}$ for which $\mu^*(|f|^p) < \infty$. We identify in $L_p(\mathscr{H})$ functions that are equal almost everywhere. The elements of $L_p(\mathscr{H})$ are equivalence classes under the relation: $f_1 \sim f_2$ if $f = f_2$ almost everywhere. $L_p(\mathscr{H})$, (denoted by L_p when no confusion is possible), is a separable Banach space, the norm being defined by:

$$\|f\|_p = \left(\int |f|^p\, d\mu \right)^{1/p}. \tag{10}$$

The Banach space $L_\infty(\mathscr{H})$ consists of all μ-essentially bounded functions. The norm is:

$$\|f\|_\infty = \mu\text{-ess sup}|f(x)| = \inf_A \sup_{x \in \mathscr{H} - A} |f(x)|, \tag{11}$$

where A ranges over the subsets of measure zero.

3.41 Proposition. *$C_0(\mathscr{H})$ is dense in $L_p(\mathscr{H})$ for all $1 \le p < \infty$.*

For $p = 1$, this is true by definition 3.27. The result is not true for L_∞, of course; otherwise every function belonging to L_∞ would be continuous. If $\mathscr{H}$ is an open set Ω of $\mathbb{R}^n$ and μ the Lebesgue measure, we have a more precise result, proved by regularization (see 3.46 below): $\mathscr{D}(\Omega)$ is dense in $L_p(\Omega)$. Here $\mathscr{D}(\Omega)$ is the set of C^∞-functions with compact support lying in Ω. Likewise, if (V, g) is a C^∞ Riemannian manifold and μ the Riemannian measure, $\mathscr{D}(V)$ is dense in $L_p(V)$. $\mathscr{D}(V)$ consists of the C^∞-functions on V, with compact support.

3.42 Proposition. *For $1 \le p < \infty$, $L_p^*(\mathscr{H})$ is isometric isomorphic to $L_q(\mathscr{H})$ with $1/p + 1/q = 1$. Hence L_p is reflexive provided $1 < p < \infty$.*

The isomorphism: $L_q \ni g \to u_g \in L_p^*$ is defined as follows:

$$L_p \ni f \to u_g(f) = \int fg \, d\mu.$$

Indeed fg, which is μ-measurable, is μ-integrable according to Hölder's inequality 3.60:

$$\|fg\|_1 \le \|f\|_p \|g\|_q. \tag{12}$$

3.43 Proposition. *Let $\{f_k\}$ be a sequence in L_p (or in L_∞) which converges in L_p to $f \in L_p$. Then there exists a subsequence converging pointwise almost everywhere to f.*

3.44 Theorem. *Let Ω be an open set of $\mathbb{R}^n$. A bounded subset $\mathscr{A} \subset L_p(\Omega)$ is precompact in $L_p(\Omega)$ if and only if for every number $\varepsilon > 0$, there exists a number $\delta > 0$ and a compact set $\mathfrak{K} \subset \Omega$, such that for every $f \in \mathscr{A}$:*

$$\int_{\Omega - \mathfrak{K}} |f(x)|^p \, dx < \varepsilon$$

and

$$\int_{\mathfrak{K}} |f(x + y) - f(x)|^p \, dx < \varepsilon \quad \text{when } \|y\| \le \delta,$$

where without loss of generality, we suppose that δ is smaller than the distance from $\mathfrak{K}$ to $\partial\Omega$, the boundary of Ω.

3.45 Theorem. *Let $1 < p < \infty$ and $\{f_k\}$ be a bounded sequence in $L_p(\mathscr{H})$, converging pointwise almost everywhere to f. Then f belongs to L_p and f_k converges to f weakly in L_p. The result does not hold for $p = 1$, of course (see below).*

5.1. Regularization

3.46 Let $\gamma \in \mathscr{D}(\mathbb{R}^n)$ be a non-negative function, whose integral equals 1. For convenience, we suppose $\operatorname{supp} \gamma \subset \bar{B}$.

For $k \in \mathbb{N}$, consider the function $\gamma_k(x) = k^n\gamma(kx)$, $\|\gamma_k\|_1 = 1$, and $\gamma_k(x) \to 0$ almost everywhere (except for $x = 0$) when $k \to \infty$. In fact $\{\gamma_k\}$ converges vaguely to the *Dirac measure* concentrated at zero. Let f be a function locally integrable on $\mathbb{R}^n$ (this means that $f\chi_{\mathfrak{K}}$ is integrable for any compact $\mathfrak{K} \subset \mathbb{R}^n$); we define the *regularization* of f by:

$$(13) \qquad f_k(x) = (\gamma_k * f)(x) = \int \gamma_k(x - y) f(y)\, dy.$$

Obviously f_k is C^∞. Moreover, if $f \in L_p(\mathbb{R}^n)$, then $\{f_k\}$ converges to f in $L_p(1 \le p < \infty)$, and

$$(14) \qquad \|f_k\|_p \le \|f\|_p.$$

When $f \in C_0(\mathbb{R}^n)$, the result is obvious, since f_k converges uniformly to f:

$$|f_k(x) - f(x)| = \left| \int \gamma_k(x - y)[f(y) - f(x)]\, dy \right| \le \sup_{\|y - x\| < 1/k} |f(y) - f(x)|,$$

so the assertion follows from the uniform continuity of f. Clearly $\|f_k - f\|_p \to 0$ when $k \to \infty$. If $\tilde{f} \in L_p(\mathbb{R}^n)$, for each $\varepsilon > 0$, there exists $f \in C_0(\mathbb{R}^n)$ such that $\|\tilde{f} - f\|_p < \varepsilon$ (Proposition 3.41). But

$$\|\tilde{f}_k - \tilde{f}\|_p \le \|\tilde{f}_k - f_k\|_p + \|f_k - f\|_p + \|f - \tilde{f}\|_p \le 2\varepsilon + \|f_k - f\|_p,$$

where we used inequality (14), which we are going to prove now. According to Hölder's inequality (21), q being defined by $1/p + 1/q = 1$:

$$\begin{aligned} |(\gamma_k * f)(x)| &\le \int [\gamma_k(x - y)]^{1/p + 1/q} |f(y)|\, dy \\ &\le \left[\int \gamma_k(x - y)\, dy \right]^{1/q} \left[\int \gamma_k(x - y) |f(y)|^p\, dy \right]^{1/p} \\ &= \left[\int \gamma_k(x - y) |f(y)|^p\, dy \right]^{1/p}. \end{aligned}$$

Hence by Fubini's theorem, 3.38:

$$\int |\gamma_k * f|^p\, dx \le \int |f|^p\, dy \int \gamma_k(x - y)\, dx = \|f\|_p^p.$$

5.2. Radon's Theorem

3.47 Theorem. *Let $\mathfrak{B}$ be a uniformly convex Banach space and $\{f_k\}$ a sequence in $\mathfrak{B}$ which converges weakly to $f \in \mathfrak{B}$.*

$$\text{If } \|f_k\| \to \|f\|, \text{ then } f_k \to f \text{ strongly } (\|f - f_k\| \to 0) \text{ as } k \to \infty.$$

Recall that $\mathfrak{B}$ is said to be uniformly convex, if $\|g_k\| = \|h_k\| = 1$ and $\|g_k + h_k\| \to 2$ implies $\|h_k - g_k\| \to 0$, when $k \to \infty$, for sequences $\{g_k\}$ and $\{h_k\}$ in $\mathfrak{B}$.

A uniformly convex Banach space is reflexive; the converse is not true. The spaces $L_p(1 < p < \infty)$ are uniformly convex. This result is due to Clarkson as a consequence of his inequality 3.63 below.
It is obvious that a Hilbert space is uniformly convex.

Proof. If $f = 0$, we have nothing to prove: $\|f_k\| \to 0$.
If $f \neq 0$, we can suppose without loss of generality that $\|f\| = 1$ and that $\|f_k\| \neq 0$ for all k. Set $g_k = \|f_k\|^{-1} f_k$ and $h_k = f$ for all k.
By the Hahn–Banach theorem 3.12, there exists $u_0 \in \mathfrak{B}^*$ such that $u_0(f) = 1$ and $\|u_0\| = 1$. Since $f_k \to f$ weakly, we have as $k \to \infty$:

$$u_0(g_k) = \|f_k\|^{-1} u_0(f_k) \to u_0(f) = 1.$$

Using $\|u_0\| = 1$ we get:

$$|1 + u_0(g_k)| \leq \|g_k + f\| \leq 2.$$

Letting $k \to \infty$, we obtain $\|g_k + f\| \to 2$. Thus $\|g_k - f\| \to 0$ and $\|f - f_k\| \to 0$, since $\|f - f_k\| \leq \|f - g_k\| + |1 - \|f_k\||$ ■

3.48 Definition. Let u be a locally integrable function in Ω, an open set of $\mathbb{R}^n$. A locally integrable function $\mathfrak{h}$ is called the *weak derivative* of u with respect to x^1 if it satisfies

$$\int_\Omega \varphi \mathfrak{h} \, dV = -\int_\Omega u \, \partial_1 \varphi \, dV \quad \text{for all } \varphi \in \mathcal{D}(\Omega).$$

By induction we define the weak derivative of u of any order if it exists.

Proposition 3.48. *Let f be a Lipschitzian function on a bounded open set $\Omega \subset \mathbb{R}^n$. Then $\partial_i f$ exists almost everywhere, belongs to $L_p(\Omega)$, and coincides with the weak derivative in the sense of the distributions.*

Proof. According to Rademacher's Theorem (3.37), $\partial_i f$ exists almost everywhere, since f is a Lipschitzian function. Moreover, $\partial_i f$ is bounded almost everywhere, since $|f(x) - f(y)| \leq k\|x - y\|$ implies $|\partial_i f| \leq k$, when $\partial_i f$ exists.

On any line segment in Ω, f is absolutely continuous. Thus by Theorem (3.36), $\partial_i f$ defines the weak derivative with respect to x^i. $\partial_i f$ is the limit function, almost everywhere, of a sequence of μ-measurable functions; hence it is measurable (see definition 3.30).
Since $|\partial_i f|^p \le k^p$, $\mu^*_{\text{on}\,\Omega}(|\partial_i f|^p) \le k^p\mu(\Omega) < \infty$. Consequently $\partial_i f \in L_p(\Omega)$, according to theorem (3.34). ∎

3.49 Proposition. *Let M_n be a C^∞ Riemannian manifold and $\varphi \in H_1^p(M)$; then almost everywhere $|\nabla|\varphi|| = |\nabla\varphi|$.*

Proof. Since a Riemannian manifold is covered by a countable set of balls, it is sufficient to establish the result for a ball B of $\mathbb{R}^n$, provided with a Riemannian metric. Denote the coordinates by $\{x^i\}$.
First of all we are going to prove the statement for C^∞ functions. Let $f \in C^\infty(B)$. When $f(x) > 0$, $\partial_i f = \partial_i |f|$, and when $f(x) < 0$, $\partial_i f = -\partial_i |f|$. Thus the result will follow, once we have shown that the set $\mathscr{A}$ of the points $x \in B$, where simultaneously $f(x) = 0$ and $|\nabla f(x)| \ne 0$, has zero measure. Since, for $x \in \mathscr{A}$, $|\nabla f(x)| \ne 0$, there exists a neighborhood Θ_x of x, such that $\Theta_x \cap f^{-1}(0)$ is a submanifold of dimension $n-1$ (see for instance Choquet-Bruhat [99] p. 12). Consequently, $\mu[\Theta_x \cap f^{-1}(0)] = 0$. As there exists a countable basis of open sets for the topology of B, $\mathscr{A}$ is covered by a countable set of $\mathscr{V}_x(x \in \mathscr{A})$. Let $\{\mathscr{V}_{x_k}\}_{k\in\mathbb{N}}$ be this set. We have $\mathscr{A} \subset \bigcup_{k=1}^\infty \mathscr{V}_{x_k}$ and $\mu[\mathscr{V}_{x_k} \cap \mathscr{A}] = 0$; thus $\mu(\mathscr{A}) = 0$.

Now let $\varphi \in H_1^p(B)$. By definition, $C^\infty(B) \cap H_1^p(B)$ is dense in $H_1^p(B)$. So using Proposition 3.43, there exists $\{f_j\}$, a sequence of C^∞ functions on B, such that $\|f_j - \varphi\|_{H_1^p} \to 0$ and such that $f_j \to \varphi$ a.e. and $\partial_i f_j \to h_i$ a.e. for all $i(1 \le i \le n)$ as $j \to \infty$, where h_i denotes the weak derivatives of φ.
Define for a function f, $f^+ = \sup(f, 0)$ and $f^- = \sup(-f, 0)$. Obviously $f_j^+ \to \varphi^+$ a.e. and $f_j^- \to \varphi^-$ a.e.
Moreover f_j^+ is a Cauchy sequence in H_1^p, which converges to an element of H_1^p, which is φ^+, since $\|f_j^+ - \varphi^+\|_p \to 0$. Thus $\varphi^+ \in H_1^p$ likewise $\varphi^- \in H_1^p$.
According to Proposition 3.43, taking a subsequence $\{f_k\}$ of $\{f_j\}$, if necessary, we have for all $i(1 \le i \le n)$:
$\partial_i f_k^+ \to h_i^+$ a.e. and $\partial_i f_k^- \to h_i^-$ a.e. as $k \to \infty$, where h_i^+ (respectively, h_i^-) denote the weak derivatives of φ^+ (respectively φ^-).
Since we have proved that almost everywhere $|\partial_i f_k| = |\partial_i |f_k||$, letting $k \to \infty$ yields $|h_i| = |h_i^+ + h_i^-|$ a.e. Likewise, $\partial_{i_1} f_k\, \partial_{i_2} f_k = \partial_{i_1}|f_k|\, \partial_{i_2}|f_k|$ a.e. gives $h_{i_1}h_{i_2} = (h_{i_1}^+ + h_{i_1}^-)(h_{i_2}^+ + h_{i_2}^-)$ a.e. Hence $|\nabla\varphi| = |\nabla|\varphi||$ a.e. ∎

3.50 Proposition. *Let $\overline{W}_n$ be a compact Riemannian manifold with boundary of class C^r. If $f \in C^{k-1}(\overline{W}) \cap \overset{\circ}{H}{}_k^p(W)$ $(1 \le k \le r)$, then f and its derivatives up to order $k-1$ vanish on the boundary ∂W.*

Proof. Let $\{\Omega_i, \varphi_i\}$ be an atlas of class C^r such that $\varphi_i(\Omega_i)$ is either the ball $B \subset \mathbb{R}^n$ or the half ball $D = B \cap \bar{E}$. Consider a C^∞ partition of unity $\{\alpha_i\}$ subordinate to the covering $\{\Omega_i\}$.

Choose any point P of ∂W. At least one of the α_i does not vanish at P. Let $\alpha_{i_0}(P) > 0$ and set $\tilde{f} = (\alpha_{i_0} f) \circ \varphi_{i_0}^{-1}$. $\tilde{f} \in C^{k-1}(\bar{D}) \cap \mathring{H}_k^p(D)$. For all u and v belonging to $C^1(\bar{D})$

$$\text{(15)} \qquad \int_D v\, \partial_1 u\, dx + \int_D u\, \partial_1 v\, dx = \int_{\partial D} uv(\vec{e}_1, \vec{v})\, d\sigma,$$

where $\vec{v}$ is the outer normal and $\vec{e}_1$ the first unit vector of the basis of $\mathbb{R}^n$.

If $u \in \mathscr{D}(D)$, the right side vanishes. Hence, by density, the left side of (15) is zero for all $u \in \mathring{H}_1^p(D)$. In particular, if $\tilde{u}$ is $\tilde{f}$ or one of its derivatives up to order $k-1$, for all $v \in C^1(\bar{D})$:

$$\int_{\partial D} \tilde{u}v(\vec{e}_1, \vec{v})\, d\sigma = \int_D v\, \partial_1 \tilde{u}\, dx + \int_D \tilde{u}\, \partial_1 v\, dx = 0.$$

Consequently $\tilde{u}$ vanishes on ∂D, and f, as well as its derivatives up to order $k-1$, vanishes at P.

Remark. It is easy to prove the converse when $f \in C^k(\bar{W})$: If f and its derivatives up to order $k-1$ vanish on ∂W, then $f \in \mathring{H}_k^p(W)$.

§6. Elliptic Differential Operators

3.51 Definition. Let M_n be a Riemannian manifold. A *linear differential operator* $A(u)$ of order $2m$ on M_n, written in a local chart (Ω, φ), is an expression of the form:

$$\text{(16)} \qquad A(u) = \sum_{\ell=0}^{2m} a_\ell^{\alpha_1\alpha_2\cdots\alpha_\ell} \nabla_{\alpha_1\alpha_2\cdots\alpha_\ell} u,$$

where a_ℓ are ℓ-tensors and $u \in C^{2m}(M)$. For simplicity, we can write $A(u) = a_\ell \nabla^\ell u$. The terms of highest order, $2m$, are called the *leading part* (a_{2m} is presumed to be nonzero).

The operator is said to be *elliptic* at a point $x \in \Omega$, if there exists $\lambda(x) \geq 1$ such that, for all vectors ξ:

$$\text{(17)} \qquad \|\xi\|^{2m} \lambda^{-1}(x) \leq a_{2m}^{\alpha_1\alpha_2\cdots\alpha_{2m}}(x) \xi_{\alpha_1} \cdots \xi_{\alpha_{2m}} \leq \lambda(x) \|\xi\|^{2m}.$$

We say that the operator is *uniformly elliptic* in Ω, if there exist λ_0 and $\lambda(x)$, $1 \leq \lambda(x) \leq \lambda_0$, such that (17) holds for all $x \in \Omega$.

3.52 Definition. A *differential operator* A of order $2m$ defined on M_n is written $A(u) = f(x, u, \nabla u, \ldots, \nabla^{2m} u)$, where f is assumed to be a differentiable function of its arguments. Then the first variation of f at $u_0 \in C^{2m}(\Omega)$ is the linear differential operator

$$(18)\quad \begin{aligned} A'_{u_0}(v) &= \sum_{\ell=0}^{2m} \frac{\partial f(x, u_0, \nabla u_0, \ldots, \nabla^{2m} u_0)}{\partial \nabla_{\alpha_1\alpha_2\cdots\alpha_\ell} u} \nabla_{\alpha_1\alpha_2\cdots\alpha_\ell} v \\ &= \frac{\partial f}{\partial \nabla^\ell u}(x, u_0, \nabla u_0, \ldots, \nabla^{2m} u_0) \nabla^\ell v. \end{aligned}$$

If A'_{u_0} is an elliptic operator, then A is called elliptic at u_0.

6.1. Weak Solution

3.53 a) Let A be a linear differential operator of order $2m$ defined on a Riemannian manifold M, with or without boundary. Until now, by a solution of the equation $A(u) = f$ we meant a function $u \in C^{2m}(M)$ such that the equation is satisfied pointwise. There are other quite natural ways that a more general function, such as an element of $H^p_{2m}(M)$ or a distribution, can be said to be a solution of $A(u) = f$.

b) If $f \in L_p$ and if the coefficients of A are measurable and locally bounded, we say that $u \in H^p_{2m}(M)$ is a strong solution in the L_p sense of $A(u) = f$ if there is a sequence $\{\varphi_i\}$ of C^∞ functions on M such that $\varphi_i \to u$ in $H^p_{2m}(M)$ and $A(\varphi_i) \to f$ in $L^p(M)$. Indeed, in this case the weak (distribution) derivatives up to order $2m$ are functions in $L^p(M)$ and $A(u) = f$ almost everywhere.

c) Let $A(u) = a_\ell \nabla^\ell u$. If the tensors $a_\ell \in C^\ell(M)$ for $0 \le \ell \le 2m$, then we define the *formal adjoint* of A by

$$A^*(\varphi) = (-1)^\ell \nabla^\ell(\varphi a_\ell).$$

We say that $u \in L_1(M)$ satisfies $A(u) = f$, in the sense of distributions, if for all $\varphi \in \mathscr{D}(M)$:

$$\int_M u A^*(\varphi)\, dV = \int_M f\varphi\, dV.$$

If the coefficients $a_\ell \in C^\infty(M)$, then a distribution u satisfies $A(u) = f$ if for all $\varphi \in \mathscr{D}(M)$:

$$\langle u, A^*(\varphi)\rangle = \langle f, \varphi\rangle.$$

Now a given distribution is some weak derivative of some order, say r, of a locally integrable function. In this case $\langle u, A^*(\varphi)\rangle$ makes sense if the coefficients satisfy $a_\ell \in C^{\ell+r}(M)$.

d) If the operator can be written in divergence form, i.e., if we can write $A(u)$ as

$$A(u) \equiv \sum_{\substack{0 \le k \le m \\ 0 \le \ell \le m}} \nabla_{\alpha_1 \cdots \alpha_k}(a_{k,\ell}^{\alpha_1 \cdots \alpha_k \beta_1 \cdots \beta_\ell} \nabla_{\beta_1 \cdots \beta_\ell} u) + \sum_{\ell=0}^{m} b_\ell \nabla^\ell u$$

where $a_{k,\ell}$ are $k + \ell$-tensors and b_ℓ ℓ-tensors, then $u \in H_m^p(M)$ is said to be a weak solution of $A(u) = f$ with $f \in L_1(M)$ if for all $\varphi \in \mathscr{D}(M)$:

$$\sum_{\substack{0 \le k \le m \\ 0 \le \ell \le m}} (-1)^k \int_M a_{k,\ell} \nabla^\ell u \nabla^k \varphi \, dV + \sum_{\ell=0}^{m} \int_M \varphi b_\ell \nabla^\ell u \, dV = \int_M f \varphi \, dV.$$

Here we need only suppose that $a_{k,\ell}$ are measurable and are locally bounded for all pairs (k, ℓ). Note these definitions of weak or generalized solutions are not equivalent; they depend on the properties of the coefficients. But the terminology is standard and does not really cause confusion.

e) For a nonlinear differential operator of the type

$$A(u) = \sum_{\ell=0}^{m} (-1)^\ell \nabla_{\alpha_1 \cdots \alpha_\ell} A_\ell^{\alpha_1 \cdots \alpha_\ell}(x, u, \nabla u, \ldots, \nabla^m u) = (-1)^\ell \nabla^\ell A_\ell,$$

where A_ℓ are ℓ-tensors on M, $u \in C^m(M)$ is said to be a weak solution of $A(u) = 0$, if for all $\varphi \in \mathscr{D}(M)$:

$$\sum_{\ell=0}^{m} \int A_\ell^{\alpha_1 \cdots \alpha_\ell}(x, u, \nabla u, \ldots, \nabla^m u) \nabla_{\alpha_1 \cdots \alpha_\ell} \varphi \, dV = 0.$$

6.2. Regularity Theorems

3.54 And now some theorems concerning the regularity of (weak) solutions in the interior, then on the boundary, of the manifold. Roughly speaking, we can hope that if we can define in some sense, for some function u, $A(u) = f$, then u will have the maximum of regularity allowed by the coefficients. This is almost the case. Precisely, we have:

Theorem. *Let Ω be an open set of $\mathbb{R}^n$ and $A = a_\ell \nabla^\ell$ a linear elliptic operator of order 2m with C^∞ coefficients ($a_\ell \in C^\infty(\Omega)$ for $0 \le \ell \le 2m$). Suppose u is a distribution solution of the equation $A(u) = f$ and $f \in C^{k,\alpha}(\Omega)$ (resp., $C^\infty(\Omega)$). Then $u \in C^{k+2m,\alpha}(\Omega)$, (resp., $C^\infty(\Omega)$) with $0 < \alpha < 1$.*

If f belongs to $H_k^p(\Omega)$, $1 < p < \infty$, then u belongs locally to H_{k+2m}^p.

Proof. Although it is basic, I did not find exactly this theorem in the literature. First of all, if $f \in C^\infty(\Omega)$, $u \in C^\infty(\Omega)$; this is the well-known result of Schwartz

(250). On the other hand, if $f \in H_k^2(\Omega)$ $(k \geq 0)$, then $u \in H_{k+2m}^2$ locally according to Bers, John and Schechter (50) p. 190. Now if $f \in C^{k,\alpha}(\Omega)$, obviously f belongs locally to H_k^2. Therefore $u \in H_{k+2m}^2$ locally and according to Morrey (204) p. 246, $u \in C^{2m+k,\alpha}(\Omega)$. Finally, let us establish the result when $f \in H_k^p(\Omega)$ $p \neq 2$. There exist continuous v and integer $r \geq 0$ such that $u = \Delta^r v$ (Bers, John and Schechter (50) p. 195). But $A\Delta^r$ is elliptic of order $2(m + r)$ and v belongs locally to L_2. Thus if $f \in H_k^p(\Omega)$ $(1 < p < \infty)$, then v belongs locally to $H_{k+2(m+r)}^p$ and $u \in H_{k+2m}^p$ locally, according to Morrey (204) p. 246. ■

3.55 Theorem (Ladyzenskaja and Uralceva [173] p. 195). *Let Ω be an open set of $\mathbb{R}^n$ and A a linear elliptic operator of order two, with $C^{k,\alpha}$ coefficients ($k \geq 0$ an integer, $0 < \alpha < 1$). If a bounded function $u \in H_2(\Omega)$ satisfies $A(u) = f$ almost everywhere, where $f \in C^{k,\alpha}(\Omega)$, then $u \in C^{k+2,\alpha}(\Omega)$. The same conclusion holds if $u \in H_1(\Omega)$ is a weak solution of $A(u) = f$.*

For this last statement, the operator must be written in divergence form. In general this requires $a^{ij} \in C^1(\Omega)$. Then, according to Ladyzenskaja (173), u is locally bounded (p. 199), and belongs locally to H_2 (p. 188). Thus we can apply the first part of theorem 3.55, on any bounded open set Θ with $\bar{\Theta} \subset \Omega$.

Remark. If the coefficients and f belong locally to H_q, with $q > 2 + n/2$, and if $u \in H_{1,\text{loc}}$, then we can prove (see Aubin (20) p. 66) that u belongs locally to H_{q+2}.

3.56 Theorem (Giraud [127], Hopf [146], and Nirenberg [216] and [217]). *Let $A(u) = F(x, u, \nabla u, \nabla^2 u)$ be à differential operator of order two, defined on Ω an open set of $\mathbf{R}^n$, F being a C^∞ differentiable function of its arguments. Suppose that A is elliptic on Ω at $u_0 \in C^2(\Omega)$, and that $A(u_0) = f \in C^{r,\beta}(\Omega)$ with $0 < \beta < 1$. Then $u_0 \in C^{r+2,\beta}(\Omega)$.*

Let Θ be a bounded subset of $C^2(\Omega)$, and suppose that A is uniformly elliptic on Ω at any $u \in \Theta$, uniformly in u (the same λ_0 is valid for all $u \in \Theta$, see definition 3.51). If $A(\Theta)$ is bounded in $C^{r,\beta}(K)$, then Θ is bounded in $C^{r+2,\beta}(K)$, for any compact set $K \subset \Omega$.

The result for $n = 2$ is due to Leray, and Nirenberg [217] established the theorem in the case $n > 2$, when there exists a modulus of continuity for the second derivatives of u_0. Previously Giraud [127] and Hopf [146] proved the result assuming that $u_0 \in C^{2,\alpha}(\Omega)$ for some $\alpha > 0$.

Remark. When A is a differential operator of order two on a compact Riemannian manifold M_n, it is possible to prove similar results: If $A(\Theta)$ is bounded in $H_q(M_n)$ with $q > 2 + n/2$, then Θ is bounded in $H_{q+2}(M_n)$, (see Aubin [20] p. 68).

3.57 Theorem (Agmon [2] p. 444). *Let Ω be a bounded open set of $\mathbb{R}^n$ with boundary of class C^{2m} and A be an elliptic linear differential operator of order $2m$ with coefficients $a_\ell \in C^\ell(\overline{\Omega})$. Let $u \in L_q(\Omega)$ for some $q > 1$, and $f \in L_p(\Omega)$, $p > 1$. Suppose that for all functions $v \in C^{2m}(\overline{\Omega}) \cap \mathring{H}^p_m(\Omega)$,*

$$\int uA(v)\,dV = \int fv\,dV.$$

Then $u \in H^p_{2m}(\Omega) \cap \mathring{H}^p_m(\Omega)$ and

$$\|u\|_{H^p_{2m}} \leq C[\|f\|_p + \|u\|_p],$$

where C is a constant depending only on Ω, A, n, and p.

Moreover, if $p > n/(m+1)$ then $u \in C^{m-1}(\overline{\Omega})$ and u is a solution of the Dirichlet problem

$$A^*u = f \quad \text{in } \Omega, \nabla^\ell u = 0 \quad \text{on } \partial\Omega, 0 \leq \ell \leq m-1$$

in the strong L_p sense.

3.58 Theorem (Gilbarg and Trüdinger [125] p. 177). *Let Ω be a bounded open set of $\mathbb{R}^n$ with C^{k+2} boundary ($k \geq 0$) and A a linear elliptic operator of order two, such that $a_2 \in C^{k+1}(\overline{\Omega})$ and $a_1, a_0 \in C^k(\overline{\Omega})$.*
Suppose $u \in \mathring{\mathsf{H}}_1(\Omega)$ is a weak solution of $A(u) = f$, with $f \in \mathsf{H}_k(\Omega)$. Then $u \in \mathsf{H}_{k+2}(\Omega)$ and

$$\text{(19)} \qquad \|u\|_{\mathsf{H}_{k+2}} \leq C(\|u\|_2 + \|f\|_{\mathsf{H}_k}),$$

where the constant C is independent of u and f.
Thus, if the coefficients and f belong to $C^\infty(\overline{\Omega})$ and if the boundary is C^∞, then $u \in C^\infty(\overline{\Omega})$.

3.59 Theorem (Gilbarg and Trüdinger [125] p. 106). *Let Ω be a bounded open set of $\mathbb{R}^n$ with $C^{k+2,\alpha}$ boundary and let A be a linear elliptic operator of order two, with coefficients belonging to $C^{k,\alpha}(\overline{\Omega})$ ($k \geq 0$ an integer and $0 < \alpha < 1$). Suppose $u \in C^0(\overline{\Omega}) \cap C^2(\Omega)$ is a solution of the Dirichlet problem $A(u) = f$ in Ω, $u = v$ on $\partial\Omega$, with $f \in C^{k,\alpha}(\overline{\Omega})$ and $v \in C^{k+2,\alpha}(\overline{\Omega})$. Then $u \in C^{k+2,\alpha}(\overline{\Omega})$.*

Now let us prove a result which will be used in Chapter 8.

Proposition 3.59. *Let Ω be a bounded open set of $\mathbb{R}^n$ with C^∞ boundary and let $A(u) = F(x, u, \nabla u, \nabla^2 u)$ be a differential operator of order two, defined on Ω, F being a C^∞ differentiable function of its arguments on $\overline{\Omega}$. Suppose that A is uniformly elliptic on $\overline{\Omega}$ at $u_0 \in C^{2,\alpha}(\overline{\Omega})$, with $0 < \alpha < 1$. If $u_{0/\partial\Omega} \in C^\infty(\partial\Omega)$ and if $A(u_0) \in C^\infty(\overline{\Omega})$, then $u_0 \in C^\infty(\overline{\Omega})$.*

Proof. By Theorem 3.56, $u_0 \in C^\infty(\Omega)$. It remains to prove the regularity up to $\partial\Omega$. Let X be a C^∞ vector field tangent to $\partial\Omega$. Differentiating $A(u_0)$ with respect to $L = X^i\,\partial_i$ yields $A'_{u_0}(Lu_0) \in C^\alpha(\overline{\Omega})$ where A'_{u_0} is a linear elliptic operator with coefficients belonging to $C^\alpha(\overline{\Omega})$. As $Lu_0 \in C^0(\overline{\Omega}) \cap C^2(\Omega)$ and as $Lu_0/\partial\Omega \in C^\infty$, by Theorem 3.59, $Lu_0 \in C^{2,\alpha}(\overline{\Omega})$. So the third derivatives of u_0 are Hölder continuous up to $\partial\Omega$, except maybe the derivatives three times normal. Now let $P \in \partial\Omega$ and ∂_ν be the normal derivative. As A is elliptic $\partial_{\partial_{\nu\nu}} A(u_0)$ is strictly positive at P. By the inverse function theorem, $\partial_{\nu\nu} u_0$ expresses itself in a neighborhood Θ of P in function of u_0 its first derivatives and its other second derivatives which belong to $C^{1,\alpha}(\overline{\Omega} \cap \Theta)$. Thus $u_0 \in C^{3,\alpha}(\overline{\Omega})$. By induction $u_0 \in C^\infty(\overline{\Omega})$.

3.60 The Neumann Problem. Until now we talked about the Dirichlet problem. But we may wish to solve an elliptic equation with other boundary conditions.

For the Neumann Problem the normal derivative of the solution at the boundary is prescribed. For this problem, and those with mixed boundary conditions, we give as references Ladyzenskaja and Uralceva [173] p. 135, Ito [152], Friedman [116], and Cherrier [97].

6.3. The Schauder Interior Estimates[1]

3.61 Let Ω be an open set of $\mathbb{R}^n$ and let $u \in C^{2,\alpha}(\Omega)(0 < \alpha < 1)$ be a bounded solution in Ω of the equation

$$a^{ij}\,\partial_{ij}u + b^i\,\partial_i u + cu = f,$$

where f and the coefficients belong to $C^\alpha(\Omega)$, a^{ij} satisfying $a^{ij}\xi_i\xi_j \geq \lambda|\xi|^2$ with $\lambda > 0$ for all $x \in \Omega$ and $\xi \in \mathbb{R}^n$. Then on any compact set $K \subset \Omega$:

$$\|u\|_{C^{2,\alpha}(K)} \leq C[\|u\|_{C^0(\Omega)} + \|f\|_{C^\alpha(\Omega)}], \tag{20}$$

where the constant C depends on K, α, λ and Λ a bound for the C^α norm of the coefficients in Ω.

§7. Inequalities

7.1. Hölder's Inequality

3.62 Let M be a Riemannian manifold. If $f \in L_p(M)$ and $h \in L_q(M)$ with $p^{-1} + q^{-1} = 1$, then $fh \in L_1(M)$ and :

$$\|fh\|_1 \leq \|f\|_p \|h\|_q. \tag{21}$$

[1] Gilbarg and Trüdinger (125) p. 85.

More generally, if $f \in L_{p_i}(M)$, $(1 \le i \le k)$, with $\sum_{i=1}^k p_i^{-1} = 1$, then $\prod_{i=1}^k f_i \in L_1(M)$ and $\|\prod_{i=1}^k f_i\|_1 \le \prod_{i=1}^k \|f_i\|_{p_i}$.

Proposition 3.62. *Let M be a Riemannian manifold. If $f \in L_r(M) \cap L_q(M)$, $1 \le r < q \le \infty$, then $f \in L_p$ for $p \in [r, q]$ and*

$$\|f\|_p \le \|f\|_r^a \|f\|_q^{1-a} \quad \textit{with } a = \frac{1/p - 1/q}{1/r - 1/q}. \tag{22}$$

The proof is just an application of Hölder's inequality.

7.2. Clarkson's Inequalities

3.63 If $u, v \in L_p(M)$, when $2 \le p < \infty$,

$$\|u + v\|_p^p + \|u - v\|_p^p \le 2^{p-1}(\|u\|_p^p + \|v\|_p^p),$$

$$\|u + v\|_p^q + \|u - v\|_p^q \ge 2(\|u\|_p^p + \|v\|_p^p)^{q-1}$$

with $p^{-1} + q^{-1} = 1$. When $1 < p \le 2$, then

$$\|u + v\|_p^q + \|u - v\|_p^q \le 2(\|u\|_p^p + \|v\|_p^p)^{q-1},$$

$$\|u + v\|_p^p + \|u - v\|_p^p \ge 2^{p-1}(\|u\|_p^p + \|v\|_p^p).$$

7.3. Convolution Product

3.64 Let $\mathfrak{u} \in L_p(\mathbb{R}^n)$, $\mathfrak{v} \in L_q(\mathbb{R}^n)$ and $p, q \in [1, \infty[$ with $p^{-1} + q^{-1} \ge 1$. Then the convolution product $(\mathfrak{u} * \mathfrak{v})(x) = \int_{\mathbb{R}^n} \mathfrak{u}(x - y)\mathfrak{v}(y)\, dy$ exists a.e., belongs to L_r with $r^{-1} = p^{-1} + q^{-1} - 1$, and satisfies

$$\|\mathfrak{u} * \mathfrak{v}\|_r \le \|\mathfrak{u}\|_p \|\mathfrak{v}\|_q \tag{23}$$

Proposition 3.64. *Let M_n, $\tilde{M}_n$ be two Riemannian manifolds and let $M_n \times \tilde{M}_n \ni (P, Q) \to f(P, Q)$ be a numerical measurable function such that, for all $P \in M_n$, $Q \to f_P(Q) = f(P, Q)$ belongs to $L_p(\tilde{M})$ with $\sup_{P \in M_n} \|f_P(Q)\|_p < \infty$, and for all $Q \in \tilde{M}_n$, $P \to \tilde{f}_Q(P) = f(P, Q)$ belongs to $L_p(M)$ with $\sup_{Q \in \tilde{M}_n} \|\tilde{f}_Q(P)\|_p < \infty$.*

If $g \in L_q(\tilde{M}_n)$ with $p^{-1} + q^{-1} \ge 1$, then $h(P) = \int_{\tilde{M}_n} f(P, Q) g(Q)\, dV(Q)$ exists for almost all $P \in M_n$ and belongs to $L_r(M_n)$ with $r^{-1} = p^{-1} + q^{-1} - 1$. Moreover:

$$\|h\|_r \le \sup_{P \in M_n} \|f_P(Q)\|_p^{1-p/r} \|g\|_q \sup_{Q \in \tilde{M}_n} \|\tilde{f}_Q(P)\|_p^{p/r}. \tag{24}$$

Proof. It is sufficient to prove inequality (24) for nonnegative C^0 functions with compact support. If $p = q = 1$, it is obvious:

$$\|h\|_1 \leq \sup_{Q \in \bar{M}} \|\tilde{f}_Q(P)\|_1 \|g\|_1.$$

In the general case we write

$$f(P, Q)g(Q) = [f^p(P, Q)g^q(Q)]^{1/r}[f^p(P, Q)]^{1/p - 1/r}[g^q(Q)]^{1/q - 1/r}.$$

Since $1/r + (1/p - 1/r) + (1/q - 1/r) = 1$, applying Hölder's inequality, we are led to

$$|h(P)| \leq \left[\int_{\bar{M}} f^p(P, Q)g^q(Q)\, dV(Q)\right]^{1/r}\left[\int_{\bar{M}} f^p(P, Q)\, dV(Q)\right]^{1/p - 1/r} \times \left[\int_{\bar{M}} g^q(Q)\, dV(Q)\right]^{1/q - 1/r},$$

and the result follows. ∎

7.4. The Calderon–Zygmund Inequality

3.65 Let $\omega \in L_\infty(\mathbb{R}^n)$ with compact support satisfy $\omega(tx) = \omega(x)$ for all $0 < t \leq 1$ and $\|x\| \leq \rho$ for some $\rho > 0$, and also satisfy $\int_{\mathbb{S}_{n-1}(\rho)} \omega(x)\, d\sigma = 0$. For all $\varepsilon > 0$ let

$$(K_\varepsilon * f)(x) = \int_{\|x\| > \varepsilon} \omega(y)\|y\|^{-n} f(x - y)\, dy \quad \text{with } f \in L_p(\mathbb{R}^n).$$

If $1 < p < \infty$, then $\lim_{\varepsilon \to 0}(K_\varepsilon * f)(x)$ exists almost everywhere and the limit function denoted by $K_0 * f$ belongs to L_p.

Moreover, $K_\varepsilon * f \to K_0 * f$ in L_p and there exists a constant C, which depends on ω and p, such that

(25) $\|K_0 * f\|_p \leq C\|f\|_p$ (*Calderon–Zygmund inequality*)

If $\omega \in L_\infty(\mathbb{R}^n)$ satisfies $\omega(tx) = \omega(x)$ for all $t > 0$ and $\int_{\mathbb{S}_{n-1}(1)} \omega(x)\, d\sigma = 0$, then $\|K_\varepsilon * f - K_0 * f\|_p \to 0$ when $\varepsilon \to 0$ and (25) holds.
In addition, if $\omega \in C^1(\mathbb{R}^n - \{0\})$, $K_\varepsilon * f \to K_0 * f$ a.e. (see Dunford and Schwartz [111]).

7.5. Korn–Lichtenstein Theorem

3.66 Theorem. *If $\omega(x)$ is a function with the properties described in* 3.65 *and K_0 is defined as above, there exists, for any $\alpha (0 < \alpha < 1)$, a constant $A(\alpha)$ such that*

$$\|K_0 * f\|_{C^\alpha} \leq A(\alpha)\|f\|_{C^\alpha},$$

for all $f \in C^\alpha(\mathbb{R}^n)$ with compact support.

3.67 Theorem. *Let M_n be a compact Riemannian manifold and p, q, and r real numbers satisfying $1/p = 1/q - 1/n$, $1 \le q < n$ and $r > n$. Define $\mathscr{A} = \{\varphi \in L_1 / \int \varphi \, dV = 0\}$.*
Then there exists a constant k, such that, for all $\alpha \ge 1$, any function $\varphi \in \mathscr{A}$ with $|\nabla |\varphi|^\alpha| \in L_q$, satisfies

$$\| |\varphi|^\alpha \|_p \le k^\alpha \| \nabla |\varphi|^\alpha \|_q. \tag{26}$$

If $\varphi \in \mathscr{A}$ with $|\nabla \varphi| \in L_r$, then $\sup|\varphi| \le \text{Const} \times \|\nabla\varphi\|_r$. If $\varphi \in L_1$ *and $\Delta\varphi \in L_q$ (in the distributional sense), then*

$$|\nabla\varphi| \in L_p \quad \textit{and} \quad \|\nabla\varphi\|_p \le \text{Const} \times \|\Delta\varphi\|_q.$$

If $\varphi \in L_1$ and $\Delta\varphi \in L_r$, then $|\nabla\varphi|$ is bounded and

$$\sup|\nabla\varphi| \le \text{Const} \times \|\Delta\varphi\|_r.$$

The constants do not depend on φ, of course.

Let $\overline{M}_n$ be a compact Riemannian manifolds with boundary. Then the theorem holds for functions $\varphi \in \mathscr{D}(M)$.

Proof. First of all, we are going to establish (26) for $\alpha = 1$.
Let $\mathsf{G}(P, Q)$ be the Green's function of the Laplacian. As $\int \varphi \, dV = 0$, in the distributional sense (Proposition 4.14):

$$\varphi(P) = \int \mathsf{G}(P, Q)\Delta\varphi(Q) \, dV(Q), \tag{27}$$

whence:

$$|\varphi(P)| \le \int |\nabla_Q \mathsf{G}(P, Q)| |\nabla\varphi(Q)| \, dV(Q) \tag{28}$$

and according to Proposition 3.64, we find

$$\|\varphi\|_q \le \|\nabla\varphi\|_q \sup_{P \in M_n} \int |\nabla_Q \mathsf{G}(P, Q)| \, dV(Q)$$

for all $\varphi \in \mathscr{A}$, such that $|\nabla\varphi| \in L_q$.

Using the Sobolev imbedding theorem 2.21, we obtain

$$\|\varphi\|_p \le K\|\nabla\varphi\|_q + A\|\varphi\|_q \le k_0\|\nabla\varphi\|_q, \tag{29}$$

with $k_0 = K + A \sup_{P \in M} \int |\nabla_Q \mathsf{G}(P, Q)| \, dV(Q)$.

Let us now prove (26) for $\alpha > 1$. Since the set of the C^∞ functions which have no degenerate critical points is dense in the spaces H_1^s (Proposition 2.16), we need only establish (26) for these functions. Let $\varphi \neq 0$ be such a function, with $\int \varphi \, dV = 0$.
Set $\check{\varphi} = \sup(\varphi, 0)$ and $\tilde{\varphi} = \sup(-\varphi, 0)$.
If the measure of the support of φ is less than or equal to ε (the ε of Lemma 3.68 below) (34) applied to $|\varphi|^\alpha$ gives (26) with $k = B$. Otherwise, let $a > 0$ be such that the measure of

$$\Omega_a = \{x \in M_n / |\varphi(x)| \geq a\} \quad \text{is equal to } \varepsilon\colon \mu(\Omega_a) = \varepsilon.$$

We have $a\varepsilon \leq \|\varphi\|_1$.

Since $|\varphi|^\alpha \leq \widecheck{|\varphi|^\alpha - a^\alpha} + a^\alpha$, then by (34) below we have

(30)

$$\||\varphi|^\alpha\|_p \leq \|\widecheck{|\varphi|^\alpha - a^\alpha}\|_p + a^\alpha\left(\int dV\right)^{1/p} \leq B\|\nabla|\varphi|^\alpha\|_q + a^\alpha\left(\int dV\right)^{1/p}.$$

Suppose that $\|\check{\varphi}\|_\alpha \leq \|\tilde{\varphi}\|_\alpha$, (otherwise, replace φ by $-\varphi$); then we write:

$$a\varepsilon \leq \|\varphi\|_1 = 2\|\check{\varphi}\|_1 \leq 2\|\check{\varphi}\|_\alpha\left(\int dV\right)^{1-1/\alpha}, \tag{31}$$

by using Hölder's inequality.
Now consider the function $\psi = (\check{\varphi})^\alpha - \|\check{\varphi}\|_\alpha^\alpha(\tilde{\varphi})^\alpha/\|\tilde{\varphi}\|_\alpha^\alpha$. ψ satisfies $\|\psi\|_1 = 2\|\check{\varphi}\|_\alpha^\alpha$ and $\int \psi \, dV = 0$. Thus applying (29), where we choose $k_0 \geq 1$, yields:

$$\|\psi\|_p \leq k_0\|\nabla\psi\|_q \leq k_0\|\nabla|\varphi|^\alpha\|_q \tag{32}$$

As $\|\psi\|_1 \leq \|\psi\|_p(\int dV)^{1-1/p}$, using (30), (31), and (32) leads to (26), with $k = B + 2k_0 \int dV/\varepsilon$. Indeed $k^\alpha > B + k_0\varepsilon^{-\alpha}2^{\alpha-1}(\int dV)^\alpha$.

One easily obtains the other results, by applying the properties of the Green's function from 4.13 below to (28) or (27) after differentiation:

$$|\nabla\varphi(P)| \leq \int |\nabla_P G(P, Q)||\Delta\varphi(Q)| \, dV(Q). \tag{33}$$ ∎

The proof for the compact manifold with boundary is similar.
Finally we must prove the following lemma which was used above.

3.68 Lemma. *Let M_n be a Riemannian manifold and p, q as above. There exist B, ε, two positive constants, such that any function $\varphi \in H_1^q$ satisfies:*

$$\|\varphi\|_p \leq B\|\nabla\varphi\|_q \tag{34}$$

when $\mu(\operatorname{supp}\varphi) = \int_{\operatorname{supp}\varphi} dV \leq \varepsilon$.

Proof. Since $\|\varphi\|_q \le \|\varphi\|_p[\mu(\operatorname{supp}\varphi)]^{1/n} \le \varepsilon^{1/n}\|\varphi\|_p$, using (29) we obtain (34) with any $\varepsilon < A^{-n}$ by setting $B = K(1 - \varepsilon^{1/n}A)^{-1}$. ∎

7.6. Interpolation Inequalities

3.69 Theorem. *Let M_n be a Riemannian manifold and q, r satisfy $1 \le q, r \le \infty$. Set $2/p = 1/q + 1/r$. Then all functions $f \in \mathscr{D}(M)$ satisfy:*

$$\|\nabla f\|_p^2 \le (n^{1/2} + |p - 2|)\|f\|_q\|\nabla^2 f\|_r. \tag{35}$$

Let $\mathfrak{F}$ denote the completion of $\mathscr{D}(M)$ under the norm $\|f\|_q + \|\nabla^2 f\|_r$.
If $f \in \mathfrak{F}$, then $|\nabla f| \in L_p(M)$ and (35) holds.
In particular, when M is compact (with boundary or without), if $f \in L_q(M)$ and $|\nabla^2 f| \in L_r$, then $f \in L_r(M)$ and (35) holds for $f \in \mathring{H}_2^r(M)$.
Moreover, if $1/q + 1/r = 1$, then all $f \in \mathscr{D}(M)$ satisfy

$$\|\nabla f\|_2^2 \le \|f\|_q\|\Delta f\|_r.$$

Proof. First of all, suppose $p \ge 2$. For $f \in \mathscr{D}(M)$:

$$\begin{aligned}\nabla^\nu(f|\nabla f|^{p-2}\nabla_\nu f) = |\nabla f|^p + f|\nabla f|^{p-2}\nabla^\nu\nabla_\nu f\\ + (p-2)|\nabla f|^{p-4} f\nabla_{\nu\mu} f\nabla^\nu f\nabla^\mu f.\end{aligned} \tag{36}$$

Integrating (36) over M leads to $\|\nabla f\|_p^p = \int f\,\Delta f\,dV$ if $p = 2$, and when $p > 2$ it yields:

$$\|\nabla f\|_p^p = \int f\Delta f|\nabla f|^{p-2}\,dV + (2-p)\int|\nabla f|^{p-4} f\nabla_{\nu\mu} f\nabla^\nu f\nabla^\mu f\,dV. \tag{37}$$

But $|\Delta f|^2 \le n|\nabla^2 f|^2$ and $|\nabla_{\nu\mu} f\nabla^\nu f\nabla^\mu f| \le |\nabla^2 f||\nabla f|^2$; thus:

$$\|\nabla f\|_p^p \le (n^{1/2} + |p-2|)\int|f||\nabla^2 f||\nabla f|^{p-2}\,dV.$$

Applying the Hölder inequality 3.62, since $1/q + 1/r + (p-2)/p = 1$, we find:

$$\|\nabla f\|_p^p \le (n^{1/2} + |p-2|)\|f\|_q\|\nabla^2 f\|_r\|\nabla f\|_p^{p-2},$$

and the desired result follows. When $1 \le p < 2$, the proof is similar, but a little more delicate (see Aubin [22]). When M is compact, if $f \in L_q(M)$ and $|\nabla^2 f| \in L_r$, then by the properties of the Green's function $f \in L_r(M)$. ∎

3.70 Theorem (See Nirenberg [220] p. 125). *M_n will be either $\mathbb{R}^n$, or a compact Riemannian manifold with or without boundary. Let q, r be real numbers $1 \le q, r \le \infty$ and j, m integers $0 \le j < m$.*

Then there exists k, a constant depending only on n, m, j, q, r, and a, and on the manifold, such that for all $f \in \mathscr{D}(M)$ (with $\int f\,dV = 0$, in the compact case without boundary):

$$\|\nabla^j f\|_p \le k\|\nabla^m f\|_r^a \|f\|_q^{1-a}, \tag{38}$$

where

$$\frac{1}{p} = \frac{j}{n} + a\left(\frac{1}{r} - \frac{m}{n}\right) + (1-a)\frac{1}{q}, \tag{39}$$

for all a in the interval $j/m \le a \le 1$, for which p is non-negative.

If $r = n/(m-j) \ne 1$, then (38) is not valid for $a = 1$.

Proof. α) The result holds also for $j = m = 0$, with $k = 1$. This is just proposition (3.62).

Once the two cases $j = 0, m = 1$, and $j = 1, m = 2$ are proved, the general case will follow by induction, by applying the inequality

$$|\nabla|\nabla^\ell f|| \le |\nabla^{\ell+1} f| \text{(see Proposition (2.11))}. \tag{40}$$

For the proof, we are going to use Hölder's inequality, Theorem (3.69), and the Sobolev imbedding theorem. It may be written (Corollary 2.12, and Theorem 3.67):

$$\|h\|_s \le \text{Const} \times \|\nabla h\|_t, \quad \text{where } \frac{1}{s} = \frac{1}{t} - \frac{1}{n} > 0, \text{ for all } h \in \mathscr{D}(M_n) \tag{41}$$

(with $\int h\,dV = 0$, when the manifold is compact without boundary).

β) *The case* $j = 0, m = 1, p < \infty$. By (41), with $t = r < n$ and Proposition (3.62):

$$\|f\|_p \le \|f\|_s^a \|f\|_q^{1-a} \le k\|\nabla f\|_r^a \|f\|_q^{1-a}, \tag{42}$$

with $1/p - 1/q = a(1/s - 1/q) = a(1/r - 1/n - 1/q)$. Thus for $j = 0$ and $m = 1$, if $r < n$, then (38) holds for $0 \le a \le 1$ and p runs from q to $s = rn/(n-r)$.

If $r \ge n$, use (41), with $1/ap = 1/\mu - 1/n$. Putting $h = |f|^{1/a}$, we find the desired result when $p < \infty$. Indeed, $\|h\|_{ap} \le C\|\nabla h\|_\mu$ becomes

$$\|f\|_p^{1/a} \le \frac{C}{a} \||\nabla f||f|^{(1/a)-1}\|_\mu \le \frac{C}{a}\|\nabla f\|_r \|f\|_q^{(1/a)-1} \tag{43}$$

by using Hölder's inequality, since $1/r + (1/a - 1)/q = 1/\mu = 1/ap + 1/n$.

γ) *The case* $j = 0, m = 1, p = +\infty$. If $r > n$, let $s \in [(n + r)/2, r]$. When $M_n \neq \mathbb{R}^n$, all $f \in \mathscr{D}(M)$ (with $\int f \, dV = 0$ in the compact case) satisfy (Theorem 3.67):

$$\|f\|_p \leq \text{Const} \times \|\nabla f\|_s$$

for all p, $1 \leq p \leq \infty$, and the constant does not depend on p and s. Thus C does not depend on p in (43). Letting $p \to \infty$ in (43), we obtain the inequality for $p = +\infty$.
If $r > n$ and $M_n = \mathbb{R}^n$, a proof similar to that of the Sobolev imbedding theorem, yields:
There exists a constant $C(v)$, such that for all $f \in \mathscr{D}(\mathbb{R}^n)$:

$$\sup|f| \leq C(v)(\|\nabla f\|_r + \|f\|_v), \quad \text{when } v > n. \tag{44}$$

Consider the function $\varphi(x) = f(tx)$, with $0 < t < \infty$. Applying (44) to φ and setting $y = tx$ lead to:

$$\sup|f| \leq C(v)(t^{1-n/r}\|\nabla f\|_r + t^{-n/v}\|f\|_v).$$

Choosing $t = (\|f\|_v \|\nabla f\|_r^{-1})^{(n/v+1-n/r)-1}$, we find:

$$\sup|f| \leq 2C(v)\|\nabla f\|_r^d \|f\|_v^{1-d}, \tag{45}$$

with $d^{-1} = 1 + v(1/n - 1/r)$.
If $q > n$, we can choose $v = q$, and the result follows for $p = +\infty$, $(d = a)$.
If $q \leq n$, since $\|f\|_v \leq \|f\|_q^{q/v}(\sup|f|)^{1-q/v}$, (45) gives:

$$(\sup|f|)^{1-(1-d)(1-q/v)} \leq 2C(v)\|\nabla f\|_r^d \|f\|_q^{(1-d)q/v},$$

which is the result for $p = +\infty$, with

$$a^{-1} = q/v\, d + 1 - q/v = 1 + q(1/n - 1/r).$$

δ) *The case* $j = 1, m = 2$. We have established (Theorem 3.69) inequality (38) for $a = j/m = 1/2$. If $r < n$, inequality (38) for $a = 1$ is just the Sobolev imbedding theorem (Corollary 2.12, Theorem 3.67). By interpolation (22), we find the inequality for $\frac{1}{2} < a < 1$. If $r \geq n$, according to (38) with $j = 0$, $m = 1$, applied to the function $|\nabla f|$:

$$\|\nabla f\|_p \leq \text{Const} \times \|\nabla^2 f\|_r^b \|\nabla f\|_s^{1-b}, \tag{46}$$

with $1/p = 1/s + b(1/r - 1/n - 1/s) > 0$ and $0 \leq b \leq 1$.
Using (35) in (46) yields the desired inequality. Indeed, $\|\nabla f\|_s^2 \leq \text{Const} \times \|\nabla^2 f\|_r \|f\|_q$ with $2/s = 1/r + 1/q$. Thus we find inequality (38) where $j = 1$, $m = 2$ and $a = (1 + b)/2$.

We can verify that $a \leq a_1 = [1 + (1/n - 1/r)/(1/n + 1/q)]^{-1}$ implies $b \leq a_0 = (1 + s/n - s/r)$. Thus (38) holds. ■

§8. Maximum Principle

8.1. Hopf's Maximum Principle[2]

3.71 *Let Ω be an open connected set of $\mathbb{R}^n$ and $L(u)$ a linear uniformly elliptic differential operator in Ω of order 2:*

$$L(u) = \sum_{i,j} a_{ij}(x) \frac{\partial^2 u}{\partial x^i \, \partial x^j} + \sum_{i=1}^{n} b_i(x) \frac{\partial u}{\partial x^i} + h(x)u$$

with bounded coefficients and $h \leq 0$.

Suppose $u \in C^2(\Omega)$ satisfies $L(u) \geq 0$.

If u attains its maximum $M \geq 0$ in Ω, then u is constant equal to M on Ω. Otherwise if at $x_0 \in \partial\Omega$, u is continuous and $u(x_0) = M \geq 0$, then the outer normal derivative at x_0, if it exists, satisfies $\partial u/\partial \nu(x_0) > 0$, provided x_0 belongs to the boundary of a ball included in Ω.

Moreover, if $h \equiv 0$, the same conclusions hold for a maximum $M < 0$.

Remark 3.71. We can state a maximum principle for weak solution (see Gilbarg and Trüdinger [125] p. 168).

Let $Lu = \partial_i(a^{ij} \partial_j u) + b^i \partial_i u + hu$ be an elliptic operator in divergence form defined on an open set Ω of $\mathbb{R}^n$, where the coefficients a^{ij}, b^i and h are assumed to be measurable and locally bounded.

$u \in \mathrm{H}_1(\Omega)$ is said to satisfy $Lu \geq 0$ *weakly* if for all $\varphi \in \mathcal{D}(\Omega)$, $\varphi \geq 0$:

$$\int_\Omega [a^{ij} \partial_i u \, \partial_j \varphi - (b^i \partial_i u + hu)\varphi] \, dx \leq 0.$$

In this case, if $h \leq 0$ then $\sup_\Omega u \leq \sup_{\partial\Omega} \max(u, 0)$.

The last term is defined in the following way: we say that $v \in \mathrm{H}_1(\Omega)$ satisfies $v/\partial\Omega \leq k$ if $\max(v - k, 0) \in \mathring{\mathrm{H}}_1(\Omega)$.

8.2. Uniqueness Theorem

3.72 *Let $\overline{W}$ be a compact Riemannian manifold with boundary and $L(u)$ a linear uniformly elliptic differential operator on $\overline{W}$:*

$$L(u) = a^{ij}(x)\nabla_i \nabla_j u + b^i(x)\nabla_i u + h(x)u$$

with bounded coefficients and $h \leq 0$.

[2] Protter and Weinberger [239].

Then, the Dirichlet problem $L(u) = f$, $u|\partial W = g$ (f and g given) has at most one solution.

Proof. Suppose $\tilde{\mathfrak{u}}$ and $\mathfrak{u}$ are solutions of the Dirichlet problem. Then $\mathfrak{v} = \tilde{\mathfrak{u}} - \mathfrak{u}$ satisfies $L\mathfrak{v} = 0$ in W and $\mathfrak{v}|\partial W = 0$. According to the maximum principle $\mathfrak{v} \leq 0$ on W. But the same result holds for $-\mathfrak{v}$. Thus $\mathfrak{v} = 0$ in W. ■

3.73 Theorem. *Let $\overline{W}$ and $L(u)$ be as above. If $\mathfrak{w} \in C^2(W) \cap C^0(\overline{W})$ is a subsolution of the above Dirichlet problem, i.e. $\mathfrak{w}$ satisfies:*

$$L\mathfrak{w} \geq f \quad \text{in } W, \mathfrak{w}/\partial W \leq g,$$

then $\mathfrak{w} \leq \mathfrak{u}$ everywhere, if $\mathfrak{u}$ is the solution of the Dirichlet problem. Likewise, if $\mathfrak{v}$ is a supersolution, i.e. $\mathfrak{v}$ satisfies $L\mathfrak{v} \leq f$ in W and $\mathfrak{v}/\partial W \geq g$ then $\mathfrak{u} \leq \mathfrak{v}$ everywhere.

8.3. Maximum Principle for Nonlinear Elliptic Operator of Order Two

3.74 Let $\overline{W}$ be a Riemannian compact manifold with boundary, and $A(\mathfrak{u}) = f(x, \mathfrak{u}, \nabla\mathfrak{u}, \nabla^2\mathfrak{u})$ a differential operator of order two defined over W, where f is supposed to be a differentiable function of its arguments. Suppose $\mathfrak{v}$, $\mathfrak{w} \in C^2(W)$ satisfy $A(\mathfrak{v}) = 0$ and $A(\mathfrak{w}) \geq 0$. Define $\mathfrak{v}_t$ by $[\mathfrak{v} + t(\mathfrak{w} - \mathfrak{v})]$.

Theorem 3.74. *Let $A(\mathfrak{u})$ be uniformly elliptic with respect to $\mathfrak{v}_t$, for all $t \in]0, 1[$. Then $\varphi = \mathfrak{w} - \mathfrak{v}$ cannot achieve a nonnegative maximum $M \geq 0$ in W, unless it is a constant, if $\partial f(x, \mathfrak{v}_t, \nabla\mathfrak{v}_t, \nabla^2\mathfrak{v}_t)/\partial\mathfrak{u} \leq 0$ on W.*

. Moreover suppose $\mathfrak{v}, \mathfrak{w} \in C^0(\overline{W})$ and $\mathfrak{w} \leq \mathfrak{v}$ on the boundary, then $\mathfrak{w} \leq \mathfrak{v}$ everywhere provided the derivatives of $f(x, \mathfrak{v}_t, \nabla\mathfrak{v}_t, \nabla^2\mathfrak{v}_t)$ are bounded (in the local charts of a finite atlas) for all $t \in]0, 1[$.

If in addition at $x_0 \in \partial W$, $\varphi(x_0) = 0$ and $\partial\varphi/\partial\nu(x_0)$ exists, then $\partial\varphi/\partial\nu(x_0) > 0$, unless φ is a constant, provided the boundary is C^2.

Proof. Consider $\gamma(t) = f(x, \mathfrak{v}_t)$. For some $\theta \in]0, 1[$ the mean value theorem shows that $0 \leq A(\mathfrak{w}) - A(\mathfrak{v}) = \gamma(1) - \gamma(0) = \gamma'(\theta)$ with

$$\gamma'(\theta) = \frac{\partial f(x, \mathfrak{v}_\theta)}{\partial\nabla_{ij}\mathfrak{u}}\nabla_{ij}\varphi + \frac{\partial f(x, \mathfrak{v}_\theta)}{\partial\nabla_i\mathfrak{u}}\nabla_i\varphi + \frac{\partial f(x, \mathfrak{v}_\theta)}{\partial u}\varphi = L(\varphi)$$

Thus $\varphi = \mathfrak{w} - \mathfrak{v}$ satisfies $L(\varphi) \geq 0$. Applying the above theorems yields the present statements. ■

3.75 As an application of the maximum principle we are going to establish the following lemma, which will be useful to solve Yamabe's problem.

Proposition 3.75. *Let M_n be a compact Riemannian manifold. If a function $\psi \geq 0$, belonging to $C^2(M)$, satisfies an inequality of the type $\Delta\psi \geq \psi f(P, \psi)$, where $f(P, t)$ is a continuous numerical function on $M \times \mathbb{R}$, then either ψ is strictly positive, or ψ is identically zero.*

Proof. According to Kazdan. Since M is compact and since ψ is a fixed non-negative continuous function, there is a constant $a > 0$ such that $\Delta\psi + a\psi \geq 0$. By the maximum principle 3.71, the result follows: $u = -\psi$ cannot have a local maximum ≥ 0 unless $u \equiv 0$. Here $L = -\Delta - a$. ∎

8.4. Generalized Maximum Principle

3.76 There is a generalized maximum principle on complete noncompact manifolds Cheng and Yau (90). Namely:

Theorem. *Let (M, g) be a complete Riemannian manifold. Suppose that for any $x \in M$ there is a C^2 non-negative function φ^x on M with support K^x in a compact neighborhood of x which satisfies $\varphi^x(x) = 1$, $\varphi^x \leq k$, $|\nabla\varphi^x| \leq k$, and $\varphi^x_{ii} \geq -kg_{ii}$ for all directions i, where k is a constant independent of x.*

If f is a C^2 function on M which is bounded from above, then there exists a sequence $\{x_j\}$ in M such that $\lim f(x_j) = \sup f$,

$$\lim |\nabla f(x_j)| = 0 \quad \textit{and} \quad \lim\sup \nabla_{ii} f(x_j) \leq 0$$

for all directions i.

Proof. Denote by L the sup of f, which we suppose not attained; otherwise the theorem is obvious by the usual maximum principle. Let $\{y_j\}$ be a sequence in M such that $\lim f(y_j) = L$. On $\mathring{K}^{y_j}$ consider the function $(L - f)/\varphi^{y_j}$. This is strictly positive and goes to ∞ when $x \to \partial\mathring{K}^{y_j}$.
Let $x_j \in \mathring{K}^{y_j}$ be a point where this function attains its minimum. We have

$$\left(\frac{L - f}{\varphi^{y_j}}\right)(x_j) \leq \left(\frac{L - f}{\varphi^{y_j}}\right)(y_j) = L - f(y_j)$$

$$\left(\frac{\nabla_i(L - f)}{L - f}\right)(x_j) = \left(\frac{\nabla_i \varphi^{y_j}}{\varphi^{y_j}}\right)(x_j)$$

$$\left(\frac{\nabla_{ii}(L - f)}{L - f}\right)(x_j) \geq \left(\frac{\nabla_{ii} \varphi^{y_j}}{\varphi^{y_j}}\right)(x_j) \quad \text{for all direction } i.$$

From these we get

$$0 < L - f(x_j) \leq k[L - f(y_j)]$$

$$|\nabla f(x_j)| \leq k[L - f(y_j)]$$

$$\nabla_{ii} f(x_j) \leq k[L - f(y_j)]g_{ii}.$$

Thus $\{x_j\}$ is a sequence having the required properties. ∎

§9. Best Constants

3.77 Theorem (Lions [188]). *Let $\mathfrak{B}_1, \mathfrak{B}_2, \mathfrak{B}_3$ be three Banach spaces and* $\mathfrak{u}, \mathfrak{v}$, *two linear operators*: $\mathfrak{B}_1 \xrightarrow{\mathfrak{u}} \mathfrak{B}_2 \xrightarrow{\mathfrak{v}} \mathfrak{B}_3$. *Suppose* $\mathfrak{u}$ *is compact and* $\mathfrak{v}$ *continuous and one to one. Then given any* $\varepsilon > 0$, *there is* $A(\varepsilon) > 0$ *such that for all* $x \in \mathfrak{B}_1$:

$$\|\mathfrak{u}(x)\|_{\mathfrak{B}_2} \leq \varepsilon\|x\|_{\mathfrak{B}_1} + A(\varepsilon)\|\mathfrak{v} \circ \mathfrak{u}(x)\|_{\mathfrak{B}_3}$$

Proof. Suppose the contrary. Then there exists $\varepsilon_0 > 0$ and a sequence $\{x_k\}$ in $\mathfrak{B}_1$, satisfying $\|x_k\|_{\mathfrak{B}_1} = 1$ such that

$$\|\mathfrak{u}(x_k)\|_{\mathfrak{B}_2} > \varepsilon_0\|x_k\|_{\mathfrak{B}_1} + k\|\mathfrak{v} \circ \mathfrak{u}(x_k)\|_{\mathfrak{B}_3}, \tag{47}$$

Since $\mathfrak{u}$ is compact, a subsequence of $\{\mathfrak{u}(x_k)\}$ converges in $\mathfrak{B}_2$, Say $\mathfrak{u}(x_{k_i}) \to y_0 \in \mathfrak{B}_2$. Rewriting (47) for this subsequence gives:

$$\|\mathfrak{u}(x_{k_i})\|_{\mathfrak{B}_2} > \varepsilon_0 + k_i\|\mathfrak{v} \circ \mathfrak{u}(x_{k_i})\|_{\mathfrak{B}_3}.$$

Whether $y_0 = 0$ or not, letting $k_i \to \infty$ yields the desired contradiction. ∎

3.78 Theorem (Aubin [17]). *Let $\mathfrak{B}_1, \mathfrak{B}_2, \mathfrak{B}_3$ be three Banach spaces and* $\mathfrak{u}, \mathfrak{w}$ *two continuous linear operators*: $\mathfrak{B}_1 \xrightarrow{\mathfrak{u}} \mathfrak{B}_2, \mathfrak{B}_1 \xrightarrow{\mathfrak{w}} \mathfrak{B}_3$.
Suppose $\mathfrak{u}$ *is not compact and* $\mathfrak{w}$ *is compact. And consider all pairs of real numbers* C, A, *such that all* $x \in \mathfrak{B}_1$ *satisfy*:

$$\|\mathfrak{u}(x)\|_{\mathfrak{B}_2} \leq C\|x\|_{\mathfrak{B}_1} + A\|\mathfrak{w}(x)\|_{\mathfrak{B}_3}. \tag{48}$$

Define $\mathsf{K} = \inf C$ *such that some A exists. Then* $\mathsf{K} > 0$.

Proof. Since $\mathfrak{u}$ is not compact, there exists a sequence $\{x_i\}$ in $\mathfrak{B}_1$ with $\|x_i\|_{\mathfrak{B}_1} = 1$, such that no subsequence of $\{\mathfrak{u}(x_i)\}$ converges in $\mathfrak{B}_2$. But $\mathfrak{w}$ is compact. Hence there exists $\{\mathfrak{w}(x_k)\}$, a subsequence of $\{\mathfrak{w}(x_i)\}$, which converges in $\mathfrak{B}_3$. Because $\{\mathfrak{u}(x_k)\}$ is not a Cauchy sequence in $\mathfrak{B}_2$, there exist $\eta > 0$ and $\{k_j\}$ an increasing sequence in $\mathbb{N}$ such that

$$\|\mathfrak{u}(y_j)\| > \eta, \quad \text{where } y_j = x_{k_{2j+1}} - x_{k_{2j}}.$$

Write (48) for y_j:

$$\|\mathfrak{u}(y_j)\|_{\mathfrak{B}_2} \leq C\|y_j\|_{\mathfrak{B}_1} + A\|\mathfrak{w}(y_j)\|_{\mathfrak{B}_3}.$$

Letting $j \to \infty$ leads to $\eta \leq 2C$, since $\mathfrak{w}(y_j) \to 0$ in $\mathfrak{B}_3$. Thus $\mathsf{K} \geq \eta/2 > 0$.
∎

9.1. Application to Sobolev Spaces

3.79 Let M_n be a compact Riemannian manifold with boundary or without. Consider the following Banach spaces $\mathfrak{B}_1 = H_1^q(M)$, $\mathfrak{B}_2 = L_p(M)$ and $\mathfrak{B}_3 = L_q(M)$ with $q < n$ and $1/p = 1/q - 1/n$.

Recall Sobolev's and Kondrakov's theorems, 2.21 and 2.34. The imbedding $\mathfrak{B}_1 \subset \mathfrak{B}_2$ is not compact (example 2.38) and the imbedding $\mathfrak{B}_1 \subset \mathfrak{B}_3$ is compact. Thus there exist constants A, C such that

$$\|f\|_p \leq C(\|\nabla f\|_q + \|f\|_q) + A\|f\|_q, \tag{49}$$

for instance $(C_0, 0)$, and $\mathrm{K} = \inf\{C \text{ such that some } A \text{ exists}\} > 0$.

Of course K depends on n, q and on the manifold. But we have proved (Theorem 2.21) that $\mathrm{K} = \mathrm{K}(n, q)$ is the same constant for all compact manifolds of dimension n and that K is the norm of the imbedding $H_1^q(\mathbb{R}^n) \subset L_p(\mathbb{R}^n)$.

Chapter 4

Complementary Material

The main aim of this book is to present some methods for solving nonlinear elliptic (or parabolic) problems and to use them concretely in Riemannian Geometry. The present chapter 4 consists in six sections. In the first two, we prove the existence of Green's function for compact Riemannian manifolds. In §3 and 4, we present some material concerning Riemannian Geometry and Partial Differential Equations, the two main fields of this book. This material, which completes the previous one (in Chapter 3), is crucially used in the sequel of this volume. Many theorems will be quoted without proof, except if they are not available in other books. Then we describe the methods and we mention the sections of the book where one finds concrete applications of them to some problems concerning curvature and also harmonic maps. For instance, to illustrate the steepest descent method, the pioneering article of Eells-Sampson is the best example. We end this chapter with a new result on the best constant in the Sobolev inequality. Its proof shows the power of the method of points of concentration.

§1. Linear Elliptic Equations

4.1 To prove the existence of Green's function, first of all we have to solve linear elliptic equations. We also need some results concerning the eigenvalues of the Laplacian. Let (M_n, g) be a C^∞ Riemannian manifold. We are going to consider equations of the type

$$(1) \qquad -\nabla^i[a_{ij}(x)\nabla^j\varphi] = f(x),$$

where $a_{ij}(x)$ are the components, in a local chart, of a C^∞ Riemannian metric (see 1.15) and where f belongs to $L_2(M)$.

1.1. First Nonzero Eigenvalue λ of Δ.

4.2 Theorem. *Let (M_n, g) be a compact C^∞ Riemannian manifold. The eigenvalues of the Laplacian $\Delta = -\nabla^\nu\nabla_\nu$ are nonnegative. The eigenfunctions, corresponding to the eigenvalue $\lambda_0 = 0$, are the constant functions. The* *first nonzero eigenvalue* λ_1 *is equal to* μ, *defined by*: $\mu = \inf\|\nabla\varphi\|_2^2$ *for all* $\varphi \in \mathscr{A}$, *with* $\mathscr{A} = \{\varphi \in \mathrm{H}_1$ *satisfying* $\|\varphi\|_2 = 1$ *and* $\int \varphi\, dV = 0\}$.

Proof. The first statement is proved in 1.77; the second in 1.71.
Let $\{\varphi_i\}_{i \in \mathbb{N}}$ be a sequence in $\mathscr{A}$, such that $\|\nabla\varphi_i\|_2^2 \to \mu$ when $i \to \infty$. $\{\varphi_i\}$ is called a minimizing sequence. Obviously $\{\varphi_i\}$ is bounded in H_1.
According to Kondrakov's theorem, 2.33, there exists a subsequence $\{\varphi_j\}$ of $\{\varphi_i\}$ and $\varphi_0 \in L_2$, such that $\varphi_j \to \varphi_0$ strongly in L_2 ($\|\varphi_j - \varphi_0\|_2 \to 0$). Hence $\|\varphi_j - \varphi_0\|_1 \to 0$ when $j \to \infty$, since the manifold is compact. Thus φ_0 satisfies $\|\varphi_0\|_2 = 1$ and $\int \varphi_0\, dV = 0$.
By Theorem 3.18, there exist $\tilde{\varphi}_0 \in \mathrm{H}_1$ and $\{\varphi_k\}$ a subsequence of $\{\varphi_j\}$ such that $\varphi_k \to \tilde{\varphi}_0$ weakly in H_1.
Furthermore, $\varphi_k \to \varphi_0$ weakly in L_2 (strong convergence implies weak convergence). Thus, since the imbedding $\mathrm{H}_1 \subset L_2$ is continuous, φ_0 and $\tilde{\varphi}_0$ are functions in L_2 which define the same distribution on $\mathscr{D}(M)$. Therefore $\varphi_0 = \tilde{\varphi}_0$, and $\varphi_0 \in \mathscr{A}$.
According to Theorem 3.17, $\|\varphi_0\|_{\mathrm{H}_1} \le \liminf_{k\to\infty} \|\varphi_k\|_{\mathrm{H}_1}$, since $\varphi_k \to \varphi_0$ weakly in H_1. This implies $\|\nabla\varphi_0\|_2^2 \le \lim_{k\to\infty} \|\nabla\varphi_k\|_2^2 = \mu$, because $\|\varphi_k\|_2 = \|\varphi_0\|_2 = 1$. But $\varphi_0 \in \mathscr{A}$, hence $\|\nabla\varphi_0\|_2^2 = \mu$, and the minimum μ is attained.

Writing Euler's equation of our variational problem (see, for instance, Berger [42], p. 123), leads to: There exist numbers α and β, such that for all $\psi \in \mathrm{H}_1$:

$$\int \nabla^\nu \varphi_0 \nabla_\nu \psi\, dV = \alpha \int \varphi_0 \psi\, dV + \beta \int \psi\, dV.$$

Picking $\psi = 1$, gives $\beta = 0$. Choosing next $\psi = \varphi_0$ leads to $\alpha = \mu$. So $\varphi_0 \in \mathrm{H}_1$ satisfies weakly

(2) $$\Delta\varphi_0 = \mu\varphi_0.$$

Thus, by the regularity theorem 3.54, $\varphi_0 \in C^\infty$. In fact, here it would be easy to prove this regularity result without using Theorem 3.54. Let us begin by observing that by (2), $\Delta\varphi_0 \in L_2$. Since

(3) $$\int \nabla^{ij}\varphi \nabla_{ij}\varphi\, dV = \int (\Delta\varphi)^2\, dV - \int R_{ij}\nabla^i\varphi\nabla^j\varphi\, dV,$$

$\varphi_0 \in \mathrm{H}_2$. Taking the Laplacian of Equation (2) gives $\Delta^2\varphi_0 = \Delta\Delta\varphi_0 = \mu\Delta\varphi_0 \in L_2$. By induction $\Delta^k\varphi_0 \in L_2$ and $|\nabla\Delta^k\varphi_0| \in L_2$ for all $k \in \mathbb{N}$. A straightforward calculation gives equalities like (3) for derivatives of higher order. So $\varphi_0 \in \mathrm{H}_k$ for all $k \in \mathbb{N}$, and the Sobolev imbedding theorem implies $\varphi_0 \in C^\infty$.

Thus μ is an eigenvalue of Δ, and φ_0 an eigenfunction. On the other hand, let γ be an eigenfunction satisfying $\Delta\gamma = \lambda_1\gamma$; then $\lambda_1 = \|\nabla\gamma\|_2^2\|\gamma\|_2^{-2} \ge \mu$, according to the definition of μ, because $\int \gamma\, dV = 0$. ∎

4.3 Corollary. *Let M be a compact C^∞ Riemannian manifold. If $\varphi \in \mathrm{H}_1$ satisfies $\int \varphi \, dV = 0$, then $\|\varphi\|_2 \le \lambda_1^{-1/2} \|\nabla\varphi\|_2$, where λ_1 is the first nonzero eigenvalue of Δ.*

4.4 Theorem. *Let $\overline{W}_n$ be a compact Riemannian manifold with boundary of class C^∞. The eigenvalues of the Laplacian Δ are strictly positive. The eigenfunctions, corresponding to the first eigenvalue λ_1, are proportional (the eigenspace is of dimension one). They belong to $C^\infty(\overline{W})$ and they are either strictly positive on W or strictly negative. Moreover λ_1 is equal to μ defined by: $\mu = \inf\|\nabla\varphi\|_2^2$ for all $\varphi \in \mathscr{A}$, with $\mathscr{A} = \{\varphi \in \mathring{\mathrm{H}}_1(W)$ satisfying $\|\varphi\|_2 = 1\}$.*

Proof. We recall that λ is said to be an eigenvalue of the Laplacian Δ if the Dirichlet problem

$$\Delta u = \lambda u \quad \text{in} \quad W, u = 0 \quad \text{on} \quad \partial W \tag{4}$$

has a nontrivial solution (in a given space—here in $C^\infty(\overline{W})$).

First of all, we are going to establish the existence of $\varphi_0 \ge 0$, a nontrivial weak solution of $\Delta u = \mu u$ in W, belonging to $\mathring{\mathrm{H}}_1(W)$. By Proposition 3.49, if $\varphi \in \mathscr{D}(W)$, $|\nabla|\varphi\| = |\nabla\varphi|$ almost everywhere. And $|\varphi| \in \mathring{\mathrm{H}}_1(W)$ by Proposition 3.48. Thus, since $\mathscr{D}(W)$ is dense in $\mathring{\mathrm{H}}_1(W)$, by definition $\mu = \inf\|\nabla\varphi\|_2^2$ for all $\varphi \in \mathscr{A}_+ = \{\varphi \in \mathring{\mathrm{H}}_1(W)$ satisfying $\|\varphi\|_2 = 1$ and $\varphi \ge 0\}$. We proceed now as in Theorem 4.2. Let $\{\varphi_i\}_{i \in \mathbb{N}}$ be a minimizing sequence in $\mathscr{A}_+$. There exist a subsequence $\{\varphi_k\}$ and a $\varphi_0 \in \mathring{\mathrm{H}}_1(W)$, such that $\|\varphi_k - \varphi_0\|_2 \to 0$ and $\varphi_k \to \varphi_0$ weakly in $\mathring{\mathrm{H}}_1$, when $k \to \infty$.
According to Proposition 3.43, there exists a subsequence of $\{\varphi_k\}$ which converges almost everywhere to φ_0. Since $\varphi_k \ge 0$ for all k, $\varphi_0 \ge 0$ and $\varphi_0 \in \mathscr{A}_+$. μ is attained by φ_0, and writing the Euler equation yields: φ_0 satisfies weakly in $\mathring{\mathrm{H}}_1(W)$, $\Delta\varphi_0 = \mu\varphi_0$. It remains to prove the regularity of the solution. According to Theorem 3.54, we have at once $\varphi_0 \in C^\infty(W)$. But the regularity at the boundary is more difficult to establish.
Let $\{\Omega_i, \psi_i\}_{i \in \mathbb{N}}$ be a C^∞ atlas of $\overline{W}$, such that $\psi_i(\Omega_i)$ is either the ball $B = B_0(1) \subset \mathbb{R}^n$ or the half ball $D = \overline{E} \cap B$. And let $\{\alpha_i\}$ be a C^∞ partition of unity subordinate to the cover $\{\Omega_i\}$. We have only to prove that the functions $f_i = (\alpha_i \varphi_0) \circ \psi_i^{-1}$ belong either to $\mathscr{D}(B)$ or to $C^\infty(\overline{D})$. Theorem 3.54 implies the result for the former. For the latter we can prove (see Nirenberg [218], p. 665) that $f_i \in \mathrm{H}_k(W)$ for all $k \in \mathbb{N}$. Hence, according to the Sobolev imbedding theorem, $f_i \in C^\infty(\overline{D})$.

So $\varphi_0 \in \mathring{\mathrm{H}}_1(W) \cap C^\infty(\overline{W})$. Consequently φ_0 is zero on ∂W, by Proposition 3.50. If μ is zero, $\|\nabla\varphi_0\|_2 = 0$ and $\varphi_0 = 0$ in W. This fact contradicts $\|\varphi_0\|_2 = 1$. Hence $\mu > 0$ and $\mu = \lambda_1$. Indeed, let $u \in C^2$ be a solution of (4). Multiplying the equation by u and integrating over W lead to $\|\nabla u\|_2^2 = \lambda\|u\|_2^2$, by means of an integration by parts of the first integral. Hence $\lambda \ge \mu$. As $\varphi_0 \not\equiv 0$ satisfies $\varphi_0 \ge 0$, $\varphi_0 \in C^\infty(\overline{W})$, $\varphi_0/\partial W = 0$, and $\Delta\varphi_0 = \mu\varphi_0$ in W, according to the maximum principle 3.71, $\varphi_0 > 0$ in W.

Let ψ_0 be an eigenfunction with $\Delta\psi_0 = \lambda_1\psi_0$. We are going to prove that ψ_0 and φ_0 are proportional. Define $\beta = \sup\{\nu \in \mathbb{R}$ such that $\varphi_0 - \nu\psi_0 > 0$ in $W\}$.
Obviously $\varphi_0 - \beta\psi_0 \geq 0$ in W. But we have more: there is a point $P \in W$ where the function $\varphi_0 - \beta\psi_0$ vanishes. Indeed, suppose $\varphi_0 - \beta\psi_0 > 0$ in W. According to the maximum principle 3.71, $(\partial/\partial\nu)(\varphi_0 - \beta\psi_0) < 0$ on ∂W, since

$$\Delta(\varphi_0 - \beta\psi_0) = \lambda_1(\varphi_0 - \beta\psi_0) \geq 0. \tag{5}$$

But the first derivatives are continuous on $\overline{W}$, so there exists an $\varepsilon > 0$ such that $\varphi_0 - (\beta + \varepsilon)\psi_0 > 0$ in W. Hence our initial supposition is false and P does exist. Applying the maximum principle yields $\varphi_0 - \beta\psi_0 = 0$ everywhere, since (5) implies that $\varphi_0 - \beta\psi_0$ cannot achieve its minimum in W, unless it is constant. ■

4.5 Remark. If $\overline{W}_n$ is only of class C^k, the preceding proof is valid, except for the regularity on the boundary.
When $k > n/2$, we can prove that $\varphi_0 \in C^r(\overline{W})$ with $r < k - n/2$. The proof is similar to the previous one, except that now the atlas $\{\Omega_i, \psi_i\}_{i\in\mathbb{N}}$ is of class C^k. The functions f_i satisfy elliptic equations on $\bar{D}$ with coefficients belonging to $C^{k-2}(\bar{D})$; then by Theorem 3.58, $f_i \in \mathrm{H}_k(\bar{D})$. Applying Theorem 2.30, we have $f_i \in C^r(\bar{D})$.

4.6 Corollary. *Let $\overline{W}$ be a compact Riemannian manifold with boundary of class C^k, $k \geq 1$, or at least Lipschitzian. There exists $\lambda_1 > 0$, such that $\|\varphi\|_2^2 \leq \lambda_1^{-1}\|\nabla\varphi\|_2^2$ for all $\varphi \in \overset{\circ}{\mathrm{H}}_1(W)$. Thus $\|\nabla\varphi\|_2$ is an equivalent norm for $\overset{\circ}{\mathrm{H}}_1(W)$.*

Proof. Since the Sobolev imbedding theorem 2.30 and the Kondrakov theorem 2.34 hold, the proof of Theorem (4.4) is valid, except for the regularity at the boundary. Thus μ is attained in $\mathscr{A}$ and consequently $\mu = \lambda_1$ is strictly positive. ■

1.2. Existence Theorem for the Equation $\Delta\varphi = f$

4.7 Theorem. *Let (M_n, g) be a compact C^∞ Riemannian manifold. There exists a weak solution $\varphi \in \mathrm{H}_1$ of (1) if and only if $\int f(x)\,dV = 0$. The solution φ is unique up to a constant. If $f \in C^{r+\alpha}$, ($r \geq 0$ an integer or $r = +\infty$, $0 < \alpha < 1$), then $\varphi \in C^{r+2+\alpha}$.*

Proof. α) if φ is a weak solution of (1) in H_1, by Definition 3.53, $\int a_{ij}\nabla^i\varphi\nabla^j\psi\,dV = \int \psi f\,dV$ for all $\psi \in \mathrm{H}_1$. Choosing $\psi = 1$, we find $\int f\,dV = 0$. This condition is necessary.

β) *Uniqueness up to a constant.* Let φ_1 and φ_2 be weak solutions of (1) in H_1. Set $\tilde{\varphi} = \varphi_2 - \varphi_1$. For all $\psi \in \mathsf{H}_1$, $\int a_{ij}\nabla^i\psi\nabla^j\tilde{\varphi}\, dV = 0$. Choosing $\psi = \tilde{\varphi}$ leads to $\int a_{ij}\nabla^i\tilde{\varphi}\nabla^j\tilde{\varphi}\, dV = 0$. Thus $\tilde{\varphi} = \text{Constant}$.

γ) *Existence of φ.* If $f \equiv 0$, the solutions of (1) are $\varphi = \text{Constant}$. Henceforth suppose $f \not\equiv 0$. Let us consider the functional $I(\varphi) = \int a_{ij}\nabla^i\varphi\nabla^j\varphi\, dV$. Define $\mu = \inf I(\varphi)$ for all $\varphi \in \mathscr{B}$, with $\mathscr{B} = \{\varphi \in \mathsf{H}_1$ satisfying $\int \varphi\, dV = 0$ and $\int \varphi f\, dV = 1\}$.
μ is a nonnegative real number, $0 \le \mu \le I(f\|f\|_2^{-2})$. Let $\{\varphi_i\}_{i\in\mathbb{N}}$ be a minimizing sequence in $\mathscr{B}$: $I(\varphi_i) \to \mu$. Since $a_{ij}(x)$ are the components of a Riemannian metric, there exists $\alpha > 0$ such that $I(\varphi_i) \ge \alpha\|\nabla\varphi_i\|_2^2$. Thus the set $\{|\nabla\varphi_i|\}_{i\in\mathbb{N}}$ is bounded in L_2. Moreover, since $\int \varphi_i\, dV = 0$, it follows by Corollary (4.3), that $\|\varphi_i\|_2 \le \lambda_1^{-1/2}\|\nabla\varphi_i\|_2$. Consequently $\{\varphi_i\}_{i\in\mathbb{N}}$ is bounded in H_1. Applying Theorems 3.17, 3.18, and 2.34 gives: There exist a subsequence $\{\varphi_k\}$ of $\{\varphi_i\}$ and $\varphi_0 \in \mathsf{H}_1$ such that $\|\varphi_k - \varphi_0\|_2 \to 0$ and such that $I(\varphi_0) \le \mu$, (see the proof of Theorem 4.2).

Hence $\varphi_0 \in \mathscr{B}$ and $I(\varphi_0) = \mu$. Since φ_0 minimizes the variational problem, there exist two constants β and γ such that for all $\psi \in \mathsf{H}_1$:

$$\int a_{ij}\nabla^i\varphi_0\nabla^j\psi\, dV = \beta\int f\psi\, dV + \gamma\int\psi\, dV.$$

Picking $\psi = 1$ yields $\gamma = 0$. Choosing $\psi = \varphi_0$ implies $\beta = \mu$. Since $\int \varphi_0 f\, dV = 1$, φ_0 is not constant and $\mu = I(\varphi_0) > 0$. Set $\tilde{\varphi}_0 = \varphi_0/\mu$. Then $\tilde{\varphi}$ satisfies Equation (1) weakly in H_1 and Theorem 3.54 implies the last statement. ∎

4.8 Theorem. *Let $\overline{W}_n$ be a compact Riemannian manifold with boundary of class C^∞. There exists a unique weak solution of* (1): $\varphi \in \mathring{\mathsf{H}}_1(W)$. *If $f \in C^\infty(\overline{W})$, then $\varphi \in C^\infty(\overline{W})$ and φ vanishes on the boundary.*

Proof. Uniqueness is obvious. Let φ_1 and φ_2 be weak solutions of (1) in $\mathring{\mathsf{H}}_1(W)$. Set $\tilde{\varphi} = \varphi_2 - \varphi_1$. For all $\psi \in \mathring{\mathsf{H}}_1$:

$$\int_W a_{ij}\nabla^i\psi\nabla^j\tilde{\varphi}\, dV = 0. \quad \text{Choosing } \psi \in \tilde{\varphi} \text{ leads to } \tilde{\varphi} \equiv 0.$$

For existence, let us consider the functional

$$J(\varphi) = \int_W a_{ij}\nabla^i\varphi\nabla^j\varphi\, dV - 2\int_W f\varphi\, dV.$$

Define $\mu = \inf J(\varphi)$ for all $\varphi \in \mathring{\mathsf{H}}_1(W)$. μ is finite. Indeed, since a_{ij} are the components of a C^∞ Riemannian metric on $\overline{W}$ which is compact, there exists an $\alpha > 0$ such that

$$\int_W a_{ij} \nabla^i \varphi \nabla^j \varphi \, dV \geq \alpha \|\nabla \varphi\|_2^2.$$

On the other hand, by Corollary 4.6, $\|\varphi\|_2^2 \leq \lambda_1^{-1} \|\nabla\varphi\|_2^2$. Thus for all $\varepsilon > 0$,

$$(6) \quad J(\varphi) \geq \alpha\|\nabla\varphi\|_2^2 - \varepsilon\|\varphi\|_2^2 - \frac{1}{\varepsilon}\|f\|_2^2 \geq (\alpha - \varepsilon\lambda_1^{-1})\|\nabla\varphi\|_2^2 - \frac{1}{\varepsilon}\|f\|_2^2.$$

Choosing $\varepsilon = \alpha\lambda_1/2$ gives $\mu \geq -2(\alpha\lambda_1)^{-1}\|f\|_2^2$.

Let $\{\varphi_i\}_{i\in\mathbb{N}}$ be a minimizing sequence of J in $\overset{\circ}{\mathrm{H}}_1(W)$. According to (6), $\{\varphi_i\}_{i\in\mathbb{N}}$ is bounded in $\overset{\circ}{\mathrm{H}}_1$. Applying Theorems 3.17, 3.18, and 2.34 yields: There exist a subsequence $\{\varphi_k\}$ of $\{\varphi_i\}$ and $\varphi_0 \in \overset{\circ}{\mathrm{H}}_1$ such that $\|\varphi_k - \varphi_0\|_2 \to 0$ and such that $J(\varphi_0) \leq \mu$. Hence $J(\varphi_0) = \mu$, and φ_0 satisfies Equation (1) weakly in $\overset{\circ}{\mathrm{H}}_1$. Theorem 3.58 and Proposition (3.50) imply the last statement. ∎

§2. Green's Function of the Laplacian

4.9 The Laplacian in a local chart can be written as follows:

$$\Delta\varphi = -\nabla_i(g^{ij}\nabla_j\varphi) = -\partial_i(g^{ij}\partial_j\varphi) - g^{kj}\partial_j\varphi\Gamma^i_{ik},$$
$$\Delta\varphi = -|g|^{-1/2}\partial_i[g^{ij}\sqrt{|g|}\partial_j\varphi], \text{ because } \Gamma^i_{ik} = \partial_k \log\sqrt{|g|}.$$

If $\varphi = f(r)$ in geodesic polar coordinates:

$$-\Delta f(r) = \frac{1}{r^{n-1}\sqrt{|g|}}\partial_r[r^{n-1}\sqrt{|g|}\partial_r f] = f'' + \frac{n-1}{r}f' + f'\partial_r \log\sqrt{|g|}.$$

By Theorem 1.53, there exists a constant A such that

$$(7) \qquad |\partial_r \log\sqrt{|g|}| \leq Ar.$$

2.1. Parametrix

4.10 In $\mathbb{R}^n$, $n \geq 3$, $\Delta_{Q\,\mathrm{distr.}}(r^{2-n}) = (n-2)\omega_{n-1}\delta_P$ and in $\mathbb{R}^2$, $\Delta_{Q\,\mathrm{distr.}} \log r = -2\pi\delta_P$, where δ_P is the Dirac function at P, $r = d(P, Q)$ and ω_{n-1} is the volume of the unit sphere of dimension $n-1$.

On a Riemannian manifold, r is only a Lipschitzian function. For this reason we must consider $f(r)$ a positive decreasing function, which is equal to 1 in a neighborhood of zero, and to zero for $r \geq \delta(P)$ the injectivity radius at P (see Theorem 1.36). We define:

(8) $\mathsf{H}(P, Q) = [(n-2)\omega_{n-1}]^{-1} r^{2-n} f(r)$, for $n > 2$ and

$\mathsf{H}(P, Q) = -(2\pi)^{-1} f(r) \log r$, for $n = 2$,

and compute when $n > 2$:

$$\Delta_Q \mathsf{H}(P, Q) = [(n-2)\omega_{n-1}]^{-1} r^{1-n}[(n-3)f' - rf'' + ((n-2)f - rf')\partial_r \log \sqrt{|g|}].$$

According to (7), there exists a constant B such that

$$|\Delta_Q \mathsf{H}(P, Q)| \leq B r^{2-n}; \tag{9}$$

B depends on P. But on a compact Riemannian manifold, we can choose $f(r)$ zero for $r \geq \delta$ the injectivity radius, and then B does not depend on P. Henceforth in the compact case, $f(r)$ will be chosen in this way.

2.2. Green's Formula

4.10

$$\psi(P) = \int_M \mathsf{H}(P, Q)\Delta\psi(Q)\, dV(Q) - \int_M \Delta_Q \mathsf{H}(P, Q)\psi(Q)\, dV(Q), \tag{10}$$

for all $\psi \in C^2$. *Recall* (8), *the definition of* $\mathsf{H}(P, Q)$.

For the proof, we compute $\int_{M - B_P(\varepsilon)} \mathsf{H}(P, Q)\Delta\psi(Q)\, dV(Q)$, integrating by parts twice. Letting $\varepsilon \to 0$ yields Green's formula.
If $\psi \in C^\infty$, by definition $\langle \Delta_{\text{distr.}} \mathsf{H}, \psi \rangle = \langle \mathsf{H}, \Delta\psi \rangle$ in the sense of distributions, and $\langle \mathsf{H}, \Delta\psi \rangle = \int_M \mathsf{H}(P, Q)\Delta\psi(Q)\, dV(Q)$. Thus:

$$\Delta_{Q\,\text{distr.}} \mathsf{H}(P, Q) = \Delta_Q \mathsf{H}(P, Q) + \delta_P(Q). \tag{11}$$

Picking $\psi \equiv 1$ in (10), gives $\int_M \Delta_Q \mathsf{H}(P, Q)\, dV(Q) = -1$.
From (10), after interchanging the order of integration, we establish that all $\varphi \in C^2$ satisfy:

$$\varphi(Q) = \Delta_Q \int_M \mathsf{H}(P, Q)\varphi(P)\, dV(P) - \int_M \Delta_Q \mathsf{H}(P, Q)\varphi(P)\, dV(P). \tag{12}$$

4.11 Definition. Let $\overline{W}_n$ be a compact Riemannian manifold with boundary of class C^∞. The *Green's function* $\mathsf{G}(P, Q)$ of the Laplacian is the function which satisfies in $W \times W$:

$$\Delta_{Q\,\text{distr.}} \mathsf{G}(P, Q) = \delta_P(Q), \tag{13}$$

and which vanishes on the boundary (for P or Q belonging to ∂W).

Let M_n be a compact C^∞ Riemannian manifold having volume V. The *Green's function* $\mathsf{G}(P, Q)$ is a function which satisfies:

$$(14) \qquad \Delta_{Q\,\text{distr.}} \mathsf{G}(P, Q) = \delta_P(Q) - V^{-1}.$$

The Green's function is defined up to a constant in this case. Recall that δ_P is the Dirac function at P.

4.12 Proposition (Giraud [126] p. 150). *Let Ω be a bounded open set of $\mathbb{R}^n$ and let $X(P, Q)$ and $Y(P, Q)$ be continuous functions defined on $\Omega \times \Omega$ minus the diagonal which satisfy*

$$|X(P, Q)| \leq \text{Const} \times [d(P, Q)]^{\alpha - n} \quad \textit{and} \quad |Y(P, Q)| \leq \text{Const} \times [d(P, Q)]^{\beta - n}$$

for some real numbers α, β belonging to $]0, n[$. Then

$$Z(P, Q) = \int_\Omega X(P, R) Y(R, Q)\, dV(R)$$

is continuous for $P \neq Q$ and satisfies:

$$|Z(P, Q)| \leq \text{Const} \times [d(P, Q)]^{\alpha + \beta - n} \quad \textit{if } \alpha + \beta < n,$$

$$|Z(P, Q)| \leq \text{Const} \times [1 + |\log d(P, Q)|] \quad \textit{if } \alpha + \beta = n,$$

$$|Z(P, Q)| \leq \text{Const} \quad \textit{if } \alpha + \beta > n;$$

in the last case $Z(P, Q)$ is continuous on $\Omega \times \Omega$.

Proof. The integral which defines $Z(P, Q)$ is less than the sum of three integrals, an upper bound of which is easily found. The integrals are over the sets $\Omega \cap B_P(\rho)$, $[B_Q(3\rho) - B_P(\rho)] \cap \Omega$, and $\Omega - \Omega \cap B_Q(3\rho)$, with $2\rho = d(P, Q)$ small enough. ∎

2.3. Green's Function for Compact Manifolds

4.13 Theorem. *Let M_n be a compact C^∞ Riemannian manifold. There exists $\mathsf{G}(P, Q)$, a Green's function of the Laplacian which has the following properties:*

(a) *For all functions $\varphi \in C^2$:*

$$(15) \quad \varphi(P) = V^{-1} \int_M \varphi(Q)\, dV(Q) + \int_M \mathsf{G}(P, Q) \Delta\varphi(Q)\, dV(Q).$$

(b) $\mathsf{G}(P, Q)$ *is* C^∞ *on* $M \times M$ *minus the diagonal (for* $P \neq Q$*).*

(c) *There exists a constant k such that:*

$$|\mathsf{G}(P, Q)| < k(1 + |\log r|) \quad \text{for} \quad n = 2 \quad \text{and}$$

$$(16) \quad |\mathsf{G}(P, Q)| < kr^{2-n} \quad \text{for} \quad n > 2, |\nabla_Q \mathsf{G}(P, Q)| < kr^{1-n},$$

$$|\nabla_Q^2 \mathsf{G}(P, Q)| < kr^{-n} \quad \text{with} \quad r = d(P, Q).$$

(d) *There exists a constant A such that* $\mathsf{G}(P, Q) \geq A$. *Because the Green function is defined up to a constant, we can thus choose the Green's function everywhere positive.*

(e) $\int \mathsf{G}(P, Q)\, dV(P) = \text{Const}$. *We can choose the Green's function so that its integral equals zero.*

(f) $\mathsf{G}(P, Q) = \mathsf{G}(Q, P)$.

Proof of existence. Define $\Gamma(P, Q) = \Gamma_1(P, Q) = -\Delta_Q \mathsf{H}(P, Q)$ and $\Gamma_{i+1}(P, Q) = \int_M \Gamma_i(P, R)\Gamma(R, Q)\, dV(R)$ for $i \in \mathbb{N}$. Pick $\mathbb{N} \ni k > n/2$ and set

$$(17) \quad \mathsf{G}(P, Q) = \mathsf{H}(P, Q) + \sum_{i=1}^{k} \int_M \Gamma_i(P, R)\mathsf{H}(R, Q)\, dV(R) + F(P, Q).$$

By (11), (12), and (14), $F(P, Q)$ satisfies

$$(18) \qquad \Delta_Q F(P, Q) = \Gamma_{k+1}(P, Q) - V^{-1}.$$

According to (9), $|\Gamma(P, Q)| \leq Br^{2-n}$. Thus from Proposition 4.12, $\Gamma_k(P, Q)$ is bounded and consequently $\Gamma_{k+1}(P, Q)$ is C^1.

Now for P fixed, there exists a weak solution of (18), (Theorem 4.7), unique up to a constant. Using the theorem of regularity 3.54, the solution is C^2. $\mathsf{G}(P, Q)$, defined by (17), satisfies (14). And $Q \to \mathsf{G}(P, Q)$ is C^∞ for $P \neq Q$ (Theorem 3.54). For the present, we choose $\mathsf{G}(P, Q)$ such that:

$$\int_M \mathsf{G}(P, Q)\, dV(Q) = 0.$$

Proof of the properties. a) (14) applied to $\varphi \in C^\infty$ leads to (15) and $\mathscr{D}(M)$ is dense in C^2.

b) We are going to prove that $P \to \mathsf{G}(P, Q)$ is continuous for $P \neq Q$. Since we know that $Q \to \mathsf{G}(P, Q)$ is C^∞ for $Q \neq P$, the result is a consequence of f): $\mathsf{G}(P, Q) = \mathsf{G}(Q, P)$, using for the derivatives a proof similar to the following. Iterating k-times ($k > n/2$) Green's formula (10) leads to:

$$\psi(P) = \int_M \mathsf{H}(P, Q)\Delta\psi(Q)\, dV(Q)$$
$$+ \sum_{i=1}^{k} \int_M \left[\int_M \Gamma_i(P, R)\mathsf{H}(R, Q)\, dV(R)\right]\Delta\psi(Q)\, dV(Q)$$
$$+ \int_M \Gamma_{k+1}(P, Q)\psi(Q)\, dV(Q).$$

Using (8), (9), and Proposition 4.12 gives:

(19a) $$|\psi(P)| \leq \text{Const} \times (\sup|\Delta\psi| + \|\psi\|_2).$$

According to Corollary 4.3, if $\int \psi\, dV = 0$:

(19b) $$\|\psi\|_2^2 \leq \lambda_1^{-1}\|\nabla\psi\|_2^2 \leq \lambda_1^{-1}\|\Delta\psi\|_2\|\psi\|_2;$$

the last inequality arises after integrating by parts and using Hölder's inequality. Hence, there exists a constant C such that the solution of $\Delta\psi = f$ with $\int \psi\, dV = 0$ and $\int f\, dV = 0$ (Theorem 4.7) satisfies:

$$\sup|\psi| \leq C \sup|f|.$$

Applying this result to (18):

$$\sup_Q\left|[F(P, Q) - F(R, Q)] - V^{-1}\int_V [F(P, Q) - F(R, Q)]\, dV(Q)\right|$$
$$\leq C \sup_Q |\Gamma_{k+1}(P, Q) - \Gamma_{k+1}(R, Q)|.$$

Since $\int_M \mathsf{G}(P, Q)\, dV(Q) = 0$, it follows from (17) that $\int_M F(P, Q)\, dV(Q)$ is a continuous function of P.
Thus $P \to F(P, Q)$ is continuous, and for $P \neq Q$, $P \to \mathsf{G}(P, Q)$ is also.

Using only this continuity of $\mathsf{G}(P, Q)$, we will shortly prove parts d)–f). Assuming this has been done, we complete the proof of part b). By f) $\mathsf{G}(P, Q) = \mathsf{G}(Q, P)$. Thus $\mathsf{G}(P, Q)$ is C^∞ in P for $P \neq Q$ and any r-derivative at P $\partial_P^r \mathsf{G}(P, Q)$ is a distribution in Q which satisfies $\Delta_Q \partial_P^r \mathsf{G}(P, Q) = 0$ on $M - \{P\}$. $\partial_P^r \mathsf{G}(P, Q)$ is then C^∞ in Q for $Q \neq P$ according to Theorem 3.54.

c) The inequalities follow from (17). When $Q \to P$, the leading part of $\mathsf{G}(P, Q)$ is $\mathsf{H}(P, Q)$.

d) From this fact, there exists an open neighborhood Ω of the diagonal in $M \times M$, where $\mathsf{G}(P, Q)$ is positive. On $M \times M - \Omega$, which is compact, $\mathsf{G}(P, Q)$ is continuous. Thus $\mathsf{G}(P, Q)$ has a minimum on $M \times M$.

e) Since $P \to \mathsf{G}(P, Q)$ is continuous for $P \neq Q$ and $|\mathsf{G}(P, Q)| \leq \text{Const}\, r^{2-n}$, we can consider $\int_M \mathsf{G}(P, Q)\, dV(P)$, and the transposition of (15):

$$\psi(Q) = V^{-1} \int_M \psi(P)\, dV(P) + \Delta_Q \int_M \mathsf{G}(P, Q)\psi(P)\, dV(P). \tag{20}$$

Picking $\psi \equiv 1$ gives $\int_M \mathsf{G}(P, Q)\, dV(P) = \text{Const}$.

f) Choosing $\psi = \Delta\varphi$ in (20) leads to

$$\Delta\varphi(Q) = \Delta_Q \int_M \mathsf{G}(P, Q)\Delta\varphi(P)\, dV(P).$$

Thus $\varphi(Q) = \int_M \mathsf{G}(P, Q)\Delta\varphi(P)\, dV(P) + \text{Const}$, by (15), and this equality yields

$$\int_M [\mathsf{G}(P, Q) - \mathsf{G}(Q, P)]\Delta\varphi(Q)\, dV(Q) = \text{Const} \tag{21}$$

for all $\varphi \in C^2$. Integrating (21) proves that the constant is zero, since

$$\int \mathsf{G}(Q, P)\, dV(P) = 0 \quad \text{and} \quad \int \mathsf{G}(P, Q)\, dV(P) = \text{Const}.$$

Thus $\mathsf{G}(P, Q) - \mathsf{G}(Q, P) = \text{Const}$. Interchanging P and Q implies the second member is zero. ∎

4.14 Proposition. *Equality* (15) *holds when the integrals make sense.*

Proof. Suppose that $\Delta\varphi \in L_1$. Since $\mathscr{D}(M)$ is dense in L_1 there exists a sequence $\{g_m\}$ in $\mathscr{D}(M)$ such that $\|g_m - \Delta\varphi\|_1 \to 0$. Thus $\int_M g_m\, dV \to 0$ and $g_m - V^{-1} \int_M g_m\, dV \to \Delta\varphi$ in L_1.

Therefore we can choose $\{g_m\}$ with $\int_M g_m\, dV = 0$ and, according to Theorem 4.7, there exists $\{f_m\}$ such that $\int_M f_m\, dV = \int_M \varphi\, dV$ and $\Delta f_m = g_m$. f_m belongs to C^∞ and satisfies

$$f_m(P) = V^{-1} \int_M f_m\, dV + \int_M \mathsf{G}(P, Q) g_m(Q)\, dV(Q).$$

According to Proposition 3.64, $f_m \to V^{-1} \int_M \varphi \, dV + \int_M \mathsf{G}(P, Q)\Delta\varphi(Q) \, dV(Q)$ in L_1. On the other hand, $f_m \to \varphi$ in the distributional sense, since $\int_M f_m \, dV = \int_M \varphi \, dV$ and $\Delta f_m \to \Delta\varphi$ in L_1. Thus φ satisfies (15) almost everywhere.

■

4.15 Remark. It is possible to define the Green's function as the sum of a series (see Aubin [12]). This alternate definition allows one to obtain estimates on the Green's function in terms of the diameter D, the injectivity radius, d, the upper bound b, of the curvature and the lower bound a of the Ricci curvature. As a consequence, Aubin ([12] p. 367) proved that λ_1 the first nonzero eigenvalue is bounded away from zero:
There exist three positive constants C, k, and ξ which depend only on n, such that $\lambda_1 \geq CD^{-2}k^{D/\delta}$, δ satisfying $-a\delta^2 \leq \xi$, $2\delta\sqrt{\sup(0, b)} \leq \pi$, and $0 < \delta \leq d$.
Other positive lower bounds were found by various authors. Let us mention only Cheeger [82] and Yau [274].

4.16 Remark. The Green's function was introduced by Hilbert [140] and the Green's form on a compact Riemannian manifold by G. De Rham [105] and Bidal and De Rham [52].

2.4. Green's Function for Compact Manifolds with Boundary

4.17 Theorem. *Let $\overline{W}_n$ be an oriented compact Riemannian manifold with boundary of class C^∞. There exists $\mathsf{G}(P, Q)$, the Green's function of the Laplacian, which has the following properties:*

(a) *All functions $\varphi \in C^2(\overline{W})$ satisfy*

$$(22) \quad \varphi(P) = \int_W \mathsf{G}(P, Q)\Delta\varphi(Q) \, dV(Q) - \int_{\partial W} v^i \nabla_{iQ} \mathsf{G}(P, Q)\varphi(Q) \, ds(Q),$$

where v is the unit normal vector oriented to the outside and ds is the volume element on ∂W corresponding to the Riemannian metric j^*g ($j: \partial W \to \overline{W}$ the canonical imbedding).

(b) *$\mathsf{G}(P, Q)$ is C^∞ on $\overline{W} \times \overline{W}$ minus the diagonal (for $P \neq Q$).*

(c) *$|\mathsf{G}(P, Q)| < kr^{2-n}$ for $n > 2$, $|\mathsf{G}(P, Q)| < k(1 + |\log r|)$ for $n = 2$,*
$|\nabla_Q \mathsf{G}(P, Q)| < kr^{1-n}$,
$|\nabla_Q^2 \mathsf{G}(P, Q)| < kr^{-n}$, with $r = d(P, Q)$ and k a constant which depends on the distance of P to the boundary.

(d) *$\mathsf{G}(P, Q) > 0$ for P and Q belonging to the interior of the manifold*

(e) *$\mathsf{G}(P, Q) = \mathsf{G}(Q, P)$.*

Proof of existence. Let $P \in W$ given. We define $\mathsf{H}(P, Q)$ as in (8), where $f(r)$ is a function equal to zero for $r > \delta(P)(k + 1)^{-1}$ with $\mathbb{N} \ni k > n/2$ and $\delta(P)$ the injectivity radius at P.
$F(P, Q)$ defined by (17) satisfies

$$\Delta_Q F(P, Q) = \Gamma_{k+1}(P, Q), F(P, Q) = 0 \quad \text{for } Q \in \partial W.$$

According to Theorem 4.8, there exists a solution in $\overset{\circ}{H}_1(W)$.
$G(P, Q)$ defined by (17) satisfies (13), is C^∞ on $\overline{W} - P$, and equals zero for $Q \in \delta W$. (We can apply the theorems of regularity to $\overline{W} - B_P(\varepsilon)$ with $\varepsilon > 0$ small enough.)

Proof of the properties. a) The result is obtained by using Stokes' formula (see 1.70).

b) The proof is similar to that of Theorem 4.13 b).

c) The leading part of $G(P, Q)$ is $H(P, Q)$

d) Let $P \in W$ given. According to the previous result $G(P, Q) > 0$ for Q belonging to a ball $B_P(\varepsilon)$ with $\varepsilon > 0$ small enough. Applying the maximum principle 3.71, $G(P, Q)$ achieves a minimum on the boundary of $W - B_P(\varepsilon)$, since $\Delta_Q G(P, Q) = 0$. Thus $G(P, Q) > 0$ for $Q \in W$.

e) Transposing (22) with φ and ψ belonging to $\mathscr{D}(W)$ yields:

$$\psi(Q) = \Delta_Q \int_W G(P, Q)\psi(P)\, dV(P).$$

Choose $\psi(Q) = \Delta\varphi(Q)$. By Theorem 4.8,

$$\varphi(Q) = \int_W G(P, Q)\Delta\varphi(P)\, dV(P).$$

Hence $G(P, Q)$ satisfies

$$\Delta_{P\,\text{distr.}}\, G(P, Q) = \delta_Q(P)$$

and $G(P, Q) = G(Q, P)$.
Indeed $\Delta_Q[G(P, Q) - G(Q, P)] = 0$ and $G(P, Q) - G(Q, P)$ vanishes for $Q \in \partial W$. Applying Theorem 4.8 yields the claimed result. ∎

4.18 Let us now prove a result similar to that of Theorem 4.7, a result which we will use in Chapter 7.

On a compact Riemannian manifold M, let $\mathfrak{Q}$ be a $C^{r+\alpha}$ section of $T^*(M) \otimes T^*(M)$, which defines everywhere a positive definite bilinear symmetric form ($\mathfrak{Q}$ is a $C^{r+\alpha}$ Riemannian metric) where $r \geq 1$ is an integer and α a real number $0 < \alpha < 1$. Consider the equation

$$-\nabla^i[a_{ij}(x)\nabla^j\varphi] + b(x)\varphi = f(x) \tag{23}$$

where $a_{ij}(x)$ are the components in a local chart of $\mathfrak{Q}$ and where $b(x)$ and $f(x)$ are functions belonging to $C^{r+\alpha}$. Moreover, we suppose that $-\nabla^i a_{ij}(x)$ belongs to $C^{r+\alpha}$.

Theorem 4.18. *If* $b(x) > 0$, *Equation* (23) *has a unique solution belonging to* $C^{r+2+\alpha}$.

Proof. Suppose at first that $a_{ij}(x)$, $b(x)$ and $f(x)$ belong to C^∞. In that case we consider the functional $I(\varphi) = \int a_{ij}\nabla^i\varphi\nabla^j\varphi \, dV + \int b\varphi^2 \, dV$ and $\mu = \inf I(\varphi)$, for all $\varphi \in \mathsf{H}_1$ satisfying $\int \varphi f \, dV = 1$. A proof similar to that of 4.7 establishes the existence of a solution, which belongs to C^∞ by the regularity theorem 3.54 and which is unique by the maximum principle 3.71.

Now in the general case we approximate in $C^{1+\alpha}$ the coefficients of Equation (23) by coefficients belonging to C^∞. We obtain a sequence of equations

$$E_k\colon -\nabla^i[a_{kij}(x)\nabla^j\varphi] + b_k(x)\varphi = f_k(x)$$

with C^∞ coefficients ($k = 1, 2, \ldots$). And we can choose E_k so that $b_k(x) > b_0$ and $a_{kij}(x)\xi^i\xi^j \geq \lambda|\xi|^2$ for some $b_0 > 0$ and $\lambda > 0$ independent of k.

By the first part of the proof, E_k has a C^∞ solution φ_k. And these solutions ($k = 1, 2, \ldots$) are uniformly bounded. Indeed, considering the maximum and then the minimum of φ_k, we get

$$\|\varphi_k\|_{C^0} \leq b_0^{-1}\|f_k\|_{C^0}.$$

Now by the Schauder interior estimates 3.61, the sequence $\{\varphi_k\}$ is bounded in $C^{2,\alpha}$. To apply the estimates we consider a finite atlas $\{\Omega_i, \psi_i\}$ and compact sets $K_i \subset \Omega_i$ such that $M = \bigcup_i K_i$.
As $\{\varphi_k\}$ is bounded in $C^{2,\alpha}$, by Ascoli's theorem 3.15, there exist $\varphi \in C^2$ and a subsequence $\{\varphi_j\}$ of $\{\varphi_k\}$ such that $\varphi_j \to \varphi$ in C^2. Thus $\varphi \in C^{2,\alpha}$ and satisfies (23). Lastly, according to Theorem 3.55, the solution φ belongs to $C^{r+2+\alpha}$ and is unique (uniqueness does not use the smoothness of the coefficients).

Remark. For the proof of Theorem 4.18, we can also minimize over H_1 the functional

$$J(\varphi) = \int a_{ij}\nabla^i\varphi\nabla^j\varphi \, dV + \int b\varphi^2 \, dV - 2\int f\varphi \, dV.$$

We considered a similar functional in the proof of Theorem 4.8.

§3. Riemannian Geometry

3.1. The First Eigenvalue

4.19 Let λ_1 be the first non-zero eigenvalue of the Laplacian on a compact Riemannian smooth manifold (M_n, g) of dimension $n \geq 2$.

Lichnerowicz's Theorem 4.19 [185]. *If the Ricci curvature of the compact manifold (M_n, g) satisfies* Ricci $\geq a > 0$, *then* $\lambda_1 \geq \frac{na}{n-1}$.

Proof. We start with the equality

$$\nabla^j \nabla_i \nabla_j f - \nabla_i \nabla^j \nabla_j f = R_{ij} \nabla^j f \tag{24}$$

valid for any $f \in C^3(M)$.

Multiplying (24) by $\nabla^i f$ and integrating leads (after integrating by parts twice) to

$$\int (\Delta f)^2 \, dV - \int \nabla^i \nabla^j f \nabla_i \nabla_j f \, dV = \int R_{ij} \nabla^i f \nabla^j f \, dV. \tag{25}$$

Choosing as f an eigenfunction of the Laplacian $\Delta = -\nabla^i \nabla_i$ related to λ_1: $\Delta f = \lambda_1 f$, we obtain at once

$$\lambda_1^2 \int f^2 \, dV \geq a \int |\nabla f|^2 \, dV = a \lambda_1 \int f^2 \, dV.$$

Thus $\lambda_1 \geq a$, but we have better, because for any $f \in C^2$:

$$\int \nabla^i \nabla^j f \nabla_i \nabla_j f \, dV \geq \frac{1}{n} \int (\Delta f)^2 \, dV. \tag{26}$$

This inequality is obtained expanding

$$\left(\nabla_i \nabla_j f + \frac{1}{n} \Delta f g_{ij}\right)\left(\nabla^i \nabla^j f + \frac{1}{n} \Delta f g^{ij}\right) \geq 0.$$

When f satisfies $\Delta f = \lambda_1 f$, (25) and (26) imply the inequality $\lambda_1 \geq \frac{na}{n-1}$ of Theorem 4.19. After this basic result, a lot of positive bounds from below and from above for λ_1 have been obtained.

4.20 For a Kähler manifold the Laplacian $\bar{\Delta}$ is one half of the real Laplacian Δ. In Chapter 7 we will write the complex Laplacian without the bar, but in this section we must have another symbol than that for the real Laplacian.

$$\bar{\Delta} f = -\nabla^\lambda \nabla_\lambda f = -\nabla^{\bar{\lambda}} \nabla_{\bar{\lambda}} f = -\frac{1}{2} \nabla^i \nabla_i f = \frac{1}{2} \Delta f,$$

$\lambda = 1, 2, \ldots, m$, where m is the complex dimension ($n = 2m$).

For a compact Kähler manifold the first non-zero eigenvalue of the Laplacian $\bar{\lambda}_1$ is equal to $\lambda_1/2$ (in Chapter 7, we write the first non-zero eigenvalue of the complex Laplacian without bar).

Theorem 4.20 (Aubin [20] p. 81). *If the Ricci curvature of the compact Kähler manifold (M_{2m}, g) satisfies* Ricci $\geq a > 0$, *then* $\bar{\lambda}_1 \geq a$.

Proof. The complex version of (24) is

$$\nabla^{\nu}\nabla_{\mu}\nabla_{\nu}f - \nabla_{\mu}\nabla^{\nu}\nabla_{\nu}f = R_{\mu\bar{\nu}}\nabla^{\bar{\nu}}f \tag{27}$$

since $\nabla^{\bar{\nu}}\nabla_{\mu}\nabla_{\bar{\nu}}f = \nabla_{\mu}\nabla^{\bar{\nu}}\nabla_{\bar{\nu}}f$. Multiplying (27) by $\nabla^{\mu}f$ and integrating yield

$$\int (\bar{\Delta} f)^2 \, dV - \int \nabla^{\nu}\nabla^{\mu} f \nabla_{\nu}\nabla_{\mu} f \, dV = \int R_{\mu\bar{\nu}}\nabla^{\mu} f \nabla^{\bar{\nu}} f \, dV.$$

Thus, for any $f \in C^2$, $\int (\bar{\Delta} f)^2 \, dV \geq \int R_{\mu\bar{\nu}}\nabla^{\mu} f \nabla^{\bar{\nu}} f \, dV$.

The inequality of Theorem 4.20 follows. This inequality will be the key for solving the problem of Einstein-Kähler metrics when $C_1(M) > 0$ (see 7.26).

Corollary 4.20. *The first non-zero eigenvalue $\bar{\lambda}_1$ of the Laplacian on a compact Einstein-Kähler manifold satisfies $\bar{\lambda}_1 \geq \bar{R}/m$, where $\bar{R}$ is the scalar curvature of (M_{2m}, g), that is one half the real scalar curvature R:*

$$\bar{R} = g^{\mu\bar{\nu}} R_{\mu\bar{\nu}} = g^{\bar{\mu}\nu} R_{\bar{\mu}\nu} = R/2.$$

We verify that $\bar{\lambda}_1 = \bar{R}/m$ for the complex projective space $P_m(\mathbb{C})$. But there are other Kähler manifolds having this property.

There is no complex version of Obata' s theorem [*260] for the sphere.

$S_2 \times S_2$ or more generally $P_m(\mathbb{C}) \times P_m(\mathbb{C})$ have this property: $\bar{\lambda}_1 = \bar{R}/m$ (see Aubin [20]).

4.21 The preceding results concern the case of positive Ricci curvature. Without this assumption we have the

Theorem 4.21 (Berard, Besson and Gallot [*36]). *Let (M_n, g) be a compact Riemannian manifold satisfying* Ricci $\geq (n-1)\varepsilon\alpha^2 D^{-2}$, *where D is the diameter and $\varepsilon = -1, 0$ or 1. Then $\lambda_1 \geq nD^{-2}a^2(n, \varepsilon, \alpha)$.*

For the value of $a(n, \varepsilon, \alpha)$, see Theorem 1.10.

3.2. Locally Conformally Flat Manifolds

4.22 Definition. The Riemannian manifold (M_n, g) is locally conformally flat if any point $P \in M$ has a neighbourhood where there exists a conformal metric $\tilde{g}$ ($\tilde{g} = e^f g$ for some function f) which is flat.

When (M_n, g) is locally conformally flat, there exists an atlas $(\Omega_i, \varphi_i)_{i \in I}$ where φ_i are conformal diffeomorphisms $(\Omega_i, g_i) \to (R^n, \mathcal{E})$, with $g_i = g/\Omega_i$.

In 1822 Gauss proved the existence of isothermal coordinates on any surface (Chern [*94] gave an easy proof of this fact). Thus, any Riemannian manifold of dimension 2 is locally conformally flat.

In dimension greater than 2, we introduce two tensor fields.

4.23 Definition. The Weyl tensor (or tensor of conformal curvature) is defined by its components in a local chart as follows

$$W_{ijkl} = R_{ijkl} - \frac{1}{n-2}\big(R_{ik}\, g_{jl} - R_{il}\, g_{jk} + R_{jl}\, g_{ik} - R_{jk}\, g_{il}\big) + \frac{R}{(n-1)(n-2)}\big(g_{jl}\, g_{ik} - g_{jk}\, g_{il}\big). \tag{28}$$

The Schouten tensor is defined by

$$S_{ij} = \frac{1}{n-2}\left[2R_{ij} - \frac{R}{(n-1)} g_{ij}\right]. \tag{29}$$

We verify that the tensor $W_i{}^j{}_{kl}$ is conformally invariant: for the metric $\tilde{g} = e^f g$, $\tilde{W}_i{}^j{}_{kl} = W_i{}^j{}_{kl}$. We verify also that, in dimension 3, $W_{ijkl} \equiv 0$.

4.24 Theorem (Schouten). *A necessary and sufficient condition for a Riemannian manifold to be locally conformally flat is that $W_{ijkl} \equiv 0$ when $n > 3$ and $\nabla_k S_{ij} = \nabla_j S_{ik}$ when $n = 3$.*

Proof. The Weyl tensor of the Euclidean metric vanishes (its curvature is zero). Since $W_i{}^j{}_{kl}$ is conformally invariant, the necessary condition follows at once when $n > 3$.

Set $\tilde{g} = e^f g$. A computation gives

$$\tilde{S}_{ij} = S_{ij} + T_{ij}, \tag{30}$$

with $T_{ij} = \nabla_i \nabla_j f - \frac{1}{2}\nabla_i f \nabla_j f + \frac{1}{4}\nabla^k f \nabla_k f g_{ij}$; thus, we have $\nabla_k S_{ij} - \nabla_j S_{ki} \equiv 0$, in particular when $n = 3$. Indeed, if $\tilde{g} = \mathcal{E}$, $\tilde{R}_{ijkl} \equiv 0$ and

$$2R_{ijkl}\nabla^j f = \nabla_k f \nabla_i \nabla_l f - \nabla_l f \nabla_i \nabla_k f + \nabla^j f\big(g_{il}\nabla_k \nabla_j f - g_{ik}\nabla_l \nabla_j f\big).$$

Thus $\nabla_k T_{ij} = \nabla_j T_{ik}$. Since $\tilde{S}_{ij} \equiv 0$, we have $\nabla_k S_{ij} - \nabla_j S_{ik} \equiv 0$, in particular when $n = 3$.

We verify that $\nabla^j W_i{}^j{}_{kl} = \frac{n-3}{2}(\nabla_k S_{il} - \nabla_l S_{ik})$. Thus $W_{ijkl} \equiv 0$ implies $\nabla_k S_{il} - \nabla_l S_{ik} \equiv 0$ when $n > 3$.

The condition is also sufficient. Assume there exists a 1-form ω with components ω_i satisfying in a local chart $\{x^k\}$:

(31) $$\partial_i \omega_j = A_{ij}(x, \omega)$$

with

$$A_{ij} = \Gamma_{ij}^k \omega_k + \frac{1}{2}\omega_i \omega_j - \frac{1}{4}\omega^k \omega_k g_{ij} - S_{ij}.$$

Since $S_{ij} = S_{ji}$ and $\Gamma_{ij}^k = \Gamma_{ji}^k$, $\partial_i \omega_j = \partial_j \omega_i$.

Thus, locally, there exists a function f such that ω = df. According to (30) and (31), for the corresponding metric $\tilde{g}$, $\tilde{S}_{ij} = 0$. This implies, $\tilde{R} = (n-1)\tilde{S}_{ij}\tilde{g}^{ij} = 0$ and then $\tilde{R}_{ij} = 0$.

So $\tilde{g}$ is flat since $\tilde{W}_{ijkl} = 0$ (by assumption when $n > 3$, in any case when $n = 3$).

The local integrability conditions of system (31) are

$$\partial_k A_{ij} + \frac{\partial A_{ij}}{\partial \omega_l} A_{kl} = \partial_i A_{kj} + \frac{\partial A_{kj}}{\partial \omega_l} A_{il}.$$

A computation shows that they are equivalent to the conditions

$$W_i{}^j{}_{kl}\omega_j = \nabla_k S_{il} - \nabla_l S_{ik}$$

which are satisfied by hypothesis (when $n > 3$, we saw that $W_{ijkl} \equiv 0$ implies $\nabla_k S_{il} - \nabla_l S_{ik} \equiv 0$).

4.25 Proposition (Hebey). *Let (M_n, g) be a locally conformally flat manifold ($n \geq 3$) and let P be a point of M. Then there exists in a neighbourhood of P a metric $\tilde{g}$ conformal to g, which is flat and invariant by any isometry σ of (M_n, g) such that $\sigma(P) = P$.*

Proof. Let us go back to the proof of Theorem 4.24. If we fix $df(P) = 0$ and $f(P) = 0$, the solution of (31) with ω = df is unique. Now $\omega \circ \sigma$ satisfies (31) and the conditions at P. Thus $f \circ \sigma = f$.

4.26 Examples. The Riemannian manifolds of constant sectional curvature are locally conformally flat. The Riemannian product of two manifolds (M_1, g_1) and (M_2, g_2) is locally conformally flat if one of them is of constant sectional curvature k and the other of dimension 1, or of constant sectional curvature - k.

We also have the

Theorem 4.26 (Gil-Medrano [*142]). *The connected sum of two locally conformally flat manifolds admits conformally flat structure.*

3.3. The Green Function of the Laplacian

4.27 Gromov [135] found a new kind of isoperimetric inequalities, which concern the compact Riemannian manifolds (M_n, g) of positive Ricci curvature. By an homothety, we can suppose that the Ricci curvature is greater than or equal to $n-1$ which is the Ricci curvature of the sphere (S_n, g_0) of radius 1 (endowed with the standard metric).

Let $\Omega \subset M$ be an open set which has a boundary $\partial\Omega$.

Gromov considers a ball $B \subset S_n$ such that

$$\text{Vol}\, B / \text{Vol}\, S_n = \text{Vol}\, \Omega / \text{Vol}\, M. \tag{32}$$

The Gromov inequality is

$$\text{Vol}(\partial\Omega)/ \text{Vol}\, M \geq \text{Vol}(\partial B)/ \text{Vol}\, S_n. \tag{33}$$

With such inequality, we can for instance obtain an estimate of the constants in the Sobolev imbedding theorem, or a positive bound from below for the first non-zero eigenvalue λ_1 of the Laplacian, see Bérard-Gallot [*37], Berard-Meyer [*38] and Gallot [*133].

However these results concerned only compact manifolds with positive Ricci curvature. This extra hypothesis has been removed.

4.28 Let (M_n, g) be a compact Riemannian manifold.

Berard, Besson and Gallot defined the *isoperimetric function* $h(\beta)$ of M as follows:

$$h(\beta) = \inf \left[\text{Vol}(\partial\Omega)/ \text{Vol}\, M \right] \tag{34}$$

for all $\Omega \subset M$ such that $\text{Vol}\, \Omega / \text{Vol}\, M = \beta$ with $\beta \in]0, 1[$ of course. Changing Ω in $M \setminus \Omega$ proves that $h(1-\beta) = h(\beta)$.

The properties of $h(\beta)$ are studied in Gallot [*133] (regularity, under-additivity).

We denote by $\text{Is}(\beta)$ the isoperimetric function of (S_n, g_0) of radius 1. Let D be an upper bound for the diameter of (M, g) and let r be the inf of the Ricci curvature of (M, g).

Theorem 4.28 (Berard, Besson and Gallot [*36], see also Gallot [*133]). *Assume*

$$rD^2 \geq \varepsilon(n-1)\alpha^2 \quad \textit{with} \quad \varepsilon \in \{-1, 0, +1\} \quad \textit{and} \quad \alpha \in R^+. \tag{35}$$

Then, for any $\beta \in]0, 1[$,

$$Dh(\beta) \geq a(n, \varepsilon, \alpha)\, \text{Is}(\beta), \tag{36}$$

with $a(n, 0, \alpha) = \left(1 + n\omega_n/\omega_{n-1}\right)^{1/n} - 1$,

$$a(n,+1,\alpha) = \alpha\big[\omega_n/\omega_{n-1}\big]^{1/n}\left(2\int_0^{\alpha/2}(\cos t)^{n-1}dt\right)^{-1/n}$$

(*in this case* $\alpha \leq \pi$)

and $a(n,-1,\alpha) = \alpha c(\alpha)$ *where* $c(\alpha)$ *is the unique positive solution* x *of the equation* $x\int_0^{\alpha}(\operatorname{ch} t + x \operatorname{sh} t)^{n-1}\,dt = \omega_n/\omega_{n-1}$.

This solution $c(\alpha)$ *satisfies* $c(\alpha) \geq b(n,\alpha) = \inf(k, k^{1/n})$ *with*

$$k = \frac{\int_0^{\pi}(\sin t)^{n-1}\,dt}{\int_0^{\alpha}(\operatorname{ch} 2t)^{(n-1)/2}\,dt} = (n-1)\omega_n/\omega_{n-1}\big(e^{(n-1)\alpha}-1\big).$$

In dimension 2, we can choose $a(2,+1,\alpha) = \alpha\,/\sin(\alpha/2)$, $a(2,0,\alpha) = 2$ *and* $a(2,-1,\alpha) = \alpha/\operatorname{sh}(\alpha/2)$.

4.29 Let $G(x,y)$ be the Green function of the Laplacian on (M,g) satisfying

$$\int G(x,y)\,dV(y) = 0.$$

In this section, we want to find a lower bound of $G(x,y)$ in terms of n,r,V and D, that is, resp., the dimension, the inf of the Ricci curvature, the volume and the diameter of the compact manifold (M_n,g).

In [*31] Bando and Mabushi gave such a lower bound

$$G(x,y) \geq -\gamma(n,\alpha)D^2V^{-1}, \tag{37}$$

where $\gamma(n,\alpha)$ is a positive constant depending only on n and $\alpha \geq 0$ a constant such that $rD^2 \geq -(n-1)\alpha^2$.

With the result of Theorem 4.28, independently Gallot found an explicit lower bound for $G(x,y)$. His proof is unpublished, we give it below.

Proposition 4.29 (Gallot). *For any* x, y,

$$G(x,y) \geq -V^{-1}\int_0^1 \beta(1-\beta)h^{-2}(\beta)d\beta, \tag{38}$$

where V *is the volume of* (M,g), h *is defined by* (34).

Proof. Note that the integral at the right side converges since $h(\beta) \sim C\beta^{1-1/n}$ when $\beta \to 0$ and $h(1-\beta) \sim C(1-\beta)^{1-1/n}$ when $\beta \to 1$.

Fix $x \in M$ and set $f(y) = G(x,y)$. Let us define the function $a\colon \mathbb{R} \to \mathbb{R}$ by

$$a(\mu) = V^{-1}\operatorname{Vol}\{y/f(y) > \mu\}$$

and the function $\tilde{f}$ of [0,1] in $\mathbb{R}$ by $\tilde{f}(\beta) = \inf\{\mu/a(\mu) < \beta\}$. Since f is harmonic on $M - \{x\}$, $\operatorname{Vol}\{y/f(y) = \mu\} = 0$ and $\mu \to a(\mu)$ is continuous. As $\mu \to a(\mu)$ decreases, $\tilde{f}$ is the inverse function of a.

According to Gallot [*133] (Lemma 5.7, p. 60),

(i) for any regular value μ of f, $\tilde{f} \circ a(\mu) = \mu$ and

$$V a'(\mu) = V / \tilde{f}'\big[a(\mu)\big] = -\int_{\{f=\mu\}} |\nabla f|^{-1}\, d\sigma,$$

where $d\sigma$ is the $(n-1)$-measure on the manifold $\{f = \mu\}$.

(ii) For any continuous function $u: \mathbb{R} \to \mathbb{R}$,

$$\int u \circ f\, dV = V \int_0^1 u \circ \tilde{f}(\beta)\, d\beta.$$

We have

$$\begin{aligned}\int_{\{f=\mu\}} |\nabla f|\, d\sigma &= \int_{\{f>\mu\}} \Delta f dV \\ &= \int_{\{f>\mu\}} \left(\delta_x - V^{-1}\right) dV = 1 - a(\mu).\end{aligned}$$

Moreover, using (i) and the Cauchy-Schwarz inequality, we have

$$\begin{aligned}(39) \qquad \left(\mathrm{Vol}\{f=\mu\}\right)^2 &\le \int_{\{f=\mu\}} |\nabla f|^{-1}\, d\sigma \int_{\{f=\mu\}} |\nabla f|\, d\sigma \\ &= -V\Big[1 - a(\mu)\Big] a'(\mu).\end{aligned}$$

Thus, by the very definition of h,

$$V h^2[a(\mu)] \le -[1 - a(\mu)] a'(\mu).$$

We can rewrite this inequality in the form

$$1 - \beta \ge -V h^2(\beta) \tilde{f}'(\beta).$$

Integrating yields

$$(40) \qquad \tilde{f}(\beta) \le \tilde{f}(1) + V^{-1} \int_\beta^1 (1-s) h^{-2}(s)\, ds.$$

Using (ii) with $u(x) = x$ gives

$$V^{-1} \int f\, dV = \int_0^1 \tilde{f}(\beta) d\beta \le \tilde{f}(1) + V^{-1} \int_0^1 \int_\beta^1 \big(1 - s\big) h^{-2}(s)\, ds\, d\beta.$$

Since $\int f\ dV = 0$ and $\tilde{f}(1) = \inf f(y) = \inf_y G(x,y)$, we get (38) after integrating by parts the last integral.

4.30 Let H be a C^1 positive function on $\big[0, 1/2\big]$. We define the function h^* by

$$h^*(\beta) = \beta^{1-1/n} H(\beta)$$

for $\beta \in [0, 1/2]$ and

$$h^*(\beta) = h^*(1-\beta)$$

for $\beta \in [1/2, 1]$.

Let us consider the function $S(\beta) = \int_\beta^{1/2} \frac{ds}{h^*(s)}$ and its inverse function $A\colon [0, L] \to [0, 1/2]$ where $L = S(0)$.

Definition 4.30. $M^* = [-L, L] \times S_{n-1}$ is the manifold endowed with the one-parameter family of metrics

$$g_t(s, x) = (ds)^2 + t^2 \Big\{ h^*\big[A\big(|s|\big)\big] \Big\}^{2/(n-1)} g_{S_{n-1}}(x),$$

where $g_{S_{n-1}}$ is the canonical metric of $S_{n-1}(1)$.

We identify all the points of $\{+L\} \times S_{n-1}$ to a pole noted x_0 (resp. all the points of $\{-L\} \times S_{n-1}$ to a pole noted x_1) of the Riemannian manifold (M^*, g_t).

$B(x_0, r)$ being the geodesic ball of (M^*, g_t) centered at x_0 of radius r, by construction,

$$\text{(41)} \qquad \operatorname{Vol}\big[\partial B(x_0, r)\big] / \operatorname{Vol} M^* = h^*\big[\operatorname{Vol} B(x_0, r) / \operatorname{Vol} M^*\big],$$

where the volumes are related to the metric g_t.

We denote by $G^*_{x_o} = G^*(x_0, \cdot)$ the Green function of the Laplacian on (M^*, g_t) with pole x_0, and $V^* = \operatorname{Vol}(M^*, g_t)$.

4.31 Proposition (Gallot). *For any compact Riemannian manifold (M, g) whose isoperimetric function h satisfies $h \geq h^*$ on $[0, 1]$,*

$$\text{(42)} \qquad G(x, y) \geq (V^*/V) G^*(x_0, x_1)$$

x, y being two points of M.

Proof. $|\nabla G^*_{x_0}|$ is constant on each hypersurface $\{G^*_{x_0} = \mu\}$, so that the Cauchy-Schwarz inequality used in (39) is an equality for $G^*_{x_0}$. Thus, according to (41), the same proof as that of Proposition 4.29 leads to (40) with equality.

$$\text{(43)} \qquad \tilde{G}^*_{x_0}(\beta) = G^*(x_0, x_1) + (V^*)^{-1} \int_\beta^1 (1-s)\big[h^*(s)\big]^{-2} ds$$

where $V^* = \operatorname{Vol}(M^*, g_t)$ and

$$\text{(44)} \qquad \inf_y G^*(x_0, y) = G^*(x_0, x_1) = -(V^*)^{-1} \int_0^1 s(1-s)\big[h^*(x)\big]^{-2} ds.$$

(38) together with (44) imply (42).

If the manifold (M_n, g) has its Ricci curvature bounded from below by $-(n-1)K^2$, according to Theorem 4.28,

$$h(\beta) \geq \gamma(KD, n)\big[\inf(\beta, 1-\beta)\big]^{1-1/n},$$

where D is the diameter of (M, g) and γ an universal function.

Set then $h^*(\beta) = \gamma(KD, n)\big[\inf(\beta, 1-\beta)\big]^{1-1/n}$.

For a suitable choice of t, (M^*, g_t) is $B_n(R)\#B_n(R)$ the union of two euclidean balls of radius $R = R(KD, n)$ glued on their boundaries by the identity. We obtain the

Corollary 4.31. *Assume* $\operatorname{Ricci}(M_n, g) \geq -(n-1)K^2$, *then*

$$G(x, y) \geq \big[2\omega_{n-1}/nV\big] R^n G^*_{Bn(R)\#Bn(R)}(x_0, x_1),$$

where $R = R\big(KD, n\big)$ *and where* x_0 *and* x_1 *are the centers of the two balls.*

4.32 Theorem (Gallot). *Assume* $\operatorname{Ricci}(M_n, g) \geq -(n-1)K^2$, *then*

$$G(x, y) \geq R^n \omega_n V^{-1} G_{S_n(R)}(x_0, x_1), \tag{45}$$

with $R = R(n, K, D) = K^{-1}b^{-1}(n, KD)$, $G_{S_n(R)}$ *being the Green function of the sphere* $S_n(R)$ *with* x_0 *and* x_1 *their two poles.* $b(n, KD)$ *comes from Theorem* 4.28.

Proof. If we choose $h^*(\beta) = Kb(n, KD)I_s(\beta)$, for a suitable choice of t, (M^*, g_t) is a canonical sphere with radius $R = K^{-1}b^{-1}(n, KD)$.

Moreover according to (13), $h(\beta) \geq h^*(\beta)$. Then (42) implies (45).

3.4. Some Theorems

4.33 The Sard Theorem [*279] (see also Sternberg [*294]). *Let* M_n *and* $\tilde{M}_p$ *be two* C^k *differentiable manifolds of dimension n and p. If f is a map of class* C^k *of* M *into* $\tilde{M}$, *then the set of the critical values of f has measure zero provided that* $k - 1 \geq \max(n-p, 0)$.

$P \in M$ is a *critical point* of f if the rank of f at P is not p. All others points of M are called regular. $Q \in \tilde{M}$ is a *critical value* of f, if $f^{-1}(Q)$ contains at least one critical point. All other points of $\tilde{M}$ are called *regular values*. Since our manifolds have countable bases, a subset $A \subset \tilde{M}$ has measure zero if for every local chart (θ, ψ) of $\tilde{M}$, $\psi(A \bigcap \theta) \subset \mathbb{R}^p$ has measure zero.

4.34 The Nash imbedding Theorem [*252]. *Any Riemannian* C^k *manifold of dimension n,* $(3 \leq k \leq \infty)$ *has a* C^k *isometric imbedding in* $(\mathbb{R}^p, \mathcal{E})$ *when* $p = (n+1)(3n+11)n/2$, *in fact in any small portion of this space. If the manifold is compact, the result holds with* $p = (3n+11)n/2$.

Previously Nash [251] had solved the C^1 isometric imbedding problem. If in the sequence of successive approximations, we keep under control only the first derivatives, Nash does not need more dimensions than Whitney (see 1.16). So for $k = 1$, the theorem holds with $p = 2n+1$ and with $p = 2n$ in the compact case.

4.35 The Cheeger Theorem [*86]. *Let (M_n, g) be a Riemannian manifold, and let d, V and H be three given real numbers, d and V positive.*

There exists a positive constant $C_n(H,d,V)$ such that if the diameter $d(M) < d$, the volume $v(M) > V$ and the sectional curvature K of M is greater than H, then every closed geodesic on M has length greater than $C_n(H,d,V)$. Thus we have a positive lower bound for the injectivity radius.

Proof. Let P be a point of the simply connected space M_H of constant curvature H, and v a non-zero vector of $\mathbb{R}^n$. We define the angle θ, $0 < \theta < \pi/2$, by $\mathrm{Vol}\,\exp_P\left[a_{d,\theta}(v)\right] = V/2$ where $a_{d,\theta}(v)$ denotes the set of vectors in $\mathbb{R}^n$ of length $\leq d$ making an angle of θ or more with both v and - v. Then we define r by

$$\mathrm{Vol}\,\exp_P\left[B_r(0) - a_{r,\theta}(v)\right] = V/2.$$

Since $\theta < \pi/2$, there exists a constant $C_n(H,d,V) > 0$ such that, if σ, τ are geodesics in M_H, $\sigma(0) = \tau(0)$, $\left(\sigma'(0), \tau'(0)\right) \leq \theta$, then the distance $d_{M_H}\left(\sigma(r), \tau(t)\right) < r$ for $0 < t \leq C_n(H,d,V)$. Suppose now there exists on (M_n, g) a closed geodesic γ of length $l < C_n(H,d,V)$, and let us prove then that $v(M) \leq V$, which is a contradiction.

By the Rauch comparison Theorem (see 1.53), since $\mathrm{K} \geq H$,

$$v\left[\exp_{\gamma(0)} a_{d,\theta}\left(\gamma'(0)\right)\right] \leq V/2$$

and

$$v\left\{\exp_{\gamma(0)}\left[B_r(0) - a_{r,\theta}\left(\gamma'(0)\right)\right]\right\} \leq V/2.$$

These inequalities imply $v(M) \leq V$ since

$$M \subset \exp_{\gamma(0)}\left\{a_{d,\theta}\left(\gamma'(0)\right) \bigcup \left[B_r(0) - a_{r,\theta}\left(\gamma'(0)\right)\right]\right\}.$$

Indeed, let σ be a geodesic with $\sigma(0) = \gamma(0)$ and $\left(\sigma'(0), \gamma'(0)\right) \leq \theta$; then $d_M(\sigma(r), \gamma(l)) < r$ since $l < C_n(H,d,V)$. But $\gamma(l) = \gamma(0)$, thus σ is not minimal between $\sigma(0) = \gamma(0)$ and $\sigma(r)$.

From this result, Cheeger proved his finiteness Theorem (see [*86]), which asserts that there are only finitely many diffeomorphism classes of compact n-dimensional manifolds admitting a metric for which an expression involving $d(M)$, $v(M)$ and $S(M)$ a bound for the sectional curvature $|K|$ (or for the norm of the covariant derivative of the curvature tensor) is bounded.

4.36 The Gromov compactness Theorem [*147] asserts that the space $m(S,V,D)$ of compact Riemannian n-manifolds of sectional curvature $|K| \leq S$, $v(M) \geq V > 0$ and $d(M) \leq D$, is precompact in the $C^{1,\alpha}$ topology.

The following theorem has the same purpose.

Theorem 4.36 (Anderson [*3]). *The space $m(\lambda, i_0, D)$ of compact Riemannian n-manifolds such that $|\,\mathrm{Ricci}\,| \leq \lambda$, $d(M) \leq D$ and injectivity radius $\geq i_0 > 0$, is compact in the $C^{1,\alpha}$ topology. More precisely, given any sequence $(M_i, g_i) \in m(\lambda, i_0, D)$, there are diffeomorphisms f_i of M_i such that a subsequence of $(M_i, f_i^* g_i)$ converges,in the $C^{1,\alpha}$ topology, to a $C^{1,\alpha}$ Riemannian manifold (M, g).*

§4. Partial Differential Equations

4.1 Elliptic Equations

4.37 Let E and F be two smooth vector bundles over a C^∞ manifold M. We consider the vector spaces of the C^∞ sections of E and F: $C^\infty(E)$ and $C^\infty(F)$.

Let (Ω_j, φ_j) be an atlas for M, $\left(x_j^1, x_j^2, \ldots, x_j^n\right)$ the coordinates in Ω_j. π being the projection $E \to M$, $\pi^{-1}(\Omega_j)$ is diffeomorphic to $\Omega_j \times \mathbb{R}^p$ if $\mathbb{R}^p$ is the fibre of E. $\left(\xi_j^1, \xi_j^2, \ldots, \xi_j^p\right)$ will be the fibre coordinates. Likewise if $\mathbb{R}^q$ is the fibre of F, $\{\eta_j^\alpha\}(\alpha = 1, 2, \ldots q)$ will be the fibre coordinates of F over Ω_j.

A C^∞ section ψ of E is represented on each Ω_j by a vector-valued C^∞ function $\psi_j(x) = \{\psi_j^i(x)\}(i = 1, 2, \ldots, p)$.

Definition 4.37. A linear partial differential operator A of order k of $C^\infty(E)$ into $C^\infty(F)$ is a linear map of $C^\infty(E)$ into $C^\infty(F)$ that can be written in the coordinate systems defined above in the form

$$A(\psi)_j^\alpha = \sum_{l=0}^{k} \left(a_{\beta l}^\alpha\right)^{i_1 i_2 \ldots i_l} \nabla_{i_1 i_2 \ldots i_l} \psi_j^\beta, \tag{46}$$

$\alpha = 1, 2, \ldots, q$ and $\beta = 1, 2, \ldots, p$. a ${}^{\alpha}_{\beta,l}$ are l-tensors and $\psi_j^\beta \in C^k(M)$.

The *principal symbol* $\sigma_\xi(A, x)$ is obtained by replacing $\partial/\partial x_j^i$ by real variables ξ_i in the leading part of A, that is the part corresponding to the highest order derivatives appearing in A:

$$\left[\sigma_\xi(A, x)\right]_\beta^\alpha = \left[a_{\beta,k}^\alpha(x)\right]^{i_1 i_2 \ldots i_k} \xi_{i_1} \xi_{i_2} \ldots \xi_{i_k}. \tag{47}$$

4.38 Definition. A linear differential operator A is elliptic at a point x $\in M$ if the symbol $\sigma_\xi(A, x)$ is an isomorphism for every $\xi \neq 0$.

A necessary condition for this is $p = q$, and we can identify E and F.

We say that $A: C^\infty(E) \to C^\infty(E)$ is *strongly elliptic* if there exists a constant $\delta > 0$ such that

(48) $$\Big[\sigma_\xi(A,x)\Big]^\alpha_\beta \eta_\alpha \eta^\beta \geq \delta |\xi|^k |\eta|^2.$$

Replacing ξ by - ξ shows that k must be even: $k = 2m$.

We have assumed here that (M,g) is a Riemannian manifold $\big(|\xi|^2 = g^{ij}(x)\xi_i\xi_j\big)$, and that a Riemannian metric $h_{\alpha\beta}(x)$ is defined on the fibres $\big(\eta_\alpha = h_{\alpha\beta}\eta^\beta\big)$.

When (M,g) is compact, we define on $C^\infty(E)$ an inner product by

$$\langle \psi, \varphi \rangle = \int h_{j\alpha\beta}(x) \psi^\alpha_j(x) \varphi^\beta_j(x)\, dV.$$

We note $L_2(E)$ the space $C^\infty(E)$ with the norm $\sqrt{\langle \psi, \psi \rangle}$.

The formal adjoint A^* of A is defined as usual by

$$\langle A\psi, \varphi \rangle = \langle \psi, A^*\varphi \rangle$$

for any ψ and φ belonging to $C^\infty(E)$.

For the strongly elliptic operator A on $C^\infty(E)$ with (M,g) compact, the *Fredholm alternative* holds: Ker A *and* Ker A^* *are finite dimensional.*

If $f \in L_2(E)$ *there is a solution* ψ *of* A $\psi = f$ *if and only if* f *is orthogonal in* $L_2(E)$ *to* Ker A^* *(there is a unique solution orthogonal to* Ker A*).*

The eigenvalues λ_j *of* A *are discrete, having a limit point only at infinity. Moreover the eigenspaces* Ker$(A - \lambda_j I)$ *are finite dimensional.*

For more details see Morrey [*243].

4.39 Definiton. A differential operator A of $C^\infty(E)$ into $C^\infty(E)$

$$A\psi = F(x, \psi, \nabla\psi, \ldots, \nabla^k\psi)$$

where F is assumed to be a differentiable map of its arguments will be elliptic (resp. strongly elliptic) with respect to ψ at x if the linearized operator at ψ is elliptic (resp. strongly elliptic).

4.40 For the equations $Au = f$, where A is a partial differential operator on scalar functions, we will find in Chapter 3, some regularity theorems. Here we mention one more.

Let Ω be an open set of $\mathbb{R}^n$ with coordinates $\{x^i\}$, and let $u(x)$ be a weak solution in $H_1(\Omega)$ of the equation

(49) $$\sum_{i=1}^n \sum_{j=1}^n \partial_i \big(a_{ij}(x)\partial_j u + a_i(x)u\big) + \sum_{i=1}^n b_i(x)\partial_i u + a(x)u = f(x) + \sum_{i=1}^n \partial_i f_i(x).$$

We suppose that there exist $\mu \geq \nu > 0$ such that

(50) $$\nu \sum_{i=1}^{n} (\xi^i)^2 \leq a_{ij}(x)\xi^i \xi^j \leq \mu \sum_{i=1}^{n} (\xi^i)^2$$

for $x \in \Omega$ and $q > n$ such that

(51) $$\|a_i\|_q, , \|b_i\|_q, \|a\|_{q/2}, \|f\|_{q/2} \quad \text{and} \quad \|f_i\|_q$$

are bounded by μ for all $1 \leq i \leq n$.

Theorem 4.40 (Ladyzenskaja-Ural'ceva [*206]). *On any open bounded subset $\theta \subset \Omega$ such that the distance $d(\theta, \partial\Omega) \geq \delta$ for some $\delta > 0$, a weak solution in $H_1(\Omega)$ of* (49) *is bounded and belongs to C^α on θ for some $\alpha > 0$, if we suppose conditions* (50) *and* (51) *satisfied. Moreover $\|u\|_{C^0(\theta)} \leq M$ a constant which depends only on n, ν, μ, q, δ and $\|u\|_{L^2(\Omega)}$. Furthermore α and k an upper bound for $\|u\|_{C^\alpha(\theta)}$ depend only on n, ν, μ, q, δ and M.*

We have a uniform estimate of $\max|\nabla u|$ *on θ depending on the same quantities if in addition $\|\partial_k a_{ij}\|_q, \|\partial_k b_i\|_q, \|a\|_q, \|f\|_q$ and $\|\partial_k f_i\|_q$ are bounded by μ for all i, j, k. According to the first part of the theorem, we have then a uniform estimate of $\|u\|_{C^{1,\beta}(\theta)}$ for some $\beta > 0$, this estimate and β depending on the same quantities.*

Indeed, differentiating (49) with respect to x^k, $v = \partial_k u$ satisfies an equation of the following form:

$$\sum_{i=1}^{n} \sum_{j=1}^{n} \partial_i \big(a_{ij} \partial_j v\big) = F(x) + \sum_{i=1}^{n} \partial_i F_i(x).$$

4.41 Let A be an elliptic linear differential operator of order two on Ω an open set of $\mathbb{R}^n$:

(52) $$Au = \sum_{i,j=1}^{n} a_{ij}(x)\partial_{ij}u + \sum_{i=1}^{n} b_i(x)\partial_i u + c(x)u$$

such that $a_{ij}(x)$ satisfy

(53) $$0 < \lambda \leq a_{ij}(x)\xi^i \xi^j \leq \Lambda$$

for $x \in \Omega$ and any $\xi \in \mathbb{R}^n$ of norm $1(|\xi| = 1)$. $b_i(x)$ and $c(x)$ are supposed to be bounded, $\|b(x)\|^2 + |c(x)| \leq k$ for $x \in \Omega$.

Theorem 4.41 *Harnack inequality* (see Krylov [*204] and Safonov). *Let $u \in H_2^n(\Omega)$ be a non-negative function* ($u \geq 0$) *which satisfies $Au = f$ in $B_r \subset \Omega$. Then, for $0 < \alpha < 1$,*

(54) $$\sup_{x \in B_{\alpha r}} u(x) \leq \frac{C}{1-\alpha} \left(\inf_{x \in B_{\alpha r}} u(x) + r\|f\|_{L_n(B_r)} \right)$$

where C depends on n, Λ/λ and kr^2/λ. B_ρ is the ball of radius ρ with center at a given point $x_0 \in \Omega$.

From (54) we can deduce uniform estimates on osc u in $B_\rho(\rho < r)$ and on the Hölder continuity of u (see Moser [*244]). In [*244], Moser gave a proof of (54) in case u ≥ 0 satisfies

$$Au \equiv \sum_{i,j=1}^{n} \partial_i\left[a_{ij}(x)\partial_j u\right] = 0 \quad \text{on} \quad \Omega. \tag{55}$$

His conclusion is: *in any compact set $K \subset \Omega$,*

$$\max_{x\in K} u(x) \leq c \min_{x\in K} u(x),$$

where c depends on K, Ω, λ and Λ only. The proof of (54), as that of (55), is given in two parts corresponding to the following two propositions.

4.42 Proposition. *Let $u \in H_2^n(\Omega)$ satisfy $Au \geq f$, with $f \in L_n(\Omega)$. Then for $B_{2r} \subset \Omega$ and $p > 0$,*

$$\sup_{x\in B_r} u(x) \leq C\left[\left(r^{-n}\int_{B_{2r}} (u^+)^p\,dx\right)^{1/p} + r\|f^-\|_{L^n(B_{2r})}\right], \tag{56}$$

where C depends on λ, Λ, n, k and p. $u^+ = \sup(u,0)$ and $f^- = \sup(-f,0)$.

For the proof we use the following Alexandrov-Bakelman-Pucci inequality.

Theorem 4.42. *Let $u \in C^0(\bar{\Omega}) \bigcap H^n_{2,\mathrm{loc}}(\Omega)$ satisfy $Au \geq f$, where A is given by (52) and (53) holds. Setting $\det\left((a_{ij})\right) = \theta^n$, we assume $c(x) \leq 0$ in Ω, $|b|/\theta$ and f/θ belonging to $L_n(\Omega)$.*

Then

$$\sup_{x\in\Omega} u(x) \leq \sup_{x\in\partial\Omega} u^+(x) + C\|f^-/\theta\|_{L_n(\Omega)}, \tag{57}$$

where C depends on n, diam Ω and $\||b|\theta^{-1}\|_{L_n(\Omega)}$ only.

4.43 Proposition. *Let $u \geq 0$ satisfy $Au \leq f$ in Q_2. Then there exists $p > 0$ so that*

$$\|u\|_{L_p(Q1)} \leq C\Big(\inf_{x\in Q_1} u(x) + \|f\|_{L_n(Q_2)}\Big),$$

where C depends on λ, Λ, n and k only.

Q_h denotes the cubes $|x^i| < h/2 (i = 1, 2, \ldots, n)$.

For Moser [*244], who studied equation (55), the two propositions are:

(i) If u is a positive subsolution of (55) in Q_4, then for $p > 1$

$$\max_{x \in Q_1} u(x) \leq C_1 \Big(\frac{p}{p-1}\Big)^2 \Big(\int_{Q_2} u^p\, dx\Big)^{1/p}.$$

(ii) If u is a positive supersolution of (55) in Q_4, then

$$\Big(\int_{Q_3} u^p\, dx\Big)^{1/p} \leq \frac{C_2}{(p_0 - p)^2} \min_{x \in Q_1} u(x)$$

for $0 < p < p_0 = n/(n-2)$, where C_1 and C_2 denote constants which depend on n, λ and Λ only.

4.2. Parabolic Equations

4.44 The heat operator L. On a compact Riemannian manifold (M_n, g), we consider the operator

$$L = \Delta + \partial/\partial t$$

on C^2-functions u: $M \times [0, \infty[\to \mathbb{R}$.

$$K(P, Q, t) = (2\sqrt{\pi})^{-n} t^{-n/2} \exp[-\rho^2(P, Q)/4t]$$

is a parametrix for L with ρ smooth, $\rho(P, Q) = d(P, Q)$ when $d(P, Q) < \delta/2$ and $\rho(P, Q) = 0$ when $d(P, Q) > \delta$ the injectivity radius.

We define $N_1(P, Q, t) = -L_P K(P, Q, t)$ and

$$N_k(P, Q, t) = \int_0^t d\tau \int_M N_{k-1}(P, R, t-\tau) N_1(R, Q, \tau)\, dV(R).$$

The *fundamental solution of the heat operator* L is

$$(58) \quad H(P, Q, t) = K(P, Q, t) + \int_0^t d\tau \int_M K(P, R, t-\tau) \sum_{k=1}^{\infty} N_k(R, Q, \tau)\, dV(R)$$

(see Milgram-Rosenbloom [*235], Pogorzelski [*265]).

$H(P, Q, t)$ is C^∞ except for $P = Q$, $t = 0$; it is positive and symmetric in P, Q. In the sense of functions, it satisfies $L_P H(P, Q, t) = 0$.

Any function $u(P, t)$ on $M \times [0, \infty[$ which is C^2 in P and C^1 in t satisfies for $t > t_0$

$$(59) \quad u(P, t) = \int_{t_0}^t d\tau \int_M H(P, Q, t-\tau) Lu(Q, \tau)\, dV(Q) + \int_M H(P, Q, t - t_0) u(Q, t_0)\, dV(Q).$$

The spectral decomposition of $H(P,Q,t)$ is

(60) $$H(P,Q,t) = V^{-1} + \sum_{i=1}^{\infty} \exp(-\lambda_i t)\varphi_i(P)\varphi_i(Q),$$

where the λ_i are the non-zero eigenvalues of Δ, the $\varphi_i(P)$ being the corresponding orthonormal eigenfunctions.

4.45 Theorem. *On a compact Riemannian manifold* (M_n, g) *let us consider the parabolic equation*

(61) $$Lu(P,t) = f(P,t), u(P,0) = u_0(P).$$

Equation (61) *has a unique solution which is given, when the integrals make sense, by*

$$u(P,t) = \int_0^t d\tau \int_M H(P,Q,t-\tau) f(Q,\tau)\, dV(Q) + \int_M H(P,Q,t) u_0(Q)\, dV(Q).$$

Assume $u_0 \equiv 0$. If f is Hölder continuous, $\frac{\partial u}{\partial t}$ and the second derivatives of u with respect to P are Hölder continuous.

If f belongs to L_p, $\frac{\partial u}{\partial t}$ and the second derivatives of u with respect to P belong to L_p; moreover

(62) $$\left\| \frac{\partial u}{\partial t} \right\|_p + \|\nabla^2 u\|_p \leq \text{Const.}\, \|f\|_p,$$

where the norm L_p is taken over $M \times [0, \infty[$.

The left hand side of (62) is the norm of $H_2^p(M \times [0,\infty[)$.

For the details on the regularity of $u(P,t)$, see Ladyzenskaja-Solonnikov-Ural'ceva [*207] and Pogorzelski [*265].

4.46. Maximum principle. *Let* $u(P,t)$ *be a continuous function on* $M \times [0, t_0]$. *Assume* $u \leq 0$ *on* $M \times \{0\}$ *and on* $\partial M \times [0, t_0]$.

If whenever $u > 0$, u *is* C^2 *in P,* C^1 *in t and satisfies*

(63) $$\partial u/\partial t \leq -\Delta u + b^i(P,t)\partial_i u + cu$$

with the b^i *bounded and c a constant, then we have always* $u \leq 0$.

Proof. Let $w = e^{-(c+1)t}u$. w and u have the same sign. Since

$$\partial w/\partial t = e^{-(c+1)t}\left[\partial u/\partial t - (c+1)u\right],$$

we have

(64) $$\partial w/\partial t \leq -\Delta w + b^i(P,t)\partial_i w - w.$$

Assume w is positive somewhere and let (Q,t) be a point where w is maximum. Then $\Delta w(Q,t) \geq 0$, $\partial_i w(Q,t) = 0$ and $\partial w(Q,t)/\partial t \geq 0$. Thus (64) implies $w(Q,t) \leq 0$, which yields a contradiction.

Remark. The usual maximum principle, when the maximum is positive, is similar to the maximum principle for elliptic equations. It holds when the coefficient of u is non-positive.

4.47 On a compact Riemannian manifold (M_n, g), let us consider a linear parabolic equation of the type

$$\partial u^\alpha/\partial t = -\Delta u^\alpha + a_\beta^{\alpha i}\partial_i u^\beta + b_\beta^\alpha u^\beta + f^\alpha \tag{65}$$

when written in a system of coordinates $\{x^i\}$.

$u^\alpha(\alpha = 1,2,\dots,k)$ are k unknown functions on $M \times [0,\infty[$, $f^\alpha(\alpha = 1,2,\dots,k)$ are k given functions on $M \times [0,\infty[$. The coefficients a $_\beta^{\alpha i}$ and b_β^α are supposed to be smooth. We write u for $(u^1,u^2,\dots,u^k)$ and f for $(f^1,f^2,\dots,f^k)$. We choose $p > n+2$.

Theorem 4.47. *For every* $f \in L_p(M \times [0,t_0])$, *there exists a unique*

$$u \in H_2^p(M \times [0,t_0])$$

satisfying (65) *for* $1 \leq \alpha \leq k$ *and* $u(P,0) \equiv 0$.

Proof. In [149] Hamilton gave a proof of Theorem 4.47 when the manifold has a boundary. His proof, written in our easier case, is the following.

Let us prove uniqueness first. We have to prove that $u \equiv 0$ is the unique solution of (65) when $f \equiv 0$.

Since $p > n+2$, u^α and $\partial_i u^\alpha$ are continuous. Using the regularity properties for a single equation (65), α fixed, by induction we show that u is smooth for $t > 0$. Let

$$\psi = \frac{1}{2}\sum_{\alpha=1}^{n}(u^\alpha)^2.$$

ψ satisfies

$$L\psi = -\sum_{\alpha=1}^{n}\nabla_i u^\alpha \nabla^i u^\alpha + \sum_{\alpha=1}^{n} u^\alpha\left(a_\beta^{\alpha i}\partial_i u^\beta + b_\beta^\alpha u^\beta\right). \tag{66}$$

Then, for an appropriate constant C, we have that the right side of (66) is smaller than $C\psi$: $L\psi \leq C\psi$. Since $\psi(P,0) \equiv 0$, the maximum principle 4.46 shows that $\psi = 0$. Thus $u \equiv 0$.

Let us prove now the existence. Denote by $\tilde{H}_2^p(M \times [0,t_0])$ the subspace of the functions of $H_2^p(M \times [0,t_0])$ which vanish for $t = 0$.

According to Theorem 4.45, $u \to$ Lu defines an isomorphism of $\tilde{H}_2^p(M \times [0,t_0])$ onto $L_p(M \times [0,t_0])$

Let Ku $= \{a_\beta^{\alpha i}\partial_i u^\beta + b_\beta^\alpha u^\beta\}$. The map $K\colon \tilde{H}_2^p \to L_p$ is compact, since the inclusion $\tilde{H}_1^p \subset L_p$ is compact.

By the theory of Fredholm mappings, the map $\tilde{H}_2^p \to L_p$ given by $u \to (Lu - Ku)$ has finite dimensional kernel and cokernel. Moreover its index is zero, since the index is invariant under compact perturbations. Since we saw that its kernel is zero, this map is an isomorphism.

4.48 Definition. A strictly parabolic equation is an equation of the type

$$\frac{\partial \Psi_t}{\partial t} = A_t \psi_t, \tag{67}$$

where $t \to \Psi_t$ belongs to $C^1([0,\infty[\ C^\infty(E))$ and $[0,\infty[.\ni t \to A_t$ is a smooth family of strongly elliptic operator ofC $^\infty(E)$ into $C^\infty(E)$, see Definition 4.39.

4.49 We now prove local existence of solutions for the non linear parabolic equation of Eells and Sampson (see 10.16).

Let $u = \{u^\alpha\}$ be k unknown functions on $M \times [0,\tau]$, and f^α be k given smooth functions on $M(\alpha = 1, 2, \dots, k)$.

Theorem 4.49. *There exists* $\varepsilon > 0$ *and* $u \in H_2^p(M \times [0,\varepsilon])$ *with* $p > n+2$ *solving the equation*

$$\begin{cases} Lu^\alpha - \Gamma^\alpha_{\beta\gamma}(u(x,t))g^{ij}(x)\partial_i u^\beta \partial_i u^\gamma = 0 \\ u^\alpha(x,0) = f^\alpha(x), \qquad \alpha = 1, 2, \dots, k. \end{cases} \tag{68}$$

Moreover, u *is unique and smooth on* $(M \times [0,\varepsilon])$.

$\Gamma^\alpha_{\beta\gamma}$ *are smooth functions on* $\mathbb{R}^k$, $u(x,t)$ *being the point of* $\mathbb{R}^k$ *whose coordinates are* $u^\alpha(x,t)$.

Proof (Hamilton [*149]). We will find u as a sum $u^\alpha(x,t) = f^\alpha(x) + v^\alpha(x,t)$ and write (45) as $P(f+v) = 0$ with $v(x,t) = 0$ when $t = 0$.

The linearized equation $A_u h$ of (45) at $u \in H_2$ has the form

$$(A_u h)^\alpha = Lh^\alpha - a^{\alpha i}_\beta(u)\partial_i h^\beta - b^\alpha_\beta(u)h^\beta, h(x,0) = 0, \tag{69}$$

with $a^{\alpha i}_\beta(u)$ and $b^\alpha_\beta(u)$ continuous since $p > n+2$. So $v \to P(f+v)$ defines a continuously differentiable map of $\tilde{H}_2^p(M \times [0,\tau])$ into $L_p(M \times [0,\tau])$. Its derivative at $v = 0$ is $A_f \colon \tilde{H}_2^p(M \times [0,\tau]) \to L_p(M \times [0,\tau])$ which is an isomorphism according to Theorem 4.47.

Therefore by the inverse function theorem the set of all $P(f+v)$ for v in a neighbourhood Ω of $0 \in \tilde{H}_2^p(M \times [0,\tau])$ covers a neighbourhood θ of $P(f)$ in

$$L_p(M \times [0,\tau]).$$

If $\varepsilon > 0$ is small enough, the function equal to 0 for t $\leq \varepsilon$ and equal $P(f)$ for $\varepsilon < t \leq \tau$ belongs to θ. Thus there exists $w \in \tilde{H}_2^p(M \times [0,\tau])$ which satisfies

$$P(f+w) = 0$$

on $M \times [0,\varepsilon]$.

4.50 Corollary. *Let (M_n, g) and $(\tilde{M}_m, \tilde{g})$ be C^∞ compact Riemannian manifolds and f_0 a smooth map $M \to \tilde{M}$. Then there exists $\varepsilon > 0$ and a map $f\colon M\times[0,\varepsilon] \ni (x,t) \to f_t(x) \in \tilde{M}$ belonging to $H_2^p\big(M \times [0,\varepsilon], \tilde{M}\big)$ satisfying the parabolic equation*

$$\partial f_t^\lambda(x)/\partial t = -\Delta f_t^\lambda(x) + g^{ij}(x)\tilde{\Gamma}^\lambda_{\mu\nu}\big(f_t(x)\big)\partial_i f_t^\mu(x)\partial_j f_t^\nu(x) \tag{70}$$

with f_0 as initial value. Moreover f is unique and smooth on $M\times[0,\varepsilon]$. $\{x^i\}(1 \le i \le n\}$ denote local coordinates of x in a neighbourhood of a point $P \in M$ and $y^\lambda(1 \le \lambda \le m)$ local coordinates of y in a neighbourhood θ of $f(P) \in \tilde{M}$. The parabolic equation is written in these systems of coordinates, $\tilde{\Gamma}^\lambda_{\mu\nu}$ are the Christoffel symbols in θ.

Proof (Hamilton [*149]). Hamilton embeds $\tilde{M}$ in $\mathbb{R}^k$, k large enough. He considers a tubular neighbourhood T of $\tilde{M}$ in $\mathbb{R}^k$ and extends the metric $\tilde{g}$ on $\tilde{M}$ smoothly to a metric on T. There is an involution $i\colon T \to T$ corresponding to multiplication by -1 in the fibres, $i(Q) = Q$ for $Q \in \tilde{M}$. We can choose the extension $\bar{g}$ of $\tilde{g}$ to T so that i is an isometry of $(T, \bar{g})$. Finally we extend $\bar{g}$ smoothly to all of $\mathbb{R}^k$.

Now we apply Theorem 4.49 with $\Gamma^\alpha_{\beta\gamma}$ the Christoffel symbols of $\big(\mathbb{R}^k, \bar{g}\big)$ and $u(x,0) = f_0$.

We have $f_0(M) \subset \tilde{M}$. If $u(x,t)$ does not always remain in $\tilde{M}$, we can suppose ε small enough so that $u(x,t) \in T$ for any $x \in M$ and $t \in \big[0,\varepsilon\big]$.

Since i is an isometry, $i \circ u$ would be another solution of (68), which is in contradiction with the uniqueness of the solution. For more details see [*149].

4.51 Theorem. *Let E be a bundle of tensors over a smooth compact Riemannian manifold (M, g). We seek a smooth family $[0,\tau[\to u_t$ of smooth tensor fields on $M\big(u_t \in C^\infty(E)\big)$ which satisfies the equation*

$$\begin{cases} \dfrac{\partial u}{\partial t} = a^{ij}(t,x,u,\nabla u)\nabla_i\nabla_j u + f(t,x,u,\nabla u) \\ u_0(x) = \varphi(x) \end{cases} \tag{71}$$

$\varphi(x)$ is given and belongs to $C^\infty(E)$, the components a^{ij} of a doubly contravariant symmetric tensor field on M are, in a local chart, smooth functions of their arguments, and f, with values in E, is smooth in its arguments. If equation (71) *is strictly parabolic at φ (the tensor field $a^{ij}(0,x,\varphi,\nabla\varphi)$ is everywhere positive definite), then there exists a unique smooth solution u on $[0,s[$ for some $s \le \tau$.*

We leave the proof to the reader. Friedman [*130] and Eidel'man [*125] deal with the local solvability of the Cauchy problem for arbitrary nonlinear parabolic systems. In [*231] Malliavin solves the Cauchy problem for linear parabolic equations on a vector bundle E when M is compact. See also Dieudonné [*113] when the coefficients of the linear equation on E do not depend on t.

§5. The Methods

4.52 In these sections, we will mention the methods used in the book for solving elliptic equations. Then we will deal with some other methods.

The *variational method*, the *continuity method* and the *method of lower and upper solutions* are studied in detail throughout this book.

The *method of successive approximations*, probably the oldest method, was used by Vaugon (see 5.15–5.17) to prove the basic theorem on the Yamabe Problem. For the Leray-Schauder fixed point theorem (3.20), see Gilbarg-Trudinger [*143] p. 228.

The *steepest descent* or the *gradient-line technique* was used by Eells-Sampson (see 10.10) to prove the existence of harmonic maps. Bahri-Coron (see 5.79) studied the gradient lines to see if they go through the critical level of the functional or if they go to infinity. The steepest descent was used by Gaveau-Mazet [*137] and by Inouë [*181] to prove the basic theorem on the Yamabe Problem.

4.53 We continue with the method which consists, instead of solving the elliptic equation directly, in studying the corresponding *parabolic equation*. Examples: Eells-Sampson (see 10.10), Hamilton (see 9.15).

The *method of points of concentration* (see 6.53). It was used by Dong [*118] to prove the basic theorem on the Yamabe Problem. The points of concentration were introduced many years ago see for instance P. L Lions [*222]. Here we use the approach developped by Vaugon [*309] ten years ago. But only recently we discovered all the possibilities of this technique.

When a group of isometries acts, Hebey (see 6.36) developed the method of *isometry-concentration*.

The *method of B-B-C* (Bahri-Brezis-Coron see 2.65–2.67): under a topological assumption, they prove that the equation cannot have no solution. There is a contradiction beetwen some algebraic-topological arguments and the study of the problem by Analysis. Bahri solved with this method the Yamabe Problem for the compact locally conformally flat manifolds.

4.54 The Mountain Pass Lemma (Ambrosetti and Rabinowitz [*2]). *Let f be a C^1 real function defined on a Banach space B and satisfying* (PS). *Assume there is an open neighbourhood $\bar{\Omega}$ of 0, and a point $x_0 \notin \Omega$, such that $f(0)$ and $f(x_0)$ are strictly less than $C_0 \leq \inf_{x\in\partial\Omega} f(x)$. Then the following number C is a critical value of f,*

$$C = \inf_{P\in\mathcal{P}} \sup_{x\in P} f(x) \geq C_0 \tag{72}$$

where $\mathcal{P}$ is the set of all continuous path P from 0 to x_0.

(PS) means Palais-Smale condition: *Any sequence $\{x_j\} \subset B$ such that $|f(x_j)| \leq M$ and $f'(x_j) \to 0$ strongly in B^* (the dual space) has a strongly convergent subsequence.*

In [*258] Nirenberg introduced the condition $(PS)_C$ *any sequence*$\{x_j\} \subset B$ *such that* $f(x_j) \to C$ *and* $f'(x_j) \to 0$ *strongly in* B^* *has a strongly convergent subsequence.* The mountain pass lemma holds under $(PS)_C$ instead of (PS). For details on the*Minimax methods* see Ni [*255], Nirenberg [*258] and Rabinowitz [*271].

4.55 The Leray Schauder Degree D. Consider a real Banach space B, θ an open set of B and a map $F: \theta \to B$ of a special form $F = I - K$ with K compact (see 3.19) and I the identity map. We consider the triplets (F, Ω, y) with Ω a bounded open set of B such that $\Omega \subset \theta$ and $y \in B$, $y \notin F(\partial\Omega)$, (here $\partial\Omega = \bar{\Omega} - \Omega$). To such a triplet (F, Ω, y) there corresponds an integer $D(F, \Omega, y)$.

The $\mathbb{Z}$-valued function D having these three basic following properties is unique

(i) $D(I, \Omega, y) = 1$ for $y \in \Omega$,
(ii) $D(F, \Omega, y) = D(F, \Omega_1, y) + D(F, \Omega_2, y)$ whenever Ω_1 and Ω_2 are disjoint, $\Omega = \Omega_1 \cup \Omega_2$ and y $\notin F(\bar{\Omega} - \Omega_1 \cup \Omega_2)$.
(iii) Let $t \to F_t$ be a continuous family of maps for $t \in [0, 1]$ of the form defined above and $t \to y_t \in B$ be continuous with $y_t \notin F_t(\partial\Omega)$ on [0,1]. Then $D(F_t, \Omega, y_t)$ is independent of $t \in [0, 1]$.

Moreover the Leray-Schauder degree has the following properties.

(73) (iv) If $D(F, \Omega, y) \neq 0$, then $F(x) = y$ has at least a solution.
(v) $D(F_1, \Omega, y) = D(F_2, \Omega, y)$ whenever $F_1/_{\partial\Omega} = F_2/_{\partial\Omega}$.
(vi) $D(F, \Omega, y) = D(F, \tilde{\Omega}, y)$ for every open subset $\tilde{\Omega}$ of Ω such that $y \notin F(\bar{\Omega} - \tilde{\Omega})$.
(vii) Suppose that a solution x of $F(x) = y$ is a regular point of F (F' is a homeomorphism). Then the local degree (index) of F at x is defined as

$$\text{ind}(x) = D(F, B_x(\varepsilon), y),$$

where $B_x(\varepsilon)$ is the ball in B with radius ε and center x.

It is independent of ε for ε small, and equals +1 or -1 according to whether the sum of the algebraic multiplicities of the negative eigenvalues of $F'(x)$ is even or odd.

For the existence of D, see for instance Leray-Schauder [180], Nirenberg [*257] and Rabinowitz [*270].

The Leray-Schauder degree was used by Chang, Gursky and Yang to prove their result on the Nirenberg problem in dimension 3 (see 6.88).

4.56 Bifurcation Theory. We will mention only one theorem, when there is *bifurcation from a simple eigenvalue.* For other results, see for instance Smoller [*293].

Let $I \subset \mathbb{R}$ be an open interval with $\lambda_0 \in I$ and B_1, B_2 be two Banach spaces, $\Omega \subset B_1$ being an open subset. We consider $f \in C^2(I \times \Omega, B_2)$ satisfying $f(\lambda, 0) \equiv 0$ when $\lambda \in I$.

Hence, for any $\lambda \in I$, $f(\lambda, x) = 0$ has a solution $x = 0$; the problem is to exhibit non trivial solutions of $f(\lambda, x) = 0$ if there are any.

A necessary condition for having non trivial solutions in a neighbourhood of $(\lambda_0, 0)$ is that $L_0 \equiv D_x f(\lambda_0, 0)$, the differential at $x = 0$ of $x \to f(\lambda_0, x)$, is not invertible according to the implicit function theorem (see 3.10). But this condition is not sufficient.

Theorem 4.56. *If* (i) Ker L_0 *is one-dimensional, spanned by* u_0,

(ii) $R(L_0)$ *the range of* L_0 *has codimension 1,*

(iii) $\big[D_\lambda D_x f(\lambda_0, 0)\big](u_0) \notin R(L_0)$*, then* $(\lambda_0, 0)$ *is a simple bifurcation point for* f*. More precisely, let* Z *be any closed complementary subspace of* u_0 *in* B_1, $(B_1 = Z \otimes \operatorname{Ker} L_0)$*, then there is a* $\delta > 0$ *and a* C^1- *curve* $]-\delta, \delta[\ni s \to \big(\lambda(s), \varphi(s)\big) \in \mathbb{R} \times Z$ *such that* $\lambda(0) = 0$, $\varphi(0) = 0$ *and* $f\big[\lambda(s), s\big(u_0 + \varphi(s)\big)\big] = 0$ *for* $|s| < \delta$*. Furthermore, there is a neighbourhood of* $(\lambda_0, 0)$ *such that any zero of* f *either lies on this curve or is of the form* $(\lambda, 0)$.

This Theorem was used by Vazquez-Veron [*312] to solve the problem of prescribing the scalar curvature in the negative case (see 6.12).

4.57 The method of Moving Planes. This method uses the maximum principle in an essential way. To understand how the method works, let us give the original proof of the

Theorem 5.57 (Gidas-Ni-Nirenberg [140]). *Let* $\Omega \subset \mathbb{R}^n$ *by a bounded open set symmetric about* $x^1 = 0$*, convex in the* x^1 *direction and with smooth boundary* $\partial\Omega$*. Suppose* $u \in C^2(\bar{\Omega})$ *is a positive solution of* $\Delta u = f(x, u)$ *in* Ω *satisfying* $u = 0$ *on* $\partial\Omega$.

Assume f *and* $\partial_u f$ *are continuous on* $\bar{\Omega}$*, and* f *is symmetric in* x^1 *with* f *decreasing in* x^1 *for* $x^1 > 0$*. Then* u *is symmetric in* x^1 *and* $\partial_1 u < 0$ *for* $x^1 > 0$.

Proof. Set $\lambda_0 = \max_{x \in \bar{\Omega}} x^1$ and let $x_0 \in \partial\Omega$ with $x_0^1 = \lambda_0$.

Since $u > 0$ in Ω and $u(x_0) = 0$, $\partial_1 u(x_0) \leq 0$. First we prove that for $x \in \Omega$ close to x_0, $\partial_1 u(x) < 0$. If $\partial_1 u(x_0) < 0$, this is obvious by continuity. If $\partial_1 u(x_0) = 0$, the proof is by contradiction. Assume there is a sequence $\{x_j\} \subset \Omega$ converging to x_0 such that $\partial_1 u(x_j) \geq 0$. Consequently $\partial_{11} u(x_0) = 0$ and hence $\Delta u(x_0) = 0$ (since $u = 0$ on $\partial\Omega$). Thus we must have $f(x_0, 0) = 0$. In that case $u > 0$ in Ω satisfies an equation of the type $\Delta u + h(x)u \geq 0$ for some function $h(x)$. By the version 1.43 of the maximum principle $\partial_1 u(x_0) < 0$, thus the contradiction.

Now we start with *the method of moving planes.*

We denote by T_λ the plane $x^1 = \lambda$. For $\lambda < \lambda_0$, λ close to λ_0, we consider the cap $\Sigma(\lambda) = \{x \in \Omega/\lambda < x^1 < \lambda_0\}$, the set of the points in Ω between T_λ and T_{λ_0}.

For any x in Ω, we use x_λ to denote its reflexion in the plane T_λ. When $\lambda \geq 0$, x_λ is defined on $\Sigma(\lambda)$ since Ω is convex in the x^1 direction and symmetric about $x^1 = 0$.

At the beginning, when λ decreases from λ_0, since $u(x)$ is strictly decreasing for x close to x_0, $w_\lambda(x) = u(x_\lambda) - u(x) > 0$ in $\Sigma(\lambda)$.

For $x \in \partial\Sigma(\lambda)$ with $x^1 > \lambda$, $w_\lambda(x) > 0$ and, for $x \in T_\lambda \cap \partial\Sigma(\lambda)$, $w_\lambda(x) = 0$ and $\partial_1 w_\lambda(x) > 0$.

Decrease λ until a critical value μ is reached, beyond which this result no longer holds: at a point $y \in T_\mu \cap \Omega$, $\partial_1 w(y) = 0$ (we drop the subscript μ in w_μ). But w satisfies in $\Sigma(\mu)$, when $\mu \geq 0$

$$\Delta w = f\big(x_\mu, u(x_\mu)\big) - f\big(x, u(x)\big) \geq f\big(x, u(x_\mu)\big) - f\big(x, u(x)\big).$$

We can write this inequality in the form

$$\Delta w \geq h(x)w. \tag{74}$$

Moreover w satisfies $w \geq 0$ in $\Sigma(\mu)$. Thus, according to Proposition 4.61, $w \equiv 0$ in $\Sigma(\mu)$ since $w(y) = 0$ and $\partial_1 w(y) = 0$. The result follows and $\mu = 0$.

We must have $\mu \geq 0$, since otherwise we start with the reflexions from $\lambda = \lambda_1 = \inf_{x\in\bar{\Omega}} x^1$ and we increase λ.

4.58 Corollary (Gidas-Ni-Nirenberg [140]). *In the ball Ω: $|x| < R$ in $\mathbb{R}^n$, let $u \in C^2(\bar{\Omega})$ be a positive solution in Ω of*

$$\Delta u = f(u) \quad \textit{with } u = 0 \textit{ on } \partial\Omega. \tag{75}$$

f is supposed to be C^1. Then u is radially symmetric and $\frac{\partial u}{\partial r} < 0$ for $0 < r < R$.

If $f'(u) \leq 0$, the Maximum Principle (see 3.71) implies that the solution u is unique. Thus u is radially symmetric, otherwise by rotations, we would get a family of solutions. In any case the result is a consequence of Theorem 4.57. We apply it for all directions. Since the paper [*140], the maximum principle was improved; it holds in narrow domains (sec Proposition 4.60), thus the hypotheses of Theorem 4.57 and Corollary 4.58 may be weakned.

Theorem 4.58 (Berestycki-Nirenberg [*39]). *Let Ω be an arbitrary bounded domain in $\mathbb{R}^n$ which is convex in the x^1 direction and symmetric with respect to the plane $x^1 = 0$. Let u be a positive solution of (75) belonging to $C(\bar{\Omega}) \cap H^n_{2\,\mathrm{loc}}$. f is supposed to be Lipschitz continuous. Then u is symmetric with respect to x^1 and $\partial_1 u < 0$ for $x^1 > 0$ in Ω.*

Proof. We can start at once with the method of moving planes. Since f is Lipschitz, w_λ satisfies (74) in a narrow band $\Sigma(\lambda)$ when λ is close to λ_0.

Moreover $w_\lambda \geq 0$ on $\partial\Sigma(\lambda)$, thus $w_\lambda > 0$ in $\Sigma(\lambda)$ (according to Proposition 4.60) and, on $T_\lambda \cap \Omega$ where $w_\lambda = 0$, we must have $\partial_1 w_\lambda > 0$, otherwise the function vanishes.

4.59 *The method of moving planes* may be used also for unbounded domains. To start with the process, we need an assumption on the asymptotic expansion of u near infinity.

Using this method in [*69], Caffarelli and Spruck proved uniform estimates for solutions of some elliptic equations.

In [*39], Berestycki and Nirenberg use with the method of moving planes a new one, *the sliding method* introduced by them. They compare translations of the function.

4.60 The Maximum Principle (see 3.71). It concerns second order elliptic operators A in a bounded domain $\Omega \subset \mathbb{R}^n$. Let $g_{ij}(x)$ be a Riemannian metric on Ω and $\xi(x)$ be a vector field on Ω. Set

(76) $$Au = g^{ij}\partial_{ij}u + \xi^i(x)\partial_i u + h(x)u.$$

A is supposed to be uniformly elliptic $\left(a^{-1}|\eta|^2 \leq g^{ij}\eta_i\eta_j \leq a|\eta|^2\right)$, and its coefficients to be bounded by b in Ω.

The maximum principle holds for A in Ω, if

$$Au \geq 0 \quad \textit{in} \quad \Omega \quad \textit{and} \quad \limsup_{x\to\partial\Omega} u(x) \leq 0$$

imply $u(x) \leq 0$ in Ω.

The usual condition for this to hold is $h(x) \leq 0$ (see 3.71).

Proposition 4.60. *The maximum principle holds if there exists a positive function $f \in H_2^\infty(\Omega) \cap C^0(\bar{\Omega})$ satisfying $Af \leq 0$, or if Ω lies in a narrow band $\alpha < x^1 < \alpha + \varepsilon$ with ε small, or (Bakelman-Varadhan) if the measure $|\Omega|$ is small enough $\left(|\Omega| < \delta\right)$More precisely, assume diam $\Omega \leq d$, there exists $\delta > 0$ depends only on n, d, a and b.*

Proof.

(i) Consider $v = \mathrm{u}\mathrm{f}^{-1}$, v satisifies

$$g^{ij}\partial_{ij}v + \left(\xi^i + 2\nabla^i log f\right)\partial_i v + vf^{-1}A(f) \geq 0.$$

Since $\limsup_{x\to\partial\Omega} v(x) \leq 0$, the usual maximum principle implies $v \leq 0$ in Ω,thus the same is true for u.

(ii) If Ω lies in a narrow band, we construct a function f as above.

(iii) We use the following theorem of Alexandroff, Bakelman and Pucci [*268] (see 4.42). *If $h(x) \leq 0$ and if u satisfies $Au \geq f$ and $\limsup_{x \to \partial\Omega} u(x) \leq 0$, then $\sup_{x \in \Omega} u(x) \leq C\|f\|_n$ where C depends only on n, d, a and b. u satisfies*

$$\left[A - h^+(x)\right]u \geq -h^+(x)u^+.$$

Thus

$$\sup_\Omega u^+ \leq C\left(\sup_\Omega h^+\right)\left(\sup_\Omega u^+\right)|\Omega|^{1/n}.$$

Choose $\delta = \left(Cb\right)^{-n}$, then $u \leq 0$.

4.61 The Maximum Principle (Second part). *Suppose there is a ball B in Ω with a point $P \in \partial\Omega \cap \partial B$ and suppose u is continuous at P and $u(P) = 0$. If $u \not\equiv 0$ in Ω and if u admits an outward normal derivative at P, then $\frac{\partial u}{\partial \nu}(P) > 0$. More generally, if Q approaches P in B along lines, then $\liminf_{Q \to P} \frac{u(P)-u(Q)}{|P-Q|} > 0$ otherwise $u \equiv 0$ in Ω. This holds for $u \in C^2(\Omega)$ satisfying (76) if $h(x) \leq 0$.*

Proposition 4.61 (Gidas-Ni-Nirenberg [*140]). *If $u \in C^2(\Omega)$, $u \leq 0$ satisfies $Au \geq 0$, the maximum principle holds. That is, if u vanishes at some point in Ω, or if u vanishes at some point $P \in \partial\Omega$ with $\frac{\partial u}{\partial \nu}(P) = 0$, then $u \equiv 0$ in Ω.*

Proof. Set $\tilde{A} = A - h^+$. $u \leq 0$ satisfies $\tilde{A}u \geq -h^+u \geq 0$. Since $-h^- \leq 0$, the usual maximum principle holds.

§6. The Best Constant

4.62 Theorem (Aubin [13], [17]). *Let (V_n, g) be a complete Riemannian manifold with positive injectivity radius and bounded sectional curvature, $n \geq 2$ the dimension. Let q be a real number satisfying $1 \leq q < n$: then for all $\varepsilon > 0$ there exists a constant $A_\varepsilon(q)$ such that any function φ belonging to the Sobolev space $H_1^q(V_n)$ satisfies*

$$\|\varphi\|_p \leq \left[K(n,q) + \varepsilon\right]\|\nabla\varphi\|_q + A_\varepsilon(q)\|\varphi\|_q \tag{77}$$

with $1/p = 1/q - 1/n$. The best constant $K(n,q)$ depends only on n and q, its value is in 2.14.

Remark. Recently, tanks to sharp estimates on the harmonic radius obtained by Anderson and Cheeger [5], Hebey [165A] was able to prove that Theorem 4.62 still holds if one replaces the bound on the sectional curvature by a lower bound on the Ricci curvature.

We can ask the question: does $A_\varepsilon(q)$ tend to ∞ when $\varepsilon \to 0$?

In [13] we made the conjecture that the best constant $K(n,q)$ is achieved $\left(A_0(q)\right)$ exists. The conjecture is proved when $n = 2$ and when $n \geq 3$ if the manifold has constant sectional curvature.

This result is obtained by choosing a nice partition of unity and by using the isoperimetric inequality (which holds when the curvature is constant).

Later Hebey and Vaugon extended this result to the locally conformally flat manifolds by a similar argument. But recently by new methods they proved

4.63 Theorem (Hebey and Vaugon [*171], [*172]). *For any complete Riemannian manifold with positive injectvity radius, bounded sectional curvature, bounded covariant derivative of the curvature tensor and dimension $n \geq 3$, the best constant $K(n,2)$ is achieved.*

The statement of Hebey and Vaugon is more precise. We still sketch the proof when the manifold (M,g) is C^∞ compact, because it is very interesting and a good illustration of new technics; but before to read it, the reader must see chapter 6 (Note that the assumption of Theorem 4.63 are obviously satisfied by compact manifolds).

Assume Proposition 4.64 below. Let (Ω_i, ψ_i) $(i = 1,2,\dots,m)$ be a finite atlas such that $\psi_i(\Omega_i) = B$ the unit closed ball of $\mathbb{R}^n$. We can choose the atlas such that B is convex for $\left(\psi_i^{-1}\right)^* g$ $(1 \leq i \leq m)$, since any point has a convex neighbourhood.

Let us consider $\{\eta_i\}$ a C^∞ partition of unity subordinated to the covering Ω_i such that $\sqrt{\eta_i}$ and $|\nabla\sqrt{\eta_i}|$ belong to $C^0(\bar{\Omega}_i)$. Setting $u_i = \sqrt{\eta_i}u$ for some $u \in C^\infty(M)$, we have by using (79)

$$\|u\|_N^2 = \|u^2\|_{N/2} \leq \sum_{i=1}^m \|u_i^2\|_{N/2} = \sum_{i=1}^m \|u_i\|_N^2$$

$$\leq K^2 \sum_{i=1}^m \|\nabla u_i\|_2^2 + C \sum_{i=1}^m \|u_i\|_2^2.$$

Since $\sum_{i=1}^m \int |\nabla u_i|^2\,dV = \int |\nabla u|^2\,dV + \sum_{i=1}^m \int u^2 |\nabla \eta_i|^2\,dV$ (indeed the additional term $\frac{1}{2}\sum_{i=1}^m \int \nabla^j \eta_i^2 \nabla_j u^2\,dV = 0$), there is a constant $\tilde{C}$ such that any $u \in H_1(M)$ satisfies

$$\|u\|_N^2 \leq K^2 \|\nabla u\|_2^2 + \tilde{C}\|u\|_2^2 \tag{78}$$

$K = K(n,2)$ is achieved.

When the manifold has constant curvature, we use the isoperimetric inequality to prove Proposition 4.64, but in the general case the proof is harder.

4.64 Proposition. *Let $B \subset \mathbb{R}^n$ be the closed ball of radius 1, and g be a C^∞ Riemannian metric on a neighbourhood of B such that B is convex for g. Then there exists a constant C such that any $\varphi \in \overset{\circ}{H}_1(B)$ satisfies (the norms are taken with the metric g):*

$$\|\varphi\|_N^2 \leq K^2 \|\nabla \varphi\|_2^2 + C\|\varphi\|_2^2. \tag{79}$$

The proof is by contradiction. For any $\alpha \geq 1$, we suppose that there exists $u_\alpha \in \overset{\circ}{H}_1(B)$ satisfying $\|u_\alpha\|_N^2 > K^2(\|\nabla u_\alpha\|_2^2 + \alpha\|u_\alpha\|_2^2)$. Thus

$$\lambda_\alpha = \inf_{u \in \overset{\circ}{H}_1(B)} [\|\nabla u\|_2^2 + \alpha\|u\|_2^2]\,\|u\|_N^{-2} < K^{-2} = \frac{n(n-2)\omega_n^{2/n}}{4}.$$

Since $\lambda_\alpha < K^{-2}$, the minimum is achieved. The proof is that of the basic theorem 5.11 on the Yamabe problem. As a consequence, there exists $\varphi_\alpha \in \overset{\circ}{H}_1(B)$, with $\|\varphi_\alpha\|_N = 1$, which satisfies the equation

(80) $$\Delta\varphi_\alpha + \alpha\varphi_\alpha = \lambda_\alpha\varphi_\alpha^{N-1} \quad \text{and} \quad \varphi_\alpha > 0 \text{ in B.}$$

Hence $\varphi_\alpha \in C^\infty(\bar{B})$, $\varphi_\alpha/\partial B = 0$ and

(81) $$\|\nabla\varphi_\alpha\|_2^2 + \alpha\|\varphi_\alpha\|_2^2 = \lambda_\alpha < K^{-2}.$$

Therefore $\lim_{\alpha\to\infty} \|\varphi_\alpha\|_2 = 0$ and there exists a sequence $q_i \to \infty$ such that $\varphi_{q_i} \to 0$ a.e.. By interpolation

(82) $$\lim_{\alpha\to\infty} \|\varphi_\alpha\|_p \to 0 \quad \text{for } 2 \leq p < N.$$

Lemma 4.64. *There exists a sequence $\{q_i\}$ such that $\{\varphi_{q_i}\}$ has a unique simple point of concentration.*

Moreover $\lambda_{q_i} \to K^{-2}$ and $q_i\|\varphi_{q_i}\|_2^2 \to 0$ when $q_i \to \infty$.

According to Theorem 6.53, there is only one point of concentration. Indeed here $f(P) = 1$ and $\mu/\mu_s = \lambda_{q_i}K^2 \leq 1$. Moreover since the energy of a point of concentration is at least K^{-2} (see 6.52, formula 64), x_0 is a simple point of concentration. Consequently $\lambda_{q_i} \to K^{-2}$ and $q_i\|\varphi_{q_i}\|_2^2 \to 0$. Remark that

(83) $$\lim_{q_i\to\infty} \int_{B_P(\delta)} \varphi_{q_i}^N \, dV = 1.$$

since P is the unique point of concentration.

4.65 *Proof of Proposition* 4.64 (continued).

For convenience set $u_i = \left(\frac{\lambda_{q_i}}{n(n-2)}\right)^{(n-2)/4} \varphi_{q_i}$. u_i satisfies

(84) $$\Delta u_i + q_i u_i = n(n-2)u_i^{N-1}.$$

Denote by x_i a point where u_i is maximum, $u_i(x_i) = m_i = \sup_B u_i \to \infty$, $x_i \to x_0$ and $u_i \to 0$ uniformly on any compact set $K \subset B - \{x_0\}$.

Let $\mu_i = (m_i)^{-2/(n-2)}$.

Now we study the speed of convergence of x_i to ∂B (if any).

α) $\liminf_{i\to\infty} \frac{d(x_i,\partial B)}{\mu_i} = 0$.

This implies $x_0 \in \partial B$. After passing to a subsequence if necessary we can suppose that the limit exists.

We do a blow-up at x_0. Define the maps ψ_i of $\mathbb{R}^n$ in $\mathbb{R}^n$: $y \to \psi_i(y) = \mu_i y + x_0$; $B_i = \psi_i^{-1}(B) = B_{-x_0/\mu_i}\left(\frac{1}{\mu_i}\right)$ and

$$v_i(y) = \frac{u_i(\mu_i y + x_0)}{m_i} = \mu_i^{(n-2)/2} u_i(\mu_i y + x_0).$$

Let us consider the metrics $h_i = \mu_i^{-2}\psi_i^* g$. On B_i, v_i satisfies $0 \le v_i \le 1$ and

(85) $$\Delta_{h_i} v_i + q_i \mu_i^2 v_i = n(n-2) v_i^{N-1}.$$

On any compact set of $\mathbb{R}^n$, $h_i \to \mathcal{E}$ uniformly in C^2 (we are able to do so that $g(x_0) = \mathcal{E}(x_0)$).

A similar proof of that of Corollary 8.36 of Gilbarg-Trudinger [143] shows that the v_i are uniformly bounded in C^1 on a neighourhood of $0 \in B_i$.

But hypothesis α) implies $\lim_{i\to\infty} d\big(\psi_i^{-1}(x_i), \partial B_i\big) = 0$ which is in contradiction with $v_i\big(\psi_i^{-1}(x_i)\big) = 1$ and $v_i = 0$ on ∂B_i. So α) is impossible.

β) $\liminf_{i\to\infty} \frac{d(x_i, \partial B)}{\mu_i} = l > 0$.

This implies also $x_0 \in \partial B$. As previously we suppose that the limit exists. Since $O(n)$ acts on B, we can suppose without loss of generality that all points x_i and x_0 are on the same ray (the $n-1$ first components of x_i and x_0 are zero).

We denote by g_i the metric corresponding to g after the action of the element of $O(n)$.

Now we do the same blow-up as previously (in α).

Then $\cup_{i=1}^{\infty} \mathring{B}_i$ is the half space

$$E = \{(y_1, y_2, \dots, y_n) \in \mathbb{R}^n / y_n < 0\}.$$

Since the sequence $\{v_i\}$ is equicontinuous, a subsequence converges uniformly on any compact set $K \subset E$ to a function v which satisfies

(86) $$\Delta_{\mathcal{E}} v = n(n-2) v^{(n+2)/(n-2)} \quad \text{in } E$$

$v/\partial E = 0$ and $v(z) = 1$ where $z = (0, 0, \dots, -l) \in E$, $\left(y_i = \frac{x_i - x_0}{\mu_i} \to z\right)$.

Indeed $q_i \mu_i^2 \to 0$. Let $K \subset E$ be a compact set. When i is large enough

$$q_i \mu_i^2 \int_K v^2 \, dV_{\mathcal{E}} \le 2 q_i \mu_i^2 \int_K v_i^2 \, dV_{h_i} \le 2 q_i \mu_i^2 \int_{B_i} v_i^2 \, dV_{h_i}$$

$$= 2 q_i \mu^{2-n} \int_{B_i} v_i^2 \, dV_{h_i} (\mu_i^2)^{n/2} = 2 q_i \int_B u_i^2 \, dV_{g_i} \to 0$$

according to Lemma 4.64.

Now such a positive function v cannot exist by Pohozahev's identity. First by the inverse of a stereographic projection, we get a function on a half sphere of pole Q. Then a stereographic projection of pole $\tilde{Q}$ (opposite to Q) yields a function $\tilde{v}$ satisfying equation (86) on a ball $\tilde{B}$ and also $\tilde{v}(0) = 1$ and $\tilde{v}/_{\partial \tilde{B}} = 0$. So β) is impossible.

γ) $\lim_{i\to\infty} \frac{d(x_i,\partial B)}{\mu_i} = +\infty$. In this case x_0 may be on ∂B or inside B. We do a blow-up at x_i. We define the maps ψ_i by $\Omega_i \ni y \overset{\psi_i}{\to} \exp_{x_i}(\mu_i y)$ with $\Omega_i = \psi_i^{-1}(B)$.
ψ_i is well defined since Ω_i is star-shaped according to the convexity of B for g.

We consider on Ω_i the metric $h_i = \mu_i^{-2}\psi_i^* g$.

$h_i(0) = \mathcal{E}(0)$ according to the properties of the exponential mapping. $h_i \to \mathcal{E}$ uniformly in C^2 on every compact set K. The function v_i on Ω_i defined by

$$v_i(y) = \frac{u_i\big(\psi_i(y)\big)}{m_i} = \mu_i^{(n-2)/2} u_i\big(\psi_i(y)\big)$$

satisfies

$$0 \le v_i \le 1, \quad v_i(0) = 1, \quad v_i/\partial\Omega_i = 0$$

and

$$\Delta_{h_i} v_i + q_i \mu_i^2 v_i = n(n-2) v_i^{(n+2)/(n-2)} \quad \text{on } \Omega_i. \tag{87}$$

The sequence $\{v_i\}$ is uniformly bounded in C_1 on every compact set K. A subsequence converges, uniformly on any K, to a function v satisfying $0 \le v \le 1$, $v(0) = 1$ and equation (86) on $\mathbb{R}^n$, since $q_i\mu_i^2 \to 0$ (see β). We know such function v, $v = \big(1 + \|y\|^2\big)^{1-n/2}$. In order to exclude the third case γ), and to establish a contradiction to the existence for any $\alpha \ge 1$ of a function φ_α satisfying (80), we need to use the Pohozahev identity

$$\begin{aligned} &\int_{\Omega_i} \nabla^k r^2 \nabla_k v_i \Delta_{\mathcal{E}} v_i \, dV \\ &\quad = -\frac{1}{2}\int_{\partial\Omega_i} \nabla^k r^2 \nabla_k v_i \partial_\nu v_i \, d\sigma - (n-2)\int_{\Omega_i} v_i \Delta_{\mathcal{E}} v_i \, dV. \end{aligned} \tag{88}$$

In (88) the metric is the euclidean metric, ∂_ν is the outside normal derivative and $r = \|y\|$.

Since Ω_i is star-shaped,

$$\begin{aligned} X &= \int_{\Omega_i} \nabla^k r^2 \nabla_k v_i \Delta_{\mathcal{E}} v_i \, dV_{\mathcal{E}} + (n-2)\int_{\Omega_i} v_i \Delta_{\mathcal{E}} v_i \, dV_{\mathcal{E}} \\ &= -\frac{1}{2}\int_{\partial\Omega_i} \partial_\nu r^2 |\partial_\nu v_i|^2 \, d\sigma \le 0. \end{aligned}$$

Define

$$\begin{aligned} Y &= \sum_{k=1}^{n} \int_{\Omega_i} \partial_k r^2 \partial_k v_i \Delta_{h_i} v_i \, dV_{\mathcal{E}} + (n-2)\int_{\Omega_i} v_i \Delta_{h_i} v_i \, dV_{\mathcal{E}} \\ &= X + Y - X \le Y - X. \end{aligned} \tag{89}$$

Using (87), we get $Y = 2q_i\mu_i^2 \int_{\Omega_i} v_i^2\, dV_\mathcal{E}$.

Now some computations lead to

$$(90) \qquad Y - X \leq C_1\mu_i^2\left[\int_{\Omega_i} v_i^2\, dV_\mathcal{E} + \int_{\Omega_i} r^2 v_i^N\, dV_\mathcal{E}\right]$$

for some constant C_1 independant of i. Moreover there exists a constant C_2 such that for any i, $v_i \leq C_2 v$.

This result is proved by different authors; it holds when the point of concentration is simple. According to the value of Y, (89) and (90) yield

$$(91) \qquad (2q_i - C_1)\int_{\Omega_i} v_i^2\, dV_\mathcal{E} \leq C_1 C_2^N \int_{\Omega_i} \frac{r^2}{(1+r^2)^n}\, dV_\mathcal{E} < \text{Const.}.$$

But $v_i \to v$ uniformly on any K, hence for i large enough

$$\int_{\Omega_i} v_i^2\, dV_\mathcal{E} \geq \frac{1}{2}\int_K v^2\, dV_\mathcal{E} > 0.$$

So we get a contradiction: v_i cannot exist for i large enough.

Chapter 5

The Yamabe Problem

Yamabe wanted to solve the Poincaré conjecture (see 9.14). For this he thought, as a first step, to exhibit a metric with constant scalar curvature. He considered conformal metrics (the simplest change of metric is a conformal one), and gave a proof of the following statement "On a compact Riemannian manifold (M, g), there exists a metric g' conformal to g, such that the corresponding scalar curvature R' is constant". The Yamabe problem was born, since there is a gap in Yamabe's proof. Now there are many proofs of this statement. We will consider some of them, but if the reader wants to see one proof, he has to read only sections 5.11, 5.21, 5.29 and 5.30.

§1. The Yamabe Problem

5.1 Let us recall the question.

Let (M_n, g) be a compact C^∞ Riemannian manifold of dimension $n \geq 3$, R its scalar curvature. The problem is:

Does there exist a metric g', conformal to g, such that the scalar curvature R' of the metric g' is constant?

In fact Yamabe [269] said that such metric always exists, but there is a gap in his proof which is impossible to overcome in general. He said that the set $\{\varphi_q\}(2 < q < N)$ is uniformly bounded (see 5.14) without proof in the positive case.

Among other things, in this chapter we give a positive answer to the problem above (see Theorems 5.11, 5.21, 5.29 and 5.30).

5.2 The differential equation. Let us consider the conformal metric $g' = e^f g$ with $f \in C^\infty$. By 1.19, if Γ'^l_{ij} and Γ^l_{ij} denote the Christoffel symbols relating to g' and g respectively:

$$\begin{aligned}\Gamma'^l_{ij} - \Gamma^l_{ij} &= \tfrac{1}{2}[g_{kj}\,\partial_i f + g_{ki}\,\partial_j f - g_{ij}\,\partial_k f]g^{kl}\\ &= \tfrac{1}{2}[\delta^l_j\,\partial_i f + \delta^l_i\,\partial_j f - g_{ij}\nabla^l f].\end{aligned}$$

According to 1.13,

$$R'_{ij} = R'^{k}_{ikj} = R_{ij} - \frac{n-2}{2}\nabla_{ij}f + \frac{n-2}{4}\nabla_i f \nabla_j f$$
$$- \frac{1}{2}\left(\nabla^\nu_\nu f + \frac{n-2}{2}\nabla^\nu f \nabla_\nu f\right) g_{ij}$$

so

$$R' = e^{-f}\left[R - (n-1)\nabla^\nu_\nu f - \frac{(n-1)(n-2)}{4}\nabla^\nu f \nabla_\nu f\right].$$

If we consider the conformal deformation in the form $g' = \varphi^{4/(n-2)}g$ (with $\varphi \in C^\infty$, $\varphi > 0$), the scalar curvature R' satisfies the equation:

(1) $4((n-1)/(n-2))\Delta\varphi + R\varphi = R'\varphi^{(n+2)/(n-2)}$, where $\Delta\varphi = -\nabla^\nu\nabla_\nu\varphi$.

So the Yamabe problem is equivalent to solving equation (1) with $R' = \text{Const}$, and the solution φ must be smooth and strictly positive.

5.3 On a C^∞ compact Riemannian manifold M_n of dimension $n \geq 3$, let us consider the differential equation

(2) $$\Delta\varphi + h(x)\varphi = \lambda f(x)\varphi^{N-1},$$

where $h(x)$ and $f(x)$ are C^∞ functions on M_n, with $f(x)$ everywhere strictly positive and $N = 2n/(n-2)$.

The problem is to prove the existence of a real number λ and of a C^∞ function φ, everywhere strictly positive, satisfying (1).

1.1. Yamabe's Method

5.4 Yamabe considered, for $2 < q \leq N$, the functional

(3) $$I_q(\varphi) = \left[\int_M \nabla^i\varphi\nabla_i\varphi \, dV + \int_M h(x)\varphi^2 \, dV\right]\left[\int_M f(x)\varphi^q \, dV\right]^{-2/q},$$

where $\varphi \not\equiv 0$ is a nonnegative function belonging to H_1, the first Sobolev space. The denominator of $I_q(\varphi)$ makes sense since, according to Theorem 2.21, $\mathsf{H}_1 \subset L_N \subset L_q$. Define

$$\mu_q = \inf I_q(\varphi) \quad \text{for all } \varphi \in \mathsf{H}_1,\ \varphi \geq 0,\ \varphi \equiv 0.$$

It is impossible to prove directly that μ_N is attained and thus to solve Equation (2). (We shall soon see why.) This is the reason why Yamabe considered the approximate equations for $q < N$:

(4) $$\Delta\varphi + h(x)\varphi = \lambda f(x)\varphi^{q-1}$$

and proved (Theorem B of Yamabe [269]):

5.5 Theorem. *For $2 < q < N$, there exists a C^∞ strictly positive function φ_q satisfying Equation* (4) *with $\lambda = \mu_q$ and $I_q(\varphi_q) = \mu_q$.*

Proof.

a) For $2 < q \leq N$, μ_q is finite. Indeed

$$I_q(\varphi) \geq \left[\inf_{x\in M}(0, h(x))\right]\left[\sup_{x\in M} f(x)\right]^{-2/q} \|\varphi\|_2^2\|\varphi\|_q^{-2} \tag{5}$$

and

$$\|\varphi\|_2^2\|\varphi\|_q^{-2} \leq V^{1-2/q} \leq \sup(1, V^{2/n}), \tag{6}$$

with $V = \int_M dV$. On the other hand,

$$\mu_q \leq I_q(1) = \left[\int_M h(x)\,dV\right]\left[\int_M f(x)\,dV\right]^{-2/q}.$$

(b) Let $\{\varphi_i\}$ be a minimizing sequence such that $\int_M f(x)\varphi_i^q\,dV = 1$:

$$\varphi_i \in \mathrm{H}_1, \varphi_i \geq 0, \lim_{i\to\infty} I_q(\varphi_i) = \mu_q.$$

First we prove that the set of the φ_i is bounded in H_1,

$$\|\varphi_i\|_{\mathrm{H}_1}^2 = \|\nabla\varphi_i\|_2^2 + \|\varphi_i\|_2^2 = I_q(\varphi_i) - \int_M h(x)\varphi_i^2\,dV + \|\varphi_i\|_2^2.$$

Since we can suppose that $I_q(\varphi_i) < \mu_q + 1$, then

$$\|\varphi_i\|_{\mathrm{H}_1}^2 \leq \mu_q + 1 + \left[1 + \sup_{x\in M}|h(x)|\right]\|\varphi_i\|_2^2$$

and

$$\|\varphi_i\|_2^2 \leq [V]^{1-2/q}\|\varphi_i\|_q^2 \leq [V]^{1-2/q}\left[\inf_{x\in M} f(x)\right]^{-2/q}$$

c) If $2 < q < N$, there exists a nonnegative function $\varphi_q \in \mathrm{H}_1$, satisfying

$$I_q(\varphi_q) = \mu_q \quad \text{and} \quad \int_M f(x)\varphi_q^q\,dV = 1.$$

Indeed, for $2 < q < N$, the imbedding $\mathrm{H}_1 \subset L_q$ is compact by Kondrakov's theorem 2.34 and, since the bounded closed sets in H_1 are weakly compact (Theorem 3.18), there exists $\{\varphi_j\}$ a subsequence of $\{\varphi_i\}$, and a function $\varphi_q \in \mathrm{H}_1$ such that:

(α) $\varphi_j \to \varphi_q$ in L_q,
(β) $\varphi_j \to \varphi_q$ weakly in H_1.
(γ) $\varphi_j \to \varphi_q$ almost everywhere.

The last assertion is true by Proposition 3.43. (α) $\Longrightarrow \int_M f(x)\varphi_q^q\,dV = 1$; γ) $\Longrightarrow \varphi_q \geq 0$, and β) implies

$$\|\varphi_q\|_{\mathsf{H}_1} \leq \liminf_{i\to\infty} \|\varphi_j\|_{\mathsf{H}_1} \qquad \text{(Theorem 3.17).}$$

Hence $I_q(\varphi_q) \leq \lim_{j\to\infty} I_q(\varphi_j) = \mu_q$ because $\varphi_j \to \varphi_q$ in L_2, according to (α) since $q \geq 2$. Therefore, by definition of μ_q, $I_q(\varphi_q) = \mu_q$.

d) φ_q satisfies Equation (4) weakly in H_1. We compute Euler's equation. Set $\varphi = \varphi_q + \nu\psi$ with $\psi \in \mathsf{H}_1$ and ν a small real number. An asymptotic expansion gives:

$$I_q(\varphi) = I_q(\varphi_q)\left[1 + \nu q \int_M f(x)\varphi_q^{q-1}\psi\,dV\right]^{-2/q}$$
$$+ 2\nu\left[\int_M \nabla^\nu\varphi_q\nabla_\nu\psi\,dV + \int_M h(x)\varphi_q\psi\,dV\right] + 0(\nu).$$

Thus φ_q satisfies for all $\psi \in \mathsf{H}_1$:

$$(7) \qquad \int_M \nabla^\nu\varphi_q\nabla_\nu\psi\,dV + \int_M h(x)\varphi_q\psi\,dV = \mu_q \int_M f(x)\varphi_q^{q-1}\psi\,dV.$$

To check that the preceding computation is correct, we note that since $\mathscr{D}(M)$ is dense in H_1 and $\varphi \not\equiv 0$, then

$$\inf_{\varphi\in\mathsf{H}_1} I_q(\varphi) = \inf_{\varphi\in C^\infty} I_q(\varphi) = \inf_{\varphi\in C^\infty} I_q(|\varphi|) \geq \inf_{\substack{\varphi\in\mathsf{H}_1\\ \varphi\geq 0}} I_q(\varphi) \geq \inf_{\varphi\in\mathsf{H}_1} I_q(\varphi).$$

$I(\varphi) = I(|\varphi|)$ when $\varphi \in C^\infty$ because the set of the point P where simultaneously $\varphi(P) = 0$ and $|\nabla\varphi(P)| \neq 0$ has zero measure (or we can use Proposition 3.49 directly).

e) $\varphi_q \in C^\infty$ *for* $2 \leq q < N$ *and the functions* φ_q *are uniformly bounded for* $2 \leq q \leq q_0 < N$.

Let $\mathsf{G}(P, Q)$ be the Green's function (see 4.13). φ_q satisfies the integral equation (see 4.14)

$$(8) \qquad \varphi_q(P) = V^{-1}\int_M \varphi_q(Q)\,dV(Q)$$
$$+ \int_M \mathsf{G}(P, Q)[\mu_q f(Q)\varphi_q^{q-1} - h(Q)\varphi_q]\,dV(Q).$$

We know that $\varphi_q \in L_{r_0}$ with $r_0 = N$. Since, by Theorem 4.13c there exists a constant B such that $|\mathsf{G}(P, Q)| \leq B[d(P, Q)]^{2-n}$, then according to Sobolev's lemma 2.12 and its corollary, $\varphi_q \in L_{r_1}$, for $2 < q \leq q_0$ with

$$\frac{1}{r_1} = \frac{n-2}{n} + \frac{q_0 - 1}{r_0} - 1 = \frac{q_0 - 1}{r_0} - \frac{2}{n}$$

and there exists a constant A_1 such that $\|\varphi_q\|_{r_1} \leq A_1\|\varphi_q\|_{r_0}^{q-1}$.

By induction we see that $\varphi_q \in L_{r_k}$ with

$$\frac{1}{r_k} = \frac{q_0-1}{r_{k-1}} - \frac{2}{n} = \frac{(q_0-1)^k}{r_0} - \frac{2}{n}\frac{(q_0-1)^k-1}{q_0-2}$$

and there exists a constant A_k such that $\|\varphi_q\|_{r_k} \leq A_k \|\varphi_q\|_{r_0}^{(q-1)^k}$.
If for k large enough, $1/r_k$ is negative, then $\varphi_q \in L_\infty$. Indeed, suppose $1/r_{k-1} > 0$ and $1/r_k < 0$. Then $(q_0-1)/r_{k-1} - 2/n < 0$, and Hölder's inequality 3.62 applied to (8) yields $\|\varphi_q\|_\infty \leq \text{Const} \times \|\varphi_q\|_{r_{k-1}}^{q-1}$.
There exists a k such that

$$\frac{1}{r_k} = (q_0-1)^k\left[\frac{1}{r_0} - \frac{2}{n(q_0-2)}\right] + \frac{2}{n(q_0-2)} < 0$$

because $n(q_0-2) < 2r_0 = 2N$, since $q_0 < N = 2n/(n-2)$.
Moreover, there exists a constant A_k, which does not depend on $q \leq q_0$, such that:

$$\|\varphi_q\|_\infty \leq A_k \|\varphi_q\|_N^{(q-1)^k}.$$

But the set of the functions φ_q is bounded in H_1 (same proof as in b)).

Thus by the Sobolev imbedding theorem 2.21 the functions φ_q are uniformly bounded. Since $\varphi_q \in L_\infty$, differentiating (8) yields $\varphi_q \in C^1$. φ_q satisfies (4); thus $\Delta\varphi_q$ belongs to C^1 and $\varphi_q \in C^2$ according to Theorem 3.54.

f) φ_q is strictly positive. This is true because $\int_M f(x)\varphi_q^q\, dV = 1$, Proposition 3.75 establishes this result since φ_q cannot be identically zero. Lastly $\varphi_q \in C^\infty$ by induction according to Theorem 3.54. ∎

5.6 Remark. The proof of Theorem 5.5 does not work for $q = N$. The problem is that if $q = N$, we cannot apply Krondrakov's theorem in c), and therefore only have

$$\int_M f(x)\varphi_N^N\, dV \leq 1.$$

Moreover, the method in e) yields nothing when $q_0 = N$. In this case $r_k = r_0 = N$ for all k. Indeed, we shall see below that if $q = N$ then Equation (4) may not have a positive solution (see Theorem 6.67).

5.7 Remark. Using the same method, one can study equations of the type $\Delta\varphi + h(x)\varphi = \lambda f(\varphi, x)$, where $f(t,x)$, a C^∞ function on $\mathbb{R} \times M$, satisfies some conditions. In particular, $|f(t,x)| \leq \text{Const} \times (1+|t|^{q_0})$ with $q_0 < N$ (see Berger [39]) or with $q_0 \leq N$ (see Aubin [16]). The idea is to consider the variational problem:

$$\text{inf of } \int_m \nabla^i\varphi\nabla_i\varphi\, dV + \int_M h(x)\varphi^2\, dV \quad \text{when} \quad \int_M F(\varphi, x)\, dV = \text{Const},$$

where $F(t,x) = \int_0^t f(u,x)\, du$.

1.2. Yamabe's Functional

5.8 To solve Equation (1), Yamabe used the variational method. He considered the functional, for $2 \leq q \leq N = 2n/(n-2)$,

$$J_q(\varphi) = \left[4\frac{n-1}{n-2}\int_M \nabla^i\varphi\nabla_i\varphi\, dV + \int_M R(x)\varphi^2\, dV\right] \|\varphi\|_q^{-2} \tag{9}$$

and defined $\mu_q = \inf J_q(\varphi)$ for all $\varphi \geq 0$, $\varphi \not\equiv 0$, belonging to H_1. Set $\mu = \mu_N$ and $J(\varphi) = J_N(\varphi)$.

Proposition. *μ is a conformal invariant.*

Proof. Consider a change of conformal metric defined by $g' = \varphi^{4/(n-2)}g$. We have $dV' = \varphi^N\, dV$ and

$$J(\varphi\psi) = \frac{4\frac{n-1}{n-2}\left[\int_M \varphi^2\nabla^\nu\psi\nabla_\nu\psi\, dV + \int_M \varphi\psi^2\Delta\varphi\, dV\right] + \int_M R\varphi^2\psi^2\, dV}{\left[\int_M \varphi^N\psi^N\, dV\right]^{2/N}}.$$

Using (1) yields $J(\varphi\psi) = J'(\psi)$ and consequently $\mu = \mu'$. J' is the functional related to g', and μ' the inf of J'. ∎

By a homothetic change of metric we can set the volume equal to one. So henceforth, without loss of generality, we suppose the volume equal to one.

1.3. Yamabe's Theorem

5.9 In his article [269], Yamabe proved Theorem 5.5 and then he claimed that the C^∞ strictly positive functions φ_q, $q \in]2, N[$ satisfying

$$4\frac{n-1}{n-2}\Delta\varphi_q + R\varphi_q = \mu_q\varphi_q^{q-1} \quad \text{and} \quad \|\varphi_q\|_q = 1 \tag{10}$$

are uniformly bounded.
But this does not hold in general. The functions $\varphi(r)$ of Theorem 5.58 are not uniformly bounded on the sphere. This counterexample shows that Yamabe's proof is wrong. Indeed (p. 35 of Yamabe [269]), the inequality $\|v^{(q)}\|_{q_n} \leq \text{Const} \times \|v^{(q)}\|_{q_1}$ must be replaced by $\|v^{(q)}\|_{q_n} \leq \text{Const} \times \|v^{(q)}\|_{q_1}^{(q-1)^{n-1}}$ and this does not yield the result.
In the negative or zero case, ($\mu \leq 0$), it is easy to overcome the mistake. But in the positive case ($\mu > 0$), it is impossible. Yamabe did not solve his problem but he proved the following.

Theorem (See Aubin [11] p. 386). *Let M_n be a C^∞ compact Riemannian manifold; there exists a conformal metric whose scalar curvature is either a nonpositive constant or is everywhere positive.*

Proof. α) the positive case ($\mu > 0$). If $\mu_{q_0} > 0$, μ_q is positive for $q \in]2, N[$. Indeed

$$\mu_q = J_q(\varphi_q) = J_{q_0}(\varphi_q)\|\varphi_q\|^2_{q_0}\|\varphi_q\|_q^{-2} \geq \mu_{q_0}\|\varphi_q\|^2_{q_0}.$$

Then consider the conformal metric $g' = \varphi_{q_0}^{4/(n-2)} g$; applying (1) and (10) leads to

$$R'(x) = \mu_{q_0}\varphi_{q_0}^{q_0-N}(x), \tag{11}$$

the scalar curvature R' is everywhere strictly positive. Moreover we can prove that $\mu > 0$. Indeed the functional J' corresponding to g' satisfies

$$J'(\psi) \geq \inf_{x\in M}\left[4\frac{n-1}{n-2}, R'(x)\right]\left[\int_M g'ij\nabla_i\psi\nabla_j\psi\, dV' + \int_M \psi^2\, dV'\right] \times \left[\int \psi^N\, dV'\right]^{-2/N}.$$

According to the Sobolev imbedding theorem, $J'(\psi) \geq \text{Const} > 0$ for all $\psi \in \mathsf{H}_1$. Thus $\mu' > 0$ and we have $\mu = \mu'$ (Proposition 5.8).

β) *The null case* ($\mu = 0$). If $\mu_{q_0} = 0$, by (11) the scalar curvature R' vanishes, and $\mu_q = 0$ for all $q \in]2, N]$, because for all ψ and q, $J_q(\psi) \geq 0$.

γ) *The negative case* ($\mu < 0$). If $\mu_{q_0} < 0$ there exists a $\psi \in C^\infty$ such that $J_{q_0}(\psi) < 0$. Hence $J_q(\psi) < 0$ for all $q \in [2, N]$ and $\mu_q < 0$. In particular, $\mu < 0$. Moreover, $\mu_q \leq J_q(\psi) = J(\psi)\|\psi\|_N^2\|\psi\|_q^{-2} \leq J(\psi)$. Thus $\mu_q (q \in [2, N])$ is bounded away from zero.
Now we are able to prove very simply that the functions $\varphi_q (q \in]q_0, N[)$ are uniformly bounded with $q_0 \in]2, N[$. At a point P where φ_q is maximum $\Delta\varphi_q \geq 0$, hence $\mu_q\varphi_q^{q-1}(P) \geq R(P)\varphi_q(P)$. We find at once that $\varphi_q^{q-2} \leq |\inf R|\,|J(\psi)|^{-1}$ and $\varphi_q \leq 1 + [|\inf R|\,|J(\psi)|^{-1}]^{1/(q_0-2)}$. By (10), φ_q satisfies:

$$\varphi_q(P) = \int_M \varphi_q(Q)\, dV(Q) + \int_M \mathsf{G}(P, Q)\frac{n-2}{4(n-1)}[\mu_q\varphi_q^{q-1}(Q) - R(Q)\varphi_q(Q)]\, dV(Q). \tag{12}$$

Differentiating (12) yields $\varphi_q \in C^1$ uniformly, and according to Ascoli's theorem 3.15, it is possible to exhibit a sequence φ_{q_i} with $q_i \to N$, such that φ_{q_i} converges uniformly to a nonnegative function φ_N.
But $0 > \mu_q \geq \inf R(x)\|\varphi_q\|_2^2 \geq \inf R(x)$. Therefore a subsequence μ_{q_i} converges to a real number ν (in fact μ_q is a continuous function of q for $q \in]2, N]$ by Proposition 5.10, so $\mu = \nu$).
Letting $q_i \to N$ in (12), shows that φ_N is a weak solution of

$$4\frac{n-1}{n-2}\Delta\varphi_N + R\varphi_N = \nu\varphi_N^{N-1}. \tag{13}$$

Since $\|\varphi_q\|_q = 1$, $\|\varphi_N\|_N = 1$. Multiplying (13) by φ_N and integrating yield $J(\varphi_N) = \nu$.

The second term in (13) is continuous; thus, by (12), $\varphi_N \in C^1$. Now apply the regularity theorem 3.54: the second member of (13) is C^1; thus $\varphi_N \in C^2$. Now according to Proposition 3.75 φ_N is strictly positive everywhere, since $\|\varphi_N\|_N = 1$ implies $\varphi_N \not\equiv 0$. We can use the regularity theorem again to prove by induction that $\varphi_N \in C^\infty$. Thus the C^∞ function $\varphi_N > 0$ satisfies (1) with $R' = \text{Const}$ (in fact, $R' = \mu$).
In the negative case it is therefore possible to make the scalar curvature constant and negative. ∎

5.10 Proposition. *μ_q is a continuous function of q for $q \in]2, N]$, which is either everywhere positive, everywhere zero, or everywhere negative. Moreover, $|\mu_q|$ is decreasing in q if we suppose the volume equal to* 1.

Proof. If the volume is equal to 1 for $\psi \in C^\infty$, $\|\psi\|_q$ is an increasing function of q. Thus $|J_q(\psi)| \le |J_p(\psi)|$ when $p < q$ and this implies $|\mu_p| \ge |\mu_q|$ since C^∞ functions are dense in H_1.
Moreover, $J_q(\psi)$ is a continuous function of q. It follows that μ_q is an upper semicontinuous function of q. Indeed, for all $\varepsilon > 0$, there exists $\psi \in C^\infty$ such that $J_p(\psi) < \mu_p + \varepsilon$ and since $\mu_q \le J_q(\psi)$, $\lim_{q\to p} J_q(\psi) = J_p(\psi)$ yields $\limsup_{q\to p} \mu_q \le \mu_p + \varepsilon$.
Let q_i be a sequence converging to $p \in]2, N]$.
In the negative case, we saw, 5.9γ, that the functions φ_q are uniformly bounded. Therefore $\|\varphi_{q_i}\|_p \to 1$ and as $\mu_p \le J_p(\varphi_{q_i}) = \mu_{q_i}\|\varphi_{q_i}\|_p^{-2}$, $\liminf_{q\to p} \mu_q \ge \mu_p$. This establishes the continuity of $q \to \mu_q$ in the negative case. Similarly we can prove that this function is continuous on $]2, N[$ in the positive case, because if $q_0 < N$, the functions φ_q are uniformly bounded for $q \in]2, q_0]$ by (5.5e). Finally $\mu_q \to \mu_N$ when $q \to N$ because the function $q \to \mu_q$ is upper semi-continuous and decreasing in the positive case. If the volume is not one, we consider a homothetic change of metric such that the volume in the new metric is equal to one.

§2. The Positive Case

5.11. Definition. Recall $\mu = \inf J(\varphi)$ for all $\varphi \in H_1$, $\varphi \not\equiv 0$, $J(\varphi)$ being the Yamabe functional.

We have the basic theorem:

Theorem 5.11 (Aubin 1976 [14]). $\mu \le n(n-1)\omega_n^{2/n}$. *If $\mu < n(n-1)\omega_n^{2/n}$, there exists a strictly positive solution $\tilde\varphi \in C^\infty$ of* (1) *with $\tilde R = \mu$ and $\|\tilde\varphi\|_N = 1$.*

Here $\tilde R$ is the scalar curvature of $(M_n, \tilde g)$ with $\tilde g = \tilde\varphi^{4/(n-2)} g$ and ω_n is the volume of the unit sphere of radius 1 *and dimension n.*

We will give below (α) to ϵ)) the proof of this Theorem. Then, to solve the Yamabe problem, we have only to exhibit a test function ψ such that $J(\psi) < n(n-1)\omega_n^{2/n}$. All subsequent work to date has centered on the discovery of appropriate test functions, except for Bahri's results obtained by algebraic-topology methods. Bahri exhibits a solution, which is not in general a minimizer of the Yamabe functional.

Conjecture (Aubin 1976 [14] p. 294). *μ satisfies $\mu < n(n-1)\omega_n^{2/n}$ if the compact Riemannian manifold (of dimension $n \geq 3$) is not conformal to (S_n, g_0).*

According to Theorems 5.21, 5.29 and 5.30, this conjecture is proved. The consequence of this conjecture is that the Yamabe Problem is proved.

Proof. α) Recall that $\mathsf{K}(n,2) = 2(\omega_n)^{-1/n}[n(n-2)]^{-1/2}$ is the *best constant* in the Sobolev inequality (Theorem 2.14). By theorem (2.21), the best constant is the same for all compact manifolds. Thus there exists a sequence of C^∞ functions ψ_i such that

$$\|\psi_i\|_N = 1, \ \|\psi_i\|_2 \to 0 \quad \text{and} \quad \|\nabla\psi_i\|_2 \to K^{-1}(n,2),$$

when $i \to +\infty$. Therefore $J(\psi_i) \to n(n-1)\omega_n^{2/n}$ and $\mu \leq n(n-1)\omega_n^{2/n}$.

β) Let us again consider the set of functions φ_q ($q \in]2, N[$) which are solutions of (10). This set is bounded in H_1 since we have $\|\varphi_q\|_2 \leq 1$ and

$$4\frac{n-1}{n-2}\|\nabla\varphi_q\|_2^2 \leq \mu_q + \sup|R| \leq \int R\,dV + \sup|R|.$$

Therefore there exists $\varphi_0 \in \mathsf{H}_1$ and a sequence $q_i \to N$ such that $\varphi_{q_i} \to \varphi_0$ weakly in H_1 (the unit ball in H_1 is weakly compact), strongly in L_2 (Kondrakov's theorem) and almost everywhere (Proposition 3.43). The weak limit in H_1 is the same as that in L_2 because H_1 is continuously imbedded in L_2, and strong convergence implies weak convergence.

γ) Since φ_{q_i} satisfies (10), then for all $\psi \in \mathsf{H}_1$:

$$4\frac{n-1}{n-2}\int \nabla^\nu\psi\nabla_\nu\varphi_{q_i}\,dV + \int R\psi\varphi_{q_i}\,dV = \mu_{q_i}\int \psi\varphi_{q_i}^{q_i-1}\,dV.$$

Letting $q_i \to N$ gives us

$$(14) \qquad 4\frac{n-1}{n-2}\int \nabla^\nu\psi\nabla_\nu\varphi_0\,dV + \int R\psi\varphi_0\,dV = \mu\int \psi\varphi_0^{N-1}\,dV.$$

Indeed, according to Theorem 3.45, $\varphi_{q_i}^{q_i-1}$ converges weakly to φ_0^{N-1} in $L_{N/(N-1)}$ since $\varphi_{q_i}^{q_i-1} \to \varphi_0^{N-1}$ almost everywhere and

$$\|\varphi_{q_i}^{q_i-1}\|_{N/(N-1)} = \|\varphi_{q_i}\|_{(q_i-1)N/(N-1)}^{q_i-1} \leq \|\varphi_{q_i}\|_N^{q_i-1} \leq \text{Const} \times \|\varphi_{q_i}\|_{H_1}^{q_i-1} \leq \text{Const}.$$

Therefore φ_0 satisfies (14) for all $\psi \in \mathrm{H}_1 \subset L_N$. According to Trudinger's theorem ([262] p. 271) $\varphi_0 \in C^\infty$ and satisfies (1) with $R' = \mu$.

δ) The problem is not solved yet because the maximum principle implies that either $\varphi_0 > 0$ everywhere or $\varphi_0 \equiv 0$, and for the moment we cannot exclude the latter case. In order to prove that φ_0 is not identically zero, we must use Theorem 2.21. We write, using (10),

$$(15) \quad 1 = \|\varphi_q\|_q^2 \leq \|\varphi_q\|_N^2 \leq [\mathrm{K}^2(n,2)+\varepsilon]\frac{n-2}{4(n-1)}\left[\mu_q - \int R\varphi_q^2\,dV\right] + A(\varepsilon)\|\varphi_q\|_2^2,$$

where $\varepsilon > 0$ is arbitrary and $A(\varepsilon)$ is a constant which depends on ε.
When $\mu < n(n-1)\omega_n^{2/n}$, if we choose ε small enough, there exist $\varepsilon_0 > 0$ and $\eta > 0$ such that for $N - q < \eta$,

$$(16) \qquad 0 < \varepsilon_0 \leq 1 - \left[\frac{\omega_n^{-2/n}}{n(n-1)} + \varepsilon\frac{n-2}{4(n-1)}\right]\mu_q$$

since $\mu_q \to \mu$ when $q \to N$.
In this case, (15) and (16) imply

$$\liminf_{q\to N} \|\varphi_q\|_2 \geq \text{Const} > 0.$$

Because φ_{q_i} converges strongly to φ_0 in L_2, $\|\varphi_0\|_2 \neq 0$. Thus $\varphi_0 \not\equiv 0$ and $\varphi_0 > 0$. Picking $\psi = \varphi_0$ in (14) gives $J(\varphi_0) = \mu\|\varphi_0\|_N^{N-2}$; thus $\|\varphi_0\|_N \geq 1$ since $J(\varphi_0) \geq \mu$. But since the sequence $\varphi_{q_i}^{q_i/N}$ of 5.11β converges weakly to φ_0 in L_N by Theorem 3.45, $\|\varphi_0\|_N \leq \liminf_{q_i\to N} \|\varphi_{q_i}\|_{q_i}^{q_i/N}$ by Theorem 3.17. Hence $\|\varphi_0\|_N = 1$ and $J(\varphi_0) = \mu$.
Moreover by Radon's theorem, 3.47, $\varphi_{q_i} \to \varphi_0$ strongly in H_1 because $\|\varphi_{q_i}\|_{\mathrm{H}_1} \to \|\varphi_0\|_{\mathrm{H}_1}$ since $\mu_{q_i} \to \mu$. Therefore by the Sobolev imbedding theorem $\varphi_{q_i} \to \varphi_0$ strongly in L_N.

ε) In fact, when $\mu < n(n-1)\,\omega_n^{2/n}$, it is possible to prove directly that the functions φ_q $q \in]2, N[$ are uniformly bounded and we can proceed as in the negative case, without using Trudinger's theorem. ■

5.12 More generally, let us consider the equation

$$(17) \qquad 4\frac{n-1}{n-2}\Delta\varphi + h(x)\varphi = \lambda f(x)\varphi^{N-1},$$

with $h \in C^\infty$, $f \in C^\infty$ given ($f > 0$), and λ (= 0, 1, or -1) to be determined as in 5.3. Let

$$I(\varphi) = \left[4\frac{n-1}{n-2}\int \nabla^i\varphi\nabla_i\varphi\,dV + \int h\varphi^2\,dV\right] \Big/ \left[\int f\varphi^N\,dV\right]^{-2/N}$$

and define $\nu = \inf I(\varphi)$ for all $\varphi \in \mathrm{H}_1$, $\varphi \not\equiv 0$. Using the same method one can prove:

Theorem. $\nu \leq n(n-1)\omega_n^{2/n}[\sup f]^{-2/N}$. *If* $\nu < n(n-1)\omega_n^{2/n}[\sup f]^{-2/N}$, *Equation* (10) *has a* C^∞ *strictly positive solution.*

5.13 Now we have to investigate when the inequalities of Theorems 5.11 and 5.12 are strictly satisfied. For this, consider the sequence of functions ψ_k ($k \in \mathbb{N}$):

$$\psi_k(Q) = \left(\frac{1}{k} + r^2\right)^{1-n/2} - \left(\frac{1}{k} + \delta^2\right)^{1-n/2}, \quad \text{for } r < \delta,$$

and $\psi_k(Q) = 0$ for $r \geq \delta$, with δ the injectivity radius, $r = d(P, Q)$, P fixed. A computation shows that $\lim_{k\to\infty} I(\psi_k) = n(n-1)\omega_n^{2/n}[f(P)]^{-2/N}$. Pick a point P where $f(P)$ has its maximum. In order to see if equality in Theorem 5.12 does not hold, we compute an asymptotic expansion.
If $n \geq 4$, the coefficient of the second term has the sign of

$$h(P) - R(P) + \frac{n-4}{2}\frac{\Delta f(P)}{f(P)}.$$

More precisely, the asymptotic expansion for $n > 4$ is

$$I(\psi_k) = n(n-1)\omega_n^{2/n}[f(P)]^{-2/N} \times \left\{1 + [n(n-4)k]^{-1}\left[h(P) - R(P) + \frac{n-4}{2}\frac{\Delta f(P)}{f(P)}\right]\right\} + o\left(\frac{1}{k}\right),$$

and for $n = 4$

$$I(\psi_k) = 12[\omega_4/f(P)]^{1/2}\left\{1 + [h(P) - R(P)]\frac{\operatorname{Log} k}{8k}\right\} + o\left(\frac{\operatorname{Log} k}{k}\right).$$

Proposition (Aubin [14] p. 286). *If, at a point* P *where* f *is maximum,* $h(P) - R(P) + ((n-4)/2)(\Delta f(P)/f(P)) < 0$, *Equation* (17) *has a* C^∞ *strictly positive solution when* $n \geq 4$.

5.14. Let us return to Yamabe's equation (1) with $R' =$ Const.. That is equation (17) with $f(P) = 1$ and $h(P) = R(P)$. We cannot apply Proposition 5.13. Hence Yamabe's equation is a limiting case in two ways: first with the exponent $(n+2)/(n-2)$ and second with the function R.

Since the original proof of Theorem 5.11 (in [14]) many new proofs of this theorem appeared which don't use the sequence φ_q of positive solutions of the approximate equations

$$4\big((n-1)/(n-2)\big)\,\Delta\varphi + R\varphi = \mu_q\varphi^{q-1} \qquad \text{with} \quad 2 < q < N.$$

Let us mention Inouë's proof [*181] (discussed also in Berger [*42]) where the steepest descent method is used and Vaugon's proof [*311] which is certainly the simplest and the fastest. This proof is an illustration of the method of successive approximations.

When $\mu \leq 0$ we can overcome the difficulty in Yamabe's proof. In the zero case, the functions φ_q are proportional , φ_q solves the Yamabe equation (1). In the negative case, the wrong term in Yamabe's proof may be removed in the inequalities (it has the good sign). So Yamabe's argument works: the functions φ_q are uniformly bounded.

In the positive case, when $\mu > 0$, the operator L= $\Delta + (n-2)R/4(n-1)$ has its first eigenvalue $\alpha > 0$. Any $\varphi \in C^\infty$ satisfies $\int \varphi L\varphi \, dV \geq \alpha \int \varphi^2 \, dV$. L is invertible with a Green function $G_L(P, Q) \geq m > 0$.

It is interesting to write up here Vaugon's proof of the following theorem which is more general than theorem 5.11.

5.15 Theorem. *Let h and f be C^∞ functions, $f > 0$ and L= $\Delta + h$ such that any $\varphi \in C^\infty$ satisfies: $\int \varphi L\varphi \, dV \geq \alpha \int \varphi^2 \, dV$ for some $\alpha > 0$. Set $\nu = \inf \int \varphi L\varphi \, dV$ for all $\varphi \in C^\infty$ such that $\int f|\varphi|^N \, dV = 1$.*

If $\nu < \nu_0 = n(n-2)\omega_n^{2/n}/4(\sup f)^{2/N}$ there exists a C^∞ strictly positive solution of the equation $L\varphi = f\varphi^{N-1}$.

Proof. Pick $\Psi_0 \in C^\infty$, $\Psi_0 > 0$ which satisfies $\int f\Psi_0^N \, dV = 1$ and $I(\Psi_0) < \nu_0$. We set

$$I(\Psi) = \int \Psi L\Psi \, dV = \int \nabla^\nu \Psi \nabla_\nu \Psi \, dV + \int h\Psi^2 \, dV.$$

Define the sequence $\{\Psi_j\}$ for $j \geq 1$ by

$$L\Psi_j = \lambda_j f|\Psi_{j-1}|^{N-1} \tag{18}$$

where the positive real numbers λ_j are fixed by the conditions $\int f|\Psi_j|^N \, dV = 1$.

If $L\Psi$ is a stricltly positive C^∞ function, Ψ is a strictly positive C^∞ function. Thus, as it is the case for $L\Psi_1$, by induction $\Psi_j \in C^\infty$ and $\Psi_j > 0$ for all j.

5.16 Lemma. *Set $I(\Psi) = \int \Psi L\Psi \, dV = \int |\nabla\Psi|^2 \, dV + \int h\Psi^2 \, dV$*

$$\lambda_{j+1} \leq I(\Psi_j) \leq \lambda_j \quad \textit{for all} \quad j \geq 1. \tag{19}$$

Indeed multiply (18) by Ψ_j and integrate, we get

$$I(\Psi_j) = \lambda_j \int f\Psi_{j-1}^{N-1}\Psi_j \, dV$$
$$\leq \lambda_j \left(\int f\Psi_{j-1}^N \, dV \right)^{1-1/N} \left(\int f\Psi_j^N \, dV \right)^{1/N} = \lambda_j$$

by the Hölder inequality used with volume element $f\,dV$.

Then multiply (18) by Ψ_{j-1}, integrating yields $\int \Psi_{j-1}L\Psi_j \, dV = \lambda_j$. But as $I(\Psi_j - \Psi_{j-1}) \geq 0$, $I(\Psi_j) + I(\Psi_{j-1}) \geq 2\int \Psi_{j-1}L\Psi_j \, dV = 2\lambda_j$. Thus $I(\Psi_{j-1}) \geq \lambda_j$.

5.17 *Proof of theorem* **5.15** (*continued*). The set $\{\Psi_j\}$ is bounded in H_1. Indeed by hypothesis $I(\Psi_j) \geq \alpha \int \Psi_j^2\,dV$, thus $\{\Psi_j\}$ is bounded in L_2, then in H_1 since by (19)

$$0 < I(\Psi_j) \leq \lambda_1 \qquad \text{for all} \quad j \geq 1.$$

According to the usual theorems (Banach's theorem, Kondrakov's theorem ...), there exists a subsequence $\{\Psi_{j_k}\}$ of $\{\Psi_j\}$ which converges weakly in H_1, strongly in L_2 and almost everywhere to a function $\tilde{\Psi} \in H_1$.

But for $j \geq 1$, $0 \leq I(\Psi_j - \Psi_{j-1}) \leq \lambda_j - 2\lambda_j + \lambda_{j-1} = \lambda_{j-1} - \lambda_j$ which goes to zero when $j \to \infty$ (the sequence λ_j is convergent, let λ its limit).

Therefore the sequence Ψ_{j_k-1} converges weakly in H_1, strongly in L_2 and almost everywhere to $\tilde{\Psi}$.

By (2) for all $\gamma \in H_1$

$$\int \nabla^i \Psi_{j_k} \nabla_i \gamma\,dV + \int h \Psi_{j_k} \gamma\,dV = \lambda_{j_k} \int f \Psi_{j_k-1}^{N-1} \gamma\,dV.$$

Letting $k \to \infty$ yields

$$\int \nabla^i \tilde{\Psi} \nabla_i \gamma\,dV + \int h \tilde{\Psi} \gamma\,dV = \lambda \int f \tilde{\Psi}^{N-1} \gamma\,dV.$$

By the Trudinger theorem of regularity [262], $\tilde{\Psi} \in C^\infty$, $\tilde{\Psi}$ satisfies $L\tilde{\Psi} = \lambda f \tilde{\Psi}^{N-1}$ and $\tilde{\Psi} \geq 0$. Now let us prove that $\tilde{\Psi} > 0$. By construction we have

$$\|\nabla \Psi_j\|_2^2 + \int h \Psi_j^2\,dV = I(\Psi_j) \leq I(\Psi_0) < \nu_0.$$

On the other hand by the Sobolev inequality

$$1 = \left(\int f \Psi_j^N\,dV\right)^{2/N} \leq (\sup f)^{2/N} \left(K^2(n,2) + \varepsilon\right) \|\nabla \Psi_j\|_2^2 + B_\varepsilon \|\Psi_j\|_2^2$$

where $\varepsilon > 0$ is chosen small enough so that $(\sup f)^{2/N}\left(K^2(n,2) + \varepsilon\right) I(\Psi_0) < 1$. This is possible since $I(\Psi_0) < \nu_0$, recall $K^{-2}(n,2) = n(n-2)\omega_n^{2/n}/4$.

We obtain $\lim_{j\to 0} \inf \|\Psi_j\|_2 > 0$. As $\Psi_{j_k} \to \tilde{\Psi}$ in L_2, $\|\tilde{\Psi}\|_2 \neq 0$ and $\tilde{\Psi} \not\equiv 0$. Then the maximum principal implies $\tilde{\Psi} > 0$.

§3. The First Results

5.18 In order to use theorem 5.11, the first idea is to choose $\Psi \equiv 1$ as test function in the Yamabe functional J.

$$J(1) = V^{-2/N} \int R\,dV \qquad \text{where } V = \int dV,$$

so we get the following

Proposition. *If $\int R\, dV \le n(n-1)\omega_n^{2/n} V^{2/N}$, there exists a conformal metric $\tilde{g}$ with constant scalar curvature $\tilde{R}$.*

When equality holds, two cases can happen:

α) $\mu < n(n-1)\omega_n^{2/n}$ and Theorem 5.11 may be applied.

β) $\mu = n(n-1)\omega_n^{2/n}$. In that case the function $\varphi \equiv 1$ minimize the functional J (φ), we have $R =$ Const.. In fact the manifold is the sphere.

5.19 To see if $\mu < n(n-1)\omega_n^{2/n}$, we can consider test functions Ψ in the Yamabe functional $\tilde{J}(\Psi)$ corresponding to a conformal metric $\tilde{g} = e^f g$, since μ is a conformal invariant (Proposition 5.8).

The components of the Ricci tensor of $\tilde{g}$ are

$$\tilde{R}_{ij} = R_{ij} - \frac{n-2}{2}\nabla_{ij} f + \frac{n-2}{4}\nabla_i f \nabla_j f + \frac{1}{2}\big(\Delta f - \frac{n-2}{2}|\nabla f|^2\big) g_{ij}.$$

At a point $P \in V$, if f satisfies $f(P) = |\nabla f(P)| = 0$ and

(20) $$\partial_{ij} f(P) = \Big[2R_{ij}(P) - R(P) g_{ij}/(n-1)\Big]/(n-2)$$

we have $\tilde{R}(P) = R(P) + (n-1)\Delta f(P) = 0$ and

(21) $$\tilde{R}_{ij}(P) = 0.$$

Moreover if we choose f such that

(22) $$\partial_{ijk} f(P) = 2\big[\nabla_k R_{ij}(P) + \nabla_i R_{kj}(P) + \nabla_j R_{ik}(P)\big]/3(n-2) - \big[\partial_k R(P) g_{ij} + \partial_i R(P) g_{jk} + \partial_j R(P) g_{ik}\big]/3(n-2)(n-1)$$

(we suppose that the coordinates are normal at P), we obtain after contraction $(\partial_k \Delta f)_P = -\partial_k R(P)/(n-1)$ according to the Bianchi identities.

Thus $|\nabla \tilde{R}(P)| = 0$ and we obtain

(23) $$\tilde{\nabla}_k \tilde{R}_{ij}(P) + \tilde{\nabla}_i \tilde{R}_{kj}(P) + \tilde{\nabla}_j \tilde{R}_{ki}(P) = 0 \qquad \text{for all } i, j, k.$$

Recall the following well known result [14], the beginning of the limited expansion of $\sqrt{|g|}$ in normal coordinates $\{x^i\}$:

$$\sqrt{|g|} = 1 - R_{ij}(P) x^i x^j/6 - \nabla_k R_{ij}(P) x^i x^j x^k/12 + O(r^4).$$

If (21) and (23) are satisfied we obtain

(24) $$\sqrt{|g|} = 1 + 0(r^4) \qquad \text{with } r = d(P, x).$$

Remark. By a suitable choice of the successive derivatives of f at P, it is possible to prove by induction (Lee and Parker Theorem 5.1 of [*208]) the existence of conformal normal coordinates at P:

5.20 Proposition (Lee and Parker [*208]). *For each $k \geq 2$ there is a conformal metric $\tilde{g}$ such that*

$$|\tilde{g}(x)| = 1 + 0(r^k) \qquad \textit{with } r = \tilde{d}(P, x). \tag{25}$$

Recently this result was improved by J. Cao [*73], then by M. Günther [*148]. They proved that, in a neighbourhood Ω of a given point P, there exists a conformal normal coordinate system such that the determinant is equal to 1 identically.

Suppose that, on Ω, $\bar{g} = \sigma g$ (σ a positive function) is such that $\bar{\gamma} = \sqrt{|\bar{g}|} = 1$ in a geodesic coordinate system $\{y^i\}$. Then u, the square of the geodesic distance to P for $\bar{g}$ $\left(u = \sum (y^i)^2\right)$, satisfies $\bar{g}^{ij} \frac{\partial u}{\partial x^i} \frac{\partial u}{\partial x^j} = 4u$ and $\bar{\Delta} u = -2n$, $\{x^i\}$ being a normal coordinate system for g). Written these two equations in the metric g, we obtain a system of two equations in the unknowns u and σ. We seek u and σ satisfying $u = r^2 + O(r^3)$ and $\sigma = 1 + O(r)$ when r, the geodesic distance to P for g, is small. J. Cao uses the Schauder fixed point theorem. As for M. Günther, he solves this system by the method of successive approximations; for that he considers the linearized equations of the system at $(r^2, 1, \delta^{ij})$.

5.21 Theorem (Aubin [14] p. 292). *If $(M_n, g)(n \geq 6)$ is a compact nonlocally conformally flat Riemannian manifold, then $\mu < n(n-1)\omega_n^{2/n}$. Hence the minimum is achieved and there exists a conformal metric g' with $R' = \mu V^{-2/n}$, V being the volume of the manifold (M_n, g').*

Proof. By hypothesis the Weyl tensor W_{ijkl} (see Definition 4.23) is not zero everywhere, there is a point $P \in M$ where $|W_{ijkl}(P)| \neq 0$. We consider a metric $\tilde{g} = e^f g$ with f satisfying (4), and we choose, as test functions for $\tilde{J}(\varphi)$, the following sequence of lipschitzian functions Ψ_k:

$$\begin{cases} \Psi_k(r) = 0 & \text{if } r = \tilde{d}(P, Q) \geq \delta > 0 \\ & \text{and} \\ \Psi_k(r) = (r^2 + \frac{1}{k})^{1-\frac{n}{2}} - (\delta^2 + \frac{1}{k})^{1-\frac{n}{2}} & \text{for } r \leq \delta, \end{cases} \tag{26}$$

where we pick δ smaller than the injectivity radius at P.

A limited expansion in k yields for $n > 6$

$$\tilde{J}(\Psi_k) = n(n-1)\omega_n^{2/n}\left[1 - k^{-2}a^2/(n-4)(n-6) + o(k^{-2})\right]$$

and

$$\tilde{J}(\Psi_k) = 30\omega_6^{1/3}\left[1 - a^2 k^{-2}(\log k)/80 + O(k^{-2})\right]$$

for $n = 6$ with $a^2 = |W_{ijkl}(P)|^2/12n$. Thus $\tilde{J}(\Psi_k) < n(n-1)\omega_n^{2/n}$ for k large enough.

5.22 Remarks. For any compact manifold $M_n (n \geq 3)$, $J(\Psi_k)$ tends to $n(n-1)\omega_n^{2/n}$ when $k \to \infty$. This implies the first part of Theorem 5.11.

In dimension 3 to 5, there are integrals on the manifold in the limited expansion of $J(\Psi_k)$ instead of a coefficient like a^2, and it is not possible to conclude a priori, but see 5.50.

For locally conformally flat manifolds, it is obvious that local test functions cannot work since for the sphere $\mu = n(n-1)\omega_n^{2/n}$ (Theorem 5.58).

5.23 Theorem ([14] p. 291). *For a compact locally conformally flat manifold M_n, ($n \geq 3$), which has a non trivial finite Poincaré's group, $\mu < n(n-1)\omega_n^{2/n}$.*

For the proof, we consider $\tilde{M}_n$ the universal covering of M_n. $\tilde{M}_n$ is compact, locally conformally flat and simply connected. Kuiper's theorem [172] then implies that $\tilde{M}_n$ is conformally equivalent to the sphere S_n. Hence Equation (1) has a solution with $R' = \mu$

5.24 Proposition. *When the minimum μ is achieved, let $J(\varphi_0) = \mu$. In the corresponding metric g_0 whose scalar curvature R_0 is constant, the first nonzero eigenvalue of the Laplacian $\lambda_1 \geq R_0/(n-1)$.*

For the proof one computes the second variation of $J(\varphi)$ (see Aubin [14] p. 292).

§4. The Remaining Cases

4.1. The Compact Locally Conformally Flat Manifolds

5.25 The effect of §4 is to prove the validity of conjecture 5.11. The results of the preceding paragraph do not concern the locally conformally flat manifolds with infinite fundamental group for which $V^{2/n-1}\int R\,dV > n(n-1)\omega_n^{2/n}$.

The known manifolds of this type are

- α) some products $\tilde{S}_{n-1} \times C$ and $\tilde{S}_p \times \tilde{H}_{n-p}$ where C is the circle and $\tilde{S}_q$ (resp. $\tilde{H}_q$) are compact manifolds of dimension q with constant sectional curvature $\rho > 0$ (resp. $-\rho < 0$).
- β) some fibre bundles with basis one of the manifolds with constant sectional curvature mentioned previously and for fibre $\tilde{S}_q$ or $\tilde{H}_q$ according to the situation.
- γ) the connected sums $V_1 \# V_2$ of two locally conformally flat manifolds (V_1, g_1), (V_2, g_2).

Most of these manifolds are endowed with a metric of constant scalar curvature by definition. But for them, according to the conjecture 5.11, the problem is to prove that the infimum of $J(\varphi)$ is achieved, and thus we shall prove

5.26 Theorem (Gil–Medrano [*142]). *The manifolds* α), β), *and* γ) *satisfy*

$$\mu < n(n-1)\omega_n^{2/n}.$$

Proof. It consists to exhibit a test function u such that $J(u) < n(n-1)\omega_n^{2/n}$.

By an homothetic change of metric, we can suppose that $\rho = 1$. Let Π be the projection

$$\tilde{S}_{n-1} \times C \to C.$$

On $\tilde{S}_{n-1} \times C$ the function u will be $u(P) = (\operatorname{ch} r)^{1-n/2}$ where r is the distance on C from $\Pi(P)$ to a fixed point $y_0 \in C$.

On $\tilde{S}_p \times \tilde{H}_{n-p}$, the same function $u(P)$ works, but here Π is the projection $\tilde{S}_p \times \tilde{H}_{n-p} \to \tilde{H}_{n-p}$ and r is the distance on $\tilde{H}_{n-p}$ from $\Pi(P)$ to a fixed point $y_0 \in \tilde{H}_{n-p}$. The proof is similar for the fibre bundles.

For the connected sums we have first to study the conformal class of the locally conformally flat metric g_0 constructed on the connected sum $V_0 = V_1 \# V_2$. Then Gil–Medrano proved that $\mu_0 \leq \inf(\mu_1, \mu_2)$, where $\mu_i (i = 0, 1, 2)$ is the μ of (V_i, g_i).

4.2. Schoen's Article [*280]

5.27 As μ is a conformal invariant, it is possible to do the computation of $J(\Psi)$, for some test function Ψ, in a particular conformal metric (as in 5.20). When the manifold is locally conformally flat, after a suitable change of conformal metric, the metric is flat in a ball B_δ of radius δ and center x_0. We saw above that locally test functions yield nothing for these manifolds. The idea of Schoen is to extend the test functions used in 5.21 by a multiple of the Green function G_L of the operator

$$L = \Delta + (n-2)R/4(n-1).$$

We are in the positive case ($\mu > 0$), L is invertible and $G_L > 0$. More precisely let $\rho < \delta/2$ and $r = d(x_0, x)$. For $\varepsilon > 0$ set

$$\Psi(x) = \begin{cases} (\varepsilon + r^2/\varepsilon)^{1-n/2} & \text{for } r \leq \rho, \\ \varepsilon_0\big[G(x) - h(x)\alpha(x)\big] & \text{for } \rho < r \leq 2\rho, \\ \varepsilon_0 G(x) & \text{for } r > 2\rho. \end{cases} \tag{27}$$

$G(x)$ is the multiple of $G_L(x_0, x)$ the expansion of which is the following in B_δ:

$$G(x) = r^{2-n} + \alpha(x) \tag{28}$$

where $\alpha(x)$ is an harmonic function in B_δ.

$h(x)$ is a C^∞ function of r which satisfies h(x)=1 for $r \leq \rho$, $h(x) = 0$ for $r > 2\rho$ and $|\nabla h| \leq 2/\rho$.

$\varepsilon_0 = (\rho^{2-n} + A)^{-1}(\varepsilon + \rho^2/\varepsilon)^{1-n/2}$ with A = $\alpha(x_0)$ in order the function Ψ is continuous hence lipschitzian, ρ will be chosen small, ε_0 infinitely small with respect to ρ, then ε is well defined and $\varepsilon \sim \varepsilon_0^{2/(n-2)}$ when $\varepsilon_0 \to 0$. Indeed the function $t \to \left[t + \rho^2/t\right]^{1-n/2}$ is increasing for t $\in]0, \rho]$ and goes from 0 to $(2\rho)^{1-n/2}$.

5.28 Proposition (Schoen [*280] 1984). *If $G(x)$ is of the form* (28) *for any $n \geq 3$ with $\alpha(x_0) = A > 0$ then $\mu < n(n-1)\omega_n^{2/n}$.*

The proof is easily understood. By an integration by parts, all the computations can be carried out in $B_{2\rho}$. They yield

$$J(\Psi) \leq n(n-1)\omega_n^{2/n} - CA\varepsilon_0^2 + O(\rho\varepsilon_0^2) \tag{29}$$

where $C > 0$ is a constant which depends on n. The result follows.

5.29 Theorem (Schoen and Yau [*289] 1988). *If (M_n, g) is a compact locally conformally flat Riemannian manifold of dimension $n \geq 3$ which is not conformal to (S_n, g_0), then $A > 0$. Hence conjecture* (5.11) *is valid and there exists a conformal metric $\tilde{g}$ with $\tilde{R} = \mu V^{-2/n}$, V being the volume of the manifold $(M_n, \tilde{g})$.*

The result follows from Proposition 5.28 combined with 5.37 for $n = 3$ and Theorem 5.48 for $n \geq 4$.

4.3. The Dimension 3, 4 and 5

5.30 Theorem (Schoen [*280]). *If (M_n, g) is any compact Riemannian manifold of dimension* 3 *to* 5, *which is not conformal to (S_n, g_0), then $\mu < n(n-1)\omega_n^{2/n}$. Hence according to theorem* 5.11, *there exists a conformal metric $\tilde{g}$ with $\tilde{R} = \mu V^{-2/n}$, V being the volume of the manifold $(M_n, \tilde{g})$.*

Proof. The result follows from Proposition 5.28 combined with the fact A > 0 to be established below §4. When $n = 3$, the Green function G_p of L at $P \in M$ has for limited expansion in a neighbourhood of P:

$$G_p(x) = \left[1/r + A + 0(r)\right]/4\pi$$

where A is a real number and $r = d(P, x)$.

This expression is the same as (28). So the method of 5.27 works. For the dimensions 4 and 5, Schoen [*280] replaces in a small ball $B_p(\rho)$ the metric g by a flat metric. He considers a C^∞ metric which is euclidean in $B_p(\rho)$ and equal to g outside the ball $B_p(2\rho)$. Thus he can use his method, but the approximation is too complicated. It is simpler to use the following fact which is one of the hypotheses of Proposition 5.28.

5.31 Proposition. *Let (M_n, g') be a compact Riemannian manifold of dimension 4 or 5, belonging to the positive case ($\mu > 0$). Pick $P \in M_n$, there exists a metric g conformal to g' such that the Green function G_p of L at P has, in a neighbourhood θ of P, the following limited expansion*

$$G_\rho(x) = (r^{2-n} + A)/(n-2)\omega_{n-1} + \alpha(x)$$

where A is a real number and $r = d(P,x)$. $\alpha(P) = 0$, $\alpha \in C^1$ for $n = 4$ and α is lipschitzian for $n = 5$.

With this proposition, the method of 5.27 works and Proposition 5.28 implies theorem 5.30.

Proof of 5.31. We consider a conformal metric g to g' which has at P the properties (21) and (23).

Thus (24) $\sqrt{|g(x)|} = 1 + 0(r^4)$ in normal coordinates. As in 4.10, consider $H(P,Q) = f(r)r^{2-n}/(n-2)\omega_{n-1}$ with $r = d(P,Q)$ and f a C^∞ function equal to 1 in a neighbourhood of zero and to zero for $r \geq \delta > 0$ (δ small enough). Recall 4.10, the singularity of $\Delta_Q H(P,Q)$ is given by $r^{1-n}\partial_r \operatorname{Log} \sqrt{|g|}/\omega_{n-1}$ which is in $0(r^{4-n})$:

$$\Delta_Q H(P,Q) = 0(r^{4-n}). \tag{30}$$

According to the Green formula (4.10), any $\varphi \in C^2$ satisfies

$$\varphi(P) = \int_V H(P,Q)L\varphi(Q)\,dV(Q) - \int_V L_Q H(P,Q)\varphi(Q)\,dV(Q).$$

Thus by induction

$$G_L(P,Q) = H(P,Q) + \sum_{i=1}^{k} \int_V \Gamma_i(P,R)H(R,Q)\,dV(R) + F_k(P,Q),$$

with $F_k(P,Q)$ continuous on $M \times M$ if $k \geq (n-1)/4$. Here $\Gamma_1(P,R) = -L_Q H(P,Q)$ and

$$\Gamma_{i+1}(P,Q) = \textstyle\int_M \Gamma_i(P,R)\Gamma_1(R,Q)\,dV(R).$$

Moreover

$$L_Q F_k(P,Q) = \Gamma_{k+1}(P,Q). \tag{31}$$

As $R(P) = 0$ and $|\nabla R(P)| = 0$ (see 2.8), $R(Q)H(P,Q) = 0(r^{4-n})$ thus

$$L_Q H(P,Q) = 0(r^{4-n}) \tag{32}$$

According to Giraud (4.12) this implies for $n \leq 5$ that $\Gamma_2(P,Q)$ is C^1, hence (31) yields $F_1(P,Q)$ is C^1 on $M \times M$.

Moreover $\int_V \Gamma_1(P,R)H(R,Q)\,dV(R)$ is a continuous function on $M \times M$. It is even C^1 when $n = 4$ and lipschitzian when $n = 5$, according to the following

5.32 Lemma. *The convolution product* $\frac{1}{r^{n-2}} * \frac{1}{r}$ *in a compact domain of* $\mathbb{R}^n (n \geq 3)$ *is lipschitzian.*

Proof. Let Q be a point of $B_P(1)$ and $r = d(P,Q)$ small. We have to compute

$$h(r) = \int_{B_P(1)} [d(P,R)]^{2-n} [d(R,Q)]^{-1} \, dV(R).$$

Set $y = d(P,R)$ and θ the angle at P of RPQ.

$$h(r) - h(0) = \omega_{n-2} \int_0^1 dy \int_0^\pi \sin^{n-2}\theta \big[y(r^2 + y^2 - 2ry\cos\theta) - 1\big]^{-1/2} \, d\theta.$$

Pick k a large integer and $k < 1/2r$. The absolute value of the integral on $B_p(kr)$ is smaller than Cr for some C (same proof as that of Giraud's theorem 4.12). With $y = rt$, the absolute value of the integral on $B_p(1) - B_p(kr)$ is smaller than

$$\omega_{n-2} \int_k^{1/r} r\,dt \int_0^{\pi/2} \sin^{n-2}\theta |(1 + 2t^{-1}\cos\theta + t^{-2})^{-1/2} + (1 - 2t^{-1}\cos\theta + t^{-2})^{-1/2} - 2| \, d\theta$$

which is smaller than Kr for some constant K. For instance we find when $n = 5$: $h(r) = 4\omega_3(1 - 3r/4 + r^2/5)/3$.

5.33 Corollary. *Let* (M_n, g) *be a compact Riemannian manifold of dimension* 3, 4 *or* 5, *such that* g *has at* P *the properties* (21) *and* (23). *Then the Green function* G *of the laplacian* Δ *satisfies:* $G(P,Q) = H(P,Q) + \beta(Q)$ *with* β *a* C^∞ *function on* $M - \{P\}$ *which, on* M, *is* C^2 *when* $n = 3$, C^1 *when* $n = 4$ *and lipschitzian when* $n = 5$.

Proof similar to that of the preceding proposition 5.31.

§5. The Positive Mass

5.34 We now prove A > 0, and hence conclude the validity of conjecture 5.11.

Definition. A C^∞ Riemannian manifold (M_n, g) is called asymptotically flat of order $\tau > 0$ if there exists a compact $K \subset M_n$ such that $M_n - K$ is diffeomorphic to $\mathbb{R}^n - B_0$ (B_0 being some ball in $\mathbb{R}^n$ with center 0), the components of the metric g satisfying in $\{y^i\}$ the induced coordinates by the diffeomorphism:

(33) $$g_{ij} = \delta_{ij} + 0(\rho^{-\tau}), \quad \partial_k g_{ij} = 0(\rho^{-\tau-1}), \quad \partial_{kl} g_{ij} = 0(\rho^{-\tau-2}).$$

Example. Let $(\tilde{M}_n, \tilde{g})$ be a compact Riemannian manifold and $\{x^i\}$ be a system of normal coordinates at $x_0 \in \tilde{M}_n$(x_0 has zero for coordinates). Set $g = r^{-4}\tilde{g}$ near x_0 with $r^2 = \sum_{i=1}^n x_i^2$ and $M = \tilde{M}_n - \{x_0\}$.

Then (M, g) is asymptotically flat of order 2 with asymptotic coordinates $y^i = r^{-2}x^i$. Indeed in polar coordinates (ρ or r, $\theta_1, \ldots, \theta_{n-1}$) with $\rho = 1/r$ we have

$$g_{\rho\rho} = \rho^{-4}r^{-4}\tilde{g}_{rr} = \tilde{g}_{rr}, \quad g_{\theta_i\rho} = \tilde{g}_{\theta_i r} = 0 \quad \text{and} \quad \rho^2 g_{\theta_i\theta_j} = \rho^4(r^2\tilde{g}_{\theta_i\theta_j}).$$

5.35 Definition. The mass $m(g)$ of the asymptotically flat manifold (M_n, g) is defined as the limit, if it exists, of

$$\omega_{n-1}^{-1}\int_{S_{n-1}(\rho)} \sqrt{|g(\rho,\theta)|}\, g^{ij}(\partial_i g_{\rho j} - \partial_\rho g_{ij})(\rho,\theta)d\tau(\theta)$$

when $\rho \to \infty$, $d\tau$ being the area element on $S_{n-1}(\rho)$.

Remark. The preceding definition depends on the asymptotic coordinates, but according to Bartnik [*32], $m(g)$ depends only on g if $\tau > (n-2)/2$.

5.36 Proposition. *Let $(\tilde{M}_n, \tilde{g})$ be as in example 5.34 with $n > 2$. Assume $(\tilde{M}_n, \tilde{g})$ belongs to the positive case ($\mu > 0$). Set $g = G^{4/(n-2)}\tilde{g}$ where $G(x) = (n-2)\omega_{n-1}G_L(x_0, x)$ and $M = \tilde{M}_n - \{x_0\}$. Suppose*

(34) $$|\tilde{g}(r,\theta)| = 1 + 0(r^k) \quad \textit{with } k > n-2$$

and

(35) $$G(x) = r^{2-n} + A + 0(r).$$

Then (M, g) is asymptotically flat of order 2 (only of order 1 if $n = 3$ and of order $n-2$ if $(\tilde{M}, \tilde{g})$ is flat near x_0) and the mass of (M, g) is $m(g) = 4(n-1)A$.

The proof of the first part is as for example 5.34,

$$g = r^{-4}\left(1 + Ar^{n-2} + 0(r^{n-1})\right)^{4/(n-2)}\tilde{g}.$$

For the computation of the mass, choose polar coordinates with $\rho = 1/r$.

$$\partial_\rho\sqrt{|g(\rho,\theta)|} = (1/2)\sqrt{|g(\rho,\theta)|}\, g^{ij}\partial_\rho g_{ij}.$$

Thus

$$m(g) = \lim_{\rho\to\infty}\omega_{n-1}^{-1}\int_{S_{n-1}(\rho)}\left(\sqrt{|g(\rho,\theta)|}\,\partial_\rho g_{\rho\rho} - 2\partial_\rho\sqrt{|g(\rho,\theta)|}\right)d\tau(\theta).$$

But

$$\sqrt{|g(\rho,\theta)|} = \rho^{-2n} G^{2n/(n-2)} \sqrt{|\tilde{g}(r,\theta)|}$$
$$= \left[1 + A\rho^{2-n} + 0(\rho^{1-n})\right]^{2n/(n-2)} \sqrt{|\tilde{g}(r,\theta)|}.$$

If k is large enough $m(g) = (4n-4)A = 4(n-1)A$.

Remark. If we choose a metric $\tilde{g}$ which satisfies properties (21) and (23) of 5.19 near x_0, $\tilde{g}$ and G satisfy (34) and (35) with $k = 4$ when $n \leq 5$. Moreover when $n = 4$ or 5, we have $(n-2)/2 < 2$ and when $n = 3$, $1/2 < 1$, thus $m(g)$ makes sense (see remark 5.35). When $(\tilde{M}, \tilde{g})$ is locally conformally flat, $m(g)$ makes sense also, as the order $\tau > (n-2)/2$.

There are the remaining cases. For them to prove $A > 0$ is equivalent to prove $m(g) > 0$ (according to the preceding proposition).

5.1. Positive Mass Theorem, the Low Dimensions

5.37 Conjecture. *If (M_n, g) is an asymptotically flat Riemannian manifold of order $\tau > (n-2)/2$ with non-negative scalar curvature belonging to $L_1(M_n)$, then $m(g) \geq 0$ and $m(g) = 0$ if and only if (M_n, g) is isometric to the euclidean space.*

In his article [*280] Schoen announced that he and Yau proved this conjecture.

Then he concluded that he proved the Yamabe problem for the remaining cases by Proposition 5.28 and 5.36 and the study of the dimensions 4 and 5. In fact at that time, the conjecture was solved without extra hypothesis only in dimension $n = 3$ (Schoen–Yau [*288], Witten [*318]).

Even now it is not known (to the Author) that a written proof of the conjecture exists.

Using the result in dimension $n = 3$, a proof by contradiction and induction on the dimension (see Lee and Parker [*208] 1987 and Schoen [*281] 1989) allows us to say that the conjecture is proved also for the dimensions 4 and 5. This proof does not work when the dimension of the manifold is greater than 7 because then a minimal hypersurface may have singularities.

It remains to consider the compact locally conformally flat manifolds. For these manifolds the proof of the positivness of the mass is quite different and appeared later on.

5.2. Schoen and Yau's Article [*289]

5.38 Let (M_n, g) be a compact locally conformally flat manifold which belongs to the positive case ($\mu > 0$). We can choose g so that the scalar curvature $R \geq R_0 > 0$. Moreover we suppose the dimension $n \geq 4$.

Consider $(\tilde{M}, \tilde{g})$ the universal Riemannian covering manifold of (M, g). Set $\pi : \tilde{M} \to M$, $\tilde{g} = \pi^* g$.

$(\tilde{M}, \tilde{g})$ is complete, locally conformally flat and simply connected. A well known theorem of Kuiper [*205] asserts that there exists Φ a conformal immersion of $(\tilde{M}, \tilde{g})$ in (S_n, g_0) where g_0 is the standard metric of S_n.

5.39 Theorem (Schoen–Yau [*289] 1988). *Φ is injective and gives a conformal diffeomorphism of $\tilde{M}$ onto $\Phi(\tilde{M}) \subset S_n$. Moreover $S_n - \Phi(\tilde{M})$ has zero Newtonian capacity and the minimal Green function of $\tilde{L}$ at $P \in \tilde{M}$ is equal to a multiple of $|\Phi'|^{\frac{n-2}{2n}} H \circ \Phi$. Where H is the Green function of L_0 at $\Phi(P)$ on (S_n, g_0) and $|\Phi'|$ is the $(\tilde{g}, g_0)$-norm of Φ'. Thus M is the quotient of a simply connected open subset Ω of S_n by some Kleinian group, $S_n - \Omega$ having zero Newtonian capacity.*

This theorem allows to prove A > 0 for manifolds of this type not conformal to (S_n, g_0). The proof (starting at the end of p. 59 of [*289]) must be completed at least at one point.

First we will give the definitions of the new words used above and explain the existence of the minimal Green function $\tilde{G}_p$ (lemma 5.44), so as the positiveness of the energy of (M_n, g), A > 0, if the manifold is not conformal to the sphere (S_n, g_0). For this we follow Vaugon (private communication) who first clearly explained the proof of 5.39.

5.40 Definition. Let (M, g) be a Riemannian manifold with scalar curvature $R \geq 0$ and dimension $n \geq 3$.

A Green function G_p of L at P is a function on $M - P$ which satisfies $LG_P = \delta_P$. Recall

(36) $$L = \Delta + (n-2)R/4(n-1).$$

G_P is the minimal Green function if any Green function G'_P satisfies $G_P \leq G'_P$.

If some Green function G'_P exists, the minimal Green function G_P exists and is obviously unique.

Let $\{\Omega_i\}$ be a sequence of open sets of M with C^∞ boundary and $\bar{\Omega}_i$ compact, such that for all i $P \in \Omega_i \subset \bar{\Omega}_i \subset \Omega_{i+1}$ and $\cup_{i=1}^{\infty} \Omega_i = M$.

Let G_i be the Green function of L at P with zero Dirichlet condition on $\partial\Omega_i$. We have $G_i > 0$ on $\Omega_i - P$. At $Q \in \Omega_{i_0}$ $(Q \neq P)$, $G_i(Q)$ is an increasing sequence for $i \geq i_0$, according to the maximum principle since $L(G_{i+1} - G_i) = 0$ and $G_{i+1} - G_i > 0$ on $\partial\Omega_i$. Likewise $G_i < G'_P$ for all i, if we extend G_i by zero outside $\bar{\Omega}_i$. So when $i \to \infty$, G_i tends to some positive function G_P which satisfies in the distributional sense $LG_P = \delta_P$ on M.

5.41 Proposition (Vaugon). *If G_P is a Green function for L at P and if $\tilde{g} = \varphi^{4/(n-2)} g$ is a conformal metric then*

(37) $$\tilde{G}_P(x) = G_P(x)/\varphi(P)\varphi(x)$$

is a Green function for the operator $\tilde{L}$ related to $\tilde{g}$. In particular if G_P is the minimal Green function for L at P, $G_P(x)/\varphi(P)\varphi(x)$ is the minimal Green function for $\tilde{L}$ at P.

Proof. For any function $f \in \mathcal{D}(V)$,

$$\int_V G_P L(f\varphi)\,dV = f(P)\varphi(P).$$

We have $d\tilde{V} = \varphi^{2n/(n-2)}\,dV$ and a computation gives

$$\Delta(\varphi f) + \frac{n-2}{4(n-1)} R\varphi f = \varphi^{(n+2)/(n-2)}\left(\tilde{\Delta} f + \frac{n-2}{4(n-1)}\tilde{R} f\right) \tag{38}$$

so

$$\int_V G_P(x)\varphi^{-1}(x)\tilde{L}f(x)\,d\tilde{V}(x) = f(P)\varphi(P).$$

Thus $G_P(x)/\varphi(x)\varphi(P)$ is a Green function for $\tilde{L}$ at P.

5.42 Definition. If g is an euclidean metric in a neighbourhood θ of P a Green function G_P at P is equal in θ to

$$G_P(x) = [d(P,x)]^{2-n}/(n-2)\omega_{n-1} + \alpha(x) \tag{39}$$

where $\alpha(x)$ is an harmonic function in θ.

When G_P is *the minimal Green function* of L at P, we call *energy* at P related to g the real number $\alpha(P)$.

5.43 Proposition. *If g and $\tilde{g} = \varphi^{4/(n-2)}g$ are euclidean metrics in a neighbourhood θ of P, $\tilde{\alpha}(P) = \alpha(P)\varphi^{-2}(P)$. In particular the sign of the energy is a conformal invariant in the set of the euclidean metrics near P.*

By 5.42 and Proposition 5.41 for $x \in \theta$:

$$\begin{aligned}\tilde{d}^{2-n}(P,x)/(n-2)\omega_{n-1} + \tilde{\alpha}(x)\\ = \left[d^{2-n}(P,x)/(n-2)\omega_{n-1} + \alpha(x)\right]/\varphi(P)\varphi(x).\end{aligned}$$

Moreover we can prove that $d^{2-n}(P,x) = \varphi(P)\varphi(x)\tilde{d}^{2-n}(P,x)$, the result follows.

5.44 Lemma. *$(\tilde{M},\tilde{g})$ being the covering manifold of (M,g) considered in* 5.38, *at each point $P \in \tilde{M}$, there exists a minimal Green function for $\tilde{L}$.*

$\bar{g} = \Phi^*(g_0)$ is conformal to $\tilde{g}$, so there exists a C^∞ function $u > 0$ such that $\tilde{g} = u^{4/(n-2)}\bar{g}$.

Set $W = \Phi^{-1}(\Phi(P))$. As $\tilde{L}(Ho\Phi) = \sum_{Q\in W} \delta_Q$, according to Proposition 5.41, there is $\tilde{H}$ a multiple of $u^{-1}Ho\Phi$ such that

$$\tilde{L}\tilde{H} = \sum_{Q\in W} a_Q \delta_Q \qquad \text{with } a_Q > 0 \text{ and } a_P = 1.$$

Let us return to the definition and to the construction of the minimal Green function (5.40). Set $\tilde{G}_i$ be the Green function of $\tilde{L}$ at P with zero Dirichlet condition on $\partial\Omega_i$. Pick $\theta_i \subset \tilde{M}$ an open set such that $(W - P) \cap \bar{\Omega}_i \subset \theta_i$ with θ_i small enough so that $\tilde{H} - \tilde{G}_i > 0$ on $\partial\theta_i$. We extend by zero $\tilde{G}_i$ on $M - \bar{\Omega}_i$. On $\Omega_i - \Omega_i \cap \theta_i$, $\tilde{L}(\tilde{H} - \tilde{G}_i) = 0$ and $\tilde{H} - \tilde{G}_i > 0$ on $\partial(\Omega_i - \Omega_i \cap \theta_i)$. Thus by the maximum principle $\tilde{G}_i < \tilde{H}$ and $\tilde{G}_P$ the minimal Green function for $\tilde{L}$ at P exists. Moreover $\tilde{G}_P \leq \tilde{H}$. We have $\tilde{H} - \tilde{G}_P > 0$ if W $\neq \{P\}$.

5.3. The Positive Energy

5.45 Definition. A compact set $F \subset S_n$ $(n \geq 3)$ has zero newtonian capacity if the constant function 1 on S_n is the limit in H_1 of functions belonging to $\mathcal{D}(S_n - F)$.

We verify that the measure of F is zero. And we can prove that the minimal Green function for L_0 at $P \in S_n - F$ on $(S_n - F, g_0)$ is the restriction to $S_n - F$ of the Green function H for L_0 at P on (S_n, g_0).

5.46 Remark *on the proof of Theorem* 5.39**.** Return to the proof of lemma 5.44. We have $\tilde{H} - \tilde{G}_P > 0$ if $W \neq \{P\}$. So if we prove that $\tilde{H} = \tilde{G}_P$, the injectivity of Φ follows. For this, define $v = \tilde{G}_P\tilde{H}^{-1}$.

We have $0 \leq v \leq 1$. After some hard computations which must be detailed, Schoen and Yau infer $v = 1$.

Set $F = S_n - \Phi(\tilde{M}_n)$, since $\tilde{H} = \tilde{G}_P$, the restriction of H to $S_n - F$ is a minimal Green function for L_0 on $(S_n - F, g_0)$. This implies F has zero Newtonian capacity.

Before the main proof, one step consists in showing that $\int_{\tilde{M}} \tilde{G}_P \, d\tilde{V} < \infty$ when $n \geq 4$ and $\int_{\tilde{M}} \tilde{G}_P^{1-\varepsilon} \, d\tilde{V} < \infty$ for some $\varepsilon < 0$ when $n = 3$.

The last inequality holds because the Ricci curvature of $(\tilde{M}, \tilde{g})$ is bounded. Let us prove the other inequality. Return to the construction of the minimal Green function $\tilde{G}_P$ (5.40).

Let u_i the unique solution of $\tilde{L}u_i = 1$, $u_i|_{\partial\Omega_i} = 0$. We extend u_i by zero outside Ω_i.

$$\int_{\tilde{M}} G_i \, d\tilde{V} = u_i(P).$$

At Q a point where u_i is maximum, $\tilde{\Delta}u_i(Q) \geq 0$ so

$$\sup u_i \leq 4(n-1)(\inf R)^{-1}/(n-2) \leq 4(n-1)/(n-2)R_0.$$

Thus $\{G_i\}$ is an increasing sequence of non-negative functions which goes to $\tilde{G}_P$. According to Fatou's theorem, $\tilde{G}_P$ is integrable and

$$\int_{\tilde{M}} \tilde{G}_P \, d\tilde{V} = \lim_{i\to\infty} \int_{\tilde{M}} G_i \, d\tilde{V} \leq 4(n-1)/(n-2)R_0.$$

5.47 Proposition. *If $(\tilde{M}_n, \tilde{g})$ is flat near a point P, the energy of $\tilde{g}$ at P is zero.*

Assuming $R \geq R_0 > 0$, we have proven the existence of the minimal Green function $\bar{G}_P$ for $\bar{L}$ corresponding to the metric $\bar{g}$. But the manifold $(\tilde{M}, \bar{g})$ is locally conformally flat, so there exists a C^∞ function $u > 0$ such that $\tilde{g} = u^{4/(n-2)}\bar{g}$ is flat near P (we can choose $u = 1$ ouside a compact neighbourhood of P). According to Proposition 5.41, $\tilde{G}_P(x) = \bar{G}_P(x)/u(x)u(P)$ is the minimal Green function of $\tilde{L}$.

Now the energy of the sphere is zero since $H^{4/(n-2)}g_0$ is the euclidean metric on $\mathbb{R}^n$ with zero mass. So by Theorem 5.39 the energy of $\tilde{g}$ is zero.

5.48 Theorem (Schoen–Yau [*289]). *Let (M_n, g) be a compact locally conformally flat manifold which belongs to the positive case ($\mu > 0$) but which is not conformal to (S_n, g_0). If g is flat in a neighbourhood of some point P then the energy of g at P is positive.*

Proof. As the Riemannian manifold is not conformal to (S_n, g_0), it is not simply connected and $(\tilde{M}_n, \tilde{g})$ is a non trivial Riemannian covering of (M_n, g). Set $\Pi : \tilde{M}_n \to M_n$, $\tilde{g} = \Pi^* g$ is flat near each point of $\Pi^{-1}(P)$, let $\tilde{P}$ one of them. We know (lemma 5.44) that the minimal Green function $\tilde{G}_{\tilde{P}}$ of $\tilde{L}$ at $\tilde{P}$ exists. In a neighbourhood θ of $\tilde{P}$.

$\tilde{G}_{\tilde{P}}(x) = \tilde{r}^{2-n}/(n-2)\omega_{n-1} + \tilde{\alpha}(x)$ with $\tilde{r} = \tilde{d}(\tilde{P}, x)$. $\tilde{\alpha}$ is an harmonic function and $\tilde{\alpha}(P) = 0$ (Proposition 5.47).

On the other hand $G_P \circ \Pi$ satisfies $\tilde{L}(G_P \circ \Pi) = \sum_{Q \in \Pi^{-1}(P)} \delta_Q$, G_P being the Green function of L at P.

Thus $G_p \circ \Pi - \tilde{G}_{\tilde{P}} > 0$ on $\tilde{M}$ (see the proof of lemma 5.44) because

$$\Pi^{-1}(P) \neq \{\tilde{P}\}.$$

But in θ $G_p \circ \Pi(x) = \tilde{r}^{2-n}/(n-2)\omega_{n-1} + \alpha \circ \Pi(x)$. So $\alpha(P) > 0$, the energy of g at P (see Definition 5.42) is positive.

§6. New Proofs for the Positive Case ($\mu > 0$)

6.1. Lee and Parker's Article [*208]

5.49 In this article Lee and Parker present, among other things, an argument which unifies Aubin and Schoen's works. They transfer the Yamabe problem from (M_n, g) to $(M_n - P, \hat{g})$ an asymptotically flat manifold, $P \in M$. If necessary, we change g by a conformal metric which has the property of Proposition 5.20 and $\hat{g} = G_P^{4/(n-2)} g$, G_p being the Green function of L at P. Then they use as test functions the well known functions φ_k.

$$\varphi_k = (k + \rho^2)^{1-n/2} \qquad \text{for } \rho > R_0, \ \varphi_k = (k + R_0^2)^{1-n/2} \text{ on } K$$

R_0 the radius of the ball B_0 (see Definition 5.34) is fixed large and we let $k \to \infty$.

In fact, after picking a good conformal metric g on M_n, Lee and Parker use in the Yamabe functional J on (M_n, g) the test functions:

$$u_\varepsilon(x) = r^{n-2} G_P(x)(\varepsilon + r^2)^{1-n/2} \qquad \text{if } r \leq \delta, \ (\delta > 0 \text{ small}),$$
$$u_\varepsilon(x) = \delta^{n-2} G_P(x)(\varepsilon^2 + \delta^2)^{1-n/2} \qquad \text{if } r > \delta, \ r = d(P, x)$$

see [*208] and they let $\varepsilon \to 0$.

6.2. Hebey and Vaugon's Article [*166]

5.50 Theorem (Hebey–Vaugon [*166]). *If the compact Riemannian manifold (M_n, g) is not conformal to (S_n, g_0), the test functions:*

$$u_\varepsilon(x) = (\varepsilon + r^2)^{1-n/2} \qquad \textit{if } r \leq \delta, \ (\delta > 0 \textit{ small})$$
$$u_\varepsilon(x) = (\varepsilon + \delta^2)^{1-n/2} \qquad \textit{if } r \geq \delta,$$

in the Yamabe functional, yield the strict inequality of the fundamental theorem 5.11.

These test functions are the simplest one. In fact the following proof is in my opinion the clearest.

Proof. First we choose a good conformal metric. When $n \geq 6$, if the Weyl tensor W_{ijkl} is not zero at P, we choose the conformal metric $\tilde{g}$ so that $\tilde{R}_{ij}(P) = 0$ (as in 5.19). $\tilde{J}(u_\varepsilon)$ has the same limited expansion than in (5.21) with $r = \tilde{d}(P, x)$ and $\varepsilon = 1/k$. Then $\tilde{J}(u_\varepsilon) < n(n-1)\omega_n^{2/n}$ for ε small enough.

When the manifold is locally conformally flat, we choose a conformal metric $\tilde{g}$ which is euclidean near a point P. We get

$$(40) \quad \tilde{J}(u_\varepsilon) = n(n-1)\omega_n^{2/n} + C\varepsilon^{n/2-1} \left[\int_V \tilde{R}\, d\tilde{V} - 4(n-1)\delta^{n-2}\omega_{n-1} + O(\varepsilon) \right]$$

which $C > 0$ a constant which does not depend on ε.

When the dimension n equals 3 to 5, we choose a conformal metric g' so that (21) and (23) hold. Then we have (24) and $R'(x) = 0(r^2)$. A limited expansion yields (40).

In the remaining cases, we will have $\tilde{J}(u_\varepsilon) < n(n-1)\omega_n^{2/n}$ if in a conformal metric as above, we have $\int_V \tilde{R}\, d\tilde{V} < 4(n-1)\delta^{n-2}\omega_{n-1}$ for some δ. This comes from the following caracterisation of the mass together with (5.37) and Theorem (5.48).

5.51 Theorem (Hebey–Vaugon [*166]). *When the compact manifold is locally conformally flat at P*

$$(41) \qquad A = \lim_{t\to 0} \sup\left[4(n-1)\left(\int \tilde{R}\, d\tilde{V}\right)^{-1} - t^{2-n}/\omega_{n-1}\right](n-2)^{-1}$$

where the sup is taken over the conformal metrics $\tilde{g} = \varphi^{4/(n-2)}g$ which are flat in $B_P(t)$ with $\varphi(P) = 1$.

When $n = 3$ to 5, (41) holds the sup being taken over the conformal metrics which are equal to g in $B_P(t)$.

Recall $m(g) = 4(n-1)A$ and $m(\tilde{g}) = m(g)$ by Proposition 5.43. The theorem holds in the low dimensions thanks to Proposition 5.31 and 5.36.

6.3. Topological Methods

5.52 In Bahri [*20] for the locally conformally flat manifolds, and in Bahri–Brezis [*23] for the dimensions 3 to 5, the authors, by using the original method of Bahri–Brezis–Coron (see 5.78 for a more complete discussion of this method) solved the Yamabe problem in the remaining cases without the positive mass theorem. They analyse the critical points at infinity of the Yamabe functional J and prove by contradiction the existence of a critical point which yields a solution of the Yamabe problem, but in general it is not a minimizer of J.

5.53 In Schoen [*282] a different approach is used. As we are in the positive case, the operator L= $\Delta + (n-2)R/4(n-1)$ is invertible, let L^{-1} its inverse. For any $\Lambda > 1$ and any $p \in [1, (n+2)/(n-2)]$, we define F_p by

$$(42) \qquad \bar{\Omega}_\Lambda \ni u \to F_p(u) = u - E(u)L^{-1}u^p \in C^{2,\alpha}$$

where

$$\Omega_\Lambda = \left\{u \in C^{2,\alpha} \;\middle|\; \|u\|_{C^{2,\alpha}} < \Lambda, \min_M u > \Lambda^{-1}\right\},$$

and

$$E(u) = \int uLu\, dV = \int |\nabla u|^2\, dV + \frac{(n-2)}{4(n-1)}\int Ru^2\, dV.$$

Theorem (Schoen [*282]). *Let (M_n, g) be a compact locally conformally flat manifold which is not conformal to (S_n, g_0). There exists a large number $\Lambda_0 > 0$ depending only on g such that $F_p^{-1}(0) \subset \Omega_{\Lambda_0}$ for all $p \in [1, (n+2)/(n-2)]$.*

Actually Schoen wrote this theorem for any Riemannian manifold not conformal to (S_n, g_0), but he gave a complete proof only for locally conformally flat manifolds. It is not known (to the Author at this time) that a general proof is written up.

5.54 When $p = 1$ it is well known that the equation $F_1(u) = 0$ has only one positive solution.

Let λ_0 be the first eigenvalue of L. Since we are in the positive case $\lambda_0 > 0$. By minimizing $E(u) = \int uLu\,dV$ on the set $A = \{u \in H_1/\|u\|_2 = 1, u \geq 0\}$ we find (as in 5.4) a positive eigenfunction $\varphi : L\varphi = \lambda_0\varphi$.

Proposition. *φ is the unique positive solution of $F_1(u) = 0$.*

First a solution of $F_p(u) = 0$ satisfies $\|u\|_{p+1} = 1$, indeed compute $\int uLF_P(u)\,dV$. So $F_1(\varphi) = 0$

Then, it is a general result for the normal-compact operators, that the eigenspace corresponding to the first eigenvalue λ_0 is one dimensional.

To see this, let $\Psi \not\equiv 0$ be such that $L\Psi = \lambda_o\Psi$. Pick $k \in \mathbb{R}$ so that $\Psi + k\varphi \leq 0$ and equals zero in some point $P \in M$. We apply the maximum principle to

$$(-\Delta)(\Psi + k\varphi) - h(\Psi + k\varphi) = [(n-2)R/4(n-1) - \lambda_0 - h](\Psi + k\varphi)$$

which is ≥ 0 when $h \in \mathbb{R}$ is chosen large enough. If the maximum of $\Psi + k\varphi$ is ≥ 0 then $\Psi + k\varphi = \text{Const}$. As $\Psi + k\varphi$ is zero at P, it is zero everywhere and Ψ is proportional to φ.

Finally if $\Psi \not\equiv 0$ is an eigenfunction of L corresponding to an eigenvalue $\lambda \neq \lambda_0$, Ψ changes of sign. Indeed multiplying $L\Psi = \lambda\Psi$ by φ and integrating on M yield

$$\lambda \int \varphi\Psi\,dV = \int \varphi L\Psi\,dV = \int \Psi L\varphi\,dV = \lambda_o \int \varphi\Psi\,dV$$

since L is self-adjoint. We get $\int \varphi\Psi\,dV = 0$.

5.55 When $1 < p < (n+2)/(n-2)$ we can prove Theorem 5.53, for any manifold, by using the following theorem:

Theorem (Gidas and Spruck [*141]). *In $\mathbb{R}^n$, $n > 2$, the equation $\Delta v = v^p$ with $1 \leq p < (n+2)/(n-2)$ has no non-negative C^2 solution except $v(x) \equiv 0$.*

Following Spruck the proof is by contradiction. On the compact manifold (M_n, g), let us suppose there exists a sequence of positive $C^{2,\alpha}$ functions u_i which satisfy:

$$Lu_i = u_i^p, \qquad \sup u_i \to \infty. \tag{43}$$

This is equivalent to $F_p(\nu u_i) = 0$ for some ν since $p > 1$. Pick z_i a point where u_i is maximum: $u_i(z_i) = m_i$ without loss of generality, since the manifold is compact, we can suppose that $z_i \to P$. We blow-up at P. In a ball $B_P(\delta)$, consider $\{x^j\}$ a system of normal coordinates at P with $x^j(P) = 0$. We suppose $z_i \in B_p(\delta/2)$. Define for $y \in \mathbb{R}^n$ with $\|y\| < \delta m_i^\alpha/2 = k_i$.

$$v_i(y) = \frac{1}{m_i} u_i(z_i + m_i^{-\alpha} y) \qquad \text{with } \alpha = (p-1)/2$$

$z_i + m_i^{-\alpha} y$ is suppose to be the point of $B_P(\delta)$ of coordinates the sum of the coordinate of z_i in $B_P(\delta)$ and the coordinates of $m_i^{-\alpha} y$ in $\mathbb{R}^n$. We set $y^j = (x^j - z_i^j) m_i^\alpha$.

The function v_i satisfies on the ball $B_0(k_{i_0})$ for $i \geq i_0$ the elliptic equation

$$g_i^{kj}(y) \frac{\partial^2 v_i}{\partial y^k \partial y^j} + a_i^j(y) \frac{\partial v_i}{\partial y^j} + a_i(y) v_i = v_i^p(y)$$

where $g_i^{kj}(y) = g^{kj}(ym_i^{-\alpha} - z_i)$, $a_i^l(y) = -m_i^{-\alpha}(g^{kj}\Gamma_{kj}^l)(ym_i^{-\alpha} - z_i)$ and

$$a_i(y) = (n-2) m_i^{-2\alpha} R(ym_i^{-\alpha} - z_i)/4(n-1).$$

When $i \to \infty$, $g_i^{kj} \to \mathcal{E}^{kj}$, $\frac{\partial g_i^{kj}}{\partial y^k} \to 0$, $a_i^j \to 0$ and $a_i \to 0$ uniformly on $\bar{B}_0(k_{i_0})$. Moreover the functions v_i are uniformly bounded $0 < v_i \leq 1$. The conditions of Theorem 4.40 are satisfied. So $\|v_i\|_{C^\alpha}$ is uniformly bounded on $\bar{B}_0(k_{i_0})$ for some $\alpha > 0$. By the Ascoli Theorem, there exists a subsequence of $\{v_i\}$ which converges uniformly to a continuous function v. v satisfies in the distributional sense on $\mathbb{R}^n$ the equation $\Delta v = v^p$, so $v \in C^2$and $v(0) = 1$ since $v_i(0) = 1$. This is in contradiction with Theorem 5.55. Thus the assumption $\sup u_i \to \infty$ is impossible and there exists a real number such that $|u_i| \leq k$.

$G_L(P,Q)$ being the Green function of L (see 5.14) we have

$$u_i(P) = \int G_L(P,Q) u_i^p(Q)\, dV(Q) \tag{44}$$

$|u_i| \leq k$ implies $\|u_i\|_{C^{1,\alpha}} <$ Const. and then $\|u_i\|_{C^{2,\alpha}} \leq \Lambda_0$ some constant. Moreover (44) implies $u_i(P) \geq m \int u_i^p\, dV$. As $\|u_i\|_{p+1}^{p-1} \geq (n-2)\mu/4(n-1) > 0$, $\int u_i^p\, dV$ is bounded away from zero since $\int u_i^{p+1}\, dV \leq k \int u_i^p\, dV$. Hence there exists a constant ν_0 such that $u_i \geq \nu_o > 0$.

Remark. For $p = (n+2)/(n-2)$ the preceding arguing yields no contradiction. The equation $\Delta v = v^{(n+2)/(n-2)}$ on $\mathbb{R}^n$ has positive solutions. The solution, with $v(0) = 1$ as maximum, is $\omega = \left[1 + \|y\|^2/n(n-2)\right]^{1-n/2}$.

New proof by Schoen [*282].

Let us return to 5.53. The map $u \to E(u) L^{-1} u^p$ is compact from $\bar{\Omega}_\Lambda$ into $C^{2,\alpha}$. So $F_p = I+$ compact and the Leray–Schauder degree makes sense. By

Theorem 5.53, $0 \notin F_p(\partial\Omega_{\Lambda_0})$ for any $p \in \left[1,(n+2)/(n-2)\right]$ thus $\deg(F_p,\Omega_{\Lambda_0},0)$ is constant for $p \in \left[1,(n+2)/(n-2)\right]$. Moreover φ the unique positive solution of $F_1(u) = 0$ (Proposition 5.54) is nondegenerate.
Thus $\deg(F_1,\Omega_{\Lambda_0},0) = \pm 1$ and $\deg(F_{(n+2)/(n-2)},\Omega_{\Lambda_0},0)$ is odd. So the Yamabe problem has at least one solution on the compact locally conformally flat manifolds.

6.4. Other Methods

5.56 In [*181] Inouë uses the steepest descent method to solve the basic theorem of the Yamabe problem. R. Ye in [*320] studies the Yamabe flow introduces by Hamilton

$$\partial g/\partial t = (s - R)g \quad \text{with } s = \int R\,dV \Big/ \int dV.$$

Ye proves that the long-time existence of the solution holds on any compact Riemannian manifold. In the positive case for the scalar curvature, if the manifold is locally conformally flat, Ye shows that the solution converges smoothly to a unique limit metric of constant scalar curvature as t tends to ∞. The estimates are obtained by using the Alexandrov reflection method.

§7. On the Number of Solutions

7.1. Some Cases of Uniqueness

5.57 In the negative and null cases ($\mu \leq 0$) two solutions of (1) with $\tilde{R} = \text{Const.}$ are proportional. Let φ_0 be a solution of (1) with $R' = \mu$. In the corresponding metric g_0 the Yamabe equation is always of the type of Equation (1), since Yamabe's problem is conformally invariant. So let φ_1 be a solution of Equation (1) with $R = R' = \mu$.

If $\mu = 0$, $\Delta\varphi_1 = 0$, thus $\varphi_1 = \text{Const.}$. If $\mu < 0$, at a point P where φ_1 is maximum, $\Delta\varphi_1 \leq 0$, thus $[\varphi_1(P)]^{N-2} \leq 1$, and at a point Q where φ_1 is minimum, $\Delta\varphi_1 \leq 0$ thus $[\varphi_1(Q)]^{N-2} \geq 1$. Consequently $\varphi_0 = 1$ is the unique solution of (1) when $R = R' = \mu < 0$.

In the positive case, we do not have uniqueness generally, neversless we have below Obata's result.

Examples. The sphere S_n (Theorem 5.58).

$M_n = T_2 \times S_{n-2}$ with T_2 the torus when $R\left(\int dV\right)^{2/n} > n(n-1)\omega_n^{2/n}$ and $n \geq 6$. Indeed in this case there exists on M_n at least two solutions of Equation (1) with $R' = R = \text{Const.}$ First, $\varphi_1 \equiv 1$ and second, φ_0 for which (Theorem 5.21)

$$J(\varphi_0) = \mu < n(n-1)\omega_n^{2/n} < J(1).$$

On the other hand, according to Obata [225], we have uniqueness for Einstein manifolds other than the sphere.

Proposition (Obata [225]). *Let $(M_n, \tilde{g})$ be a compact Einstein manifold not isometric to (S_n, g_0). Then the conformal metrics g to $\tilde{g}$ with constant scalar curvature are proportional to $\tilde{g}$.*

Proof. Let us consider the conformal metric g on the form $g = u^2\tilde{g}$.
Set $T_{ij} = R_{ij} - (R/n)g_{ij}$. We have (see 5.2):

$$\tilde{T}_{ij} = T_{ij} + (n-2)u^{-1}\left[\nabla_i\nabla_j u + (\Delta u/n)g_{ij}\right]. \tag{45}$$

As $T_{ij}g^{ij} = 0$

$$\int ug^{ik}g^{jl}\tilde{T}_{ij}\tilde{T}_{kl}\,dV \geq \int uT_{ij}T^{ij}\,dV + 2(n-2)\int T^{ij}\nabla_{ij}u\,dV.$$

But $\nabla_i T^i_j = \left(\frac{1}{2} - \frac{1}{n}\right)\nabla_j R$ by the second Bianchi identity (see 1.23). If g has constant scalar curvature, we get $\int uT_{ij}T^{ij}\,dV = 0$ since $\tilde{T}_{ij} = 0$.

Thus if $R =$ Const., g is Einstein. According to (45), if $u \not\equiv$ Const. there exists a non-trivial solution of

$$\nabla_i\nabla_j u + (\Delta u/n)g_{ij} = 0.$$

In that case Obata proves that $(M_n, \tilde{g})$ is isometric to (S_n, g_0).

Remark. When g is Einstein, μ the inf. of the Yamabe functional J is attained by the constant function. $J(1) = \mu$. So $\lambda_1 \geq R/(n-1)$ (see [14] p. 292), which is the inequality of Theorem 1.78.

But we have more for Einstein manifolds:

$$J(1) = R\left(\int dV\right)^{\frac{2}{n}} = \mu \leq n(n-1)\omega_n^{2/n},$$

with equality only for the sphere.

7.2. Particular Cases

5.58 The case of the Sphere.

Theorem 5.58 (Aubin [13] p. 588 and Aubin [14] p. 293). *For the sphere S_n, $(n \geq 3)$, $\mu = n(n-1)\omega_n^{2/n}$ and Equation* (1) *with $R' = R$ has infinitely many solutions. In fact, the functions $\varphi(r) = (\beta - \cos\alpha r)^{1-n/2}$, with $1 < \beta$ a real number and $\alpha^2 = R/n(n-1)$, are solutions of* (1) *with $R' = R(\beta^2 - 1)$.*

Moreover on the sphere S_n with $\int dV = 1$, all $\varphi \in H_1$ satisfy:

$$\|\varphi\|_N^2 \leq K^2(n,2)\|\nabla\varphi\|_2^2 + \|\varphi\|_2^2. \tag{46}$$

Proof. Recall r is the distance to a fixed point P. According to Theorem 6.67, Equation (1) has no solution on the spere when $R' = 1 + \varepsilon \cos \alpha r$, $\varepsilon \neq 0$. Indeed, $F = \cos \alpha r$ are the spherical harmonics of degree 1: $\Delta F = \lambda_1 F$, with $\lambda_1 = R/(n-1)$ the first nonzero eigenvalue.

If $\mu < n(n-1)\omega_n^{2/n}$, we can choose ϵ small enought to apply Theorem 5.12 with $h = R$ and $f = 1 + \varepsilon \cos \alpha r$ since $\nu \leq \mu(1-\varepsilon)^{-2/N}$. This contradicts the nonsolvability of Theorem 6.67.

Writing $\mu = n(n-1)\omega_n^{2/n}$ yields (46) when $\int dV = 1$. (46) is an improvement of the inequality of Theorem 2.28. Both constants are optimal, the second, $A_q(0) \geq 1$, since the inequality must be satisfied by the function $\varphi \equiv 1$.

For the unit Sphere (S_n, g_0) the solutions of (1) with $\tilde{R} = R = n(n-1)$ are

$$\varphi_{\beta,P}(Q) = \left[(\beta^2 - 1)/(\beta - \cos r)^2\right]^{(n-2)/4} \tag{47}$$

with $\beta \in]1, \infty]$, $P \in S_n$ and $r = d(P, Q)$, (see [14] p. 293). All solutions are minimizing for J : $J(\varphi_{\beta,P}) = n(n-1)\omega_n^{2/n}$.

There is no other solution. To see this, let π be the stereographic projection at P, π is a conformal map from $S_n \setminus \{P\}$ onto $\mathbb{R}^n$. Consider (ρ, θ_i) $i = 1, 2, \ldots, (n-1)$ polar coordinates in $\mathbb{R}^n$ and set $g = (\pi^{-1})^* g_0$. As $\rho = \operatorname{cotg}(r/2)$, $g = 4\sin^4(r/2) = 4(1+\rho^2)^{-2}\mathcal{E}$.

By virtue of (38), $L(\varphi_{\beta,P}) = \frac{n(n-2)}{4}\varphi_{\beta,P}^{N-1}$ yields

$$\Psi = \left[(1+\rho^2)/2\right]^{1-n/2}\left[n(n-2)/4\right]^{(n-2)/4}\varphi_{\beta,P} \tag{48}$$

with $\cos r = (\rho^2 - 1)/(\rho^2 + 1)$ is a solution of

$$\sum_{j=1}^{n} \partial_{jj}\Psi + \Psi^{N-1} = 0 \qquad \text{on } \mathbb{R}^n. \tag{49}$$

According to Gidas–Ni–Nirenberg [*140], the positive solutions of (49) are radial symmetric. The solutions $\Psi(r)$ satisfy a second order equation, moreover $\Psi'(0) = 0$. So there is only one positive radial solution such that $\Psi(0) = k$ a given real positive number. This solution is

$$\Psi(\rho) = k\left[1 + k^{4/(n-2)}\rho^2/n(n-2)\right]^{1-n/2}.$$

It is a solution found in (48) with $\beta \in]-\infty, -1[\, \cup \,]1, \infty]$.

It is of the kind (47) ; indeed $\varphi_{-\beta,P} = \varphi_{\beta,\bar{P}}$ with $\bar{P}$ the opposite point to P on the sphere.

5.59 Schoen [*281] found all solutions of the Yamabe problem for $C \times S_{n-1}$ the product of the circle of radius τ with the sphere of radius 1. Set $\tau_0 = (n-2)^{-1/2}$. The result is:

If $\tau \leq \tau_0$ the unique solution of (1) with $\tilde{R} = R$ is $\varphi \equiv 1$. If $\tau \in]\tau_0, 2\tau_0]$ there are two inequivalent solutions, the constant solution and the minimizers of J which are a C-parameter family of solutions with fundamental period $2\pi\tau$.

For $\tau \in](k-1)\tau_0, k\tau_0]$ there are k inequivalent solutions, $(k-1)$ C-parameter families of solutions and the constant solution for which J has the greatest critical value.

7.3. About Uniqueness

5.60 In the positive case there is no uniqueness in general. It is very easy to construct manifolds for which equation (1) with $\tilde{R} = R$ has more than one solution. Let us consider two compact manifolds (M_1, g_1), (M_2, g_2) with dimension n_1 (resp. n_2) volume V_1 (resp. V_2) and constant scalar curvature R_1 (resp. $R_2 > 0$). Pick k large enough so that $(R_1+kR_2)(V_1V_2)^{2/n} > n(n-1)\omega_n^{2/n}$. In that case the functional J for the manifold $(M_1, kg_1)\times(M_2, g_2)$ satisfies $J(1) > n(n-1)\omega_n^{2/n}$. The constant function is not a minimizer for J. Hence there are at least two solutions.

Now we will discuss another method to exhibit examples with several solutions.

7.4. Hebey–Vaugon's Approach

5.61 Let us consider (M_n, g) be a compact Riemannian manifold of dimension $n \geq 3$ and G be a compact subgroup of $C(M, g)$ the group of conformal diffeomorphisms.

Theorem (Hebey–Vaugon [*168]). *The inf, on the set of G-invariant metrics g' conformal to g, of $J(g') = \left[\int dV'\right]^{-(n-2)/n} \int R'\,dV'$ is achieved.*

The case $G = \{\mathrm{Id}\}$ is the Yamabe problem. For most cases the proof consists in two steps. First they prove that the inf. is attained if it is strictly less than $\beta = n(n-1)\omega_n^{2/n}\left(\inf_{x\in M} \operatorname{Card} O_G(x)\right)^{2/n}$ (when $G = \{\mathrm{Id}\}$ this is Theorem 5.11). Then they prove that the inf is smaller than β (we have always $\beta \leq 0$).

Now if, under some conditions,

$$\inf_{u>0, u\ G\text{-invariant}} J(u) > \mu = \inf J(u) > 0$$

the theorem above yields two solutions to the Yamabe problem. The corresponding critical values of the Yamabe functional J are not the same.

A more general case is that of the Riemannian covering manifolds $\pi : (\tilde{M}, \tilde{g}) \to (M, g)$ with $\tilde{g} = \pi^* g$. The question is: find conditions so that

$$\inf_{u\in H_1(M), u\not\equiv 0} \tilde{J}(u\circ\pi) > \inf_{u\in H_1(\tilde{M}), u\not\equiv 0} \tilde{J}(u),$$

$\tilde{J}$ being the Yamabe functional of $(\tilde{M}, \tilde{g})$.

5.62 Theorem (Hebey–Vaugon [*167]). *Let (M_0, g_0) be a compact Riemannian manifold not conformally equivalent to the standard sphere. We suppose there* *exist m Riemannian manifolds (M_i, g_i) $(i = 1, 2, \dots, m)$ such that (M_0, g_0) is a Riemannian covering manifold of (M_i, g_i) with b_i sheets $(\pi_i : (M_0, g_0) \to (M_i, g_i)$, $1 = b_0 < b_1 < \dots < b_m)$. If for each i there exists $k_i \in [0, 1]$ such that*

$$C_{k_i}(M_i, g_i)\left(\int_{M_0} dV_0\right)^{2/n} \le n(n-2)\omega_n^{2/n}\left[(1-k_i)b_i^{2/n} - b_{i-1}^{2/n}\right]/4$$

then on (M_0, g_0) there exist at least $m + 1$ metrics conformal to g_0 with same constant scalar curvature (and different critical values of J).

$C_k(M, g)$ is the smallest positive real number C such that any $u \in H_1(M)$ satisfies

$$(1-k)n(n-2)\omega_n^{2/n}2^{-2}\int |u|^{2n/(n-2)}\,dV$$

$$\le \int |\nabla u|^2\,dV + \frac{(n-2)}{4(n-1)}\int Ru^2\,dV + C\int u^2\,dV$$

$C_k(M, g)$ always exists when $0 < k \le 1$ (see the value of the best constant $K(n, 2)$). It is proven that $C_0(M, g)$ exists when the manifold has constant curvature (Aubin [14]), more generally when the manifold is locally conformally flat (Hebey–Vaugon [*166]), and recently by the same authors for any compact manifolds (see 4.63).

For applications of the Theorem above see Hebey–Vaugon [*167].

7.5. The Structure of the Set of Minimizers of J

5.63 Theorem (Y. Li [*Thesis, Univ. of Paris VI]). *In a conformal class the set of the metrics with volume 1 and constant scalar curvature μ is analytic compact of finite dimension bounded by a constant which depends only on n.*

Moreover for a generic conformal class, the minimal solution is unique.

§8. Other Problems

8.1. Topological Meaning of the Scalar Curvature

5.64 We have seen that μ is a conformal invariant (Proposition 5.8), but does its sign have a topological meaning? Considering a change of metric (a non-conformal one, obviously) it is possible to prove:

Theorem (Aubin [11] p. 388). *A compact Riemannian manifold M_n $(n \ge 3)$ carries a metric whose scalar curvature is a negative constant.*

Proof. According to Theorem 5.9 if we are not in the negative case, there exists a metric g with $R \ge 0$. Then we consider a change of metric of the kind:

$\tilde{g}_{ij} = \psi g_{ij} + \partial_i\psi\,\partial_j\psi$ with $\psi > 0$ a C^∞ function. It is possible to determine ψ such that the corresponding functional $\tilde{J}(u)$ is negative for some u. Hence the result follows by Theorem 5.9. ∎

Since on every compact manifold M_n ($n \geq 3$) there is some metric with $\mu < 0$, there is no topological significance to having negative scalar curvature. In contrast to this, Lichnerowicz [186] has proved that there are topological obstructions to admitting a metric with $\mu > 0$, that is, to positive scalar curvature. He showed that if there is a metric with nonnegative scalar curvature (not identically zero), then the Hirzebruch Â-genus of M must be zero. This work was extended by Hitchin [145], who proved that certain exotic spheres do not admit metrics with positive scalar curvature—and hence certainly have no metrics with positive sectional curvature.

In a related work, Kazdan and Warner [161] proved that there are also topological obstructions to admitting metrics with identically zero scalar curvature, that is, to $\mu = 0$. Thus there are obstructions to $\mu > 0$ and $\mu = 0$, but not to $\mu < 0$.

More recently, Gromov and Lawson [136] and [137] proved that every compact simply-connected manifold M_n ($n \geq 5$), which is not spin, carries a metric of positive scalar curvature. For the spin manifolds they generalize Hirzebruch's Â-genus in order to obtain almost necessary and sufficient conditions for a compact manifold to carry a metric of positive scalar curvature. In particular, the tori T_n, $n \geq 3$, do not admit metrics of positive scalar curvature. For the details see the article in references [136] and [137] or Bourbaki [34].

8.2. The Cherrier Problem

5.65 It concerns the C^∞ compact orientable Riemannian manifolds (M, g) with boundary and dimension $n \geq 2$. Denote by ξ the unit vector field defined on the boundary ∂M, normal to ∂M and oriented to the outside.

When $n \geq 3$, let h be the mean curvature of ∂M. h is the trace of the following endomorphism of the vector fields X on $\partial M : X \to \nabla_X \xi/(n-1)$.

If we consider as previously the change of conformal metric $\tilde{g} = \varphi^{\frac{4}{n-2}} g$, $\varphi \in C^\infty$, $\varphi > 0$, the new scalar curvature $\tilde{R}$ is given by

$$(1) \qquad 4((n-1)/(n-2))\Delta\varphi + R\varphi = \tilde{R}\varphi^{(n+2)/(n-2)}.$$

and the new mean curvature $\tilde{h}$ by

$$(50) \qquad (2/(n-2))\partial_\xi\varphi + h\varphi = \tilde{h}\varphi^{n/(n-2)}.$$

5.66 The problem is [97]: *given $R' \in C^\infty(M)$ and $\tilde{h} \in C^\infty(\partial M)$ does there exist a Riemannian metric g' conformal to g such that R' and h' are respectively the scalar curvature of (M, g') and the mean curvature of ∂M in (M, g').*

The problem is equivalent to solve the Neumann problem constituted by (1) and (50) with $\tilde{R} = R'$ and $\tilde{h} = h'$.

First, Cherrier [*97] proves the existence of best constants in the Sobolev inequalities including norms of the trace and he proves inequalities with norms of the trace in the exceptional case of the Sobolev theorem. Then he shows that a variational problem I with constraint Γ has a minimizer. Writting the Euler equation yields a weak solution for (1) and (50).

Finally, and it is not the simplest, he proves that the solution is regular. For the geometrical problem the functional

$$I = \int_M \left(|\nabla\varphi|^2 + \frac{n-2}{4(n-1)} R\varphi^2 \right) dV + \frac{n-2}{2} \int_{\partial M} h\varphi^2 \, d\sigma, \tag{51}$$

and the constraint

$$\Gamma = \int_M R'|\varphi|^{\frac{2n}{n-2}} \, dV + \frac{n}{n-1} \int_{\partial M} h'|\varphi|^{2\frac{n-1}{n-2}} \, d\sigma. \tag{52}$$

5.67 Theorem (Cherrier [*97]). *If the inf μ_0 of the functional I under the constraint $\Gamma = 1$ is smaller than an explicit positive constant, then the problem has a solution. The constant depends on the data and the best constants.*

For instance, if we want to find a conformal metric with constant scalar curvature ($R' = 1$), such that the boundary is minimal ($h' = 0$), the condition is $\mu_0 < \tilde{K}^{-2}(n, 2)$.

This last part is the equivalent of Theorem 5.11 for the Yamabe problem. $\tilde{K}(n, 2) = 2^{1/n} K(n, 2)$ is the best constant for manifolds with boundary.

The same problem occurs in dimension 2. In this case h is the geodesic curvature of ∂M and the equation to solve is

$$\Delta\varphi + R = R'e^{\varphi} \qquad \text{on } M \tag{53}$$

$$\partial_\xi \varphi + 2h = 2h'e^{\varphi/2} \qquad \text{on } \partial M. \tag{54}$$

5.68 Hamza [*155] studied the particular cases of a hemisphere and of an euclidean ball. For these manifolds, there are obstructions similar to those of Kazdan and Warner for the Nirenberg problem (see 6.66 and 6.67).

5.69 Theorem (Escobar [*126]). *Define*

$$E = \left\{ (x, t) \in \mathbb{R}^n / x \in \mathbb{R}^{n-1}, t > 0 \right\}$$

when $n \geq 3$. Any $\varphi \in \mathcal{D}(E)$ satisfies

$$\left[\int_{\partial E} |\varphi|^{2(n-1)/(n-2)} \, dx \right]^{(n-2)/(n-1)} \leq k(n, 2) \int_E |\nabla\varphi|^2 \, dx \, dt \tag{55}$$

where $k(n, 2) = (n/2 - 1)\omega_{n-1}^{1/(n-1)}$. The equality holds in (55) only if φ is proportional to a function of the form:

$$\varphi_\varepsilon(x, t) = \left[(\varepsilon + t)^2 + |x - x_0|^2 \right]^{1-n/2}. \tag{56}$$

Finding these functions is the key point. They act the part of the functions $(\varepsilon + r^2)^{1-n/2}$ for the Yamabe problem.

5.70 Theorem (Escobar [*127] and [*128]). *Let (M_n, g) be a compact Riemannian manifold with C^∞ boundary and $n \geq 3$, there exists a conformal metric of constant scalar curvature such that ∂M is minimal, except if the manifold satisfies these four properties all together: $n \geq 6$, non-locally conformally flat but with a Weyl tensor vanishing on ∂M which is umbilic.*

The case of these manifolds is still open. For the proof Escobar considers test functions, constructed from the functions (56), in the functional $I\Gamma^{(2-n)/n}$ [see (51) and (52) for the definitions of I and Γ] after a change of conformal metric as for the Yamabe problem. A limited expansion yields in most cases the inequality $\mu_0 < \tilde{K}^{-2}(n, 2)$ which allows to use theorem 5.67.

8.3. The Yamabe Problem on CR Manifolds

5.71 Let M be an orientable manifold of odd dimension $2n+1$. A CR structure on M is given by a complex n–dimensional subbundle E of the complexified tangent bundle CTM satisfying $E \cap \bar{E} = \{0\}$.

A CR manifold is such manifold with an integrable CR structure (the Frobenius condition $[E, E] \subset E$ is satisfied). $G = Re(E + \bar{E})$ is a real $2n$– dimensional subbundle of TM which carries the complex structure $J : G \to G$ defined by $J(X + \bar{X}) = i(X - \bar{X})$ for $X \in E$. As M is orientable there is a 1-form θ which is zero on G . Now we define the Levi form L_θ of θ by

$$L_\theta(X, Y) = 2\, d\theta(X, JY) \qquad \text{for } X, Y \in G, \tag{57}$$

and we suppose L_θ positive definite. Then θ defines a contact structure and we say that M is strictly pseudoconvex.

Example. A strictly pseudoconvex hypersurface in $\mathbb{C}^{n+1}$ is a *strictly pseudoconvex* CR manifold.

Associated to the Levi form, Webster [*315] has defined a curvature, thus a scalar curvature S.

The CR Yamabe problem *is: given a compact, strictly pseudoconvex CR manifold, find a contact form θ for which the Webster scalar curvature S is constant.*

5.72 Theorem (Jerison & Lee [*187]). *Let M be a compact, orientable, strictly pseudoconvex CR manifold of dimension $2n + 1$. Define the functional on the contact forms θ:*

$$F(\theta) = \left[\int S(\theta)\theta \wedge d\theta^n\right] \left[\int \theta \wedge d\theta^n\right]^{n/(n+1)}.$$

$\lambda(M) = \inf_\theta F(\theta)$ depends only on the CR structure, $\lambda(M) \leq \lambda(S_{2n+1})$.

If $\lambda(M) < \lambda(S_{2n+1})$ then the infimum is attained by a contact form $\tilde{\theta}$ which has constant Webster scalar curvature $S(\tilde{\theta}) = \lambda(M)$.

Given θ a contact form, any contact form $\tilde{\theta}$ is of the form $\tilde{\theta} = u^{2/n}\theta$ with u a C^∞ positive function. The transformation law for the Webster scalar curvature S is

$$\tilde{S} = u^{-(1+2/n)}[2(1+1/n)\Delta_b u + Su]. \tag{58}$$

So there is a C^∞ function $v > 0$ (given by the theorem) which satisfies the equation

$$2(1+1/n)\Delta_b u + Su = \lambda(M)u^{1+2/n}. \tag{59}$$

Here Δ_b is the sublaplacian operator defined on the C^∞ function by

$$\int (\Delta_b u) w\theta \wedge d\theta^n = \int L^*_\theta(du, dw)\theta \wedge d\theta^n$$

for all $w \in C^\infty(M)$ where L^*_θ is the dual form on G^* of L_θ. L^*_θ extends naturally to T^*M. For $\omega \in T^*M$

$$L^*_\theta(\omega, \omega) = 2\sum_{j=1}^{n} |\omega(Z_j)|^2$$

whenever $Z_1, \ldots, Z_n$ form an orthonomal basis for E.

5.73 Theorem (Jerison & Lee [*188]). *Let $z \in \mathbb{C}^{n+1}$ and $\hat{\theta} = \frac{i}{2}(\bar{d} - d)|z|^2$. The restriction to TS_{2n+1} of $\hat{\theta}$ is a contact form for S_{2n+1} which minimizes the functional $F(\theta)$ on the sphere. The corresponding Webster scalar curvature $\hat{S} = n(n+1)$ and $\lambda(S_{2n+1}) = 2\pi n(n+1)$. Any contact form with constant scalar curvature is obtained from a constant multiple of the standard form $\hat{\theta}$ by a CR automorphism of the sphere.*

Now with the extremal contact forms of $F(\theta)$ on the sphere, Jerison and Lee [*188] can prove that most CR manifolds M satisfy $\lambda(M) < \lambda(S_{2n+1})$.

8.4. The Yamabe Problem for Noncompact Manifolds

5.74 In [11] Aubin proved that we can decrease (until negative values) the local average of the scalar curvature only by local changes of metrics.

Then we can exhibit a sequence of metrics g_i each having negative scalar curvature on Ω_i with $\Omega_i \subset \Omega_{i+1}$ and $\sum_{i=1}^{\infty} \Omega_i = M$.

As the manifold M is denumberable at infinity there is no problem of converging, since each point has a neighbourhood where the sequence g_i is constant from some rank. For such Riemannian manifold (M, g) with negative scalar curvature, we can ask if there exists a conformal metric g' such that the scalar curvature R' = Const., and if (M, g') may be complete.

Contrary to the compact case, the Yamabe problem on complete Riemannian manifolds has not always a solution. In [*190] Jin Z. gives some counterexamples. Let us consider a Riemannian compact manifold (M, g) with scalar curvature $R = -1$ and dimension $n \geq 3$. Let P be a point of M. On $M - \{P\}$ there does not exist a complete Riemannian metric $g' \in [g]$.

Indeed equation (1) has no positive solution if R' equals 0 or 1. If $u > 0$ is a solution of (1) with $R' = -1$, according to a result of Aviles [16] u can be extended to a C^1 function on M. Thus $u \equiv 1$.

5.75 Theorem (Aviles–Mc Owen [*18]). *Let (M, g) be a complete Riemannian manifold. Assume the Yamabe functional (see 5.8) is negative for some function belonging to $\mathcal{D}(M)$, then there is a conformal metric with scalar curvature equal to -1.*

There is a complete conformal metric $\tilde{g}$ with $\tilde{R} = -1$ if the scalar curvature R of (M, g) is non-positive and bounded away from zero on $M \setminus M_0$ for some compact set M_0 or if on $M \setminus M_0$ $R(x) \leq -c_1[r(x)]^{-l}$ and the Ricci curvature at x greater than $-c_2[r(x)]^{-2\alpha}$ on M where $0 \leq \alpha < 1$ and $2\alpha \leq l < 1 + \alpha$ (c_1 and c_2 are two constants and $r(x)$ is the distance of x to a given point x_0 in the interior of M_0).

For the proof Aviles and Mc Owen use the method of upper and lower solutions.

5.76 In [*105] Delanoë studies the following problem:

Given a compact Riemannian manifold (M, g) with dimension n, a closed d-submanifold Σ and a real function f, is there a complete metric on $M \setminus \Sigma$ conformal to g with scalar curvature f?

Among other results he gives the proof of

Theorem. *There exists on $M \setminus \Sigma$ a complete conformal metric $\tilde{g}$ with scalar curvature*

$$\tilde{R} = -1 \quad \text{if and only if } d > n/2 - 1,$$
$$\tilde{R} = 0 \quad \text{if } d \leq n/2 - 1 \text{ and } \mu(g) > 0.$$

There is no complete conformal metric on $M \setminus \Sigma$ with $R \geq 0$ if $\mu(g) \leq 0$.

For instance if $M = S_n$ and $d > n/2 - 1$, we have the standard metric on S_n restricted to $S_n \setminus \Sigma$ with $R = +1$, a conformal metric g' with $R' = 0$ (obtained by some stereographic projection) and a conformal complete metric $\tilde{g}$ with $\tilde{R} = -1$.

8.5. The Yamabe Problem on Domains in $\mathbb{R}^n$

5.77 We will consider this problem on smooth bounded domain Ω with Dirichlet data. If the Dirichlet data are zero we have to solve the following equation for $n > 2$:

$$(60) \qquad \Delta u = u^{(n+2)/(n-2)}, \; u > 0 \qquad \text{on } \Omega \text{ ts with } u|_{\partial\Omega} = 0.$$

We know by the Pohozaev identity (see 6.58) that (60) has no solution if Ω is star-shaped.

On the other hand if Ω is an annulus, i.e.

$$\Omega = \{x \in \mathbb{R}^n / 0 < a < |x| < b\}$$

Kazdan and Warner pointed out that (60) has a solution. Seeking a solution depending only on r, we have to minimize the functional $\int_a^b (u')^2 r^{n-1}\, dr$ on the set

$$A = \left\{u \in \overset{\circ}{H}_1([a,b]) / u \geq 0 \quad \text{and} \quad \int_a^b u^N r^{n-1}\, dr = 1\right\}.$$

As $A \subset C^{1/2}([a,b])$, it is easy to prove the existence of minimizer.

5.78 Theorem (Bahri–Coron [*25] see also Brezis [*57]). *If there exists a positive integer d such that $H_d(\Omega, \mathbb{Z}/2\mathbb{Z}) \neq 0$, then (60) has a solution.*

The proof is difficult and of a new type. Analysis and algebraic topology arguments are used. The best is to read the article. Nevertheless we give below some steps of the proof to have an idea on it.

Define $\Sigma = \{u \in \overset{\circ}{H}_1(\Omega)/u \geq 0 \text{ and } \|\nabla u\|_2 = 1\}$ and $J(u) = 1/\int u^N dV$ for $u \in \Sigma$. According to the Sobolev imbedding theorem, for $u \in \mathcal{D}(\Omega) \subset \mathcal{D}(\mathbb{R}^n)$, $\|u\|_N \leq K(n,2)\|\nabla u\|_2$, thus J(u) $\geq [K(n,2)]^{-N}$.

If u is a critical point of J in Σ, then $u\big[J(u)\big]^{1/(N-2)}$ is a solution of (60). The proof of the existence of a solution proceed by contradiction.

We assume henceforth that (60) has no solution. First this implies the following

Lemma. *Let $u_i \in \Sigma$ be a sequence such that $J(u_i)$ converges to a real number ν and such that $J'(u_i) \to 0$, then $\nu = b_k = k^{2/(n-2)}[K(n,2)]^{-N}$ for some $k \in N^*$.*

We can suppose without loss of generality that $u_i \in \mathcal{D}(\Omega)$. The sequence $\{u_i\}$ is bounded in H_1, thus there is a subsequence which converges weakly in H_1 strongly in L_2 and a.e.. As the differential of J on $\Sigma : J'(u_i) \to 0$,

$$(61) \qquad \Delta u_i - u_i^{N-1} J(u_i) = w_i \quad \text{with } w_i \to 0 \text{ in } H_{-1},$$

H_{-1} is the dual of H_1. The assumption that (60) has no solution implies that any converging subsequence converges to zero. Thus $u_i \to 0$ a.e..

In §5 of Chapter 6 we will study a similar situation. There is a subsequence $\{u_j\}$ which has only points of concentration, in the sense of definition 6.38 when at x $u_i(x)$ does not converge to 0. Let $\mathfrak{E} = \{P_1, P_2, \dots, P_m\} \subset \Omega$ be the set of the points of concentration. $\mathfrak{E}$ is finite and nonempty.

Pick $\delta > 0$ small enough so that the balls $B(P_l, \delta) \subset \Omega$ are disjoint $(l = 1, 2, \dots, m)$. We have $\eta_l = \lim_{j\to\infty} \int_{B(P_l,\delta)} u_j^N \, dV > 0$ and $\sum_{l=1}^m \eta_l = 1/\nu$.

Moreover if we blow-up at a point of concentration (see Chapter 6, §5.5), we find that the sequence $v_j \to \omega > 0$ which satisfies on $\mathbb{R}^n$ $\Delta\omega = \nu\omega^{N-1}$. Hence $\int_{\mathbb{R}^n} \omega^N \, dV = [K(n,2)]^{-n}\nu^{-n/2}$. Thus $\eta_l = q_l[K(n,2)]^{-n}\nu^{-n/2}$ where q_l is the order of multiplicity of P_l as a point of concentration.

So $\nu^{(n-2)/2} = k[K(n,2)]^{-n}$ with $k = \sum_{l=1}^m q_l$ a positive integer.

5.79 The proof of Theorem 5.78 proceeds as follows.

Define $J_c = \{u \in \Sigma / J(u) \le c\}$ and set $W_k = J_{b_{k+1}}$. When c_1 and c_2 belong to $]b_k, b_{k+1}]$ for some k, the topologies of J_{c_1} and J_{c_2} are the same. The change in topology across the level b_k is described in the article. For this we consider the lines of the flow associated to - J' starting from $u_0 \in W_k - W_{k-1}$. Let $f : [0, \infty[\times \Sigma \to \Sigma$ be the solution of

(62) $$\frac{\partial}{\partial t} f(t,u) = -J'(f(t,u)), \;\; f(0,u) = u_0.$$

The solution of (62) is in Σ according to the maximum principle.

Recall $J'(u_0) \ne 0$ for any $u_0 \in \Sigma$ since we suppose that (60) has no solution. When k is large enough ($k \ge k_0$ for some integer k_0),Bahri–Coron prove that the solution of (62) with $b_k < J(u_0) \le b_{k+1}$ laies in W_{k-1} for large t. *Thus there is no change in topology across the level* b_k *for* $k \ge k_0$.

However Bahri–Coron prove the following.

Lemma. *If* $H_d(\Omega, \mathbb{Z}/2\mathbb{Z}) \ne 0$ *for some* $d > 0$, *then for each* $k \ge 1$ *the pair* (W_k, W_{k-1}) *is nontrivial, assuming that* $J'(u_0) \ne 0$ *for any* $u_0 \in \Sigma$.

X being a topological space and $A \subset X$, the pair (X, A) is trivial if there is a continuous map $[0,1] \times X \ni (t,x) \to f_t(x) \in X$ such that $f_t(x) = x$ for $x \in A$ and all t, $f_1(x) \subset A$ and $f_0(x) = x$ for all $x \in X$. The proof of the Lemma is by induction and uses algebraic topology arguments. The Lemma is in contradiction with the analysis of the lines of the flow solutions of (62) for large k, thus $J(u)$ has a critical point in Σ.

5.80 When $n = 3$, if $H_k(\Omega, \mathbb{Z}/2\mathbb{Z}) = 0$ for $k = 1, 2$ then Ω is contractible.

Moreover if Ω is star-shaped (60) has no solution. Thus we could think that (60) has a solution only when Ω is not contractible. This is not true (see Ding [*114]), there are examples of contractible bounded regular open sets Ω with solutions of (60).

5.81 Let us consider now the same equation, but with non zero Dirichlet data ($\varphi \not\equiv 0$):

$$\Delta u = \lambda u^{(n+2)/(n-2)}, \; u > 0 \quad \text{on } \Omega \text{ with } u|_{\partial\Omega} = \varphi \geq 0 \tag{63}$$

for some constant $\lambda > 0$. This problem is quite different to the former one.

Let h be the harmonic function on Ω such that $h|_{\partial\Omega} = \varphi$ and consider the following variational problem.

$$\inf_{u \in A} \int_\Omega |\nabla u|^2 \, dx \tag{64}$$

with

$$A = \left\{ u \in H_1(\Omega) \mid u - h \in \mathring{H}_1(\Omega), u \geq 0 \quad \text{and} \quad \int_\Omega u^N \, dx = \gamma \right\}.$$

If u is a solution of (63), u is a supersolution of the equation

$$\Delta v = 0, \quad v|_{\partial\Omega} = \varphi. \tag{65}$$

Thus $u > h$ on Ω (see 3.73), h being the solution of (65). Moreover $h \geq \inf \varphi \geq 0$.

So when $\gamma > \int_\Omega h^N \, dx$, a minimizer of (64), if it exists, satisfies (63) with $\lambda > 0$. For $\gamma = \int_\Omega h^N \, dx$ the minimizer is h, $\lambda = 0$. A solution of (63) with $\lambda < 0$ is a subsolution of (65) and so it is smaller than h.

5.82 Theorem (Caffarelli–Spruck [*69]). *Suppose $\partial\Omega \in C^2$ and $\varphi \in C^{1+\beta}(\partial\Omega)$ ≥ 0 positive somewhere. If $\gamma > \int_\Omega h^N \, dx$, there exists a minimizer $u \in C^\infty(\Omega) \cap C^{1+\beta}(\bar{\Omega})$ of* (64) *which satisfies* (63) *with some positive constant λ.*

8.6. The Equivariant Yamabe Problem

5.83 Let (M_n, g) be a compact Riemannian manifold of dimension $n \geq 3$ and $I(M, g)$ be its group of isometries. The problem is:

Given $G \subset I(M, g)$ a subgroup of isometries, does there exist a G-invariant metric $\tilde{g}$ conformal to g which realizes the infimum $\nu(G)$ of

$$J(g') = \left(\int dV' \right)^{-(n-2)/n} \int R' \, dV' \tag{66}$$

on the set of the G-invariant metrics g' conformal to g. J on the set of conformal metrics to g, is the Yamabe functional when we write g' on the form $g' = \varphi^{4/(n-2)} g$. When the infimum $\nu(G)$ is attained at $\tilde{g}$, the scalar curvature $\tilde{R}$ of $\tilde{g}$ is constant.

5.84 Theorem (Hebey–Vaugon [*168]).

$$\nu(G) \leq n(n-1)\omega_n^{2/n} \Big[\inf_{x \in M} \operatorname{Card} O_G(x) \Big]^{2/n}. \tag{67}$$

If the inequality is strict $\nu(G)$ is achieved.

$O_G(x)$ is the orbit of x under G. When all orbits are infinite this Theorem implies immediately the existence of a minimizer for J. There is equality on (S_n, g_0) when G has a fixed point. But for the other manifolds, Hebey and Vaugon prove that the inequality (67) is strict in most cases

In particular

5.85 Theorem (Hebey–Vaugon [*168]). *The inequality* (67) *is strict in each of these cases*

1) *All the orbits of G are infinite.*
2) $3 \leq n \leq 11$.
3) *The manifold is locally conformally flat.*
4) *There exists a point P of some minimal orbit of G such that* $W_{ijkl}(P) \neq 0$ *or* $|\nabla W_{ijkl}(P)| \neq 0$ *or* $|\nabla^2 W_{ijkl}(P)| \neq 0$.
5) *There exists a point P of some minimal orbit of G such that* $|\nabla^m W_{ijkl}(P)| = 0$ *for all* m *satisfying* $0 \leq m \leq (n-6)/2$.

When $3 \leq n \leq 5$ or when the manifold is locally conformally flat the proof is "classic" if we consider as well known the proof of the Yamabe problem. But before it is necessary to prove, for a point P of some minimal orbit of G , the existence of a G-invariant metric g' conformal to g which is euclidean on a neighbourhood of P.

In the general case the computations are done in a special conformal metric $\tilde{g}$, the analogue of Proposition 5.20 is proved but with $\tilde{g}$ *a* G-invariant metric. The proof of 5) assumes the strong form of the positive mass conjecture. Hebey–Vaugon conjecture that the inequality (67) is strict except if the manifold is (S_n, g_0) and G has a fixed point. But they solve the problem (5.83) in all cases.

5.86 Theorem (Hebey–Vaugon [*168]). *Let* (M_n, g) *be a compact Riemannian manifold of dimension* $n \geq 3$ *and let G be a compact subgroup of* $C(M, g)$ *the set of conformal transformations. There exists a conformal metric to g which realizes the infimum of* $J(g')$ *on the set of G-invariant metrics conformal to g.*

Corollary (Hebey–Vaugon [*168]). *Let* (M_n, g) *be a compact Riemannian manifold of dimension* $n \geq 3$, *which is not conformal to* (S_n, g_0). *There exists a conformal metric* $\tilde{g}$ *for which* $\tilde{R} = \text{Const.}$ *and* $I(M, \tilde{g}) = C(M, g)$.

8.7. An Hard Open Problem

5.87 Let (M, g) be a compact Riemannian manifold of dimension $n > 2$. Consider the functional (66) $J(g') = \left(\int dV'\right)^{-\frac{n-2}{n}} \int R'\, dV'$ on the set of the C^∞ Riemannian metrics g' on M.

Let us prove the well known result: *The critical points of J are Einstein-metrics.*

Let h be a symmetric twice-covariant smooth tensor field, h_{ij} its components in a local chart. We consider for t small the family of metrics $g_t = g + th$. In a local chart set

$$C^j_{ik} = \Gamma'^j_{ik} - \Gamma^j_{ik} = \frac{t}{2} g_t^{j\lambda} (\nabla_i h_{k\lambda} + \nabla_k h_{i\lambda} - \nabla_\lambda h_{ik})$$

where g_t^{ik} is the inverse matrix of $(g_t)_{ik}$ and Γ'^j_{ik} the Christoffel symbols of g_t.

A computation gives (see [*7] p. 396):

$$R_{tij} = R_{ij} + \nabla_\alpha C^\alpha_{ij} - \nabla_i C^\alpha_{\alpha j} + C^\alpha_{\alpha\beta} C^\beta_{ij} - C^\beta_{\alpha i} C^\alpha_{\beta j}.$$

The first derivatives with respect to t, at $t = 0$, of R_{tij} the components of the Ricci tensor of g_t are

$$2\left(\frac{d}{dt} R_{tij}\right)_{t=0} = \nabla^k(\nabla_i h_{jk} + \nabla_j h_{ik} - \nabla_k h_{ij}) - \nabla_i(\nabla_j h^k_k).$$

Thus we have (R_t the scalar curvature of g_t):

$$\left(\frac{d}{dt} R_t\right)_{t=0} = \nabla^i \nabla^j h_{ij} - \nabla^j \nabla_j h^i_i - R^{ij} h_{ij}.$$

Moreover $\left(\frac{d}{dt}\sqrt{|g_t|}\right)_{t=0} = \frac{1}{2}\sqrt{|g|} g^{ij} h_{ij}$.

These two results give

$$\left(\frac{dJ(g_t)}{dt}\right)_{t=0} = \left[\int dV\right]^{-2\frac{n-1}{n}} \left[\int (Rg^{ij}/2 - R^{ij}) h_{ij}\, dV \int dV \right.$$
$$\left. - (1/2 - 1/n) \int R\, dV \int h_{ij} g^{ij}\, dV\right].$$

If g is a critical point of J

$$(68) \qquad [R_{ij} - (R/2) g_{ij}] \int dV + (1/2 - 1/n)\left(\int R\, dV\right) g_{ij} = 0.$$

Multiply by g^{ij} yields $R = \text{Const.}$. Thus (68) implies $R_{ij} = (R/n) g_{ij}$.

5.88 In a conformal class $[g]$ the functional J is the Yamabe functional. We know that $\mu_{[g]}$ the inf of J on $[g]$ is attained (Theorems 5.11, 5.21, 5.29 and 5.30). For the sphere (S_n, g_0) $\mu_{[g_0]} = n(n-1)\omega_n^{2/n}$ (see 5.58). It is the unique manifold having this property.

Define $\bar\mu = \sup \mu_{[g]}$ on the set of the conformal classes.

We can ask the questions

1) *If $\bar\mu$ is achieved by a metric $\tilde g$, is $\tilde g$ an Einstein metric?*
2) *When $\bar\mu < n(n-1)\omega_n^{2/n}$, is $\bar\mu$ achieved?*

There are partial answers to the first question.

5.89 Proposition. *Assume $\tilde{\mu}$ is achieved by a metric $\tilde{g}$. If $\tilde{\lambda}_1$ the first nonzero eigenvalue of $\tilde{\Delta}$ satisfies $\tilde{\lambda}_1 > \tilde{R}/(n-1)$ then $\tilde{g}$ is an Einstein metric.*

Remark the functional J (66) is invariant by homotheties, for $\alpha > 0$ $J(\alpha g') = J(g')$. So we can suppose that the volume is constant, equal to 1: $\tilde{V} = 1$. Consider again the family of metrics $g_t = \tilde{g} + th$. The scalar curvature R_t of g_t is not constant in general, so we have to solve the equation

$$4\frac{n-1}{n-2}\Delta_t \varphi_t + R_t \varphi_t = \mu_t \varphi_t^{N-1}$$

where μ_t is the inf of J on $[g_t]$, the set of the metrics conformal to g_t. We have $\mu_0 = \tilde{\mu} = \tilde{R} = R_0$ and $\varphi_0 \equiv 1$.

Let Ω be a neighbourhood of φ_0 in $C^{r,\alpha}$ $(0 < \alpha < 1, r > 2)$ such that any function in Ω is positive. Now consider the following map:

$$]-\varepsilon, \varepsilon[\times \Omega \ni (t, \gamma) \xrightarrow{\Gamma} 4\frac{n-1}{n-2}\Delta_t \gamma + R_t \gamma - \mu_\circ \gamma^{N-1} \in C^{r-2}.$$

Γ is continuously differentiable and the differential of Γ with respect to γ at $(0, \varphi_0)$ is

$$D_\gamma \Gamma(0, \varphi_0)(\Psi) = 4\frac{n-1}{n-2}\left(\tilde{\Delta}\Psi - \frac{R_0}{n-1}\Psi\right).$$

As $\tilde{\lambda}_1 > R_0/(n-1)$, $D_\gamma \Gamma(0, \varphi_0)$ is invertible.

By the implicite function theorem, for t small ($|t| < \tilde{\varepsilon}$), there is a unique function $\tilde{\varphi}_t$ in Ω which satisfies

$$4\frac{n-1}{n-2}\Delta_t \tilde{\varphi}_t + R_t \tilde{\varphi}_t = \mu_0 \tilde{\varphi}_t^{N-1}$$

and $]-\tilde{\varepsilon}, \tilde{\varepsilon}[\ni t \to \tilde{\varphi}_t$ is smooth.

Moreover $\mu_t = \mu_0 (\int \tilde{\varphi}_t^N \, dV)^{2/n}$ is a smooth function of t and $\varphi_t = \tilde{\varphi}_t \|\tilde{\varphi}_t\|_N^{-1}$. Then the family of metrics $\tilde{g} = \varphi_t^{4/(n-2)} g_t$ is smooth as a function of t.

$\mu_t = J(\tilde{g}_t)$ is the scalar curvature of $\tilde{g}_t$. Writting $\left(\frac{d}{dt}\mu_t\right)_{t=0} = 0$ implies $\tilde{g}$ is an Einstein metric.

Application. Remind there are three types of compact manifolds according to they carry a Riemanian metric whose scalar curvature has a given sign. Consider those which has a metric with zero scalar curvature and which carry no metric with positive scalar curvature (examples: T_n the torus, $K3$ surfaces). Those manifolds with zero scalar curvature are Ricci flat ($\tilde{\mu} = 0$ is achieved).

5.90 Remarks. We know that $\tilde{\lambda}_1 \geq \tilde{R}/(n-1)$, but in the case $\tilde{\lambda}_1 = \tilde{R}/(n-1)$ we cannot conclude. For instance we can find a family φ_t of t which has a derivative with respect to t from the left at $t = 0$ and a different one from the right (at $t = 0$). In this situation we cannot conclude.

When $\bar{\mu} = n(n-1)\omega_n^{2/n}$, there are two cases. Either the manifold is conformal to the sphere with its canonical metric and $\bar{\mu}$ is attained, or it is not conformal to the sphere and in this case $\bar{\mu}$ is not achieved. As an example, for the manifold $C \times S_{n-1}(n > 2)$, Gil–Medrano [*142] proved that $\bar{\mu} = n(n-1)\omega_n^{2/n}$.

An other result. Aviles and Escobar [*17] proved that there exists $\varepsilon(n) > 0$ such that $\mu_{[g]} < n(n-1)\omega_n^{2/n} - \varepsilon(n)$ if (M_n, g) is any compact Einstein manifold which is not conformal to (S_n, g_0).

In fact we known very few on $\bar{\mu}$. For instance it would be interesting to prove that $\bar{\mu} = n(n-1)(\omega_n/2)^{2/n}$ for the real projective space $P_n(\mathbb{R})$, in other words that the metric with constant curvature has the greatest μ.

8.8 Berger's Problem

5.91 The problems concerning scalar curvature turn out to be very special when the dimension is two, the scalar curvature is then twice the Gaussian curvature.

Let (M, g) be a compact Riemannian manifold of dimension two.
It is well known that there exists a metric on M whose curvature is constant. Considering conformal metrics g' of g, Melvyn Berger ([40]) wanted to prove this result by using the variational method.

Set $g' = e^{\varphi} g$. Then the problem is equivalent to solving the equation

$$\Delta\varphi + R = R' e^{\varphi} \tag{69}$$

with R' some constant (see 5.2). Here R denotes the scalar curvature of (M, g) (in dimension two R is twice the sectional curvature) and R' is the scalar curvature of (M, g').

By Theorem 4.7 we can write $R = \tilde{R} + \Delta\gamma$ with $\tilde{R} = \text{Const}$ and $\gamma \in C^\infty(M)$ satisfying $\int \gamma\, dV = 0$. Setting $\psi = \varphi + \gamma$, Equation (69) becomes

$$\Delta\psi + \tilde{R} = R' e^{-\gamma} e^{\psi}. \tag{70}$$

Note that we know the sign of R'; it is that of χ, the Euler–Poincaré characteristic. Indeed, by the Gauss–Bonnet theorem $4\pi\chi = \int R\, dV$, so integrating (69) over M gives $R' \int e^{\varphi}\, dV = \int R\, dV$, which is equal to $\tilde{R} \int dV$.

5.92 *Solution for* $\chi \leq 0$. If $\chi = 0$, $R' = \tilde{R} = 0$, $\psi = 0$ is a solution of (70), so the metric $g' = e^{-\gamma} g$ solves the problem.

If $\chi < 0$, $\tilde{R} < 0$ and R' will be negative. We are going to use the variational method to solve Equation (70).
Consider the functional

$$I(\psi) = \frac{1}{2} \int \nabla^\nu \psi \nabla_\nu \psi\, dV + \tilde{R} \int \psi\, dV \tag{71}$$

and set $\nu = \inf I(\psi)$ for all $\psi \in \mathsf{H}_1$ satisfying $\int e^{\psi - \gamma}\, dV = 1$.

a) ν *is finite*. Since the exponential function is convex

$$\int (\psi - \gamma)\, dV \leq \log \int e^{\psi-\gamma}\, dV = 0.$$

Thus $\int \psi\, dV \leq 0$ and $\nu \geq 0$.

b) ν *is attained*. Let $\{\psi_i\}$ be a minimizing sequence. We can choose it such that $I(\psi_i)$ is smaller than $\nu + 1$. Then

$$\int \nabla^\nu \psi_i \nabla_\nu \psi_i\, dV < 2(1+\nu) \quad \text{and} \quad \tilde{R} \int \psi_i\, dV < 1 + \nu.$$

Thus the set $\{\psi_i\}_{i\in\mathbb{N}}$ is bounded in H_1, since $\|\nabla\psi_i\|_2 \leq$ Const and $(1+\nu)/\tilde{R} < \int \psi_i\, dV \leq 0$. Indeed, by Corollary 4.3 $\|\psi_i\|_2 \leq |\int \psi_i\, dV|[\int dV]^{-1/2} + \lambda_1^{-1/2}\|\nabla\psi_i\|_2$ where λ_1 is the first nonzero eigenvalue. Using Theorems 3.18, 2.34, and 2.46 we see that there exist $\tilde{\psi} \in \mathsf{H}_1$ and a subsequence $\{\psi_j\}$ such that $\psi_j \to \tilde{\psi}$ weakly in H_1, strongly in L_1, and such that $e^{\psi_j} \to e^{\tilde{\psi}}$ strongly in L_1. Therefore $\int e^{\tilde{\psi}-\gamma}\, dV = 1$ and according to Theorem 3.17, $I(\tilde{\psi}) \leq \nu$. But $\tilde{\psi}$ satisfies the constraint. Thus by definition of ν we have $I(\tilde{\psi}) = \nu$. Hence ν is attained.

c) Writing Euler's equation yields

$$\int \nabla^\nu \tilde{\psi} \nabla_\nu h\, dV + \tilde{R} \int h\, dV = K \int h e^{\tilde{\psi}-\gamma}\, dV$$

for any $h \in \mathsf{H}_1$. Picking $h = 1$ gives the value of K, the Lagrange multiplier $K = \tilde{R} \int dV = 4\pi\chi$. Thus $\tilde{\psi}$ is a weak solution of

(72) $$\Delta\tilde{\psi} + \tilde{R} = 4\pi\chi e^{\tilde{\psi}-\gamma}$$

By Theorem 2.46 the right-hand side belongs to L_p for all p. Therefore $\Delta\tilde{\psi} \in L_p$ for all p. Differentiating Equation (15) in 4.13 and using the properties of the Green's function show that $\tilde{\psi} \in C^{1+\alpha}$ for $0 < \alpha < 1$. Thus $\Delta\tilde{\psi} \in C^{1+\alpha}$ and according to Theorem 3.54, $\tilde{\psi} \in C^{3+\alpha}$. By induction $\tilde{\psi} \in C^\infty$.

5.93 The Positive Case $\chi > 0$.

There are only two compact manifolds which are involved $\mathbb{S}_2$ if $\chi = 2$ and the real projective space $\mathbb{P}_2$ if $\chi = 1$. We suppose M is one of these manifolds.

From now on $\tilde{R}$ is a positive real number. More generally than previously, we will consider the equation

(73) $$\Delta\varphi + \tilde{R} = f e^\varphi,$$

where $f \in C^\infty$ is a function positive at least at one point.
This property of f is necessary in order that Equation (73) have a solution, since $\tilde{R} \int dV = \int f e^\varphi\, dV$.

Henceforth, without loss of generality, we suppose the volume equals 1. Set $\nu = \inf I(\varphi)$ for all $\varphi \in H_1$ satisfying $\int f e^\varphi \, dV = \tilde{R}$, where $I(\varphi)$ is the functional (71).

Theorem 5.93. *Equation* (73) *has a* C^∞ *solution if* $\tilde{R} < 8\pi$.

Proof

a) ν *is finite*. First of all there are functions satisfying $\int f e^\varphi \, dV = \tilde{R}$ since f is positive somewhere.
On the other hand, according to Theorem 2.51 or 2.53

(74) $$\tilde{R} = \int f e^\varphi \, dV \leq \sup f \int e^\varphi \, dV$$
$$\leq C(\varepsilon) \sup f \exp\left[(\mu_2 + \varepsilon) \|\nabla\varphi\|_2^2 + \int \varphi \, dV\right].$$

Thus

(75) $$I(\varphi) \geq \left[\tfrac{1}{2} - (\mu_2 + \varepsilon)\tilde{R}\right] \|\nabla\varphi\|_2^2 + \tilde{R} \log(\tilde{R}/C(\varepsilon) \sup f).$$

$\mu_2 = 1/16\pi$, if $\tilde{R} < 8\pi$ we can choose $\varepsilon = \varepsilon_0 > 0$ small enough so that $2(\mu_2 + \varepsilon_0)\tilde{R} < 1$. Therefore ν is finite, $\nu \geq \tilde{R} \log(\tilde{R}/C(\varepsilon_0) \sup f)$.

b) ν *is attained*. Let $\{\varphi\}_{i \in \mathbb{N}}$ be a minimizing sequence; (75) implies

$$\left[\tfrac{1}{2} - (\mu_2 + \varepsilon_0)\tilde{R}\right] \|\nabla\varphi_i\|_2^2 \leq \text{Const}, \text{ thus } \|\nabla\varphi_i\|_2 \leq \text{Const}.$$

Moreover, $I(\varphi_i) \leq$ Const implies $\int \varphi_i \, dV \leq$ Const, and by (74) $\int \varphi_i \, dV \geq$ Const. Therefore $\{\varphi_i\}_{i \in \mathbb{N}}$ is bounded in H_1 (Corollary 4.3).
As in 5.92b it follows that ν is attained. There exists $\tilde{\varphi} \in H_1$ such that $I(\tilde{\varphi}) = \nu$ and $\int f e^\varphi \, dV = \tilde{R}$.

c) Writing Euler's equation yields

$$\int \nabla^\nu \tilde{\varphi} \nabla_\nu h \, dV + \tilde{R} \int h \, dV = K \int h f e^\varphi \, dV$$

for any $h \in H_1$. Picking $h = 1$ gives $K = 1$. $\tilde{\varphi}$ satisfies Equation (73) weakly and by the proof in 5.92c, $\tilde{\varphi} \in C^\infty$. ∎

Chapter 6

Prescribed Scalar Curvature

§1. The Problem

1.1. The General Problem

6.1 Let (M_n, g) be a C^∞ Riemannian manifold of dimension $n \geq 2$. Given f a smooth function on M_n, the Problem is:

Does there exist a metric g' on M such that the scalar curvature R' of g' is equal to f?

This problem was solved entirely by Kazdan and Warner [*195] [*198] [*200]. Since the equations are different for $n = 2$ and $n \geq 3$, the proofs are different as are the results. When $n = 2$ the scalar curvature R has strong topological meaning because the sectional curvature is fully determined by R (At a point where the coordinates are chosen orthonormal $R = 2R_{1212}$).

So more often than not, we will present the proofs when $n \geq 3$.

6.2 Theorem (Kazdan and Warner [*198]). *Let M be a C^∞ compact manifold of dimension $n \geq 3$. If $f \in C^\infty(M)$ is negative somewhere, then there is a C^∞ Riemannian metric on M with f as its scalar curvature.*

Proof. Using Aubin [11], on any M we can choose a smooth metric g whose scalar curvature $R = -1$. This shows that negative scalar curvature has no topological meaning.

In a pointwise conformal metric $g' = u^{4/(n-2)}g$ with $u > 0$, we use formula (1) in 5.2 for R'. If $u \in H_2^p(M)$ with $p > n$, then $u \in C^1(M)$ according to the Sobolev imbedding theorem. Set $\Omega = \{u \in H_2^p(M)/u > 0\}$. Now we consider the map

$$\Omega \times L_p \ni (u, K) \xrightarrow{\Gamma} 4(n-1)(n-2)^{-1}\Delta u - u - Ku^{N-1} \in L_p,$$

where $N = \frac{2n}{n-2}$. Γ is continuously differentiable and its partial differential with respect to u is

$$D_u\Gamma(v) = 4(n-1)(n-2)^{-1}\Delta v - v - (N-1)Ku^{N-2}v.$$

At (1,-1), $D_u\Gamma(v) = 4(n-1)(n-2)^{-1}\Delta v + (N-2)v$ is invertible as an operator acting on $\mathcal{L}(H_2^p, L_p)$. Hence, by the implicit function theorem, there exists $\varepsilon > 0$

such that if K satisfies $\|K+1\|_p < \varepsilon$ the equation $\Gamma(u, K) = \Gamma(1, -1) = 0$ has a solution in a neighbourhood in H_2^p of the constant function $u \equiv 1$. We can choose ε small enough so that the functions u in this neighbourhood are positive everywhere (this follows from the Sobolev theorem).

Since f is negative somewhere there exist $\alpha > 0$ and φ a C^∞ diffeomorphism of M such that $K = \alpha f o \varphi$ satisfies $\|K+1\|_p < \varepsilon$. So αf is the scalar curvature of the metric $\tilde{g} = (\varphi^{-1})^*(u^{4/(n-2)}g)$, where u is the solution founded above of the equation $\Gamma(u, K) = 0$. $u \in C^\infty$ by the bootstrap method, since $u \in C^1$ is a solution of $\Delta u = (n-2)(u + Ku^{N-1})/4(n-1)$ with $K \in C^\infty$. Therefore a C^∞ metric (homothetic to $\tilde{g}$) has f as scalar curvature.

6.3 Theorem (Kazdan and Warner [*198]). *Let M be a C^∞ compact manifold of dimension $n \geq 3$ which admits a metric whose scalar curvature is positive. Then any $f \in C^\infty(M)$ is the scalar curvature of some C^∞ Riemannian metric on M.*

Proof. We know that there are compact manifolds, such as the torus T^n which have metrics with zero scalar curvature but no metric with positive scalar curvature. Here by hypothesis there is a metric with positive scalar curvature, hence the manifold admits a metric with zero scalar curvature. Indeed we can pass continuously from a metric with positive scalar curvature to a metric with negative scalar curvature. So we get a metric which is in the zero case: $\mu = 0$ (μ is defined in 5.8).

Thus we have to consider only the case f positive somewhere. By the theorems which solve the Yamabe problem, there exists a metric g with scalar curvature equal to +1 which minimizes the Yamabe functional in the conformal class $[g]$. Then we procede as for Theorem 6.2. We consider on $\Omega \times L_p$,

$$\Gamma(u, K) = 4(n-1)(n-2)^{-1}\Delta u + u - Ku^{N-1}.$$

At (1,1), $D_u\Gamma(v) = 4(n-1)[\Delta v - v/(n-1)]/(n-2)$ is invertible only if $\lambda_1(g) > 1/(n-1)$ which is not always true (we can have $\lambda_1(g) = 1/(n-1)$, for instance on the sphere with the standard metric satisfying $R = 1$). If $\lambda_1(g) = 1/(n-1)$, we choose a metric $\tilde{g}$ close to g (so that $R(\tilde{g}) > 0$) not belonging to $[g]$.

For $\tilde{g}$ well chosen, a minimizing metric in $[\tilde{g}]$ for the Yamabe functional, with scalar curvature equal to 1, will have its $\lambda_1 > 1/(n-1)$. With this metric the proof of Theorem 6.2 will work, $K = \alpha f o \varphi$ satisfying $\|K-1\|_p < \varepsilon$.

Using this result, the problem of describing the set of scalar curvature functions on M_n is completely solved if $n \geq 3$. To see this, note that the topological obstructions mentioned above show that there are tree cases.

The first case: M does not admit any metrics with $\mu \geq 0$. Then $\mu < 0$ for every metric, so the scalar curvature functions are precisely those which are negative somewhere.

The second case: M does not admit a metric with $\mu > 0$, but does admit metrics with $\mu = 0$ and $\mu < 0$. This is identical with the first case except that the zero function is also a scalar curvature.

The third case: M has some metric g with $\mu > 0$. Any function is scalar curvature.

6.4 Theorem (Kazdan and Warner [*198]). *Let M be a non compact manifold of dimension $n \geq 3$ diffeomorphic to an open submanifold of some compact manifold $\tilde{M}$. Then, every $f \in C^\infty(M)$ is the scalar curvature of some Riemannian metric on M.*

Proof. Without loss of generality, we can suppose that $\tilde{M} - M$ contains an open set. On $\tilde{M}$ we pick a metric g with scalar curvature equal to -1. Consider a diffeomorphism φ of M such that $fo\varphi \in L_p(M)$, and an extension $\tilde{f}$ of $fo\varphi$ on $\tilde{M}$ by defining it to be identically equal to -1 on $\tilde{M} - M$. Therefore given $\varepsilon > 0$ there exists a diffeomorphism Ψ of $\tilde{M}$ such that $\|\tilde{f}o\Psi + 1\|_p < \varepsilon$. Now we can apply the proof of Theorem 6.2.

1.2. The Problem with Conformal Change of Metric

6.5 Henceforth on this chapter we will deal with the following problem:

Let (M_n, g) be a C^∞ Riemannian manifold of dimension $n \geq 2$. Given $f \in C^\infty(M)$ does there exist a metric $\tilde{g}$ conformal to g ($\tilde{g} \in [g]$), such that the scalar curvature of $\tilde{g}$ equals f?

We suppose $f \not\equiv$ Const., otherwise we would be in the special case of the Yamabe problem. The problem turns out to be very special when the Riemannian manifold is (S_n, g_0) the sphere endowed with its canonical metric. This comes from the fact that (S_n, g_0) is the unique Riemannian manifold for which the set of conformal transformations is not compact. This result was a conjecture of Lichnerowicz solved by Lelong–Ferrand [175].

Thus the problem on (S_n, g_0) is especially hard. It was raised by Nirenberg on S_2 in the sixties. Chapter 4 will deal with the Nirenberg Problem. In this chapter we suppose that (M_n, g) is not conformal to (S_n, g_0).

6.6 Recall the equations to solve.

When $n = 2$, we write the conformal change of metric on the form $\tilde{g} = e^\varphi g$. The problem is equivalent to finding a C^∞ solution of

$$\Delta\varphi + R = fe^\varphi \tag{1}$$

where R is the scalar curvature of (M_n, g).

When $n \geq 3$, we consider the change of conformal metric $\tilde{g} = \varphi^{4/(n-2)} g$. The problem is equivalent to finding a positive C^∞ solution of

$$4(n-1)(n-2)^{-1}\Delta\varphi + R\varphi = f\varphi^{N-1}, \tag{2}$$

where $N = 2n/(n-2)$. For simplicity set $\tilde{R} = (n-2)R/4(n-1)$. Then (2) becomes

(3) $$\Delta\varphi + \tilde{R}\varphi = f\varphi^{N-1}, \varphi > 0,$$

where we have written f for $(n-2)f/4(n-1)$ without loss of generality.

As the problem concerns a given conformal class of metrics, in writing equations (1) and (2) we may use in any metric in this conformal class. So when M is compact, we choose g the (or one of the) metric minimizing the Yamabe functional, accordingly $R = $ Const..

§2. The Negative Case when M is Compact

6.7 In this section we consider (1)–(3) when R (or $\tilde{R}$) are negative. The first result is in Aubin [11].

Theorem 6.7 *Let (M_n, g) be a compact C^∞ Riemannian manifold with $\mu < 0$ and $n \geq 2$. Given a C^∞ function $f < 0$, there is a unique conformal metric with scalar curvature f. μ is defined in* 2.1.

When $n \geq 3$ we consider the functional

$$I(\varphi) = \int |\nabla\varphi|^2 \, dV + \int \tilde{R}\varphi^2 \, dV.$$

Set $\nu_q = \inf I(\varphi)$ for all

$$\varphi \in \mathcal{A}_q = \left\{\varphi \in H_1/\varphi \geq 0, \int f\varphi^q \, dV = -1\right\}$$

with $2 \leq q < N$. Consider a minimizing sequence $\{\varphi_i\}$.

Since $\int \varphi_i^q \, dV \leq \frac{1}{\sup f} \int f\varphi^q \, dV = \frac{-1}{\sup f}$, $\|\varphi_i\|_2 \leq$ Const.; and $\|\varphi_i\|_{H_1} \leq$ Const. because $I(\varphi_i) \to \nu_q$. The proof proceeds now as for the Yamabe problem. In the negative case, a uniform bound in C^0 for the minimizers φ_q satisfying $\Delta\varphi_q + \tilde{R}\varphi_q = -\nu_q f\varphi_q^{q-1}$ is very easy to find. At a point P where φ_q is maximum $\Delta\varphi_q(P) \geq 0$, thus $\varphi_q^{q-2}(P) \leq -\tilde{R}/\nu_q f(P) \leq$ Const.. Uniqueness is proved by Proposition 6.8 and the solution $\Psi = \lim_{q\to N} \varphi_q$.

When $n = 2$ we consider the functional

$$I(\varphi) = \frac{1}{2}\int |\nabla\varphi|^2 \, dV + \int R\varphi \, dV.$$

Set $\nu = \inf I(\varphi)$ for all

$$\varphi \in \mathcal{A} = \left\{\varphi \in H_1/\int fe^\varphi \, dV = \int R \, dV\right\}.$$

$\nu \leq I(0) = 0$ since $\varphi \equiv 0$ is not a solution of (1) when $R = $ Const. and $f \not\equiv$ Const..

Consider a minimizing sequence $\{\varphi_i\}$, $0 > I(\varphi_i) \to \nu$.

First step. $|\int \varphi_i \, dV| <$ Const.. Obviously $\int \varphi_i \, dV > 0$. Furthermore the result follows from

$$\int \varphi_i \, dV \leq V \log \left(\int e^{\varphi_i} \, dV/V \right)$$

and

$$\int e^{\varphi_i} \, dV \leq [\inf(-f)]^{-1} \int (-f) e^{\varphi_i} \, dV = \int R \, dV / \sup f$$

where $V = \int dV$.

Second step. $\|\varphi_i\|_{H_1} <$ Const..

$I(\varphi_i) < 0$ implies $||\nabla_i||_2^2 < -2R \int \varphi_i \, dV <$ Const. and

$$\|\varphi_i\|_2^2 \leq \|\varphi_i - \bar{\varphi}_i\|_2^2 + V\bar{\varphi}_i^2 \leq \|\nabla \varphi_i\|_2^2 \lambda_1^{-1} + V\bar{\varphi}_i^2$$

where $\bar{\varphi}_i = \int \varphi_i \, dV | V$.

Third step. ν is attained by a C^∞ function.

The map $H_1 \ni \varphi \to e^\varphi \in L_p$ is compact for any p (Theorem 2.46), so a subsequence $\{\varphi_j\}$ of $\{\varphi_i\}$ tends to a weak solution of (1) in H_1. By the bootstrap method together with the regularity theorem, the solution is smooth.

Remark 6.7. Kazdan pointed out that one does not need any assumption on the exponent $q > 2$ in the negative case when f is negative. He proved that the equation

$$\Delta u = g(x, u) \tag{3b}$$

has a solution when the continuous function $g(x,t)$: $M \times \mathbb{R} \to \mathbb{R}$ has the property that there exists numbers $a < b$ so that if $t > b$ then $g(x,t) < 0$, and if $t < a$ then $g(x,t) > 0$.

When $g(x,u) = f(x)u|u|^{q-2} - \tilde{R}u$, we get a positive solution. Indeed we can use the method of lower and upper solutions with $b > a > 0$. We verify that $a > 0$ is a lower solution of (3b) if a is small enough:

$$g(x,a) \geq a(-\tilde{R} + \inf_{x \in M} f(x) a^{q-2}) > 0 = \Delta a$$

when $a < [\tilde{R}/\inf_{x \in M} f(x)]^{\frac{1}{q-2}}$. Moreover $b > [\tilde{R}/\sup_{x \in M} f(x)]^{\frac{1}{q-2}}$ is an upper solution of (3b).

6.8 Proposition (Aubin [14], Kazdan and Warner [*198]). *If $f \leq 0$ on M, equation* (1) (*resp. equation* (2)) *has at most one solution (resp. one positive solution).*

We suppose $f \not\equiv 0$ otherwise the problem has no solution. Set $\Omega = \{x \in M / f(x) = 0\}$ and let Ψ be a solution of (1) when $n = 2$ (resp. a positive solution of (2) when $n \geq 3$). Consider $\tilde{g} = e^{\psi} g$ when n=2 (resp. $\tilde{g} = \psi^{4/(n-2)} g$ when $n \geq 3$).

When $n \geq 3$, if there is another solution, equation (2) (written in the metric $\tilde{g}$)

$$(4) \qquad 4(n-1)(n-2)^{-1}\tilde{\Delta} u + fu = fu^{N-1}, \qquad (u = \varphi/\psi)$$

would have a solution not equal to the constant function $u \equiv 1$. First suppose $\Omega = \emptyset$ then $u \equiv 1$ is the unique solution of (4) indeed at a point P where u is maximum $\tilde{\Delta} u(P) \geq 0$ thus $u(P) \leq 1$; and at a point Q where u is minimum $\tilde{\Delta} u(Q) \leq 0$ thus $u(Q) \geq 1$.

If $\Omega \neq \emptyset$, $\tilde{\Delta} u = 0$ on Ω and u cannot reach a maximum or a minimum on $\mathring{\Omega}$. Therefore, if $u > 1$ somewhere on M, u attains its maximum at a point $P \notin \mathring{\Omega}$. Accordingly there is a sequence $\{P_i\} \subset M - \Omega$ which tends to P. $\tilde{\Delta} u(P) = 0$ and for P_i near enough to P, $\tilde{\Delta} u(P_i) \geq 0$. Thus $u(P_i) \leq 1$ and $u(P) \leq 1$. Likewise if Q is a point where u is minimum, $u(Q) \geq 1$.

Similarly when $n = 2$, we prove that $u \equiv 0$ is the unique solution of equation (1) written in the metric $\tilde{g}$

$$\tilde{\Delta} u + f = f e^{u} \qquad (u = \varphi - \Psi).$$

6.9 Proposition (Kazdan and Warner [*198]). *A necessary condition for a solution of* (3) *to exist is that the unique solution of*

$$(5) \qquad \Delta u - (N-2)(\tilde{R} u - f) = 0$$

is positive.

A necessary condition for a solution of (1) *to exist is that the unique solution of*

$$(6) \qquad \Delta u - Ru + f = 0$$

is positive. In both cases this implies the weaker necessary condition $\int f \, dV < 0$.

Proof. If $\varphi > 0$ satisfies (3), multiplying both members by φ^{1-N} and integrating yields $\int f \, dV < 0$. Since $u > 0$, integrating (5) gives $\int f \, dV = \tilde{R} \int u \, dV < 0$.

As $\tilde{R} < 0$, the operator $\Gamma = \Delta - (N-2)\tilde{R}$ is invertible (in the space of C^∞ functions for instance). We have to prove that if (3) has a solution $\varphi > 0$ then the unique solution u of (5) is positive. For this we compute $\Gamma(\varphi^{2-N})$ and find

$$\Gamma(\varphi^{2-N}) = -(N-2)f - (N-2)(N-1)\varphi^{-N} \nabla^i \varphi \nabla_i \varphi \leq -(N-2)f.$$

Thus $-\Gamma(\varphi^{2-N} - u) \geq 0$. According to the maximum principle $\varphi^{2-N} - u < 0$ and $u > 0$ (we have $\varphi^{2-N} - u \neq$ Const.).

Similarly, when $n = 2$, we prove $(-\Delta + R)(e^{-\varphi} - u) \geq 0$. This yields $u > e^{-\varphi}$ which is positive.

Remark. With Proposition 6.9 it is easy to find functions f satisfying $\int f\,dV < 0$ such that equations (1) and (3) have no solution.

For instance $f = -\Delta u/(N-2) + \tilde{R}u$ when $n \geq 3$, and $f = Ru - \Delta u$ when $n = 2$, where u is a function changing sign and satisfying $\int u\,dV > 0$.

6.10 Proposition (Kazdan and Warner [*198]). *If $f \in C^\infty$ is the scalar curvature of a conformal metric, any $h \in C^\infty$, satisfying $h \leq \alpha f$ for some real number $\alpha > 0$, is the scalar curvature of a conformal metric. More generally, if* (3) *has a positive solution for some $f \in C^0$, the equation*

$$(7) \qquad \Delta u + au = hu^{N-1} \qquad \textit{with } \tilde{R} \leq a < 0$$

will have a positive solution for any $h \in C^0$ satisfying $h \leq \alpha f$ with $\alpha > 0$. If (1) *has a solution for some $f \in C^0$, equation*

$$(8) \qquad \Delta u + a = he^u \qquad \textit{with } \tilde{R} \leq a < 0$$

will have a solution for any $h \in C^0$ satisfying $h \leq \alpha f$ with $\alpha > 0$.

Proof. As equation (3) has a solution for αf, we have to prove Proposition 6.10 when $\alpha = 1$. Let φ be a solution of (3), $u^+ = \varphi$ is an upper solution of (7). Indeed

$$\Delta\varphi + a\varphi - h\varphi^{N-1} = (a - \tilde{R})\varphi + (f - h)\varphi^{N-1} \geq 0$$

Pick a positive real number β small enough $\beta \leq \left[\frac{-a}{\sup(-h)}\right]^{\frac{1}{N-2}}$ and $\beta \leq \inf \varphi$, the constant function $u_- \equiv \beta$ is a lower solution of (8) which satisfies $0 < u_- \leq u^+ = \varphi$. Indeed $\Delta\beta + a\beta - h\beta^{N-1} = \beta(a - h\beta^{N-2}) \leq 0$.

We are in position to use the method of upper and lower solutions, and conclude that (7) has a solution. Similarly (8) has a solution.

Here $\beta = u_- \leq u^+ = \varphi$ with $\beta \leq \log[(-a)/\sup(-h)]$.

If h is only C^0, the solution of (7) (resp. (8)) is in $C^{1,\beta}$ for any $\beta \in]0,1[$. In this case, for the proof we use the Maximum Principle for weak solution (Remark 3.71). Kazdan and Warner state Proposition 6.10 for $h \in L_p(M)$ with $p > n$, the solution is then in C^β for some $\beta > 0$ and also in H_2^p.

6.11 Theorem. *Assume f is the scalar curvature of a conformal metric. If $f \leq 0$, there exists a neighbourhood $\mathcal{V}$ of f in C^α (any $\alpha \in]0,1[$) such that each $h \in \mathcal{V}$ is the scalar curvature of a conformal metric.*

A necessary condition for (3) *to have a solution is $-\tilde{R} < \lambda$, where*

$$\lambda = \inf\left[\|\nabla u\|_2^2 \|u\|_2^{-2}\right] \qquad \textit{for all } u \in \mathcal{D}(\mathring{\Omega}).$$

If $\mathring{\Omega} = \emptyset$, $\lambda = +\infty$. Here

$$\Omega = \{x \in M / f(x) \geq 0\},$$

f may have positive values.

Proof. Writting equation (2) in the conformal metric $\tilde{g}$ which has f as scalar curvature, the equation to solve is

$$(9) \qquad 4(n-1)(n-2)^{-1}\tilde{\Delta}u + fu - hu^{N-1} = 0, \quad u > 0.$$

Consider the map $(u,h) \xrightarrow{\Gamma} 4(n-1)(n-2)^{-1}\tilde{\Delta}u + fu - hu^{N-1}$ from

$$\{u \in C^{2,\alpha}, u > 0\} \times C^\alpha \qquad \text{in } C^\alpha.$$

$$D_u\Gamma(v) = 4(n-1)(n-2)^{-1}\tilde{\Delta}v + fv - (N-1)hu^{N-2}v.$$

At $(1, f)$, $D_u\Gamma = 4(n-1)(n-2)^{-1}\tilde{\Delta} - (N-2)f$ is invertible. Indeed,

$$\nu = \inf\left[(n-1)\int|\tilde{\nabla}v|^2 d\tilde{V} - \int fv^2 d\tilde{V}\right]\left[\int v^2 d\tilde{V}\right]^{-1}$$

is achieved by a smooth positive function, thus $\nu > 0$. Recall $N - 2 = \frac{4}{n-2}$. As Γ is continuously differentiable, we can apply the implicit function theorem. This proves the existence of $\mathcal{V}$. Similarly when $n = 2$, we write (1) in the metric $\tilde{g}$ and we consider the map

$$(u,h) \xrightarrow{\Gamma} \tilde{\Delta}u + f - he^u.$$

At $(0, f)$, $D_u\Gamma = \tilde{\Delta} - f$ is invertible, the spaces being well chosen.

Remark. According to Proposition 6.10, the neighbourhood $\mathcal{V}$ of f may be in C^0.

Proof of the second part. Let $W \subset \Omega$ be a submanifold with smooth boundary ($\overset{\circ}{W} \neq \emptyset$), and let $\tilde{\lambda}$ be the first eigenvalue for the Laplacian Δ on W with zero Dirichlet data. On W consider Ψ the eigenfunction such that $\Psi \leq \varphi$ on W with $\Psi = \varphi$ somewhere in $\overset{\circ}{W}$ (recall φ is supposed to be a solution of (3)). On $\overset{\circ}{W}$, $\Delta\Psi = \tilde{\lambda}\Psi$, $\Psi \in C^\infty$ and

$$(10) \qquad \Delta(\varphi - \Psi) = \tilde{\lambda}(\varphi - \Psi) - (\tilde{R} + \tilde{\lambda})\varphi + f\varphi^{N-1}.$$

If $\tilde{\lambda} \leq -\tilde{R}$, as $f \geq 0$ on W, (10) implies $\Delta(\varphi - \Psi) \geq 0$. Thus $\varphi > \psi$ everywhere on W since $\varphi > \Psi$ on ∂W. This contradicts $\Psi = \varphi$ somewhere.

So $-\tilde{R} < \tilde{\lambda}$, but $\tilde{\lambda}$ is as close as one wants to λ and we obtain $-\tilde{R} \leq \lambda$.

To get the strict inequality, consider $f^- = \inf(0, f)$ which is Lipschitz continuous. According to Proposition 6.10, equation (3) with $f^-\varphi^{N-1}$ in the right hand side has a positive solution in $C^{o,\alpha}$. Then using the first part of the Theorem proved above, there exists a neighbourhood of f^- in C^α, where we can choose $h \in C^\infty$ having zero as regular value and satisfying $h(x) \geq 0$ when $x \in \Omega$.

Now, if (3) has a solution, $\lambda = -\tilde{R}$ yields a contradiction. Indeed equation (3) with $h\varphi^{N-1}$ has a positive solution $\varphi_0 \in C^\infty$. Let λ_0 be the first eigenvalue of

Δ (with zero Dirichlet data) on $\Omega_0 = \{x \in M/h(x) \geq 0\}$. Ω_0 is a submanifold, thus $-\tilde{R} < \lambda_\circ$. As $\lambda_0 \leq \lambda$, we get the desired inequality.

6.12 Theorem (Ouyang [*262], Rauzy [*273], Tang [*298], Vazquez–Veron [*312]). *On (M_n, g) a C^∞ compact Riemannian manifold of dimension $n \geq 3$, let $f \leq 0$ be a C^α function ($\alpha \in]0,1]$). Define $K = \{x \in M/f(x) \geq 0\}$ and $\lambda = \inf[\|\nabla u\|_2^2 \|u\|_2^{-2}]$ for all $u \in \mathcal{D}(\overset{\circ}{K})$. Then Equation* (3) *has a solution if and only if*

$$-\tilde{R} < \lambda. \tag{11}$$

When $f \in C^\infty$, f is the scalar curvature of a conformal metric, if (11) *holds.*

This theorem is a particular case of Theorem 6.13 below. For the proof Ouyang, Vazquez and Veron use the method of bifurcation. They study equation

$$\Delta u - \lambda u = f u^p, \qquad u > 0 \tag{12}$$

with $\lambda \geq 0$ and $p > 1$. For this they consider

$$C^{2,\alpha} \times \mathbb{R} \ni (u, \lambda) \to f(u, \lambda) = \Delta u - \lambda u - f|u|^{p-1}u \in C^\alpha. \tag{13}$$

(0,0) is a point of bifurcation and there exists a C^1 bifurcated branch issuing from (0,0).

Recently Tang gave a simple proof of Theorem 6.12, using the method of upper and lower solutions, advocated by Kazdan and Warner. If we exhibit a positive upper solution u^+ of (3), a positive constant β, small enough, is a lower solution of (3) and we can take $u^- \leq u^+$. Indeed $\Delta\beta + \tilde{R}\beta \leq f\beta^{N-1}$ as soon as $\beta \leq [\tilde{R}/\inf f]^{(n-2)/4}$.

So we can choose $u^- = \beta \leq u^+$ and we are in position to use the method (see §12 of Chapter 7).

We saw in 6.11 that condition (11) is necessary. Let us prove it is sufficient. Assume $-\tilde{R} < \lambda$, there exists a neighbourhood of K : $\bar{W}$ which is a manifold with boundary whose first eigenvalue $\tilde{\lambda}$ satisfies $-\tilde{R} < \tilde{\lambda} < \lambda$ (for Δ with zero Dirichlet data). Since K is compact, W has a finite number of connected components W_i $(1 \leq i \leq k)$. On each W_i, pick $\varphi_i > 0$ an eigenfunction satisfying $\varphi_i/\partial W_i = 0$ and on W_i $\Delta\varphi_i = \lambda_i\varphi_i$, with $\lambda_i \geq \tilde{\lambda}$ the first eigenvalue for Δ on $\overline{W_i}$.

Now consider φ a positive C^∞ function on M which is equal to φ_i on a neighbourhood θ_i of $\overline{K_\cap W_i} \subset W_i$. For α large enough, let us verify that $u^+ = \alpha\varphi$ is an upper solution of (3). On any θ_i

$$\Delta u^+ + \tilde{R}u^+ = (\lambda_i + \tilde{R})u^+ > 0 \geq f(u^+)^{N-1}.$$

And on $M - \cup_{i=1}^k \theta_i$, as $f \leq -\varepsilon$ for some $\varepsilon > 0$, we will have

$$\Delta\varphi + \tilde{R}\varphi \geq -\varepsilon\alpha^{N-2}\varphi^{N-1} \geq \alpha^{N-2} f \varphi^{N-1},$$

if α is large enough.

6.13 Theorem (Rauzy [*273]). *On (M_n, g) a C^∞ compact Riemannian manifold of dimension $n \geq 3$, let f be a C^∞ function satisfying* (11) *where λ is the first eigenvalue for Δ on Ω with zero Dirichlet data (as defined in Theorem* 6.12).

There exists a positive constant C which depends only on $f^- = \sup(-f, 0)$ such that if f satisfies

$$\sup f < C \tag{14}$$

then equation (3) *has a solution (f is the scalar curvature of a conformal metric). Assume* $\sup f > 0$. *Equation* (3) *has more than one positive solution when $6 \leq n < 10$ if at a point P where f is maximum $\Delta f(P) = 0$, and when $n \geq 10$ if in addition $\|W_{ijkl}(P)\| \neq 0$ and $\Delta\Delta f(P) = 0$.*

The first part of the theorem is proved by using the mini-max method. Condition (14) means that, when f^- is given, equation (3) has a solution for any f^+ on Ω satisfying (14). For the proof of the second part of Theorem 6.13, Rauzy uses the method of points of concentration.

Remark 6.13. We can ask how C depends on f^-. The answer is given by Aubin–Bismuth in [*13].

Set $K = \{x \in M_n / f(x) \geq 0\}$, K must satisfy $\lambda(K) > -\tilde{R}$. Condition (14) is

$$\sup f \leq C(K) \inf_{(M-\Omega)} [-f(x)].$$

where Ω is a neigbourhood of K such that $\lambda(\Omega) > -\tilde{R}$, $\lambda(\Omega)$ being the first eigenvalue of Δ on Ω with zero Dirichlet data.

6.14 Theorem. *When $n = 2$, if f a C^∞ function on (M_2, g) satisfies $f \leq 0$ and $f \not\equiv 0$, there is a conformal metric with scalar curvature f.*

If we consider $f^- = \sup(-f, 0) \not\equiv 0$ as given, there exists a positive constant C such that the same conclusion holds whenever $\sup f \leq C$.

Proof. Assume $f \leq 0$ and set $\Omega = \{x \in M / f(x) = 0\}$.

Let $\overline{W}$ be a manifold with boundary which is a neighbourhood of Ω. $\overline{W}$ exists since $\Omega \neq M$. On $\bar{W}$ let w be a solution of $\Delta w + R = 0$, for instance with zero Dirichlet data. If k is large enough let us verify that $w^+ = \gamma + k$ is an upper solution of (1) when $\gamma = w$ on a neighbourhood θ of Ω, with $\bar{\theta} \subset W$.

On θ, $\Delta(\gamma + k) + R = 0 \geq f e^{\gamma + k}$. And on $M - \theta$, as $f \leq -\varepsilon$ for some $\varepsilon > 0$, we will have $\Delta(\gamma + k) + R \geq -\varepsilon e^{\gamma + k} \geq f e^{\gamma + k}$. On the other hand when $\tilde{k}$ is large enough $w^- = -\tilde{k}$ is a lower solution of (1) satisfying $w^- \leq w^+$.

Indeed $\Delta w^- + R = R \leq f e^{-\tilde{k}} = f e^{w^-}$ for large $\tilde{k}$. The method of lower and upper solutions yields a solution of (1).

For the proof of the second part of Theorem 6.14, we use Theorem 6.11. According to the proof above, $-f^-$ is the scalar curvature of a conformal metric, so there exists a neighbourhood $\mathcal{V}$ of $-f^-$ in C^α such that each function in $\mathcal{V}$

is the scalar curvature of a conformal metric. In $\mathcal{V}$ there are functions $h \geq -f^-$ which are equal on Ω to a positive constant C if C is small enough.

Now if $\sup f \leq C$, $f \leq h$ on M and by Proposition 6.10, f is the scalar curvature of some metric conformal to g.

Remark. The necessary condition of Proposition 6.9 is satisfied under the hypothesis $f \leq 0$, $f \not\equiv 0$. Indeed, G_R beeing the Green function of $\Delta - R$, the solution of (6) is $u(P) = -\int G_R(P,Q)f(Q)\,dV(Q)$.

We know that G_R satisfies $G_R \geq \varepsilon$ for some $\varepsilon > 0$. Thus $u \geq -\varepsilon \int f\,dV > 0$.

Similarly when $n \geq 3$, if $f \leq 0$ and $f \not\equiv 0$, the solution of (5) is positive. In case f changes sign, if Theorem 6.12 can be apply to the function $-f^-$ (i.e. (11) is satisfied), there is a positive constant $C(f^-)$ such that f is the scalar curvature of a conformal metric whenever $\sup f \leq C(f^-)$.

The proof is similar to that of the second part of Theorem 6.14. It is an alternative proof to the first part of Rauzy's Theorem.

§3. The Zero Case when M is Compact

6.15 In this case, the manifold carries a metric with zero scalar curvature. In this metric equations (1) and (3) are

$$\Delta\varphi = fe^{\varphi}, \qquad \text{when } n = 2. \tag{15}$$

$$\Delta\varphi = f\varphi^{N-1}, \qquad \varphi > 0 \quad \text{when } n \geq 3. \tag{16}$$

Obviously there are two necessary conditions:

$$f \text{ changes sign} \tag{17}$$

$$\int f\,dV < 0. \tag{18}$$

$\int \Delta\varphi\,dV = 0$ implies (17). Multiplying (15) by $e^{-\varphi}$, (16) by φ^{1-N} and integrating yield (18).

The zero case is not different than the positive case, we can use the variational method. Whereas the negative case, when f changes sign, is very peculiar.

6.16 Theorem. *When $n = 2$, $f \in C^\infty$ is the scalar curvature of a conformal metric (equation* (15) *has a solution) if and only if* (17) *and* (18) *hold.*

Proof. Define $\nu = \inf \|\nabla u\|_2$ for all

$$u \in \mathcal{A} = \left\{\varphi \in H_1 \Big/ \int fe^{\varphi}\,dV = 0\right\}.$$

To see that $\mathcal{A} \neq \emptyset$, let $\varphi \in C^\infty$ a function which is positive in a ball Ω where $f > 0$ and which is zero outside Ω, $\int fe^{\alpha\varphi}\,dV = 0$ for some $\alpha > 0$ since $\int f\,dV < 0$.

Consider a minimizing sequence $u_i \in \mathcal{A}$. Set $u_i = v_i + \bar{u}_i$ where $\bar{u}_i = V^{-1}\int u_i\,dV$, $\nu = \lim_{i\to\infty} \|\nabla v_i\|_2$ and $\int fe^{v_i}\,dV = 0$. As $\bar{v}_i = 0$ the set $\{v_i\}$ is bounded in H_1.

So there exist $\Psi \in H_1$ and a subsequence $\{v_j\}$ such that $v_j \to \Psi$ weakly in H_1, strongly in L_2 and such that $e^{v_j} \to e^{\psi}$ in L_1 since the map $H_1 \ni \varphi \to e^{\varphi} \in L_1$ is compact (Theorem 2.46).

This implies $\int fe^{\psi}\,dV = \lim_{j\to\infty}\int fe^{v_j}\,dV = 0$, thus $\psi \in \mathcal{A}$ and $\|\nabla\|_2 = \nu$ since $\|\nabla\|_2 \leq \lim \|\nabla v_j\|_2 = \nu$. We cannot have $\nu = 0$, otherwise $\psi \equiv 0$ which contradicts $\int f\,dV < 0$. Hence ψ satisfies

$$\Delta\psi = kfe^{\psi} \qquad \text{with } k \in \mathbb{R}\,. \tag{19}$$

Multiplying both members by $e^{-\psi}$ and integrating implies $k\int f\,dV < 0$. Thus $k > 0$ and $\varphi = \psi + \log k$ is a solution of (15). Regularity follows by a standard bootstrap argument.

6.17 Proposition. *When $n \geq 3$, if* (17) *and* (18) *hold, there is a positive C^∞ solution φ_q of the equation $\Delta\varphi = f\varphi^{q-1}$ for $2 < q < N$.*

Proof. Define $\nu_q = \inf \|\nabla u\|_2^2$ for all

$$u \in \mathcal{A}_q = \left\{ u \in H_1/u \geq 0, \int fu^q\,dV = 1 \right\}.$$

$\mathcal{A}_q \neq \emptyset$ (see the proof of 6.16). Consider a minimizing sequence $\{u_i\}$. If no subsequence of the sequence $\{\|u_i\|_2\}$ is bounded, set $v_i = u_i/\|u_i\|_2$.
The functions v_i satisfy $\|v_i\|_2 = 1$, $\|\nabla v_i\|_2 \to 0$ and $\int fv_i^q\,dV \to 0$. Thus $\{v_i\}$ is bounded in H_1 and $v_i \to V^{-1/2}$ in H_1 (V is the volume). This implies a contradiction with (18) since we would have $\int f\,dV = V^{q/2}\lim\int fv_i^q\,dV = 0$.

Similarly $\nu_q \neq 0$, otherwise as we know now that $\{u_i\}$ is bounded in H_1, the constant function is a minimizer and this implies a contradiction.

Consequently $\nu_q > 0$ and there exists a subsequence $\{u_j\}$ which is bounded in H_1. As $H_1 \subset L_q$ is compact, we prove (as we did many times) the existence of a positive solution $\varphi_q \in C^\infty$ satisfying

$$\Delta\varphi_q = \nu_q f\varphi_q^{q-1} \quad \text{and} \quad \int f\varphi_q^q\,dV = 1, \tag{20}$$

where $\nu_q > 0$ is the inf of the variational problem considered above.

6.18 Lemma. *The set of the functions φ_q satisfying* (20) *is bounded in H_1.*

First of all let us prove that the set of the ν_q is bounded. Thus we will have

$$\|\nabla\varphi_q\|_2 \leq \text{Const.}.$$

For this we may pick any function $u \geq 0$, $u \not\equiv 0$, with support in the subset of M where $f \geq 0$. But in order to have a proof useful in a more general context we choose $u = f^+ = \sup(f, 0)$. $\gamma = f^+\left[\int (f^+)^{q+1}\, dV\right]^{-1/q} \in \mathcal{A}_q$ (see the proof of 6.17 for the definition of $\mathcal{A}_q$).

Thus,

$$\begin{aligned}
\nu_q &\leq \|\nabla\gamma\|_2^2\, \|\nabla f^+\|_2^2 \left[\int (f^+)^{q+1}\, dV\right]^{-2/q} \\
&\leq \|\nabla f^+\|_2^2 \left(\int f^+\, dV\right)^{1-2/q} \left[\int (f^+)^3\, dV\right]^{-1} \\
&\leq \sup\left(\int f^+\, dV, 1\right) \|\nabla f^+\|_2^2 \left[\int (f^+)^3\, dV\right]^{-1}.
\end{aligned}$$

Now the proof is by contradiction. Suppose $\|\varphi_q\|_2$ is not uniformly bounded. There exists a sequence q_i such that $\|\varphi_{q_i}\|_2 \to \infty$ when $i \to \infty$. Then, $v_i = \varphi_{q_i}\|\varphi_{q_i}\|_2^{-1}$ is a sequence of C^∞ functions which satisfies $\|\nabla v_i\|_2 \to 0$, $\|v_i\|_2 = 1$ and $\int f v_i^{q_i}\, dV \to 0$. Thus $v_i \to V^{-1/2}$ in H_1 when $i \to \infty$ and by the Sobolev Theorem $\|v_i - V^{-1/2}\|_N \to 0$. This implies

$$\begin{aligned}
\int |v_i^{q_i} - V^{-q_i/2}|\, dV &\leq q_i \int |v_i - V^{-1/2}| |v_i^{q_i-1} + V^{-(q_i-1)/2}|\, dV \\
&\leq \text{Const.}\, \|v_i - V^{-1/2}\|_N \to 0.
\end{aligned}$$

So $\int f v_i^{q_i}\, dV \to 0$ yields $\int f\, dV = 0$ which contradicts with (18) the necessary assumption $\int f\, dV < 0$.

6.19 Now we need to use a Theorem which will also be useful later on, which is why we consider a more general situation.

Let (M, g) be a C^∞ compact riemannian manifold. Consider the operator

$$Lu = \Delta u + hu \tag{21}$$

where $h \in C^\infty$.

Theorem 6.19 (Aubin [14] p. 280). *Assume there exists a sequence, bounded in H_1, of positive C^∞ functions φ_{q_i} $(2 < q_i < N, \lim q_i = N)$ satisfying $\int f \varphi_{q_i}^{q_i}\, dV = 1$ and*

$$L\varphi_{q_i} = \mu_i f \varphi_{q_i}^{q_i-1} \tag{22}$$

where f is a C^∞ function with $\sup f > 0$ and μ_i positive real numbers. If

$$0 < \mu = \lim_{i\to\infty} \mu_i < n(n-2)\omega_n^{2/n}/4[\sup f]^{1-2/n} \tag{23}$$

then a subsequence $\{\varphi_{q_j}\}$ converges weakly in H_1 to a positive C^∞ function ψ which satisfies

$$L\psi = \mu f \psi^{N-1}. \tag{24}$$

Proof. Since $\|\varphi_{q_i}\|_{H_1} <$ Const., we can proceed as for the Yamabe problem. There exists a subsequence $\{\varphi_{q_j}\}$ and $\psi \in H_1$ such that $\varphi_{q_j} \to \psi$ weakly in H_1, strongly in L_2 and almost everywhere. Then $\varphi_{q_j}^{q_j-1} \to \psi^{N-1}$ weakly in $L_{N/(N-1)}$ and ψ satisfies (24) weakly in H_1.

According to Trudinger [262], $\psi \in C^\infty$. Then the maximum principle implies either $\psi > 0$ or $\psi \equiv 0$. In order to show that the last case cannot occur, we must prove that $\|\psi\|_2 \neq 0$.

For this we will use $K(n,2) = 2\omega_n^{-1/n}[n(n-2)]^{-1/2}$ the best constant in the Sobolev inequality (Proposition 2.18). Given $\varepsilon > 0$ there exists $A(\varepsilon)$ such that all $\varphi \in H_1$ satisfy:

$$\|\varphi\|_N^2 \leq [K^2(n,2)+\varepsilon]\|\nabla\varphi\|_2^2 + A(\varepsilon)\|\varphi\|_2^2 \tag{25}$$

We now write

$$1 = \int f\varphi_{q_i}^{q_i}\, dV \leq (\sup f)\|\varphi_{q_i}\|_{q_i}^{q_i} \leq (\sup f)V^{1-\frac{q_i}{N}}\|\varphi_{q_i}\|_N^{q_i}. \tag{26}$$

From (21) and (22) we get

$$\|\nabla\varphi_{q_i}\|_2^2 + \int h\varphi_{q_i}^2\, dV = \mu_i \tag{27}$$

(25) and (26) together with (27) yield

$$1 - (\sup f)^{2/q_i}V^{2(N-q_i)/Nq_i}[K^2(n,2)+\varepsilon]\mu_i \leq \text{Const.}\, \|\varphi_{q_i}\|_2^2.$$

Taking the lim inf of both sides when $i \to \infty$ we find

$$1 - (\sup f)^{2/N}[K^2(n,2)+\varepsilon]\mu \leq \text{Const.}\, \liminf_{i\to\infty}\|\varphi_{q_i}\|_2^2.$$

Since ε is as small as one wants, if μ satisfies (23) then $\|\psi\|_2^2 > 0$.

6.20 Remark. When $f \geq 0$ there is an alternative proof of Theorem 6.19 which doesn't use Trudinger's Theorem. For this see (6.39) below. We do the same computations without the function $\eta(\eta \equiv 1)$. In the left hand side of (45), the limit when $i \to \infty$ of the term in brackets is

$$l = 1 - k^2(2k-1)^{-1}C\mu\varepsilon_0^{2/n}(\sup f)^{2/N}.$$

The assumption (23) $\mu(\sup f)^{2/N}K^2(n,2) < 1$ allows us to pick C enough near $K^2(n,2)$ and $k > 1$ enough near 1 so that $l > 0$, since $\varepsilon_0 = 1$ according to the hypothesis.

So we get for some $k > 1$, $\|\varphi_{q_i}\|_{kN} \leq$ Const.. This is sufficient to prove that $\{\varphi_{q_i}\}$ is bounded in C^r for any $r \geq 0$. The proof begins by using the Green function of the Laplacian as in 5.5: we prove that the functions φ_{q_i} are uniformly bounded in C^0 then in C^1.

The bound in C^r is obtained by induction thanks to the regularity theorems (§6.2 of Chapter 3).

Hence a subsequence of $\{\varphi_{q_i}\}$ converges in C^{r-1} to a smooth positive solution of (24).

6.21 Let's return to our problem, the existence of a positive C^∞ solution of (16). Of course we assume (17) and (18). According to Proposition 6.17, Lemma 6.18 and Theorem (6.19), the problem is reduced to find sufficient conditions so that

$$\text{(28)} \qquad \nu = \lim_{q\to N} \nu_q < n(n-2)\omega_n^{2/n}/4[\sup f]^{1-2/n}.$$

Moreover, ν is equal to $\bar\nu = \inf$ of $\|\nabla\varphi\|_2^2$ for all

$$\varphi \in \mathcal{A} = \left\{\varphi \in H_1/\varphi \geq 0, \textstyle\int f\varphi^N\,dV = 1\right\}.$$

Let us prove by contradiction that ν_q has a limit when $q \to N$ and that $\nu = \bar\nu$. Consider a sequence ν_{q_i} with $q_i \to N$ such that $\lim_{i\to\infty} \nu_{q_i} = \nu \neq \bar\nu$.

We will prove in 6.39 (see Corollary 6.40) that there exists a subsequence $\{q_j\}$ of $\{q_i\}$ such that the sequence $\{\varphi_{q_j}\}$ converges uniformly on the set $K = \{x \in M/f(x) \leq 0\}$.

Moreover

$$\int_{M\setminus K} f\varphi_{q_j}^{q_j}\,dV \leq \left(\int_{M\setminus K} f\varphi_{q_j}^N\,dV\right)^{q_j/N}\left(\int_{M\setminus K} f\,dV\right)^{1-q_j/N}.$$

Thus, given $\varepsilon > 0$, if q_j is enough close to N

$$\int_{M\setminus K} f\varphi_{q_j}^N\,dV > \int_{M\setminus K} f\varphi_{q_j}^{q_j}\,dV - \varepsilon \quad \text{and} \quad \left|\int_K \left(\varphi_{q_j}^N - \varphi_{q_j}^{q_j}\right)dV\right| < \varepsilon.$$

This implies $a_j^N = \int f\varphi_{q_j}^N\,dV > 1 - 2\varepsilon$. Hence

$$\bar\nu \leq \|\nabla\varphi_{q_j}\|_2^2 a_j^{-2} = \nu_{q_j} a_j^{-2} < (1-\varepsilon)^{-2/N}\nu_{q_j}.$$

We find $\bar\nu \leq (1-\varepsilon)^{-2/N}\nu$ for any $\varepsilon > 0$. So $\bar\nu < \nu$ according to our assumption.

In that case there exists a positive C^∞ function u such that $\int fu^N\,dV = 1$ and $\|\nabla u\|_2^2 < \nu_{q_j}$ for all $q_j > N - \eta$ ($\eta > 0$ some real number). Now this is impossible since

$$\int fu^{q_j}\,dV \to 1 \qquad \text{when } q_j \to N.$$

Remark. We have always $\nu \leq n(n-2)\omega_n^{2/n}/4[\sup f]^{1-2/n}$.

The proof is similar to that for the Yamabe problem. We take the standard test-functions centered at a point where f is maximum. As results of existence we have

6.22 Theorem (Escobar and Schoen [*129]). *Suppose (M, g) of dimension $n \geq 3$ is locally conformally flat with zero scalar curvature. Assume $f(P) > 0$ at a point P, where the C^∞ function f attains its maximum. If all its derivatives of order less than or equal to $n-3$ vanish at P and if $\int f\,dV < 0$, then f is the scalar curvature of a conformal metric.*

When $n = 3, 4$ the locally conformally flat assumption on M can be removed. So, in these cases,the result is optimal.

We have the same result for the dimension $n = 5$, according to Bismuth [*54B].

6.23 Theorem (Aubin–Hebey [*15]). *Let f be a C^∞ function satisfying $\int f\,dV < 0$ and $\sup f > 0$. If at a point P where f is maximum the Weyl tensor is not zero, then f is the scalar curvature of a conformal metric in the following cases:*

when $n = 6$ if $\Delta f(P) = 0$
when $n > 6$ if $\Delta f(P) = 0$ and $|\Delta\Delta f(P)|/f(P)$ is small enough.

§4. The Positive Case when M is Compact

6.24 We write the equation in the metric which minimizes the Yamabe functional: $R = \text{Const.} > 0$. When $n \geq 3$ we set $\tilde{R} = (n-2)R/4(n-1)$. We have to solve

$$\text{(29)} \qquad \Delta\varphi + R = fe^{\varphi}, \qquad \text{when } n = 2$$

$$\text{(30)} \qquad \Delta\varphi + \tilde{R}\varphi = f\varphi^{N-1}, \varphi > 0 \qquad \text{when } n \geq 3.$$

Since we deal with the sphere in §7 to 10 below, when $n = 2$ the manifold is the projective space. If it comes from the unit sphere, $R = 2$ and its volume $V = 2\pi$. If $\varphi \in C^\infty$ satisfies (29) or (30), at a point P where f is maximum, $\Delta\varphi(P) \geq 0$ and we get $f(P) > 0$. The only necessary condition for f to be the scalar curvature of a conformal metric is f is positive somewhere.

6.25 Theorem (M.S Berger [40] and J. Moser [*245]). *On the projective space $P_2(\mathbb{R})$, any $f \in C^\infty$ with $\sup f > 0$ is the scalar curvature of a conformal metric.*

Proof. Define $I(\varphi) = \frac{1}{2}\|\nabla\varphi\|_2^2 + 2\int \varphi\,dV$ and set $\lambda = \inf I(\varphi)$ for all

$$\varphi \in \mathcal{A} = \left\{u \in H_1 \Big/ \int fe^u\,dV = 1\right\}, \quad \mathcal{A} \neq \emptyset.$$

Recall the following inequality:
For any $\varepsilon > 0$ there exists $C(\varepsilon)$ such that all $\varphi \in H_1$ satisfy

$$\text{(31)} \qquad \int e^{\varphi}\,dV \leq c(\varepsilon)\exp\left[(\varepsilon + 1/16\pi)\|\nabla\varphi\|_2^2 + (1/2\pi)\int \varphi\,dV\right].$$

For $u \in \mathcal{A}$, $1 = \int fe^u\,dV \leq \sup f \int e^u\,dV$. Thus

$$\text{(32)} \qquad \int u\,dV \geq -(2\pi\varepsilon + 1/8)\|\nabla u\|_2^2 - \text{Const.}$$

This implies

$$(33) \qquad I(u) \geq (1/4 - 4\pi\varepsilon)\|\nabla u\|_2^2 - \text{Const.}$$

As ε is as small as one wants, λ is finite. Let $\{u_i\}$ be a minimizing sequence. From (33) we get $\|\nabla u_i\|_2 \leq$ Const. since $I(u_i) \to \lambda$ when $i \to \infty$.

$I(u_i) \leq$ Const. and (32) imply $|\int u_i \, dV| \leq$ Const..

So $\|u_i\|_{H_1} \leq$ Const.. The method used many times yields a minimizer $\Psi \in H_1$ for our variational problem. For this we use the compactness of the map $H_1 \ni \varphi \to e^\varphi \in L_1$. Thus Ψ satisfies (29) weakly in H_1. By a bootstrap argument $\Psi \in C^\infty$.

Berger's Problem is described in Chapter 5 (§8.8).

6.26 Proposition. *When $n \geq 3$, if $\sup f > 0$ there is a positive C^∞ solution $\varphi_q (2 < q < N)$ of the equation $\Delta\varphi + \tilde{R}\varphi = f\varphi^{q-1}$, $\varphi > 0$.*

Proof. By the variational method as for the Yamabe problem (see 5.5). Define $I(u) = \|\nabla u\|_2^2 + \tilde{R}\|u\|_2^2$ and $\lambda_q = \inf I(u)$ for all

$$u \in \mathcal{A}_q = \left\{\varphi \in H_1/\varphi \geq 0, \ \int f\varphi^q \, dV = 1\right\}.$$

$\mathcal{A}_q \neq \emptyset$ since we have supposed $\sup f > 0$. Consider $\{u_i\}$ a minimizing sequence. As $I(u_i) \to \lambda_q$, $\{u_i\}$ is bounded in H_1. A subsequence $\{u_j\}$ converges to φ_q weakly in H_1, strongly in L_q (since $q_i < N$) and a.e.. Hence, $\varphi_q \geq 0$ satisfies weakly in H_1

$$(34) \qquad \Delta\varphi_q + \tilde{R}\varphi_q = \lambda_q f \varphi_q^{q-1}.$$

Moreover $\int f\varphi_q^q \, dV = 1$ implies $\varphi_q \not\equiv 0$ and $\lambda_q > 0$.

According to the maximum principle, $\varphi_q > 0$, and by the regularity theorems $\varphi_q \in C^\infty$.

6.27 Proposition. *The variational problem, considered above, has a minimizer $\varphi_q \in C^\infty$. $\varphi_q > 0$ satisfies (34) and $\int f\varphi_q^q \, dV = 1$. The set of the functions φ_q ($q \in [2, N[$) is bounded in H_1.*

If $\lambda = \inf I(u)$ for all $u \in \mathcal{A} = \{\varphi \in H_1/\varphi \geq 0, \int f\varphi^N \, dV = 1\}$, $\lim_{q\to N} \lambda_q = \lambda$. Since $\lambda_q \leq$ Const. (see $\nu_q \leq$ Const. in Lemma 6.18), $\|\varphi_q\|_{H_1} \leq \lambda_q / \inf(1, \tilde{R}) \leq$ Const.. The proof of the last assertion is in (6.21). Then Theorem 6.19 implies:

6.28 Theorem (Aubin [14] p. 280). *If*

$$(35) \qquad \lambda < n(n-2)\omega_n^{2/n}/4[\sup f]^{1-2/n} = \Lambda,$$

equation (30) has a C^∞ solution which minimizes $I(u)$ over $\mathcal{A}$. Therefore f is the scalar curvature of a conformal metric.

With this theorem, the problem is reduced to finding a test-function u_0 satisfying $I(u_0) < \Lambda$. As $\lambda \leq \Lambda$, an asymptotic expansion can produce the desired inequality.

6.29 Corollary (Aubin [14] p. 286). *When $\int f\,dV > 0$, if*

$$\int R\,dV \leq n(n-1)\omega_n^{2/n}\left[\int f(x)\,dV/\sup f\right]^{1-2/n}, \tag{36}$$

f is the scalar curvature of a conformal metric.

Taking $\left[\int f\,dV\right]^{-1/N}$ as test-function yields

$$\lambda \leq I\left(\left[\int f\,dV\right]^{-1/N}\right) = \left[\int f\,dV\right]^{-2/N}(n-2)\int R\,dV/4(n-1).$$

Then applying (36) gives

$$\lambda \leq n(n-2)\omega_n^{2/n}/4[\sup f]^{1-2/n}$$

If $\lambda = \Lambda$, the constant function is a minimizer, f is proportional to $\tilde{R}$. Otherwise $\lambda < \Lambda$ and we can apply Theorem 6.28.

6.30 Theorem (Aubin [14] p. 289). *If (M, g') is not conformal to the sphere with the standard metric, there exists a constant $k > 1$ (which depends on the manifold) such that any $f \in C^\infty$ satisfying*

$$0 < \sup f \leq k \inf f \tag{37}$$

is the scalar curvature of a conformal metric.

Proof. The hypothesis on (M, g') implies (see Aubin's conjecture 5.11) that μ the inf of the Yamabe functional, achieved by the metric $g \in [g']$, satisfies:

$$\mu = RV^{2/n} < n(n-1)\omega_n^{2n}.$$

Pick $k = [n(n-1)/R]^{n/(n-2)}(\omega_n/V)^{2/(n-2)}$, (37) implies

$$\left[\sup f/\inf f\right]^{1-2/n} \leq k^{1-2/n} = n(n-1)(\omega_n/V)^{2/n}/R$$

and

$$R\left[V\sup f\Big/\int f\,dV\right]^{1-2/n} \leq R\left[\frac{\sup f}{\inf f}\right]^{1-2/n} \leq n(n-1)\omega_n^{2n}V^{-\frac{2}{n}}$$

which is inequality (36). The result follows from Corollary (6.29).

6.31 Theorem (Escobar–Schoen [*129]). *Let f be a C^∞ function with* $\sup f > 0$ *on a compact riemannian manifold (M_n, g) not conformal to the sphere with the* *standard metric. Then f is the scalar curvature of a conformal metric when $n = 3$. The same conclusion holds for the locally conformally flat manifolds when $n \geq 4$ if, at a point P where f is maximal, all its derivatives up to order $n - 2$ vanish.*

When the manifold is locally conformally flat there exists a metric g' which is flat in a neighbourhood of P. Escobar–Schoen use the test-functions centered at P that Schoen constructed for the Yamabe Problem. In the limited expansion in $r = d(P, Q)$, the first term after the constant $\Lambda = n(n-2)\omega_n^{2/n}/4[\sup f]^{1-2/n}$ will be that with $-\alpha(P)\big($ see 5.28, $A = \alpha(P)\big)$ if the function f is enough flat at P. It is the reason of the hypothesis on the derivatives of f at P. This result is improved by Hebey–Vaugon.

6.32 Theorem (Aubin–Hebey [*15]). *Define*

$$W = \{Q \in M / |W_{ijkl}(Q)| \neq 0\}.$$

Let f be a C^∞ function with $\sup f > 0$. If at a point P where f attains its maximum $P \in W$ and $\Delta f(P) = 0$ and $|W_{ijkl}(P)| \neq 0$, then f is the scalar curvature of a conformal metric when $n = 6$. When $n \geq 7$, the same conclusion holds if in addition $\frac{|\Delta\Delta f(P)|}{|f(P)|}$ is small enough.

We use the test-functions $(\varepsilon + r^2)^{1-n/2}$ of the original proof of the Yamabe problem. When $n \geq 6$ the first term after the constant Λ, in the limited expansion in $r = d(P, Q)$, will be that with $-|W_{ijkl}|^2$ if $\Delta f(P) = 0$. When $n \geq 7$ a term with $|\Delta^2 f(P)|/|f(P)|$ is of the same order of that with $-|W_{ijkl}|^2$. We set $\Delta^1 f = \Delta f$ and $\Delta^k f = \Delta\Delta^{k-1} f$ for $k > 1$ entire.

6.33 Theorem (Hebey–Vaugon [*168]). *Let f be a C^∞ function satisfying $\sup f > 0$ and $\Delta f(P) = 0$ at a point P where f is maximum.Then f is the scalar curvature of a conformal metric when $n = 4$ or 5. When $n \geq 6$ we suppose $|W_{ijkl}(P)| = 0$. The same conclusion holds if $|\Delta^2 f(P)| = 0$, when $n = 6$ or 7, and when $n = 8$ if in addition $|\nabla W_{ijkl}(P)| \neq 0$ or $\Delta^3 f(P) = 0$.*

When $n > 8$ the same conclusion holds if $|\nabla W_{ijkl}(P)| \neq 0$, $\Delta^2 f(P) = 0$ and $\Delta^3 f(P) = 0$, or when $|\nabla W_{ijkl}(P)| = 0$ if $|\nabla^2 W_{ijkl}(P)| \neq 0$, $\Delta^2 f(P) = 0$, $\Delta^3 f(P) = 0$ and $\Delta^4 f(P) = 0$, or when all derivatives of W_{ijkl} vanish at P if $\Delta^k f(P) = 0$ for all k satisfying $1 \leq k \leq n/2 - 1$.

For other results when $|\nabla^2 W_{ijkl}| = 0$ and $|\nabla^k W_{ijkl}(P)| \neq 0$ for some $k > 2$ see [*168]. For the proof they use their test-functions (see 5.50) and the positive mass theorem. From Theorems 6.32 and 6.33 we get

6.34 Corollary. *When $n \geq 4$, the set of the functions which are the scalar curvature of some conformal metric is C^1 dense in the set of the C^∞ functions which are positive somewhere.*

Given $f \in C^\infty$ with $\sup f > 0$, for any $\varepsilon > 0$ there exists $\tilde{f}$ satisfying $\|f - \tilde{f}\|_{C^1} < \varepsilon$, $\sup \tilde{f} = \sup f$ and $\tilde{f} = \sup f$ in a neighbourhood $\mathcal{V}$ of P a point where f(P) = $\sup f$. At a point $Q \in \mathcal{V}$ we can apply Theorem 6.32 if (M, g) is not locally conformally flat at P. Otherwise we can apply theorem 6.33.

6.35 Remark. All the results obtained are proved by using Theorem 6.28: Find sufficient conditions which imply that λ, the inf of the functional, is smaller than Λ. When the function f is neither close to the constant function nor flat enough at a point where f attains its maximum, we suspect that $\lambda = \Lambda$ and that there is no minimizer. In this case we must use other methods (those used for the Nirenberg problem), for instance the method of isometry-concentration (*Hebey's method*) studied in the next paragraph or algebraic-topology methods (*Bahri–Coron's method*) studied in Chapter 5 (see 5.78). With this method, we can prove many results of the following type:

Theorem 6.35 (Aubin–Bahri [*11]). *On a compact manifold (M_n, g), of dimension $n > 4$, let f be a C^2 function with only non-degenerate critical point $y_0, y_1, \dots y_k$.*

We suppose $\Delta f(y_i) > 0$ for $0 \leq i \leq l$, $\Delta f(y_j) < 0$ for $l < j \leq k$ and

$$f(y_0) \geq \dots \geq f(y_l) > f(y_{l+1}) \geq \dots \geq f(y_k).$$

Let Z be a pseudo-gradient for f which has the Morse–Smale property. For this pseudo-gradient we define $X = \overline{\cup_{i \leq l} W_s(y_i)}$, where $W_s(y_i)$ is the stable manifold of y_i. Assume X is non-contractible, but contractible in K^c for some positive real number $c < f(y_l)$. There exists a constant c_0 independent of f such that if $f(y_0)/c \leq 1 + c_0$ then f is the scalar curvature of some metric conformal to g.

Here $K^c = \{x \in M / f(x) \geq c\}$. The constant c_0 is of the order of 1. Let us remark that we do not assume that f is positive everywhere. On the sphere (S_n, g_0), this theorem can be seen as a generalization of Chang and Yang's theorem (see 6.81). In [*7] Aubin–Bahri generalized this result.

For compact manifolds of dimension four, (the sphere (S_4, g_0) included), Ben Ayed–Chtioui–Hammami (E.N.I.T. of Tunis) and Y. Chen (Rutgers University) obtained a nice result in [*34]. It is not a generalisation of the Bahri–Coron Theorem 6.87. The result is of a new type and the proof technically still more difficult. The hypothesis is different, although it is of the same kind.

§5. The Method of Isometry-Concentration

5.1. The Problem

6.36 Let (M_n, g) be any compact C^∞ riemannian manifold of dimension $n \geq 3$ and scalar curvature $R \geq 0$. The manifold may have a smooth boundary ∂M. We consider a group of isometries G, which can be reduced to the identity.

Given f a G-invariant C^∞ function on M, the problem is: Find a G-invariant metric g' conformal to g such that the scalar curvature $R' = f$.

More precisely we want to find a G-invariant C^∞ solution of the equation

$$\Delta\varphi + \tilde{R}\varphi = f\varphi^{N-1}, \qquad \varphi > 0. \tag{38}$$

When $\partial M \neq \emptyset$, φ must vanish on the boundary. Here $\tilde{R} = \frac{(n-2)}{4(n-1)} R$ and $N = \frac{2n}{n-2}$. Of course we suppose that f satisfies the necessary condition: f positive somewhere (see 6.24) and when $R \equiv 0$ with $\partial M = \emptyset$ the second necessary condition $\int f\,dV < 0$ (see 6.15).

6.37 We rewrite the proofs of (6.17) and (6.26) with G-invariant functions. So there exists for any $q \in [2, N[$ a G-invariant function $\varphi_q \geq 0$ satisfying for any Ψ G-invariant:

$$\int \varphi(\Delta\Psi + \tilde{R}\Psi)\,dV = \mu_q \int f\varphi_q^{q-1}\Psi\,dV,$$

where $\mu_q = \inf I(\varphi)$ for all

$$\varphi \in \mathcal{A}_q = \left\{u \in \mathring{H}_1 / u \geq 0,\ u\ G\text{-invariant},\ \int fu^q\,dV = 1\right\}.$$

Recall $I(\varphi) = \int |\nabla\varphi|^2\,dV + \int \tilde{R}\varphi^2\,dV$.

When $R \not\equiv 0$, $\Delta + \tilde{R}$ is invertible. Let Ψ_q be the solution of the equation

$$\Delta\Psi_q + \tilde{R}\Psi_q = \mu_q f\varphi_q^{q-1}.$$

As the right hand side is G-invariant, the equation has a G-invariant solution. So we have for all G-invariant functions Ψ

$$\int (\varphi_q - \Psi_q)(\Delta\Psi + \tilde{R}\Psi)\,dV = 0.$$

Pick $\Psi = \varphi_q - \Psi_q$, we find $\varphi_q = \Psi_q$. Thus φ_q is a positive C^∞ function satisfying

$$\Delta\varphi_q + \tilde{R}\varphi_q = \mu_q f\varphi_q^{q-1} \quad \text{and} \quad \int f\varphi_q^q\,dV = 1. \tag{39}$$

When $R \equiv 0$, the proof is similar and φ_q satisfies (39). The set $\{\varphi_q\}$ $q \in [2, N[$ is bounded in H_1, this is already proved in (6.18) when $R \equiv 0$. Indeed as f

is G-invariant, $f^+ = \sup(f,0)$ is G-invariant. So $\gamma = f^+\left[\int (f^+)^{q+1}\,dV\right]^{-1/q} \in \mathcal{A}_q$ thus

$$\|\varphi_q\|_{H_1} \le I(\varphi_q)/\inf(1,\tilde{R}) = \frac{\mu_q}{\inf(1,\tilde{R})} \le I(\gamma)/\inf(1,\tilde{R}) \le \text{Const.}.$$

Rewritting the proof of Theorem (6.19), there exists a sequence $q_i \to N$ such that either φ_{q_i} converges in L_2 to a G-invariant positive C^∞ solution of (38) or $\varphi_{q_i} \to 0$ in L_2 and almost everywhere.

In this case $\varphi_{q_i} \to 0$ in all L_q with $q < N$. Indeed as $\{\varphi_{q_i}\}$ is bounded in H_1, it is bounded in L_N and the result follows from the inequality (when $2 < q < N$):

$$\int \varphi_{q_i}^q\,dV \le \left(\int \varphi_{q_i}^2\,dV\right)^{(N-q)/(N-2)} \left(\int \varphi_{q_i}^N\,dV\right)^{(q-2)/(N-2)}.$$

We will study such a sequence.

6.38 Definition. P is a point of concentration for the sequence $\{\varphi_{q_j}\}$ which is supposed tending to zero a.e. and in L_q for all $q < N$, if for any $\delta > 0$

(40) $$\lim_{q_j \to N} \int_{B(P,\delta)} f\varphi_{q_j}^{q_j}\,dV > 0.$$

We take a subsequence for which (40) holds in case we only have that the lim sup of the integral in (40) is positive. As there is only a finite number of points of concentration (see below), without loss of generality we suppose that either the limit of the integral is zero or (40) holds.

Remark. Let us consider a sequence of C^∞ functions $u_i > 0$ which satisfy $\int f u_i^N\,dV = 1$ and

(39b) $$\Delta u_i + \tilde{R} u_i - \nu_i f u_i^{N-1} = w_i$$

with $w_i \to 0$ in H_{-1} and $\nu_i \to \nu > 0$ when $i \to \infty$.

Assume the sequence $\{u_i\}$ tends to zero a.e. and in L_2. Thus $u_i \to 0$ in L_q for any $q < N$.

For such a sequence, P is a point of concentration if for any $\delta > 0$

(40*b*) $$\lim_{i \to \infty} \int_{B(P,\delta)} f u_i^N\,dV > 0.$$

5.2. Study of the Sequence $\{\varphi_{q_j}\}$

6.39 Theorem. *Suppose the C^∞ positive functions $\varphi_{q_j}(q_j \to N)$ satisfy* (39) *and are uniformly bounded in H_1. There exist $\varepsilon_0 > 0$ and $\delta_0 > 0$ such that if $\int_{B(P,\delta)} f\varphi_{q_j}^{q_j}\,dV \leq \varepsilon_0$ for all q_j and all $\delta \leq \delta_0$, then we can exhibit a subsequence $\{\varphi_{q_i}\}$ of $\{\varphi_{q_j}\}$ so that $\{\varphi_{q_i}\}$ converges to zero in $C^r(r \geq 0$ given) on a neighbourhood of P.*

Proof. Multiply (39), with $\varphi_q = \varphi_{q_i}$, by $\eta^2\varphi_{q_i}^{2k-1}$ and integrate. $k > 1$ is a real number, and $\eta \geq 0$ a C^∞ function with support in $B(P,\delta)$, is equal to 1 on $B(P,\delta/2)$.

For simplicity we drop the subscript q_i. The first term is

$$\int \eta^2\varphi^{2k-1}\Delta\varphi\,dV = (2k-1)\int \eta^2\varphi^{2k-2}|\nabla\varphi|^2\,dV + 2\int \eta\varphi^{2k-1}\nabla^i\eta\nabla_i\varphi\,dV. \tag{41}$$

When we compute $\int|\nabla(\eta\varphi^k)|^2\,dV$ we find the first term in the right hand side of (41):

$$\int|\nabla(\eta\varphi^k)|^2\,dV = \frac{k^2}{2k-1}\left[\mu\int f\eta^2\varphi^{2k+q-2}\,dV - \int \tilde{R}\eta^2\varphi^{2k}\,dV + \frac{1}{k}\int \nabla^i(\eta\nabla_i\eta)\varphi^{2k}\,dV\right] + \int|\nabla\eta|^2\varphi^{2k}\,dV - 2\int \nabla^i(\eta\nabla_i\eta)\varphi^{2k}\,dV. \tag{42}$$

At first suppose $f(P) > 0$. We pick δ so that f is positive on $B(P,\delta)$. Applying the Hölder inequality gives

$$\int f\varphi^q\eta^2\varphi^{2k-2}\,dV \leq \left(\int_{B(P,\delta)} f\varphi^q\,dV\right)^{1-2/q}\left(\int f\varphi^q\eta^q\varphi^{(k-1)q}\,dV\right)^{\frac{2}{q}} \leq \varepsilon_0^{1-2/q}\Big(\sup_{B(P,\delta)} f\Big)^{2/q}\|\eta\varphi^k\|_q^2. \tag{43}$$

Recall that for any $C > K^2(n,2) = 4\omega_n^{-2/n}/n(n-2)$, there exists $A(C)$ such that

$$\|\eta\varphi^k\|_N^2 \leq C\|\nabla(\eta\varphi^k)\|_2^2 + A(C)\|\eta\varphi^k\|_2^2. \tag{44}$$

Using (43) and (44) in C times (42) yields

$$\|\eta\varphi_{q_i}^k\|_N^2\left(1 - \frac{k^2}{2k-1}C\mu_{q_i}\varepsilon_0^{1-\frac{2}{q_i}}(\sup f)^{\frac{2}{q_i}}V^{2(\frac{1}{q_i}-\frac{1}{N})}\right) \leq \text{Const.}\int \varphi_{q_i}^{2k}\,dV \tag{45}$$

since $\|\eta\varphi^k\|_q^2 \leq \|\eta\varphi^k\|_N^2 V^{2(1/q-1/N)}$.

We suppose q_i close to N, for instance $2+2/(n-2) < q_i < N$. Then, it is possible to choose ε_0 small enough so that the left hand side in (45) is positive for some $k_0 > 1+n/2$. ε_0 is independent of P and q_i since

$$0 < \mu_{q_i} \leq \sup(1, \tilde{R})\|\varphi_{q_i}\|_{H_1}^2 \leq \text{Const.}.$$

For $k < k_0$ (45) gives $\|\eta\varphi_{q_i}^k\|_N^2 \leq \text{Const.} \int \varphi_{q_i}^{2k}\, dV$.

As $\|\varphi_{q_i}\|_{H_1} \leq$ Const., by the Sobolev Theorem $\|\varphi_{q_i}\|_N \leq$ Const.. Choose $2k = N$, we find $\|\eta\varphi_{q_i}^{N/2}\|_N \leq$ Const..

Then pick $k = (N/2)^l$ with $l = 2, 3, \dots$ until $(N/2)^l > 1+n/2$. If $\rho < \delta/2$ is small enough, we obtain

$$\int_{B(P,2\rho)} \varphi_{q_i}^p\, dV \leq \text{Const.} \qquad \text{for some } p > \tfrac{n(n+2)}{n-2}. \tag{46}$$

Using the properties of the Green function of $\Delta + \tilde{R}$:

$$\begin{aligned}|\nabla\varphi_{q_i}(y)| &\leq \mu_{q_i} \int |\nabla_y G(y,x)||f(x)|\varphi_{q_i}^{q_i-1}(x)\, dV(x)\\ &\leq \text{Const.}\left[\int_{B(y,\rho)} \frac{\varphi_{q_i}^{q_i-1}(x)}{[d(y,x)]^{n-1}}\, dV(x) + \rho^{1-n}\int \varphi_{q_i}^{q_i-1}\, dV\right]\\ &\leq \text{Const.}\end{aligned}$$

since the integral on $B(y,\rho)$ is smaller than Const. $\int_{B(P,2\rho)} \varphi_{q_i}^p\, dV$ which is bounded by virtue of (46).

Then we obtain a uniform estimate of the functions φ_{q_i} in C^{r+1} near P (Theorem 4.40). Thus a subsequence of $\{\varphi_{q_i}\}$ converges uniformly in C^r on a neighbourhood of P.

Now if $f(P) < 0$ we pick δ so that f is negative on $B(P,\delta)$. From (42) and (44) we get immediately:

$$\|\eta\varphi_{q_i}^k\|_N^2 \leq \text{Const.} \int \varphi_{q_i}^{2k}\, dV$$

where the constant does not depend on $k > 1$. The proof continues as above.

If $f(P) = 0$, for any $\xi > 0$ there exists a ball $B(P,\delta)$ such that $f < \xi$ on $B(P,\delta)$. In (42) we write $f < \xi$. The Hölder inequality (43) without f yields

$$\int \eta^2\varphi_{q_i}^{2k+q_i-2}\, dV \leq B\|\eta\varphi_{q_i}^k\|_N^2$$

where B is a constant since $\|\varphi_{q_i}\|_{q_i} \leq \text{Const.}\, \|\varphi_{q_i}\|_{H_1}$ which is bounded. Thus instead of (45) we obtain:

$$\|\eta\varphi_{q_i}^k\|_N(1 - k^2(2k-1)^{-1}C\mu_{q_i}\xi B) \leq \text{Const.} \int \varphi_{q_i}^{2k}\, dV.$$

As we can choose ξ as small as one wants the proof is completed just as above.

Corollary 6.39. *Let $\{u_i\}$ be as in Remark 6.38. There exist $\varepsilon_0 > 0$ and $\delta_0 > 0$ such that if $\int_{B(P,\delta)} f u_i^N \, dV \leq \varepsilon_0$ for all u_i and all $\delta \leq \delta_0$, then there is a subsequence $\{u_j\}$ of $\{u_i\}$, so that $u_j \to 0$ in L_N on a neighbourhood of P.*

The proof is similar. $\|\eta u_i^k\|_{H_1} \leq$ Const. implies by the Kondrakov Theorem that a subsequence $\{u_j^k\}$ converges in $L_{N/k}$ on a neighbourhood of P. The limit may be only zero since $u_j \to 0$ a.e.

6.40 Corollary. *Let h, f be C^∞ functions and $0 < \mu_j <$ Const.. Assume the C^∞ positive functions φ_j satisfy $\Delta\varphi_j + h\varphi_j = \mu_j f \varphi_j^{q_j-1}$, $2 \leq q_j < N$. Define $K = \{x \in M / f(x) \leq 0\}$.*

If the set $\{\varphi_j\}$ is bounded in H_1, then there is a subsequence of $\{\varphi_j\}$ which converges in C^r ($r \geq 0$ given) on a neighbourhood of K.

5.3. The Points of Concentration

6.41 Proposition. *Assume P is a point of concentration for the sequence of functions φ_{q_i} satisfying (39), $\|\varphi_{q_i}\|_{H_1} \leq$ Const. and the conditions of Definition 6.38, then for any $\delta > 0$*

$$\lim_{q_i \to N} \int_{B(P,\delta_0)} f \varphi_{q_i}^{q_i} \, dV \geq \varepsilon_0. \tag{47}$$

Assume there exists some $\delta_0 > 0$ for which

$$\lim_{q_i \to N} \int_{B(P,\delta_0)} f \varphi_{q_i}^{q_i} \, dV < \varepsilon_0.$$

Therefore there exists $\tilde{q}_0 < N$ such that $\int_{B(P,\delta_0)} f \varphi_{q_i}^{q_i} \, dV < \varepsilon_0$ if $q_i > \tilde{q}_0$.

Since $\varphi_{q_i} \to 0$ almost everywhere when $q_i \to N$, $\varphi_{q_i} \to 0$ uniformly on a neighbourhood of $K = \{x \in M / f(x) \leq 0\}$, according to the end of the proof of (6.39). Thus there exists $\tilde{q}_1 < N$ such that $\int_{B(P,\delta)} f \varphi_{q_i}^{q_i} \, dV < \varepsilon_0$ for all $\delta \leq \delta_0$ if $q_i > \tilde{q}_1$.

Then we can apply Theorem 6.39,

$$\lim_{q_i \to N} \int_{B(P,\delta)} f \varphi_{q_i}^{q_i} \, dV = 0$$

and P is not a point of concentration.

6.42 Proposition. *The set $\mathfrak{G}$ of the points of concentration is finite and non-empty: $\mathfrak{G} = \{P_1, P_2, \ldots, P_m\}$. A subsequence of $\{\varphi_{q_i}\}$ tends to zero in C^r_{loc} ($r \geq 0$) on $M - \mathfrak{G}$.*

Moreover $f(P_j) > 0$ for all j.

Proof. Let P be a point of concentration. If $f(P) = 0$, pick $\delta > 0$ so that $f(Q) < \eta$ on $B(P,\delta)$. $\eta > 0$ will be chosen small enough in order to get a contradiction. We can write

$$\int_{B(P,\delta)} f\varphi_{q_i}^{q_i}\,dV \le \eta \int_{B(P,\delta)} \varphi_{q_i}^{q_i}\,dV \le \eta\|\varphi_{q_i}\|_{q_i}^{q_i}$$
$$\le \eta\,\text{Const.}\,\|\varphi_{q_i}\|_{H_1}^{q_i} \le \eta C$$

with C a constant. Choose $\eta < \varepsilon_0/C$, by (47) P cannot be a point of concentration.

If $f(P) < 0$ the proof is simpler, in this case we have only to choose δ so that f is negative on $B(P,\delta)$.

According to Theorem 6.39 if $P \in M - \mathfrak{G}$, a subsequence of $\{\varphi_{q_i}\}$ tends to zero in C^r on a neighbourhood of P.

Since M has a denumberable basis of neighbourhoods, after taking subsequences, we can find a subsequence of $\{\varphi_{q_i}\}$ which tends to zero in C^r_{loc} on $M - \mathfrak{G}$. Hence we find again

(48) $$\int_{f\le 0} f\varphi_{q_i}^{q_i}\,dV \to 0 \qquad \text{when } q_i \to N.$$

For simplicity we write all subsequence $\{\varphi_{q_i}\}$.

Consider m points of concentration $P_j (j = 1, 2, \dots, m)$ and choose δ small enough so that the balls $B(P_j,\delta)$ are disjoint. Applying (48) together with Proposition 6.41 gives

$$1 = \int f\varphi_{q_i}^{q_i}\,dV \ge \lim_{q_i\to N} \sum_{j=1}^{m} \int_{B(P_j,\delta)} f\varphi_{q_i}^{q_i}\,dV \ge m\varepsilon_0.$$

Thus, there are at most $1/\varepsilon_0$ points of concentration, $\mathfrak{G}$ is finite and $\mathcal{E} \ne \emptyset$ since

(49) $$\lim_{q_i\to N} \sum_{P_j\in\mathcal{E}} \int_{B(Pj,\delta)} f\varphi_{q_i}^{q_i}\,dV = 1.$$

Remark. The sequence $\{u_i\}$, introduced in Remark 6.38, satisfies Proposition 6.42 except there is a subsequence $\{u_j\}$ which tends to zero in $L_{N loc}$ on $M - \mathfrak{G}$ (see Corollary 6.39). Indeed we know only that $w_i \to 0$ in H_{-1}.

6.43 Proposition. *Assume P is a point of concentration (see definition 6.38) for the sequence of functions φ_{q_j} such that $\varphi_{q_j} \to 0$ in C^2_{loc}on $M - \mathfrak{G}$.*

Then, in the sense of measures on a neighbourhood Ω of P such that $\bar{\Omega} \subset M - \mathfrak{G} + \{P\}$

$$\varphi_{q_j}^{q_j} \to [l/f(P)]\delta_P \quad \textit{and} \quad |\nabla\varphi_{q_j}|^2 \to l\mu\delta_P$$

where $\mu = \lim_{q_i\to N} \mu_{q_i}$ and $l = \lim_{q_j\to N} \int_\Omega f\varphi_{q_j}^{q_j}\,dV$.

Proof. Let h be a continuous function with supp $h \subset \Omega$ and $B = B(P,\delta) \subset \Omega$.

$$\left|\int_\Omega h\varphi_{q_j}^{q_j}\, dV - \frac{h(P)}{f(P)}\int_B f\varphi_{q_j}^{q_j}\, dV\right|$$
$$\leq \int_{\Omega - B} |h|\varphi_{q_j}^{q_j}\, dV + \sup{}_B |h - h(P)f/f(P)| \int \varphi_{q_j}^{q_j}\, dV$$

$\|\varphi_{q_j}\|_{H_1} \leq$ Const. thus $\int \varphi_{q_j}^{q_j}\, dV < C_0$ some constant. Given $\varepsilon > 0$, we choose δ small enough so that $\sup_B |h - h(P)f/f(P)| < \frac{\varepsilon}{3C_0}$.

Then, there exists $\tilde{q} < N$ such that

$$\frac{|h(P)|}{f(P)}\left|\int_B f\varphi_{q_j}^{q_j}\, dV - l\right| < \frac{\varepsilon}{3} \quad \text{and} \quad \int_{\Omega - B} |h|\varphi_{q_j}^{q_j}\, dV < \frac{\varepsilon}{3}$$

for $q_j > \tilde{q}$. So $|\int_\Omega h\varphi_{q_j}^{q_j}\, dV - lh(P)/f(P)| < \varepsilon$.

For the proof of the second assertion we suppose $h \in C^2$.

$$\int_\Omega h\nabla^\nu\varphi_{q_j}\nabla_\nu\varphi_{q_j}\, dV = \int_\Omega h\varphi_{q_j}\Delta\varphi_{q_j}\, dV - \int_\Omega \varphi_{q_j}^2 \Delta h\, dV/2$$
$$= \mu_{q_j}\int_\Omega hf\varphi_{q_j}^{q_j}\, dV - \int_\Omega (h\tilde{R} + \Delta h/2)\varphi_{q_j}^2\, dV.$$

So

$$\lim_{q_j \to N}\int_\Omega h\nabla^\nu\varphi_{q_j}\nabla_\nu\varphi_{q_j}\, dV = \mu l h(P).$$

Because the C^2 functions are dense in C^0, the proof is complete.

Corollary. *Assume P is a point of concentration for the sequence of functions u_i introduced in Remark* 6.38. *Consider a subsequence $\{u_j\}$ such that $u_j \to 0$ in $L_{N\,\mathrm{loc}}$ (see Remark* 6.42).

Then in the sense of measures on a neighbourhood Ω of P with $\bar{\Omega} \subset M - \mathfrak{G} + \{P\}$ we have: $u_j^N \to [l/f(P)]\delta_P$ and $|\nabla u_j|^2 \to l\nu\delta_P$ where

$$l = \lim_{j\to\infty}\int_\Omega fu_j^N\, dV.$$

6.44 Proposition. *If P is a point of concentration for the sequence of functions φ_{q_j} satisfying* (39), *P is a critical point of f(i.e. $|\nabla f(P)| = 0$) when $P \notin \partial M$. When $P \in \partial M$ the result holds if $\partial_n f(P) \geq 0$.*

This result was proved by Bahri–Coron [*26] and very easily by Hebey [*162] on the sphere by using the conditions of Kazdan and Warner (see 6.67).

In fact this is a general result, that we prove below, assuming $P \notin \partial M$ when $\partial M \neq \emptyset$. But in most cases, it appears that a point of ∂M cannot be a point of concentration (for instance when $\partial_n f(P) > 0$).

Let Ψ be a C^∞ function, with support in a neighbourhood Ω of P, such that $\partial_i\Psi(P) = \partial_i f(P)$ and $\partial_{ij}\Psi(P) = 0$ for all i, j in a system of normal coordinates at P. From (39) and an integration by parts we obtain

(50)
$$\begin{aligned}
&\int \varphi_{q_j}^{q_j}\nabla^\nu f \nabla_\nu \Psi \, dV \\
&= \int f\varphi_{q_j}^{q_j}\Delta\Psi \, dV - q_j\int f\varphi_{q_j}^{q_j-1}\nabla_\nu\Psi\nabla^\nu\varphi_{q_j}\, dV \\
&= \int f\varphi_{q_j}^{q_j}\Delta\Psi \, dV - \left(q_j/\mu_{q_j}\right)\int \Delta\varphi_{q_j}\nabla_\nu\Psi\nabla^\nu\varphi_{q_j}\, dV \\
&\quad + \left(q_j/2\mu_{q_j}\right)\int \varphi_{q_j}^2\nabla^\nu\left(\tilde{R}\nabla_\nu\Psi\right) dV.
\end{aligned}$$

Integrating again by parts gives

(51)
$$\begin{aligned}
&\int \Delta\varphi_{q_j}\nabla_\nu\Psi\nabla^\nu\varphi_{q_j}\, dV \\
&= \int \nabla_\mu\varphi_{q_j}\nabla_\nu\Psi\nabla^{\nu\mu}\varphi_{q_j}\, dV + \int \nabla^\mu\varphi_{q_j}\nabla_{\nu\mu}\Psi\nabla^\nu\varphi_{q_j}\, dV \\
&= \frac{1}{2}\int |\nabla\varphi_{q_j}|^2\Delta\Psi\, dV + \int \nabla^\mu\varphi_{q_j}\nabla_{\nu\mu}\psi\nabla^\nu\varphi_{q_j}\, dV.
\end{aligned}$$

According to Proposition 6.43 and taking in account the properties of Ψ at P we get

$$l|\nabla f(P)|^2/f(P) = \lim_{q_j\to N}\int \varphi_{q_j}^{q_j}\nabla^\nu f\nabla_\nu\Psi\, dV = 0$$

indeed the limit of each term in the right hand side of (51), then of (50), is zero.

Integrating by parts (50) is valid if $P \in \partial M \neq \emptyset$ since $\varphi_{q_j}|_{\partial M} = 0$. But the right hand side of (51) contains an additional term $-\frac{1}{2}\int_{\partial M}|\nabla\varphi|^2\partial_n\Psi d\sigma$. Thus this computation yields only $\partial_n f(P) \geq 0$, where ∂_n means the normal derivative oriented to the outside.

Corollary 6.44. *If P is a point of concentration for the sequence of functions u_i introduced in Remark 6.38, P is a critical point of f.*

5.4. Consequences

6.45 From Proposition 6.42 we get immediatly some consequences.

Examples. Consider the unit ball $B \subset \mathbb{R}^n (n \geq 3)$ endowed with the euclidean metric. If f is a C^∞ radial function positive somewhere with $f(0) \leq 0$, equation (38) with zero Dirichlet condition ($\varphi|_{\partial B} = 0$) has a C^∞ solution.

Indeed a sequence $\left\{\varphi_{q_i}\right\}$ of radial functions cannot have point of concentration. 0 is not possible since $f(0) \leq 0$. The same is true for other points P of B, otherwise all the points of the sphere centered at 0 with radius $r = d(0, P)$ would be points of concentration which is impossible since the points of concentration are isolated.

Likewise equation (38) on the sphere S_n with the standard metric has a C^∞ positive solution if the C^∞ function f, positive at some point, depends only on the distance to a point $P \in S_n$ and if $f(P) \leq 0$ and $f(\tilde{P}) \leq 0$ where $\tilde{P}$ is the antipodal point to P.

These results will be improved below.

6.46 Lemma. *Assume P is a point of concentration for the sequence of functions φ_{q_j} satisfying (39) with $\mu = \lim \mu_{q_j}$ when $q_j \to N$. Then*

$$[f(P)]^{2/N} \mu \left[\lim_{q_j \to N} \int_{B(P,\delta)} f \varphi_{q_j}^{q_j} \, dV \right]^{2/n} \geq \mu_s \tag{52}$$

where $\mu_s = K^{-2}(n,2) = n(n-2)\omega_n^{2/n}/4$.

For simplicity write B for $B(P,\delta)$. Pick $\delta_0 > 0$ so that $\bar{B} \subset M - \mathfrak{G}$.

For $\delta \leq \delta_0$ we saw that $\lim \int_B f \varphi_{q_j}^{q_j} \, dV$ when $q_j \to N$ does not depend on δ. Set this limit equals to l.

Return to the proof of Theorem 6.39. If

$$(\mu/\mu_s) l^{2/n} [f(P)]^{2/N} < 1, \tag{53}$$

it is possible to use inequality (45) for some k(l is defined in 6.43).

Indeed we can choose C near $1/\mu_s$, δ small enough and j large enough so that

$$C \mu_{q_j} \left[\int_B f \varphi_{q_j}^{q_j} \, dV \right]^{1-2/q_j} (\sup_B f)^{2/q_j} \leq a < 1.$$

Thus for some $\alpha > 0$ and some $\rho < \delta$

$$\int_{B(P,\rho)} \varphi_{q_j}^{N+\alpha} \, dV < \text{Const. for all } q_j \text{ large enough.} \tag{54}$$

Using the Hölder inequality, for any open set θ with $\bar{\theta} \subset M - \mathfrak{G} + \{P\}$

$$\text{Const.} \left(\int_\theta \varphi_{q_j}^N \, dV \right)^{q_j/N} \geq \int_\theta \varphi_{q_j}^{q_j} \, dV \to l/f(P) > 0 \tag{55}$$

according to Proposition (6.43) and

$$\left(\int_\theta \varphi_{q_j}^N \, dV \right)^2 \leq \left(\int_\theta \varphi_{q_j}^{N-\alpha} \, dV \right) \left(\int_\theta \varphi_{q_j}^{N+\alpha} \, dV \right). \tag{56}$$

Pick $\theta = B(P,\rho)$, (54) and (55) together with (56) imply

$$\liminf_{q_j \to N} \int_\theta \varphi_{q_j}^{N-\alpha} \, dV > 0.$$

This contradicts a property of $\{\varphi_{q_j}\}$ (see the end of (6.37)), $\{\varphi_{q_j}\}$ or a subsequence converges to zero in $L_{N-\alpha}$.

6.47 Proposition. *Assume P is a point of concentration. Then*

$$\mu[f(P)]^{1-2/n} \geq \mu_s[\operatorname{Card} O(P)]^{2/n} \tag{57}$$

where $O(P)$ is the orbit of P under G and $\mu = \liminf_{q_j \to N} \mu_{q_j}$.

Proof. Set $k = \operatorname{Card} O(P)$. There are at least k points of concentration P_j which are the points of $O(P)$. Choose δ small enough so that the balls $B(P_j, \delta)$ are disjoint and without other point of concentration. We have

$$\lim_{q_i \to N} \sum_{j=1}^{k} \int_{B(Pj,\delta)} f\varphi_{q_i}^{q_i}\, dV \leq \lim_{q_i \to N} \int f\varphi_{q_i}^{q_i}\, dV = 1$$

since $\varphi_{q_i} \to 0$ uniformly on a neighbourhood of the set K of the points where $f \leq 0$ (Corollary 6.40). Put in (52) the inequality

$$\lim_{q_i \to N} \int_{B(Pj,\delta)} f\varphi_{q_i}^{q_i}\, dV \leq 1/k$$

we get (57).

So we have proved the

6.48 Theorem. *The equation* (38) *has a G-invariant C^∞ positive solution if*

$$f(P) < [\operatorname{Card} O(P)]^{2/(n-2)}[\mu_s/\mu]^{n/(n-2)} \tag{58}$$

at any critical point P of f satisfying the necessary condition: f is positive somewhere, and in the case $R \equiv 0$, M without boundary, the second necessary condition $\int f\, dV < 0$.

Recall $\mu = \inf I(\varphi)$ for all

$$\varphi \in \mathcal{A} = \left\{ u \in \overset{\circ}{H}_1 / u \geq 0, u \text{ G-invariant}, \int f u^N\, dV = 1 \right\}$$

where $I(\varphi) = \|\nabla\varphi\|_2^2 + \int \tilde{R}\varphi^2\, dV$. Moreover $\mu = \lim \mu_{q_j}$ when $q_j \to N$. The proof is written up in (6.21). $O(P)$ is the orbit of P under G. $\mu_s = K^{-2}(n, 2)$.

We will see, in §6 through 10, some applications of this theorem of Hebey.

5.5. Blow-up at a Point of Concentration

6.49 Assume P is a point of concentration for the sequence φ_{q_i} which is supposed tending to zero a.e. and in L_q for all $q < N$. φ_{q_i} satisfies (39), $\mu_{q_i} \to \mu$ when $q_i \to N$. So we suppose for all $\delta > 0$,

$$\lim_{q_i \to N} \int_{B(P,\delta)} \varphi_{q_i}^{q_i}\, dV \geq \varepsilon_0 > 0.$$

Let $\delta_i > 0$ with $\delta_i < \delta/2$ be a sequence tending to zero.

After passing to a subsequence, if necessary, we can suppose that $\varphi_{q_i} < \eta$ on $B(P,\delta) - B_P(\delta_i)$ for some small constant $\eta > 0$ and all i, since $\{\varphi_{q_i}\}$ tends to zero in $C^r_{\text{loc}}(r \geq 0)$ on $M - \mathfrak{G}$ (Proposition 6.42). Our hypothesis implies that $m_i = \sup \varphi_{q_i}$ on $B(P,\delta_i)$ tends to $+\infty$ when i $\to \infty$. Pick $z_i \in B(P,\delta_i)$ a point where $m_i = \varphi_{q_i}(z_i)$. Consider $\{x^j\}$ a system of normal coordinates at P with $x^j(P) = 0$. Set

$$(59) \qquad v_i(y) = \frac{1}{m_i}\varphi_{q_i}(z_i + m_i^{-\alpha_i} y) \qquad \text{with } \alpha_i = q_i/2 - 1.$$

$y \in B_{k_i}$ the ball in $\mathbb{R}^n$ of radius $k_i = \delta m_i^{\alpha_i}/2$.

Fix k large in $\mathbb{N}$. For $k_i > k$ let us prove that the functions v_i are bounded in $H_1(B_k)$, B_k being endowed with the euclidean metric.

On $B(P,\delta)$ there exists $\lambda \geq 1$ such that

$$\lambda^{-1}\sum_{j=1}^{n}(\xi_j)^2 \leq g_{jk}\xi^j\xi^k \leq \lambda \sum_{j=1}^{n}(\xi^j)^2, \qquad \text{for any vector } \xi.$$

We have $0 \leq v_i \leq 1$ thus $\int_{B_k} v_i^2\, dV \leq$ Const.. Moreover

$$\int_{B_k} \mathcal{E}^{\alpha\beta}\frac{\partial v_i}{\partial y^\alpha}\frac{\partial v_i}{\partial y^\beta}\, d\mathcal{E} \leq m_i^{(n-2)\alpha_i - 2}\lambda^{1+\frac{n}{2}}\int g^{\alpha\beta}\frac{\partial \varphi_{q_i}}{\partial x^\alpha}\frac{\partial \varphi_{q_i}}{\partial x^\beta}\, dV \leq C$$

some constant, since $(n-2)\alpha_i - 2 = (n-2)q_i/2 - n < 0$. Here $\mathcal{E}^{\alpha\beta} = \delta^\alpha_\beta$ and $d\mathcal{E}$ is the euclidean measure.

6.50 After passing to a subsequence if necessary, using the Banach Theorem, we can suppose without loss of generality that:

$$(60) \qquad v_i \to \omega \qquad \text{weakly in } H_1(B_k) \text{ for any large } k \in \mathbb{N}.$$

Let us seek the equation satisfied by ω on $\mathbb{R}^n$.

Let $\Psi \in \mathcal{D}(B_k)$ and set $\tilde{\Psi}_i(x) = \Psi\left[m_i^{\alpha_i}(x - z_i)\right]$ whose support is included in $B(z_i, km_i^{-\alpha_i}) \subset B(P,\delta)$ since i is large enough. Since

$$\int g^{\alpha\beta}\frac{\partial \varphi_{q_i}}{\partial x^\alpha}\frac{\partial \tilde{\Psi}_i}{\partial x^\beta}\, dV + \int \tilde{R}\varphi_{q_i}\tilde{\Psi}_i\, dV = \mu_{q_i}\int f\varphi_{q_i}^{q_i-1}\tilde{\Psi}_i\, dV$$

in coordinates y^i we get

$$(61) \qquad m_i^{2\alpha_i}\int g^{\alpha\beta}\frac{\partial v_i}{\partial y^\alpha}\frac{\partial \Psi}{\partial y^\beta}\sqrt{|g|}\, dy + \int \tilde{R}v_i\Psi\sqrt{|g|}\, dy$$
$$= m_i^{q_i-2}\mu_{q_i}\int f v_i^{q_i-1}\Psi\sqrt{|g|}\, dy.$$

Now there exists a constant C_0 such that

$$\left|(g^{\alpha\beta}\sqrt{|g|} - \mathcal{E}^{\alpha\beta})\xi_\alpha\xi_\beta\right| \leq C_0\|y\|^2 m_i^{-2\alpha_i}\|\xi\|^2 \leq C_0 k^2 m_i^{-2\alpha_i}\|\xi\|^2$$

for all vectors ξ(‖ ‖ is the euclidean norm) and

$$\left|f\sqrt{|g|} - f(P)\right| \leq C_0 \|y\| m_i^{-\alpha_i}.$$

These two inequalities suggest writting (61) in the form:

$$\begin{aligned}\int \mathcal{E}^{\alpha\beta} \frac{\partial v_i}{\partial y^\alpha}\frac{\partial \Psi}{\partial y^\beta}\,dy - \mu_{q_i} f(P)\int v_i^{q_i-1}\Psi\,dy \\ = \int \left(\mathcal{E}^{\alpha\beta} - g^{\alpha\beta}\sqrt{|g|}\right)\frac{\partial v_i}{\partial y^\alpha}\frac{\partial \Psi}{\partial y^\beta}\,dy - m_i^{-2\alpha_i}\int \tilde{R} v_i \Psi\sqrt{|g|}\,dy \\ + \mu_{q_i}\int \left(f\sqrt{|g|} - f(P)\right) v_i^{q_i-1}\Psi\,dy.\end{aligned}$$

Where the right hand side tends to zero when $i \to \infty$. Since $v_i \to \omega$ weakly in $H_1(B_k)$ and $\mu_{q_i} \to \mu$, we get

$$\int \mathcal{E}^{\alpha\beta}\frac{\partial \omega}{\partial y^\alpha}\frac{\partial \Psi}{\partial y^\beta}\,dy - \mu f(P)\int \omega^{N-1}\Psi\,dy = 0.$$

That is, ω satisfies weakly in H_1 on $\mathbb{R}^n$:

$$\sum_{j=1}^{n} \partial_{jj}\omega + \mu f(P)\omega^{N-1} = 0. \tag{62}$$

The functions $\{v_i\}$ for i large satisfies an equation E_i on B_k. The equations E_i are uniformly elliptic, the coefficients in the left hand side and in the right hand side are bounded. Thus according to Theorem 4.40, there exist β and k_o such that $\|v_i\|_{C^\beta(B_k)} \leq k_o$. By Ascoli's theorem, $\{v_i\}$ or some subsequence tends uniformly on any compact set. This implies that ω is non trivial since $\omega(0) = 1$. Moreover $\omega \in C^\infty$ by the regularity theorems.

6.51 When ω is maximum in y=0, which is the case here, we claim that all positive solutions of (62) are of the form $C^2(1 + \|y\|^2/\varepsilon)^{1-n/2}$ with $\varepsilon > 0$ a real number. As $\omega(0) = 1$ the solution of (62) is

$$\omega_P = (1 + \|y\|^2/\varepsilon)^{1-n/2} \qquad \text{with } \varepsilon = n(n-2)/\mu f(P). \tag{63}$$

Indeed as equation (62) is radial symmetric, according to Gidas, Ni and Nirenberg [*140], the positive solution is radial symmetric: $\omega = h(r)$. Thus $h(r)$ satisfies a second order equation, so by the Cauchy Theorem the solutions depend on two constants $h(0)$ and $h'(0)$. Now in our problem $h(0) = 1$ and $h'(0) = 0$. Hence (62) has only one radial solution which achieves its maximum 1 at $y = 0$. We can verify that this solution is ω_P given in (63).

6.52 According to Bliss (Lemma 2.19), ω is a minimizer for the Yamabe functional. Thus

$$(64)\qquad \int |\nabla\omega|^2\,dy\Big(\int \omega^N\,dy\Big)^{-2/N} = K^{-2}(n,2) = \mu_s.$$

Using (62) yields

$$(65)\qquad \Big(\int \omega_P^N\,dy\Big)^{2/n} = \mu_s/\mu f(P).$$

Furthermore

$$\lim_{i\to\infty}\int_{B(P,\delta)} f\varphi_{q_i}^{q_i}\,dV = \lim_{i\to\infty} m_i^{q_i-n\alpha_i}\int_{B_{k_i}} f v_i^{q_i}\sqrt{|g|}dy.$$

As $q_i-n\alpha_i = q_i(1-\frac{n}{2})+n>0$ and since $v_i\to\omega_P$ uniformly on all compact set

$$(66)\qquad f(P)\int\omega_P^N\,dy \le \lim_{i\to\infty}\int_{B(P,\delta)} f\varphi_{q_i}^{q_i}\,dV.$$

Now $\varphi_{q_i}\to 0$ uniformly on any compact set included in $V-\mathfrak{G}$ thus

$$(67)\qquad \lim_{i\to\infty}\sum_{P\in\mathfrak{G}}\int_{B(P,\delta)} f\varphi_{q_i}^{q_i}\,dV = 1.$$

if δ is chosen small enough so that the balls $B(P_j,\delta)$ are disjoint.

From (66) and (67) we get $\sum_{P\in\mathfrak{G}} f(P)\int\omega_P^N\,dy\le 1$.

Applying (65) yields

6.53 Theorem. *Let $\mathfrak{G}$ be the set of the points of concentration. Then*

$$(68)\qquad \sum_{P\in\mathfrak{G}}[f(P)]^{1-n/2}\le(\mu/\mu_s)^{n/2}.$$

See (6.48) to recall the definitions.

Inequality (68) is valid for the sequence $\{u_i\}$ introduced in 6.38 with $\mu=\nu$. For $\mu=\operatorname{Inf}_{\varphi\in\mathcal{A}}\int(|\nabla\varphi|^2+\tilde{R}\varphi^2)\,dV$ with

$$\mathcal{A}=\Big\{\varphi\in\mathring{H}_1/\varphi\ \ G\text{-invariant},\ \int f|\varphi|^N\,dV=1\Big\},$$

we can prove that actually $\mathfrak{G}$ is the orbit of some point P $(\mathfrak{G}=O(P))$. In that case (68) is not other than (57).

On the one hand, considering in the functional the test functions $u_k=\sum_{Q\in O(P)}\Psi_k(r_Q)$, where r_Q is the distance to Q and Ψ_k is defined in 5.21 with 2δ smaller than the distance of two points in $O(P)$, we get

$$\mu\le\mu_s\big[\operatorname{Card}O(P)\big]^{1-2/N}\big[f(P)\big]^{-2/N}.$$

On the other (52) implies

$$[f(P)]^{\frac{2}{N}}\mu\Big(\lim_{q_j\to N}\sum_{Q\in O(P)}\int_{B(Q,\delta)} f\varphi_{q_j}^{q_j}\,dV\Big)^{\frac{2}{n}} \ge \mu_s\big[\operatorname{Card} O(P)\big]^{\frac{2}{n}}.$$

If $\mathfrak{G} \neq O(P)$, $\lim_{q_j\to N}\sum_{Q\in O(P)}\int_{B(Q,\delta)} f\varphi_{q_j}^{q_j}\,dV < 1 = \int f\varphi_{q_j}^{q_j}\,dV$ and the inequalities above yield a contradiction.

Example. In the Yamabe Problem $f(P) = 1$. If $\mu < \mu_s$, $\mathfrak{G} = \emptyset$ according to (68). There is no point of concentration, φ_{q_j} cannot tend to zero almost everywhere, thus the Yamabe Problem has a solution.

When $\mu = \mu_s$ we are on the sphere, where there exist sequences $\{u_j\}$ of solutions of (38) with $f = 1$ such that $u_j \to 0$ a.e. and there is one point of concentration. See the proof of the Yamabe Problem in [*118] where R. Dong uses the idea above.

§6. The Problem on Other Manifolds

6.1. On Complete Non-compact Manifolds

6.54 On $\mathbb{R}^n (n \ge 3)$ endowed with the euclidean metric $\mathcal{E}$, the equation to solve reduces to

(69) $$\Delta u = f u^{N-1}, \qquad u > 0 \qquad \text{with } \Delta = -\textstyle\sum_{i=1}^n \partial_{ii}.$$

There are many results on this equation and also on the more general equation in $(\mathbb{R}^n, \mathcal{E})$:

(70) $$\Delta u = f u^p, \qquad u > 0 \qquad \text{with } p > 1.$$

Theorem 6.54 (Ni [*256]). *Let $x = (x_1, x_2) \in \mathbb{R}^3 \times \mathbb{R}^{n-3}$. If $|f(x)| \le C|x_1|^l$ for some $l < -2$, uniformly in x_2 when $x_1 \to \infty$, then equation* (70) *has infinitely many bounded positive solutions which are bounded below by positive constants.*

If $f(x) \le 0$ and $|f(x)| \ge C|x|^l$ at ∞ for some $l \ge -2$, equation (69) *does not have positive solution.*

If we seek solutions of (70) in H_1, of course f must be positive somewhere since $\int f u^N\,dx > 0$. There is another nonexistence result proved by using the Pohozaev identity in Li and Ni [*212]. In [*256] the asymptotic behavior of radial solutions of (70) is studied in case f is radially symmetrical and decaies at infinity.

Bianchi and Egnell (in [*52]) seek radial solutions of (69) satisfying $u(x) = 0(|x|^{2-n})$ as $x \to \infty$. They have results of existence and nonexistence, that we can compare with those of 4.32. Indeed when u satisfies the preceding assumption at infinity, the problem is similar to the Nirenberg problem on the sphere.

When u is bounded from below by some positive constant (as in the Theorem above), we are guaranted that the conformal metric is complete. In [*77] A. Chajub–Simon proves some existence results of solutions of (69) such that $u-1$ belong to some weighted Sobolev spaces.

In [*213] Yan-Yan Li studies equation (69) on $\mathbb{R}^3$, especially when f is periodic in one of its variables. For more results see the articles in references an their bibliographies.

On a manifold which is not $(\mathbb{R}^n, \mathcal{E})$, let us mention the two following results.

6.55 Ratto, Rigoli and Veron [*272] studied the problem of prescribed scalar curvature on the hyperbolic space (H_n, g) of sectional curvature -1. Let B be the unit ball in $\mathbb{R}^n$ endowed with the Poincaré metric $g_H = 4(1-|x|^2)^{-2}\mathcal{E}$. Given $K \in C^\infty(B)$, they seek a complete metric conformal to g_H whose scalar curvature is K. Among results of existence and non-existence they prove the following

Theorem 6.55. *Let $a(r)$ ba a nondecreasing positive function on $[0,1[$ satisfying $\int_0^1 a(r)\,dr < \infty$. If for some $\delta \in]0,1[$, $-a^2(|x|) \le K(x) < 0$ when $1-\delta \le |x| < 1$, then there exists $\sigma > 0$ such that if $K(x) \le \sigma$ in B, $K(x)$ is the scalar curvature of a metric conformal to g_H. The metric is complete if in addition*

$$\int_0^1 \left(\int_r^1 a(s)\,ds \right)^{-1} dr = +\infty.$$

6.56 Theorem (Le Gluher [*209]). *Let (M_n, g) be a complete Riemannian manifold ($n \ge 3$) with injectivity radius $\delta > 0$, bounded curvature and $I(\varphi)$ coercice on H_1. Given f a C^∞ function on M, positive somewhere and satisfying* $\limsup f(x) \le 0$ *at infinity, then equation* (2) *has a C^∞ positive solution in H_1 if*

$$\inf_{\varphi \in \mathcal{A}} I(\varphi) < \mu_s \inf_{x \in K} \left\{ [\operatorname{Card} 0(x)]^{2/n} [f(x)]^{-2/N} \right\} \tag{71}$$

f is supposed to be invariant under G a group of isometries of (M_n, g) which can be reduced to the identity.

$$\mathcal{A} = \left\{ \varphi \in H_1,\ \varphi \ \ G\text{-invariant},\ \varphi \ge 0 \Big/ \int f\varphi^N\,dV = 1 \right\}.$$

$$K = \left\{ x \in M / f(x) > 0 \ \ and \ \ |\nabla f(x)| = 0 \right\}.$$

μ_s is defined in 6.46 *and* $I(\varphi) = \int \left(|\nabla\varphi|^2 + \tilde{R}\varphi^2 \right) dV$.

The proof uses the method of Bahri–Coron.

6.2. On Compact Manifolds with Boundary

6.57 Let Ω be a bounded domain of $\mathbb{R}^n$ $(n \geq 3)$ with C^∞ boundary. We consider the following equation:

$$(72) \qquad \Delta u + a(x)u = f(x)u^{N-1}, \quad u > 0 \text{ on } \Omega, \quad u(x) = 0 \text{ on } \partial\Omega.$$

$a(x) \geq 0$, $f(x)$ are given functions in $C^\infty(\bar{\Omega})$ and $\Delta = -\sum_{i=1}^n \frac{\partial^2 u}{(\partial x^i)^2}$.

If $f \leq 0$ on Ω, equation (72) has no solution. Indeed multiplying (72) by u and integrating yield $\int_\Omega f(x)u^N\,dx \geq 0$.

According to Kazdan and Warner, equation (72) has also no solution, when Ω is star-shaped with respect to 0, if

$$(73) \qquad \frac{\partial a}{\partial r} \geq 0, \qquad f \geq 0 \quad \text{and} \quad \frac{\partial f}{\partial r} \leq 0 \text{ on } \Omega.$$

This result is an improvement of Pohozahev's identity below. If a = Const. and $f \equiv 1$, conditions (73) are satisfied and we get Corollary 6.58.

6.58 Pohozahev identity [*267]. Let Ω be a star-shaped open set of $\mathbb{R}^n$ with $\partial\Omega$ differentiable. f being a continuous function on $\mathbb{R}$, we set $F(v) = \int_0^v f(t)dt$. If $u \in C^2(\bar{\Omega})$ satisfies $\Delta u = f(u)$ on Ω, $u/\partial\Omega = 0$, then

$$(74) \qquad (1 - n/2)\int_\Omega uf(u)\,dx + n\int_\Omega F(u)\,dx = \frac{1}{2}\int_{\partial\Omega} \partial_\nu h(\partial_\nu u)^2 d\sigma$$

where $h(x) = \|x\|^2/2$, $\Delta = -\sum_{i=1} \partial_{ii}$ and ∂_ν denotes the outer normal derivative on $\partial\Omega$.

For the proof we compute $A = \int_\Omega \nabla^i(\nabla^j h \nabla_j u \nabla_i u)\,dx$ in two different ways. At first $A = \int_{\partial\Omega} \nabla^j h \nabla_j u \partial_\nu u d\sigma = \int_{\partial\Omega} \partial_\nu h(\partial_\nu u)^2 d\sigma$ since $u/\partial\Omega = 0$. Then, as $\nabla_{ij} h = \delta_i^j$, $A = \int \nabla^i u \nabla_i u\,dx + \frac{1}{2}\int \nabla^j h \nabla_j |\nabla u|^2\,dx - \int \nabla^j h \nabla_j u f(u) d\sigma$, and (74) follows after integrating by parts.

Corollary 6.58. *On Ω a star-shaped open set of $\mathbb{R}^n$ with $\partial\Omega$ differentiable, the equation $\Delta u = u^{N-1}$, $u > 0$ on Ω, $u/\partial\Omega = 0$ has no solution.*

As Ω is star-shaped $\partial_\nu h > 0$. So the right hand side of (74) is strictly positive since $\partial_\nu u > 0$ on $\partial\Omega$ according to the Maximum Principle (see Chapter 3, §8). But this is impossible, since the left hand side of (74) is zero when $f(u) = u^{N-1}$.

6.59 Let G be a group of isometries of $(\Omega, \mathcal{E})$. We suppose $a(x)$ and $f(x)$ are G-invariant. Let's denote the orbit of x $\in \Omega$ by O(x) $= \{\sigma(x), \sigma \in G\}$ and consider the functional

$$I(v) = \int_\Omega |\nabla v|^2\,dx + \int_\Omega a(x)v^2\,dx.$$

We define $\mu(G) = \inf I(v)$ for $v \in \mathcal{A}(G) = \{v \in \overset{\circ}{H}_1(\Omega),\ v \geq 0,\ v\ G\text{-invariant}$ and $\int_\Omega f(x)v^N\,dx = 1\}$.

Theorem 6.59 (Hebey [*163]). *Let* $\mathfrak{G} = \{x \in \bar{\Omega}/f(x) > 0$ *and* $|\nabla f(x)| = 0\}$. *Equation* (72) *has a smooth solution if*

$$\mu(G)\big[f(x)\big]^{1-2/n} < \mu_s\big[\operatorname{Card} O(x)\big]^{2/n} \text{ for all } x \in \mathfrak{G}.$$

The proof uses the method of isometry-concentration.

Corollary 6.59 (Hebey [*163]). *When* Ω *is a ball in* $\mathbb{R}^n (n > 4)$, $a(x)$ *and* $f(x)$ *are radial functions, equation* (72) *has a smooth solution if* $f(0) \leq 0$. *The same conclusion holds when* $f(0) > 0$, *if*

$$(n-2)(n-4)\Delta f(0) + 8(n-1)a(0)f(0) < 0.$$

6.60 On a smooth compact orientable Riemannian manifold (M_n, g) with boundary, the *Cherrier Problem* consists in finding g' conformal to g such that the scalar curvature of (M_n, g') and the mean curvature of ∂M in (M_n, g') are given functions. The equation to solve is equation (2) (resp. (1) when $n = 2$) with non-linear Neumann boundary condition.

We studied this problem in Chapter 5.

§7. The Nirenberg Problem

6.61 In 1969–70 Nirenberg posed the following problem: Given a (positive) smooth function f on (S_2, g_0) ("close" to the constant function, if we want), is it the scalar curvature of a metric g conformal to g_0 (g_0 is the standard metric whose sectional curvature is 1).

Recall that if we write g in the form $g = e^\varphi g_0$, the problem is equivalent to solving the equation

$$\Delta\varphi + 2 = fe^\varphi. \tag{75}$$

Since the radius $(1/\alpha)$ of the sphere is chosen equal to 1, the scalar curvature $R = 2\alpha^2 = 2$.

Consider the operator $\Gamma : \varphi \to e^{-\varphi}(\Delta\varphi + 2)$. It is well known that the differential of Γ at $\varphi_0 = 0$, $D\Gamma_{\varphi_o}(\Psi) = \Delta\Psi - 2\Psi$ is not invertible. The kernel of $D\Gamma_{\varphi_o}$ is the three dimensional eigenspace corresponding to the first non zero eigenvalue of Δ. Indeed the functions $\cos r$, where r is the distance to a given point of S_2, satisfy

$$\Delta \cos r = -(\cos r)'' - \operatorname{cotg} r(\cos r)' = 2\cos r.$$

Or if we consider $S_2 \subset \mathbb{R}^3$, the traces of the coordinates x^i $(i = 1, 2, 3)$ satisfy $\Delta x^i = 2x^i$.

6.62 The same problem can be posed on (S_n, g_0) with $n > 2$. Given f a smooth function on (S_n, g_0), is it the scalar curvature function of a conformal metric g to g_0.

If we write g on the form $g = \varphi^{4/(n-2)} g_0$ the problem is equivalent to exhibiting a positive solution of the equation

$$4\frac{n-1}{n-2}\Delta\varphi + n(n-1)\varphi = f\varphi^{(n+2)/(n-2)}. \tag{76}$$

As before the differential of the operator

$$\tilde{\Gamma} : \varphi \to \varphi^{-(n+2)/(n-2)}\left[4\frac{n-1}{n-2}\Delta\varphi + n(n-1)\varphi\right]$$

is not invertible at $\varphi_1 = 1$, and the kernel of

$$d\tilde{\Gamma}_{\varphi_1}(\Psi) = 4\frac{n-1}{n-2}[\Delta\Psi - n\Psi]$$

is the $n+1$ dimensional eigenspace corresponding to the first non zero eigenvalue of Δ.

§8. First Results

6.63 Let us try to solve the Nirenberg problem by a variational method. We consider the functional

$$I(\varphi) = \int |\nabla\varphi|^2 \, dV + 4\int \varphi \, dV \tag{77}$$

and the constraint $\int f e^{\varphi} \, dV = 8\pi$, where 4π is the volume of (S_2, g_0). Set $\nu = \inf I(\varphi)$ for $\varphi \in \mathcal{A} = \{\varphi \in H_1 / \int f e^{\varphi} \, dV = 8\pi\}$.

First we have to prove that if $\varphi \in H_1$, e^{φ} is integrable, and for the sequel, that the mapping $H_1 \ni \varphi \to e^{\varphi} \in L_1$ is compact (see Theorem 2.46).

So if f is positive somewhere $\mathcal{A}$ is non empty.

Then we must see if ν is finite. For this we need an inequality of the type (see 2.46 and Theorem 2.51):

$$\int e^{\varphi} \, dV \le C(\mu) \exp\left[\mu \int |\nabla\varphi|^2 \, dV + V^{-1} \int \varphi \, dV\right] \tag{78}$$

which holds, on a compact manifold of dimension 2, for all $\varphi \in H_1$ when $\mu > \mu_2 = 1/16\pi$. Here V is the volume and $C(\mu)$ a constant. On (S_2, g_0), (78) is valid with $\mu = 1/16\pi$ ($C(\mu_2)$ exists) and $V = 4\pi$. Thus

$$8\pi \le \sup f \int e^{\varphi} \, dV \le C \sup f \exp\bigl[I(\varphi)/16\pi\bigr].$$

So ν is finite. Unfortunately, the value of μ_2 does not enable us to prove that a minimizing sequence $\{\varphi_i\}$ is bounded in H_1. Indeed, $I(\varphi_i) \to \nu$ but we can have

$$\|\nabla\varphi_i\|_2 \to +\infty \quad \text{and} \quad \int \varphi_i \, dV \to -\infty.$$

6.64 In higher dimensions the variational method breaks down immediately. Consider the functional

$$J(\varphi) = \left[4\frac{n-1}{n-2}\int |\nabla\varphi|^2 \, dV + n(n-1)\int \varphi^2 \, dV\right]\left[\int f\varphi^N \, dV\right]^{-2/N}$$

for $\varphi \in H_1$. By using Aubin's test function (see 5.10) centered at P, a point where f is maximum, it is easy to show that

$$\inf J(\varphi) = n(n-1)\omega_n^{2/n}[\sup f]^{-2/N}.$$

On the other hand if $f \equiv 1$, we know the functions Ψ for which $J(\Psi) = n(n-1)\omega_n^{2/n}$ (see 5.58). For these functions if $f \not\equiv$ Const., $\int f\varphi^N \, dV < \sup f \int \varphi^N \, dV$. Thus if f is not constant, for any $\varphi \in H_1$, $\varphi \not\equiv 0$, $J(\varphi) > \inf J(\varphi)$. So the inf cannot be achieved.

Nevertheless, J. Moser succeeded in solving the Nirenberg problem in the particular case when the function f is invariant under the antipodal map $x \to -x$ (S_2 is considered imbedded in R^3).

8.1. Moser's Result

6.65 Theorem (Moser [*245]). *On (S_2, g_0) let $f \in C^\infty$ be a function invariant under the antipodal map $x \to -x$. If $\sup f > 0$, f is the scalar curvature of a metric conformal to g_0.*

If φ satisfies (1), $\int fe^\varphi \, dV = 8\pi$. So the condition $\sup f > 0$ is both necessary and sufficient.

As f is antipodally symmetric, we can pass to the quotient on $P_2(\mathbb{R})$. Now on $P_2(\mathbb{R})$ the problem of prescribed scalar curvature is entirely solved. The proof is written up in 6.25. The variational method works on $P_2(\mathbb{R})$. The reason is that the volume of $P_2(\mathbb{R})$ is half of that of the sphere. With $V = 2\pi$ in (78), it is possible to prove that a minimizing sequence is bounded in H_1.

Remark. For $n \geq 3$, we can consider the same problem as Moser. We will deal with this subject in a more general situation when f is invariant under a group of isometries (see §9), not only under the antipodal map.

8.2. Kazdan and Warner Obstructions

6.66 Theorem (Kazdan and Warner [*195]). *Let F be the eigenspace corresponding to the first non zero eigenv $\lambda_1 = 2$ of the laplacian of the unit sphere (S_2, g_0).*

If φ satisfies (75) *then for all $\xi \in F$*

$$\int_{S_2} \nabla^\nu \xi \nabla_\nu f e^\varphi \, dV = 0. \tag{79}$$

Proof. Differentiating $f = e^{-\varphi}[\Delta\varphi + 2]$ gives

$$\nabla_\nu f = e^{-\varphi}\nabla_\nu \Delta\varphi - [\Delta\varphi + 2]e^{-\varphi}\nabla_\nu\varphi.$$

Multiplying by $e^\varphi \nabla^\nu \xi$ and integrating yields

$$\int e^\varphi \nabla^\nu \xi \nabla_\nu f \, dV$$
$$= \int \nabla^\nu \xi \nabla_\nu \Delta\varphi \, dV - 2\int \nabla^\nu \xi \nabla_\nu \varphi \, dV - \int \Delta\varphi \nabla^\nu \xi \nabla_\nu \varphi \, dV.$$

Any $\xi \in F$ satisfies $\nabla_{ij}\xi = -\xi g_{ij}$ and $\Delta\xi = 2\xi$. Thus integrating by parts twice gives $\int \nabla^\nu \xi \nabla_\nu \Delta\varphi \, dV = 2\int \nabla^\nu \xi \nabla_\nu \varphi \, dV$. Moreover

$$\int \Delta\varphi \nabla^\nu \xi \nabla_\nu \varphi \, dV = \int \nabla^\mu \varphi \nabla_{\nu\mu}\varphi \nabla^\nu \xi \, dV + \int \nabla^\mu \varphi \nabla_{\nu\mu}\xi \nabla^\nu \varphi \, dV$$
$$= \frac{1}{2}\int \nabla_\nu |\nabla\varphi|^2 \nabla^\nu \xi \, dV - \int \xi |\nabla\varphi|^2 \, dV = 0.$$

6.67 Theorem (Kazdan and Warner [*198] p. 130). *Let F be the eigenspace corresponding to the first non zero eigenvalue $\lambda_1 = n$ of the laplacian on the unit sphere (S_n, g_0) $n \geq 3$. If φ satisfies* (76), *then for all $\xi \in F$*

$$\int_{S_n} \nabla^\nu \xi \nabla_\nu f \varphi^N \, dV = 0 \qquad \text{with } N = 2n/(n-2). \tag{80}$$

The proof is similar to those of 6.66. We differentiate

$$f = (n-1)\big[4\Delta\varphi/(n-2) + n\varphi\big]\varphi^{1-N}.$$

Then, after multiplying by $\varphi^N \nabla^\nu \xi$, integration over S_n yields

$$\int \nabla^\nu \xi \nabla_\nu f \varphi^N \, dV = \frac{4(n-1)}{n-2}\left[\int \varphi \nabla^\nu \xi \nabla_\nu \Delta\varphi \, dV \right. \tag{81}$$
$$\left. + (1-N)\int \Delta\varphi \nabla_\nu \varphi \nabla^\nu \xi \, dV - n\int \varphi \nabla_\nu \varphi \nabla^\nu \xi \, dV \right].$$

As ξ satisfies $\nabla_{\nu\mu}\xi = -\xi g_{\nu\mu}$ and $\Delta\xi = n\xi$, integrating by parts many times yields

$$\int \varphi\nabla^\nu\xi\nabla_\nu\Delta\varphi\, dV = -\int \Delta\varphi\nabla^\nu\xi\nabla_\nu\varphi\, dV + n\int \xi\varphi\Delta\varphi\, dV$$

$$\int \Delta\varphi\nabla^\nu\xi\nabla_\nu\varphi\, dV = \int \nabla^\nu\xi\nabla_{\nu\mu}\varphi\nabla^\mu\varphi\, dV + \int \nabla^\mu\varphi\nabla_{\nu\mu}\xi\nabla^\nu\varphi\, dV$$

$$= (n/2-1)\int \xi|\nabla\varphi|^2\, dV,$$

$$\int \xi\varphi\Delta\varphi\, dV = \int \xi|\nabla\varphi|^2\, dV + \int \varphi\nabla^\nu\varphi\nabla_\nu\xi\, dV.$$

Thus the right hand side in (81) is zero.

Consequences. Many smooth functions on (S_n, g_0) are not scalar curvature of any metric conformal to g_0. If $\nabla^\nu\xi\nabla_\nu f \geq 0$ for instance, for some $\xi \in F$ ($\xi \not\equiv 0$), equation 75, if $n = 2$ or equation 76 if $n \geq 3$ has no solution. But we have more. The set of functions f, which are scalar curvature of some metric conformal to g_0, is stable under $C(S_n)$ the conformal group of (S_n, g_0).

So if there exist $u \in C(S_n)$ and $\xi \in F$ ($\xi \not\equiv 0$) such that $\nabla^\nu(\xi o u)\nabla_\nu f \geq 0$, then f is not scalar curvature of any metric conformal to g_0. In this way we have the following

6.68 Theorem (Bourguignon–Ezin [*56]). *Let X be a conformal vector field on a compact Riemannian manifold (M, g). Then*

(82) $$\int X(R)\, dV = 0 \qquad \textit{where } R \textit{ is the scalar curvature of } g.$$

For $n \geq 3$ the identity is obtained by integrating the formula of Lichnerowicz [185] p. 134.

$$\Delta(\nabla_i X^i) = R\nabla_i X^i/(n-1) + nX^i\nabla_i R/2(n-1).$$

For $n = 2$ the proof is in [*56], where the authors exhibit a function f such that $\nabla^\nu\xi\nabla_\nu f$ does not keep a fixed sign for any $\xi \in F$, but such that $X(f)$ keeps a fixed sign for some conformal vector field X on S_2.

Note that the integral condition (82) provides examples of functions f which are not scalar curvature of any conformal metric only on (S_n, g_0). Indeed by the Lelong–Ferrand theorem [175], the connected component of the identity of $C_0(M, [g])$, the conformal group, is compact, except if the manifold is (S_n, g_0). If $C_0(M, [g])$ is compact, there exists $\tilde{g} \in [g]$ such that $C_0(M, [g])$ is the group of isometries of $(M, \tilde{g})$: $\tilde{\nabla}_\nu X^\nu = 0$.

Thus on $(M, \tilde{g})$ (11) is trivial:

$$\int X(f)d\tilde{V} = \int X^\nu\tilde{\nabla}_\nu f d\tilde{V} = \int f\tilde{\nabla}_\nu X^\nu d\tilde{V} = 0.$$

Examples. In [*56] the authors exhibit a function R which cannot be excluded by (79), but for which there exists a conformal vector field X such that $X(R) \geq 0$.

Let us mention also the example of Xu and Yang in [*319], of a rotationally symmetric function R on S_2 for which the obstruction (79) is satisfied but equation (1) has no rotationally symmetric solution.

Chen and Li [*89] generalized this result: if R is rotationally symmetric and monotone in the region where $R > 0$, then equations (1) and (2) have no rotationally symmetric solution.

6.69 We saw in 6.67 that the necessary conditions for equations (75) and (76) to have a solution are

α) f is positive somewhere

β) f satisfies the Kazdan–Warner conditions (i.e. there does not exist $u \in C(S_n)$ and $\xi \in F$ ($\xi \not\equiv 0$) such that $\nabla^\nu(\xi \circ u)\nabla_\nu f \geq 0$).

Are these conditions sufficient? The answer is no. Chen and Li [*90] produced functions f which satisfy α) and β), but are not the scalar curvature of any metric $g \in [g_0]$.

Theorem 6.69 (Chen and Li [*90]). *If f is rotationally symmetric and monotone in the region where $f > 0$, then equations* (75) *and* (76) *have no solution.*

Under these hypotheses, in order to satisfy β, it is essential that f changes sign. When $n = 2$ we have more. According to Xu and Yang's result [319] (see 6.85), for the class of positive nondegenerate rotationally symmetric functions, β) is a necessary and sufficient condition.

To go further, Han and Li [*158] produced, when $2 \leq n \leq 4$, a family of positive functions f satisfying β) which are not the scalar curvature of any metric $g \in [g_0]$.

8.3. A Nonlinear Fredholm Theorem

6.70 On the unit sphere (S_2, g_0), any $\varphi \in H_1$, satisfies (78) with $\mu = \mu_2 = 1/16\pi$ and $V = 4\pi$. The constant $C(\mu_2)$ can be taken equal to 1 according to Onofri [*261]. But we can improve the best constant μ_2:

Theorem 6.70 (Aubin [21]). *Let F be the eigenspace corresponding to the first non zero eigenvalue for Δ. The functions $\varphi \in H_1$ satisfying $\int \varphi\, dV = 0$ and $\int \xi e^\varphi\, dV = 0$ for all $\xi \in F$ satisfy*

$$\int e^\varphi\, dV \leq C(\mu) \exp\left(\mu \|\nabla\varphi\|_2^2\right) \quad \text{with any } \mu > \frac{\mu_2}{2} = \frac{1}{32\pi} \tag{83}$$

$C(\mu)$ being some constant.

Chang and Yang pointed out that (83) is valid with $\mu = 1/32\pi$ for any functions φ which, in addition of the hypothesis of Theorem 6.70, satisfies equation (1) with $f \geq 0$. In fact if we look at the proof of Theorem 6.70 when we are on the sphere S_2, we can write μ_2 instead of $\mu_2 + \varepsilon$ (p. 158 of [21]) and we have to bound a term in $\|\varphi\|_2$.Thus

Corollary 6.70. *If in addition to the hypothesis of Theorem* 6.70, $\|\varphi\|_2 \leq k$, *there exists a constant* $C(k)$ *such that* φ *satisfies:*

$$\int e^{\varphi}\, dV \leq C(k) \exp\left(\|\nabla\varphi\|_2^2/32\pi\right). \tag{84}$$

When φ satisfies (75) with $f \geq 0$, $\|\Delta\varphi\|_1 < 16\pi$. Moreover, as $\int \varphi\, dV = 0$, $\|\varphi\|_2 \leq C_1$, $\|\Delta\varphi\|_1 \leq C_2$ and (84) holds (C_1 and C_2 two constants).

6.71 On the unit sphere (S_n, g_0) $n \geq 3$ we know that any function $\varphi \in H_1$ satisfies

$$\|\varphi\|_N^2 \leq K^2(n,2)\|\nabla\varphi\|_2^2 + \omega_n^{-2/n}\|\varphi\|_2^2, \tag{85}$$

where $K(n,2)$ is the best constant in the Sobolev imbedding theorem

$$K(n,2) = 2\omega_n^{-1/n}[n(n-2)]^{-1/2}, \qquad N = 2n/(n-2).$$

But we can improve the best constants (see Theorem 2.40).

Theorem 6.71 (Aubin [21]). *Let* $\xi_i (i = 1, 2, \ldots, n+1)$ *be a basis of* F *on* (S_n, g_0). *Then all* $\varphi \in H_1$ *satisfying* $\int \xi_i |\varphi|^N\, dV = 0$ *for all* i *satisfy*

$$\|\varphi\|_N^2 \leq \left[2^{-2/n}K^2(n,2) + \varepsilon\right] \|\nabla\varphi\|_2^2 + A(\varepsilon)\|\varphi\|_2^2 \tag{86}$$

where $A(\varepsilon)$ *is a constant which depends on* $\varepsilon > 0$, ε *as small as one wants.*

Recall F is the space of the eigenfunctions corresponding to $\lambda_1 = n$ (the space of the spherical harmonics of degree 1).

We saw in 6.64 that the variational method breaks down. But if we consider

$$\nu = \inf J(\varphi) \text{ for all } \varphi \in \mathcal{A} = \left\{ \varphi \in H_1 \Big/ \int \xi_i |\varphi|^N\, dV = 0 \text{ for all } i \right\},$$

then ν may be achieved (for the definition of $J(\varphi)$ see (6.64)).

6.72 Theorem (Aubin [21]). *Given a smooth function* $\tilde{f}$ *on* S_2 *satisfying* $\int \tilde{f}\, dV > 0$, *there exists* $h(\tilde{f}) \in F$ *such that Equation* (75) *with* $f = \tilde{f} - h(\tilde{f})$ *has a solution* $\Psi \in C^\infty$, e^Ψ *being orthogonal to* F *in the* L_2 *sense. Moreover* Ψ *minimizes* $I(\varphi)$ *on* $\tilde{\mathcal{A}}$.

Proof. As in 6.63, consider the functional (77):

$$I(\varphi) = \int |\nabla\varphi|^2\, dV + 4 \int \varphi\, dV.$$

Here we will consider $\tilde{\nu}$, the inf of I (φ) for $\varphi \in \tilde{\mathcal{A}}$ with

$$\tilde{\mathcal{A}} = \left\{ \varphi \in H_1 \Big/ \int \tilde{f} e^{\varphi}\, dV = 8\pi \quad \text{and} \quad \int \xi e^{\varphi}\, dV = 0 \text{ for all } \xi \in F \right\}.$$

Similarly we prove that $\tilde{\nu}$ is finite. Let $\{\varphi_i\}$ be a minimizing sequence. Pick μ satisfying $1/2 < 16\pi\mu < 1$, by (83) any $\varphi \in \tilde{\mathcal{A}}$ satisfies:

$$8\pi \le \sup \tilde{f} C(\mu) \exp\left[\mu \|\nabla\varphi\|_2^2 + \int \varphi\, dV/4\pi \right].$$

Thus, for some constants C_j,

$$(1 - 16\pi\mu)\|\nabla\varphi_i\|_2^2 + C_1 \le I(\varphi_i) \le C_2.$$

Hence $\|\nabla\varphi_i\|_2^2 \le C_3$ and $|\int \varphi_i\, dV| \le C_4$. As the map $H_1 \ni \varphi \to e^{\varphi} \in L_1$ is compact, there is a subsequence $\{\varphi_j\}$ of $\{\varphi_i\}$ and $\Psi \in \mathcal{A}$ such that $\varphi_j \to \Psi$ weakly in H_1. So Ψ minimizes $I(\varphi)$ on $\mathcal{A}$. Consequently Ψ satisfies weakly in H_1

$$\Delta\Psi + 2 = \left[\tilde{f} - h(\tilde{f})\right] e^{\Psi} \qquad \text{where } h(\tilde{f}) \in F.$$

Bootstrap method then implies $\Psi \in C^{\infty}$.

Corollary 6.72. *On (S_2, g_0) a necessary condition for solving the Nirenberg problem is that the candidate function is positive somewhere. This condition is also sufficient modulo a vector space of dimension at least three.*

Proof. Suppose f is positive at P, and consider a conformal transformation of the sphere with pole at P. The new metric is of the form $\tilde{g}(Q) = (\beta - \cos\alpha r)^{-2} g(Q)$ for some $\beta > 1$, $1/\alpha$ being the radius of the sphere ($R = 2\alpha^2$). The new scalar curvature is constant $\tilde{R} = R(\beta^2 - 1)$. Then on the sphere we have to solve an equation like (75) in the metric $\tilde{g}$ with $\tilde{R}$ instead of 2 ($\tilde{\Delta}\varphi + \tilde{R} = f e^{\varphi}$).

Since we can choose β so that $\int f\, d\tilde{V} > 0$, we can apply Theorem 6.72 in the metric $\tilde{g}$.

6.73 Theorem (Aubin [21]). *Given a smooth function $\tilde{f}$ on (S_n, g_0) $(n \ge 3)$, satisfying $\sup \tilde{f} < 4^{1/(n-2)} \inf \tilde{f}$, there exists $h(\tilde{f}) \in F$ such that Equation* (76) *with $f = \tilde{f} - h(\tilde{f})$ has a solution $\varphi \in C^{\infty}$. So f is the scalar curvature of some metric in $[g_0]$.*

Since $H_1 \subset L_N$ is not compact the proof is harder than that of Theorem 6.72. We must consider the approximation equation.

$$\text{(87)} \qquad 4\frac{n-1}{n-2}\Delta\varphi + n(n-1)\varphi = f\varphi^{q-1}, \qquad \text{for } 2 < q < N.$$

First of all we prove the existence of functions $\Psi_q \in C^{\infty}$, $\Psi_q > 0$ satisfying

$$\int \xi \Psi_q^q\, dV = 0$$

for all $\xi \in F$ which solves (87) with f = $\tilde{f} - h_q(\tilde{f}), h_q(\tilde{f})$ belonging to F. Moreover, Ψ_q is a minimizer of the functional

$$J_q(\varphi) = \left[4\frac{n-1}{n-2}\int |\nabla\varphi|^2\,dV + n(n-1)\int \varphi^2\,dV\right]\left[\int \tilde{f}|\varphi|^q\,dV\right]^{-2/q}$$

over the set of the functions φ of H_1 satisfying $\int \xi|\varphi|^q\,dV = 0$ for all $\xi \in F$.

Then we consider a sequence $q_i \to N$. As $\Psi_{q_i}^{q_i}$ is orthogonal to F in the L_2 sense, we can apply Theorem 6.71 to the functions $\Psi_q^{q/n}$ (the inequality (86) instead of (85)). This allows us to complete the proof of Theorem 6.76.

This result was improved recently by Hebey.

6.74 Theorem (Hebey [*164A]). *Given a smooth function $\tilde{f}$ satisfying* $\sup \tilde{f} > 0$ *on (S_n, g_0) $n \geq 3$, there exists $h(\tilde{f}) \in F$ and a conformal diffeomorphism $u \in C(S_n)$ such that $\tilde{f} - h(\tilde{f}) \circ u$ is the scalar curvature of some metric in $[g_0]$ the conformal class of g_0. On (S_3, g_0) if there exists a point $x \in S_3$ such that* $\tilde{f}(x) = \tilde{f}(-x) = \sup f > 0$, *then there exists $h(\tilde{f}) \in F$ such that $f = \tilde{f} - h(\tilde{f})$ is the scalar curvature of some metric in* $[g_0]$.

§9. G-invariant Functions f

6.75 In this section, we suppose that f is invariant under a non trivial group G of isometries and we seek a solution of (76) invariant by G. If the group acts freely, $M_n = S_n/G$ is a manifold.

Let f' be the quotient of f on M_n. The problem then becomes the problem of prescribed scalar curvature on M_n studied above. The advantage of this approach is that the inf of the functional may be attained on M_n and there is no longer any obstruction on M_n like those of Kazdan and Warner. As M_n is locally conformally flat, Theorem 6.31 can be applied.

On (S_n, g_0) $n \geq 3$, any C^∞ function, positive somewhere and invariant under a nontrivial group of isometries acting freely, is the scalar curvature of a metric conformal to g_0 when $n = 3$, and when $n > 3$ if, at a point where f is maximal, all its derivatives up to order $n-2$ vanish.

Actually, the hypothesis G acts freely is quite restrictive. When n is even, there is only one such group, the group with two isometries, the identity and the antipodal map. Thus when n is even, only antipodally symmetric functions are covered by the result of Escobar–Schoen (see Theorem 6.31).

More general groups of isometries were deal with by Hebey. For $n = 3$ Hebey established the best result possible in this context.

6.76 Theorem (Hebey [*162]). *On (S_3, g_0), any C^∞ function, positive somewhere and invariant under a group G of isometries acting without fixed point, is the scalar curvature of some metric in* $[g_0]$.

We saw that the hypothesis "positive somewhere" is necessary by integrating (76). Moreover, if there is a point of S_3 fixed by G, functions like ξ + Const. with $\xi \in F$ are not excluded. So the second hypothesis cannot be weakened too much. For the proof Hebey used the method of isometry-concentration studied in §5.

First, for any q satisfying $2 < q < N = 2n/(n-2)$, there exists a G-invariant C^∞ function $u_q > 0$ solution of the equation obtained from (76) by substituting $q-1$ for the exponent $(n+2)/(n-2)$ in the right hand side of (2). The existence of such sub-critical sequence $\{u_{q_i}\}$ with $q_i \to N$ is proved without difficulty since the Kondrakov Theorem $\big(H_1 \subset L_q$ is compact $\big)$ holds for $q < N$. If any subsequence of $\{u_{q_i}\}$ does not converge (the bad case), there exists a subsequence which converges to zero except at a finite number of points, the points of concentration.

Then the main idea is to estimate from above the number of points of concentration and to obtained a contradiction. Using this method Hebey established many results; let us mention some of them.

6.77 Set $\mu(G)$ equal the inf of $J(\varphi)$ (see 6.64 for the definition) for all G-invariant functions $\varphi \in H_1$ such that $\int f|\varphi|^N \, dV = 1$. Define $C(f) = \big\{x \in S_n / |\nabla f(x)| = 0$ and $f(x) > 0$ for $f \in C^\infty(S_n)\big\}, O(x) = \big\{u(x)/u \in G\big\}$ for $x \in S_n$ and $\mu_0 = n(n-1)\omega_n^{2/n}$, which is the inf of $J(\varphi)$ for all $\varphi \in H_1$ when $f \equiv 1$.

Theorem 6.77 (Hebey [*162], [*163]). *Let $f \in C^\infty$ be a G-invariant function on (S_n, g_0) which is positive somewhere. If*

$$f(x) < \big[\mu_0/\mu(G)\big]^{n/(n-2)} \big[\mathrm{Card}\, O(x)\big]^{2/(n-2)}$$

holds for all $x \in C(f)$, then f is the scalar curvature of a metric in $[g_0]$.

Corollary (Hebey [*162]). *Let $f \in C^\infty$ be a G-invariant function on (S_n, g_0) such that $\int f \, dV > 0$. If for all $x \in C(f)$*

$$f(x) \leq \big[\mathrm{Card}\, O(x)\big]^{2/(n-2)} \int f \, dV / \omega_n, \tag{88}$$

f is the scalar curvature of a metric in $[g_0]$.

From the theorem above, corollary follows directly by writing

$$\mu(G) \leq J(1) = n(n-1)\omega_n \Big(\int f \, dV\Big)^{-2/N}.$$

For details on the proof see also Aubin [10].

6.78 Theorem (Hebey [*162]). *Let f be a G-invariant C^∞ function on (S_n, g_0). At P a point with $f(P) = \sup f > 0$, we suppose that all derivatives of f of order less than or equal to $n-3$ vanish. Moreover we suppose* Card $O(P) \geq 2$.

If for all $x \in C(f)$

(89) $$f(x) \le \left[\operatorname{Card} O(x)/\operatorname{Card} O(P)\right]^{2/(n-2)} \sup f,$$

then f is the scalar curvature of some metric in $[g_0]$, when $n \ge 5$ is odd, or when $n = 4$ if in addition $\Delta f(P)/\sup f$ is smaller than an explicit positive constant. The same result holds when $n \ge 6$ is even, if $|\nabla^{n-2} f|/\sup f$ at P is smaller than an explicit positive constant.

We dealt with the case $n = 3$ in 6.76. When $n \ge 4$ we need some condition of flatness of f at P for the conclusion.

Remark. There is no obstruction (seen in §1.2) for a G-invariant function f to be the scalar curvature of some metric in $[g_0]$, when G acts without fixed point. Indeed for all $x \in S_n$ $\operatorname{Card} O(x) \ge 2$. So there exists a constant k enough large such that $f + k$ satisfies (88). Corollary 6.77 then implies that $f + k$ is the scalar curvature of some metric in $[g_0]$.

6.79 Theorem (Hebey [*162]). *Let f be a G-invariant function which is positive somewhere on (S_n, g_0). Define*

$$C_0 \subset C(f) = \left\{x \in S_n / |\nabla f(x)| = 0 \text{ and } f(x) > 0\right\},$$

the set of the points where the function $[\operatorname{Card} O(x)]^{\frac{2}{n}} [f(x)]^{\frac{2}{n-1}}$ is minimum on $C(f)$.

Assume all derivatives up to order $n-3$ vanish at a point $x_0 \in C_0$ where $\operatorname{Card} O(x_0) \ge 2$. Then f is the scalar curvature of some metric in $[g_0]$ when n is odd or when $n = 4$ if

$$\Delta f(x_0) < 6\left[\operatorname{Card} O(x_0) - 1\right] f(x_0).$$

The same result holds when $n \ge 6$ is even if

$$|\nabla^{n-2} f(x_0)| < 2^{3-n}(n-1)!(\operatorname{Card} O(x_0) - 1) f(x_0)/(n-2).$$

With this result, when G acts without fixed point, we can prove, on (S_4, g_0) for instance, the following: There exists $\varepsilon > 0$ such that any G-invariant function f, positive somewhere and satisfying $\|f - 1\|_{C^2} < \varepsilon$, is the scalar curvature of some metric in $[g_0]$.

§10. The General Case

10.1. Functions f Close to a Constant

6.80 We begin with a result of Chang and Yang. For any $n \geq 2$ they solve the Nirenberg Problem (or its extension for $n \geq 3$) in a neighbourhood of the constant under the two following weak hypotheses

1) *f has only critical points non degenerate of order at most n, when n is even and at most $n-1$ when n is odd.*

This has the following meaning. Let P be the south pole of $S_n \subset \mathbb{R}^{n+1}$ with coordinates $x_1, x_2, \ldots, x_{n+1}$. $P(0,0,\ldots,-1)$. We let $y_1, y_2, \ldots y_n$ be the stereographic coordinates, with respect to the north pole Q, of $y \in S_n - Q$.

$$x_i = 2y_i/\left(1+|y|^2\right) \qquad \text{for } 1 \leq i \leq n, \quad x_{n+1} = 1 - 2/\left(1+|y|^2\right).$$

The limited expansion of f at P of order α can be written

$$f_k(y) = f(P) + \sum_{k=1}^{\alpha} R_k(y)$$

where R_k is a homogenous polynomial in y_i of degree k.

Definition. f is nondegenerate at P of order α if R_k vanish for $k < \alpha$ and $G(P,t)$ defined by (90) satisfies $|G(P,t)| \geq ct^{-\alpha}$ if $\alpha < n$ and , if $\alpha = n$ $|G(P,t)| \geq ct^{-n}\log t$ for some $c > 0$ when t is large.

We verify that a critical point P is nondegenerate of order 2 if $\Delta f(P) \neq 0$. With this definition the critical points need be not isolated.

2) *The map G defined below has* $\deg(G, B, 0) \neq 0$.

Let $\varphi_{Q,t}$ be the conformal map of S_n defined by $\varphi_{Q,t}(y) = ty$. G is the map from the unit ball $B \subset \mathbb{R}^{n+1}$ given by:

$$B \ni q = (1-1/t)Q \overset{G}{\to} G(Q,t) = \int (f \circ \varphi_{Q,t})\vec{x}. \tag{90}$$

$q = 0$ $(t = 1)$ being identified with the identity map, the set of conformal transformations is homeomorphic to $B = \left\{q \in \mathbb{R}^{n+1}/|q| < 1\right\}$. For $q \neq 0$, $|q| = 1 - 1/t$ and $Q = q/|q|$.

6.81 Theorem (Chang and Yang [*85]). *Let f be a C^∞ function on (S_n, g_0) whose critical points are nondegenerate of order at most n when n is even and at most $n-1$ when n is odd. Assume that the map G defined by* (90) *has* $\deg(G, B, 0) \neq 0$. *Then there exists some constant $\varepsilon(n)$ such that if f satisfies* $\sup|f-1| \leq \varepsilon(n)$, *$f$ is the scalar curvature of some metric conformal to g_0.*

For the proof, Chang and Yang start by applying Aubin's result (Theorem 6.72 when $n = 2$ and Theorem 6.73 when $n \geq 3$), to the family of functions $f_p = f \circ \varphi_{P,t}$ with $p = (1 - 1/t)P \in B$.

There exists $\vec{\Lambda}_p \in \mathbb{R}^{n+1}$ and $w_p \in C^\infty(S_2)$ (resp. $u_p > 0$ smooth on S_n $(n \geq 3)$) satisfying

$$\Delta w_p + 2 = \left(f_p - \vec{\Lambda}_p \cdot \vec{x}\right)e^{w_p} \qquad \text{when n=2 and} \tag{91}$$

$$4\frac{n-1}{n-2}\Delta u_p + n(n-1)u_p = \left(f_p - \vec{\Lambda}_p \cdot \vec{x}\right)u_p^{N-1} \text{ when } n \geq 3. \tag{92}$$

Indeed we can choose $\varepsilon(n)$ such that $\sup f < 4^{1/(n-2)} \inf f$ when $n \geq 3$. Moreover w_p minimizes $I(w) = \|\nabla w\|_2^2 + 4\int w\, dV$ on

$$\mathcal{A}_2 = \left\{ w \in H_1 / \int f_p e^w\, dV = 8\pi \quad \text{and} \quad \int e^w \vec{x}\, dV = 0 \right\},$$

and if $\varepsilon(n)$ is small enough, u_p minimizes

$$J(u) = \left[4\frac{n-1}{n-2}\|\nabla u\|_2^2 + n(n-1)\|u\|_2^2\right]\left[\int f_p u^N\right]^{-2/N}$$

on $\mathcal{A}_n = \left\{u \in H_1 / \int |u|^N \vec{x}\, dV = 0, u \neq 0\right\}$.

Given $p \in B$, it is proven by contradiction in [*85] that w_p (resp. u_p) is uniquely determined if $\varepsilon(n)$ is small enough. Join two distinct minima u_p and $\tilde{u}_p$ by a 1-parameter family $u_\lambda^N = \lambda \tilde{u}_p^N + (1 - \lambda)u_p^N$ and show that $\lambda \to J(u_\lambda)$ is convex. Similarly in the case $n = 2$. By the implicit function theorem it is proved that w_p, u_p and Λ_p are continuous in p. In particular $\Lambda : B \to \mathbb{R}^{n+1}$ is a continuous map.

If $\Lambda_p = 0$ at some $q \in B$ with $|q| < 1$, equation (75) (resp. (76)) has a solution, and f is the scalar curvature of some metric in $[g_0]$. Indeed w_q satisfies (91) with $\Lambda_q = 0$:

$$\Delta w_q + 2 = f_q e^{w_q}. \tag{93}$$

Thus $\tilde{w} = (w_q - \log|\det D\varphi_q|) \circ \varphi_q^{-1}$ is a solution of Equation (75), where $\varphi_q = \varphi_{Q,t}$ with $q = (1 - 1/t)Q$. Similarly when $n \geq 3$, u_q satisfies (92) with $\Lambda_q = 0$:

$$4\frac{n-1}{n-2}\Delta u_q + n(n-1)u_q = f_q u_q^{N-1}. \tag{94}$$

Thus $v = \left(u_q |\det D\varphi_q|^{-1/N}\right) \circ \varphi_q^{-1}$ is a solution of Equation (76).

To finish the proof, suppose Λ does not vanish. Under the non-degeneracy condition 1 in 6.80, it is shown in [*85] that $G(P, t)$ does not vanish for t large enough and that

$$\deg(\Lambda, B, 0) = \deg(G, B, 0).$$

Thus the condition $\deg(G, B, 0) \neq 0$ implies the contradiction and Λ vanishes somewhere in B.

10.2. Dimension Two

6.82 Theorem (A. Chang and Yang [*81]). *Let $f > 0$ be a C^∞ function on (S_2, g_0) with only nondegenerate critical points, where Δf does not vanish. Suppose f has $p+1$ local maxima and $q \neq p$ saddle points where $\Delta f > 0$; then f is the scalar curvature of a metric in $[g_0]$.*

Recently Xu and Yang [*319] pointed out that we can remove the hypothesis $f > 0$.

Set $\Omega = \{x \in S_2/f(x) > 0\} \neq \emptyset$. Suppose f has only nondegenerate critical points where $\Delta f(x) \neq 0$ when $x \in \Omega$. If, on Ω, f has $p+1$ local maxima and $q \neq p$ saddle points where $\Delta f > 0$, then f is the scalar curvature of a metric in $[g_0]$.

The critical points where $f \leq 0$ do not matter. This is not surprising, since concentration phenomena can happen only at points where $f > 0$ (see 6.42).

Before these theorems, there were partial results in Chang and Yang [*81] and Chen and Ding [*88]. The proofs are quite different than that of Theorem 4.21 which was recently improved by removing the condition "close to constant".

6.83 Theorem (A. Chang, Gursky and Yang [*78]). *Let $f > 0$ be a C^∞ function on (S_2, g_0), such that $\Delta f(Q) \neq 0$ whenever Q is a critical point of f.*

If $\deg(G, B, O) \neq 0$, *then f is the scalar curvature of a metric in $[g_0]$.*

This result generalizes Theorem 6.82: f may have degenerate critical points. Moreover the assumption is weaker. Indeed, when f has only nondegenerate critical points, the hypothesis $q \neq p$ (or $p+1-q \neq 1$) is equivalent to the index counting condition:

(95) $$\sum_{Q \text{ critical}, \Delta(Q)>0} (-1)^{k(q)} \neq (-1)^n,$$

where $k(Q)$ denotes the Morse index of f at Q, and it is shown in [*78] that (95) implies the hypothesis $\deg(G, B, 0) \neq 0$ in any dimension.

For the proof of Theorem 6.83, consider the family of functions:

(96) $$f_s = sf + 2(1-s).$$

If $s_0 > 0$ is small enough, we can apply Theorem 6.81.

So there exists a C^∞ function w_{so} solution of (75) with $f = f_{so}$. Moreover it is shown in [*85] that this solution is unique if s_0 is small enough. Now we will solve for $s \in [s_0, 1]$ the following continuous family of equations

(97) $$\Delta w + 2 = f_s e^w$$

by using the method of topological degree.

The critical points Q of f_s are those of f and when $s \in [s_0, 1]$, $|\Delta f_s(Q)| = s|\Delta f(Q)| \geq s_0|\Delta f(Q)| \geq \varepsilon$ for some $\varepsilon > 0$. Indeed suppose there is a sequence

Q_i of critical points of f such that $\Delta f(Q_i) \to 0$. By passing to a subsequence, $Q_i \to Q$ which is a critical point of f where $\Delta f(Q) = 0$. This is in contradiction with the hypothesis.

Moreover $f > 0$ implies $0 < m \leq f_s \leq M$ for some m and M independent of $s \in [s_0, 1]$. Thus we can apply Proposition 6.84 below to the solution of (97). These solutions satisfy $\|w\|_{2,\alpha} < C$ for some constant C. Set

$$\Omega = \left\{ w \in C^{2,\alpha}(S_2) \Big/ \int w\, dV = 0 \quad \text{and} \quad \|w\|_{2,\alpha} < C \right\},$$

and consider the map:

$$w \to \Psi_s(w) = w - \Delta^{-1}\left(f_s e^{w-\rho_s}\right) \tag{98}$$

where $\rho_s = \log\left[\int f_s e^w\, dV/8\pi\right]$.

We verify that $\Psi_s(w) = 0$ implies $w_s = w - \rho_s$ is a solution of (97). Conversely if w_s is a solution of (97), $w = w_s - \int w_s\, dV/4\pi$ satisfies $\Psi_s(w) = 0$.

Now as $w \to \Delta^{-1}\left(f_s e^{w-\rho_s}\right) + \bar{w}$ is a Fredholm map $\Omega \to C^{2,\alpha}$, continuous in s and $0 \notin \Psi_s(\partial\Omega)$ for $s \geq s_0$, $\deg(\Psi_s, \Omega, 0)$ is well defined and independent of s for $s \geq s_0$. Equation (97) has a unique solution for $s = s_0$; thus (97) has a solution for $s = 1$. For more details and the proof of the following proposition, see [*78] and [*85].

6.84 Proposition (A. Chang, Gursky and Yang [*78]). *Let f be a C^∞ function on S_2 and let $\mathfrak{G}$ be the set of its critical points. Assume $\Delta f(Q) \neq 0$ when $Q \in \mathfrak{G}$ and $0 < m \leq f \leq M$ for some m and M, then there exists a constant C which depends only on m, M and $\inf_{Q\in\mathfrak{G}} |\Delta f(Q)|$, such that any solution w of* (75) *satisfies $|w| \leq C$.*

First if $f \leq M$, by (4) $I(\varphi) = \|\nabla\varphi\|_2^2 + 4\int \varphi\, dV \geq$ Const., and under the hypothesis $m \leq f \leq M$, Chang and Yang proved that $I(\varphi)$ is bounded from above. Then the proof is by contradiction. A limited expansion in a neighbourhood of a point of concentration yields the contradiction by using the Kazdan and Warner condition (79).

For this, the hypothesis $|\Delta f(Q)| \geq \varepsilon > 0$ for $Q \in \mathfrak{G}$ is crucial.

6.85 When f is rotationally symmetric, we could hope that the problem would be easier. Indeed, if we seek for rotationally symmetric solutions, solving Equation (1) is equivalent in this case to solve an ordinary differential equation. Actually the difficulties are almost the same. Let us mention the following

Theorem 6.85 (Xu and Yang [*319]). *Let f be a rotationally symmetric C^∞ function on $(S_2, g_0) : f(x) = K(r)$ where r is the distance of x to a given point. Assume $K''(r) \neq 0$ when $K'(r) = 0$. If K' has both positive and negative values in the set where $K > 0$, then f is the scalar curvature of some metric in $[g_0]$.*

We complete this set of results on the Nirenberg Problem with the following

6.86 Theorem (K.C. Chang and Lin [*79]). *On* (S_2, g_0) *let* f *be a* C^∞ *function which is positive somewhere. Set* $\Omega = \{x \in S_2/f(x) > 0 \text{ and } \Delta f(x) > 0\}$. *Assume* $|\nabla f| \neq 0$ *when* $\Delta f = 0$ *or when* $f = 0$. *If* $\deg(\Omega, \nabla f, 0) \neq 1$, *then* f *is the scalar curvature of some metric in* $[g_0]$.

10.3. Dimension $n \geq 3$

6.87 Theorem (Bahri and Coron [*26]). *On* (S_3, g_0), *let* f *be a positive* C^∞ *function which has only non degenerate critical points where* $\Delta f \neq 0$. *If* (95) *holds, then* f *is the scalar curvature of some metric in* $[g_0]$.

We talked about the method used for the proof in Chapter 5. Bahri and Coron consider the functional $H(u) = \left(\int f(x)u^6\, dV\right)^{-\frac{1}{2}}$ on the set

$$\Sigma^+ = \left\{u \in H_1/u \geq 0 \quad \text{and} \quad 8\int |\nabla u|^2\, dV + 6\int u^2\, dV = 1\right\}.$$

They study the flow solution in Σ^+ of du/ds $= -H'(u)$, $u(0) \in \Sigma^+$. When the integral lines go to infinity, there is a lack of compactness. They introduce a pseudo-gradient near infinity and concentration phenomena occur. It appears that a point of concentration is a critical point where $\Delta f > 0$.

6.88 Theorem (S-Y. Chang, Gursky and Yang [*78]). *On* (S_3, g_0), *let* f *be a positive* C^∞ *function such that* $\Delta f \neq 0$ *at its critical points. If* $\deg(G, B, 0) \neq 0$, *then* f *is the scalar curvature of a metric in* $[g_0]$.

This result is proved by removing the condition "close to constant" of Theorem 6.81 as for Theorem 6.86. G is defined by (90). The authors showed that, if the critical points are nondegenerate, the hypothesis (95) of Theorem 6.87 implies $\deg(G, B, 0) \neq 0$.

The proof is similar to that of Theorem 6.86. We consider a family of equations

$$8\Delta u + 6u = f_s u^5, \qquad u > 0, \tag{99}$$

where $f_s = sf + 6(1 - s)$.

By Theorem 6.81, for $s = s_0 > 0$ small enough, (99) has a unique positive solution. On $\Omega = \left\{u \in C^{2,\alpha}(S_3)/\|u\|_{2,\alpha} < C \text{ and } C^{-1} < u < C\right\}$, where $C > 1$ is large and $0 < \alpha < 1$, define the map

$$\Omega \ni u \to \psi_s(u) = u - L^{-1}\left(f_s u^5\right) \in C^{2,\alpha}(S_3), \quad \text{where } L = 8\Delta u + 6u.$$

Equation (99) is rewritten in the form $\psi_s(u) = 0$. According to Proposition 6.89 below, for C large enough $0 \notin \psi_s(\partial\Omega)$. Thus $\deg(\psi_s, \Omega, 0)$ is well defined and independent of s for $s \geq s_0$, since $u \to L^{-1}(f_s u^5)$ is a Fredholm map continuous in s. In $\mathbb{Z}/2\mathbb{Z}$, $\deg(\psi_s, \Omega, 0) = 1$.

The hard part of the proof is to establish the a priori estimates of the following Proposition.

6.89 Proposition (S-Y Chang, Gursky and Yang [*78], see also Y-Y Li [*214]). *Suppose u is some positive solution on (S_3, g_0) of*

$$8\Delta u + 6u = fu^5$$

where $f \in C^\infty(S_3)$ satisfies $0 < m \le f$ and

$$\min_{\{x\in S_3, |\nabla K(x)|\le d\}} |\Delta K(x)| \ge d$$

for some $d > 0$. Then there exists a constant k, which depends only on m, d, $\|K\|_{C^2(S_3)}$, α and the modulo of continuity of $\nabla^2 K$ on S_3 such that

$$\|u\|_{C^{3,\alpha}(S_3)}, \qquad \|u^{-1}\|_{C^{3,\alpha}(S_3)} \le k.$$

6.90 We can say that Bahri–Coron's result (6.87) and then Theorem 6.88 solve the problem of the existence of a positive solution of Equation (76) when $n = 3$ and $f > 0$.

Of course, we can hope to find some improvements as for dimension 2, in the case where f is not always positive. But in some sense, the hypothesis $\deg(G, B, 0) \neq 0$ or (95) is optimal, except if we find some more general topological assumption. Such hypothesis cannot be removed, since there are the Kazdan–Warner obstructions.

When $n > 3$, there is Theorem 6.81, and until recently, only partial results such as that of Bahri–Coron [*24].

In [*214] and [*215] Yan-Yan Li states existence results of positive solutions of Equation (76) when f is some positive function on (S_n, g_0). When $n = 3$ Li's result is similar to that of Bahri–Coron. But when $n > 3$, we have a new answer to the problem. As in 6.80, Li considers the leading part of $f(y) - f(q)$ in a neighbourhood of some critical point q of f. He supposes that for any $q \in \mathfrak{G}$ (the set of the critical points of f), there exists some real number $\beta = \beta(q) \in]n-2, n[$ such that the leading part $R_\beta(y)$ of $f(y) - f(q)$ expresses, in some geodesic normal coordinate system centered at q, in the form

$$(100) \qquad R_\beta(y) = \sum_{j=1}^{n} a_j |y_j|^\beta, \quad \text{where } a_j \neq 0 \text{ and } A(q) = \sum_{j=1}^{n} a_j \neq 0.$$

6.91 Theorem (Yan-Yan Li [*215]). *On (S_n, g_0), $n \ge 3$, let f be a positive C^1 function which satisfies* (100) *at any $q \in \mathfrak{G}$. Then Equation* (76) *has a positive solution if*

$$\sum_{q\in\mathfrak{G} \text{ with } A(q)<0} (-1)^{i(q)} \neq (-1)^n,$$

where $i(q)$ is the number of negative $a_j(q)$, $1 \le j \le n$.

The main ingredients in the proof are some blow up analysis and some a priori estimates of positive solutions of (76).

Let J be the functional of the problem (see 5.78) and

$$[\delta(x,\lambda)](y) = \left[\lambda^2/2\right]^{(n-2)/4}\left[1+\lambda^2-\lambda^2\cos d(x,y)\right]^{1-n/2},$$

where $d(x,y)$ is the distance on the sphere of the two points x and y. When we compute a limited expansion in λ of $J\left[\sum_{i=1}^{p}\alpha_i\delta(x_i,\lambda_i)\right]$, we find that the interaction of two masses is in λ^{2-n}, whereas the self-interaction is in general in λ^{-2}. When the self-intersection is smaller than the interaction of two masses, the critical points at infinity are points where there is only one mass. Hence the Bahri–Coron Theorem 6.87 in dimension $n = 3$.

The assumptions of Li's Theorem 6.91 imply that we are in the same situation, the interaction of two masses predominates. Thus the points of concentration are simple.

10.4. Rotationally Symmetric Functions

6.92 Theorem (Hebey [*162]). *On (S_n, g_0), $n \geq 3$, let f be a rotationally symmetric C^∞ function which is positive somewhere. Denote by P and $\tilde{P}$ the poles of f. Then f is the scalar curvature of a metric in $[g_0]$, if*

$$\max\left[f(P), f(\tilde{P})\right] \leq \int f\,dV/\omega_n. \tag{101}$$

The same conclusion holds when $n = 3$ if

$$\max\left[f(P), f(\tilde{P})\right] \leq \sup f/4$$

or for $n \geq 4$, if $\Delta f(P) < 0$ when $f(P) \geq f(\tilde{P})$.

The results are proved by the method of Isometry-Concentration. Only P and $\tilde{P}$ may be points of concentration. An hypothesis like (101) does not allow that P or $\tilde{P}$ be point of concentration.

§11. Related Problems

11.1. Multiplicity

6.93 Theorem (Hebey and Vaugon [*167]). *On (S_3, g_0), let f be a positive C^∞ function invariant under two distinct finite groups of isometries G_1 and G_2. Assume G_2 acts freely, its cardinality $b > a$ the cardinality of G_1, and G_1 acts without fixed point. If*

$$(b/a)^{2/3} > 1 + b^3\left(\int f\,dV/\omega_3 \sup f\right)^{1/6}, \tag{102}$$

then f is the scalar curvature of at least two distinct metrics in $[g_0]$ which are respectively G_1-invariant and G_2-invariant. Their energies are different.

Set $g_1 = \varphi_1^{4/(n-2)} g_0$, $J(\varphi_1)$ is the energy of g_1 ($J(\varphi)$ is defined in 6.64). We present here this theorem on (S_3, g_0), but Hebey and Vaugon proved similar results on (S_n, g_0) for $n \geq 3$.

We can obtain as many metrics in $[g_0]$ with scalar curvature f as one wants. Suppose a finite group of isometries G_3, with cardinality $c > b$, acts freely. If $(c/b)^{2/3} > 1 + c^3\left(\int f\,dV/\omega_3 \sup f\right)^{1/6}$, then there exists g_3 in $[g_0]$ with scalar curvature f. As the energy of g_3 is different than those of g_1 and g_2, the three metrics are distinct. And so on. It is very easy to find functions f satisfying (102), $\sup f / \int f\,dV$ must be large enough.

The main ingredient in the proof of Theorem 6.93 is the value of the second best constant in the Sobolev inbedding theorem for the quotient of the sphere.

Let (M_n, g) be a compact Riemannian manifold, $n \geq 3$. If the manifold has constant sectionnal curvature (Aubin [14]) or if the manifold is only locally conformally flat (Hebey and Vaugon [*166]), there exists a constant C such that any $\varphi \in H_1$ satisfies

$$\|\varphi\|_N^2 \leq \frac{4\omega_n^{-2/n}}{n(n-2)} \|\nabla\varphi\|_2^2 + C\|\varphi\|_2^2.$$

Recently Hebey–Vaugon [*171] and [*172] proved that such constant C exists on any compact manifold. The proof is very different, it proceeds by contradiction. Blow-up technics are used (see 4.63).

6.94 Yan-Yan Li [*215] proved that any given somewhere positive continuous function may be perturbed in any C^0-neighbourhood of any given point on $S_n (n \geq 3)$ such that there exist many solutions for the perturbed function.

11.2. Density

6.95 The result of Li, just above, shows that the functions which are scalar curvature of some metric in $[g_0]$ on $(S_n, g_0)\, n \geq 3$, are dense in the set $\Omega \subset C^0(S_n)$ of the functions positive somewhere. Before this new result, we had the following L_p density theorem:

Theorem 6.95 (Bourguignon and Ezin [*56]). *Any smooth function on (S_2, g_0) which is positive somewhere belongs to the L_p-closure of the set of the functions which are scalar curvature of some metric in $[g_0]$.*

With Hebey's results, the same proof works on $(S_n, g_0)\, n \geq 3$. In fact the condition f positive somewhere is unnecessary since in any L_p-neighbourhood there are functions positive somewhere.

Actually with the results of §10 we have the following C^1-density theorem:

6.96 Theorem. *Let f be a smooth function positive somewhere on (S_2, g_0), or a smooth positive function on (S_n, g_0) when $n \geq 3$. In any $C^{1,\alpha}$-neighbourhood of $f (0 < \alpha < 1)$, there are smooth functions which are scalar curvature of some metrics in $[g_0]$.*

We can suppose without loss of generality that f has only nondegenerate critical points. For the proof, when $n = 2$, use for instance the improvement of Xu and Yang of Theorem 6.82. In case $q = p \neq 0$, it is easy to see that we can approximate in $C^{1,\alpha}$ the function f by a function $\tilde{f}$ for which $\Delta\tilde{f} < 0$ at some saddle point where $\Delta f > 0$. Thus $\tilde{q} = q - 1 \neq p = \tilde{p}$. In case $q = p = 0$, it is easy to see that we can approximate in C^1 the function f by a function $\tilde{f}$ which has a second maximum near the maximum of f. Thus $\tilde{p} = 1$.

When $n = 3$, use for instance Bahri–Coron's Theorem 6.87 and argue as above. When $n > 3$, use Li's Theorem 6.91 when n is odd, and when n is even, use Li's Theorem 0.13 in [*215].

11.3. The Problem on the Half Sphere

6.97 H. Hamza studied the Cherrier Problem (see §8.2 of Chapter 5) in the particular case of the hemisphere W_n endowed with g_0 the canonical metric on the sphere.

When $n = 2$, the equation to solve is (see 5.67)

$$\Delta\varphi + R = R'e^{\varphi} \text{ on } W_2, \partial_\xi\varphi + 2h = 2h'e^{\varphi/2} \text{ on } \partial W_2 = S_1. \tag{103}$$

When $n \geq 3$, the equation to solve is (see 5.65)

$$4\frac{n-1}{n-2}\Delta\varphi + R\varphi = R'\varphi^{\frac{n+2}{n-2}}, \varphi > 0 \text{ on } \quad W_n, \tag{104}$$

$$\frac{2}{n-2}\partial_\xi\varphi + h\varphi = h'\varphi^{\frac{n}{n-2}} \text{ on } \partial W_n = S_{n-1}.$$

Let us consider

$$\Lambda_n = \{F \in C^\infty(W_n)/\Delta F = nF \text{ on } W_n, \partial_\xi F = 0 \text{ on } \partial W_n\}.$$

If $W_n = \{x \in \mathbb{R}^{n+1}/|x| = 1,\ x^{n+1} \geq 0\}$, Λ_n is the set of the traces on W_n of the coordinate functions $x^i (1 \leq i \leq n)$, $\dim \Lambda_n = n$. Any $F \in \Lambda_n$ satisfies $\nabla_{ij}F = -Fg_{0ij}$ on W_n and of course $\tilde{\Delta}F = (n-1)F$ on $\partial W_n = S_{n-1}$ where $\tilde{\Delta}$ is the laplacian on (S_{n-1}, g_0), $\tilde{\nabla}$ will denote the covariant derivative on (S_{n-1}, g_0).

6.98 Theorem (Hamza [*155]). *A solution of* (103) *satisfies for any* $F \in \Lambda_2$*:*

$$\int_{W_2} e^{\varphi}\nabla^i R'\nabla_i F\, dV + 4\int_{S_1} e^{\varphi/2}\tilde{\nabla}^i h'\tilde{\nabla}_i F d\sigma = 0.$$

A solution of (104) *satisfies for any* $F \in \Lambda_n$*:*

$$\int_{W_n} \varphi^{\frac{2n}{n-2}}\nabla^i R'\nabla_i F\, dV + 2n\int_{S_{n-1}} \varphi^{2\frac{n-1}{n-2}}\tilde{\nabla}^i h'\tilde{\nabla}_i F d\sigma = 0.$$

These conditions are similar to the integrability conditions of Kazdan–Warner (see 6.66 and 6.67). On S_n there is one more independent condition, corresponding to the trace of the coordinate x^{n+1}.

A consequence of these conditions is that equations (103) and (104) have no solution if for some $F \in \Lambda_n$ $\nabla^i R' \nabla_i F > 0$ on W_n and $\tilde{\nabla}^i h' \tilde{\nabla}_i F > 0$ on ∂W_n.

For the euclidean ball, H. Hamza established also some integrability conditions (see [*155]).

Chapter 7

Einstein–Kähler Metrics

7.1 Introduction. In this chapter we shall use the continuity method and the method of upper and lower solutions to solve *complex Monge–Ampère equations*.

But they can also be solved by the variational method. The difficulty is to obtain the *a priori estimates*; either method can be used indiscriminately.

These equations arise in some geometric problems which will be explained. The results and proofs appeared in Aubin [11], [18] and [20], and Yau [277]. An exposition can also be found in Bourguignon [59] and [60].

We introduce some notation. Let g, ω, Ψ (respectively, g', ω', Ψ') denote the metric, the first fundamental form 7.2, and the Ricci form 7.4. For a compact manifold, $V = \int dV$. In complex coordinates, d' and d'' are defined by $d'\varphi = \partial_\lambda \varphi \, dz^\lambda$ and $d''\varphi = \partial_{\bar\mu}\varphi \, dz^{\bar\mu}$. Also, let $d^c\varphi = (d' - d'')\varphi$. Then $dd^c\varphi = -2\partial_{\lambda\bar\mu}\varphi \, dz^\lambda \wedge dz^{\bar\mu}$.

First definitions. Let M_{2m} be a manifold of real even dimension $2m$. We consider only local charts (Ω, φ), where Ω is considered to be homeomorphic by a map φ to an open set of $\mathbb{C}^m$: $\varphi(\Omega)$.

The complex coordinates are $\{z^\lambda\}$, $(\lambda = 1, 2, \ldots, m)$. We write $z^{\bar\lambda} = \overline{z^\lambda}$. A *complex manifold* is a manifold which admits an atlas whose changes of coordinate charts are holomorphic. A complex manifold is analytic. A *Hermitian metric* g is a Riemannian metric whose components in a local chart satisfy for all ν, μ:

$$g_{\nu\mu} = g_{\bar\nu\bar\mu} = 0, \qquad g_{\nu\bar\mu} = g_{\bar\mu\nu} = \bar g_{\mu\bar\nu}$$

The *first fundamental form* of the Hermitian manifold is $\omega = (i/2\pi) g_{\lambda\bar\mu} \, dz^\lambda \wedge dz^{\bar\mu}$, where g is a Hermitian metric.

§1. Kähler Manifolds

7.2 A Hermitian metric g is said to be *Kähler*: if the first fundamental form is closed: $d\omega = 0$. A necessary and sufficient condition for g to be Kähler is that its components in a local chart satisfy, for all λ, μ, ν,

$$\partial_\lambda g_{\nu\bar\mu} = \partial_\nu g_{\lambda\bar\mu}.$$

On a Kähler manifold we consider the Riemannian connection (Lichnerowicz [184] and Kobayashi–Nomizu [167]).

It is easy to verify that Christoffel's symbols of mixed type vanish. Only $\Gamma^{\nu}_{\lambda\mu} = \overline{\Gamma^{\bar{\nu}}_{\bar{\lambda}\bar{\mu}}}$ may be nonzero. Thus, if $f \in C^2$, then $\nabla_{\lambda\bar{\mu}} f = \partial_{\lambda\bar{\mu}} f$. On a Kähler manifold we will write $\Delta f = -g^{\lambda\bar{\mu}}\,\partial_{\lambda\bar{\mu}} f$, which is half of the real Laplacian (warning!).

Only the components of mixed type $R_{\alpha\bar{\beta}\lambda\bar{\mu}}$ of the curvature tensor may be nonzero. It is easy to verify that the components of the Ricci tensor satisfy $R_{\lambda\mu} = R_{\bar{\lambda}\bar{\mu}} = 0$ and

$$(*)\qquad R_{\lambda\bar{\mu}} = -\partial_{\lambda\bar{\mu}} \log |g|,$$

where $|g|$ is the determinant of the metric,

$$g = \begin{vmatrix} g_{1\bar{1}} & \cdots & g_{1\bar{m}} \\ \vdots & & \vdots \\ g_{m\bar{1}} & \cdots & g_{m\bar{m}} \end{vmatrix}.$$

In the real case we used the square of this determinant.

$\eta = (i/2)^m |g|\, dz^1 \wedge dz^{\bar{1}} \wedge \cdots \wedge dz^{\lambda} \wedge dz^{\bar{\lambda}} \wedge \cdots \wedge dz^m \wedge dz^{\bar{m}}$ defines a global $2m$-form. A complex manifold is orientable.

1.1. First Chern Class

7.3 $\Psi = (i/2\pi) R_{\lambda\bar{\mu}}\, dz^{\lambda} \wedge dz^{\bar{\mu}}$ is called the *Ricci form*. According to $(*)$, Ψ is closed: $d\Psi = 0$. Hence Ψ defines a cohomology class called the *first Chern class*: $\mathrm{C}_1(M)$. Recall that the cohomology class of Ψ is the set of the forms homologous to Ψ. Chern [91] defined the classes $\mathrm{C}_r(M)$ in an intrinsic way. For our purpose we only need to verify that $\mathrm{C}_1(M)$, so defined, does not depend on the metric. Indeed, let g' be another metric and Ψ' the corresponding Ricci form, let us prove that $\Psi' - \Psi$ is homologous to zero.

Since η and η' are positve $2m$-forms, there exists f, a strictly positive function, such that $\eta' = f\eta$. Hence, according to $(*)$,

$$\Psi' - \Psi = -\frac{i}{2\pi} \partial_{\lambda\bar{\mu}} \log f\, dz^{\lambda} \wedge dz^{\bar{\mu}},$$

and the result follows from the following:

Lemma. *A 1-1 form $\gamma = a_{\lambda\bar{\mu}}\, dz^{\lambda} \wedge dz^{\bar{\mu}}$ is homologous to zero if and only if there exists a function h such that $a_{\lambda\bar{\mu}} = \partial_{\lambda\bar{\mu}} h$. For the necessity we suppose the manifold is compact.*

Proof. The sufficiency is established at once:

$$\gamma = \partial_{\lambda\bar{\mu}} h\, dz^{\lambda} \wedge dz^{\bar{\mu}} = dd''h, \qquad \text{where } d''h = \partial_{\bar{\mu}} h\, dz^{\bar{\mu}}.$$

Now let us consider γ, a 1-1 form homologous to zero.

Pick a function h such that $\Delta h = -g^{\lambda\bar{\mu}}a_{\lambda\bar{\mu}}+\text{Const}$ (in fact the constant is zero), and define $\tilde{\gamma} = \tilde{a}_{\lambda\bar{\mu}}\, dz^{\lambda} \wedge dz^{\bar{\mu}}$ with $\tilde{a}_{\lambda\bar{\mu}} = \partial_{\lambda\bar{\mu}}h$.

$$g^{\lambda\bar{\mu}}(\tilde{a}_{\lambda\bar{\mu}} - a_{\lambda\bar{\mu}}) = \text{Const}, \text{ so } \nabla_{\nu}\left[g^{\lambda\bar{\mu}}(\tilde{a}_{\lambda\bar{\mu}} - a_{\lambda\bar{\mu}})\right] = g^{\lambda\bar{\mu}}\nabla_{\nu}(\tilde{a}_{\lambda\bar{\mu}} - a_{\lambda\bar{\mu}}) = 0$$

and $\delta''(\tilde{\gamma}-\gamma) = g^{\lambda\bar{\mu}}\nabla_{\lambda}(\tilde{a}_{\nu\bar{\mu}} - a_{\nu\bar{\mu}})\, dz^{\nu} = 0$, since $d\tilde{\gamma} = d\gamma = 0$ implies $\nabla_{\lambda}a_{\nu\bar{\mu}} = \nabla_{\nu}a_{\lambda\bar{\mu}}$ and $\nabla_{\lambda}\tilde{a}_{\nu\bar{\mu}} = \nabla_{\nu}\tilde{a}_{\lambda\bar{\mu}}$. Likewise, $\delta'(\tilde{\gamma}-\gamma) = -g^{\lambda\bar{\mu}}\nabla_{\bar{\mu}}(\tilde{a}_{\lambda\bar{\nu}}-a_{\lambda\bar{\nu}})\, dz^{\bar{\nu}} = 0$.

$\tilde{\gamma} - \gamma$ is homologous to zero and coclosed, so it vanishes (de Rham's theorem 1.72). On p-forms, the operators δ' and δ'' are defined by $\delta' = (-1)^{p-1} *^{-1} d' *$ and $\delta'' = (-1)^{p-1} *^{-1} d'' *$, (see 1.69); they are, respectively, of type $(-1, 0)$ and $(0, -1)$. ■

1.2. Change of Kähler Metrics. Admissible Functions

7.4 Let us consider the change of Kähler metric:

(1) $$g'_{\lambda\bar{\mu}} = g_{\lambda\bar{\mu}} + \partial_{\lambda\bar{\mu}}\varphi,$$

where $\varphi \in C^{\infty}$ is said to be admissible (so that g' is positive definite). Obviously g' is a Kähler metric, since (1) is satisfied.

Let $\mathsf{M}(\varphi) = |g'|\,|g|^{-1}$. Then $dV' = \mathsf{M}(\varphi)\, dV$. Since

$$\mathsf{M}(\varphi) = |g' \circ g^{-1}| = \begin{vmatrix} 1 + \nabla^1_1\varphi & \nabla^1_2\varphi & \cdots & \nabla^1_m\varphi \\ \nabla^2_1\varphi & 1 + \nabla^2_2\varphi & & \\ \vdots & & \ddots & \\ \nabla^m_1\varphi & & & 1 + \nabla^m_m\varphi \end{vmatrix},$$

by expanding the determinant we find

(1a)

$$\mathsf{M}(\varphi) = 1 + \nabla^{\nu}_{\nu}\varphi + \frac{1}{2}\begin{vmatrix} \nabla^{\nu}_{\nu}\varphi & \nabla^{\nu}_{\mu}\varphi \\ \nabla^{\mu}_{\nu}\varphi & \nabla^{\mu}_{\mu}\varphi \end{vmatrix} + \cdots + \frac{1}{m!}\begin{vmatrix} \nabla^{\nu}_{\nu}\varphi & \nabla^{\nu}_{\mu}\varphi & \cdots & \nabla^{\nu}_{\lambda}\varphi \\ \nabla^{\mu}_{\nu}\varphi & \nabla^{\mu}_{\mu}\varphi & & \\ \vdots & & \ddots & \\ \nabla^{\lambda}_{\nu}\varphi & & & \nabla^{\lambda}_{\lambda}\varphi \end{vmatrix}.$$

where the last determinant has m rows and m columns.

Remark. The first fundamental forms ω' corresponding to the metrics g' defined by (1) belong to the same cohomology class (Lemma 7.3). Conversely, if two first fundamental forms belong to the same cohomology class, there exists a function φ such that the corresponding metrics satisfy (1).

A cohomology class γ is said to be positive definite if there exists in γ a Hermitian form $(i/2\pi)C_{\lambda\bar{\mu}}\, dz^{\lambda} \wedge dz^{\bar{\mu}} \in \gamma$ such that everywhere $C_{\lambda\bar{\mu}}\xi^{\lambda}\xi^{\bar{\mu}} > 0$ for all vectors $\xi \neq 0$. A Kähler manifold M has at least one positive definite cohomology class defined by ω. Thus the second Betti number, $b_2(M)$, is nonzero.

If a Kähler manifold has only one positive definite cohomology class up to a proportionality constant, in particular if $b_2(M) = 1$, the all Kähler metrics are proportional to one of the form (1).

7.5 Lemma. *The Kähler manifolds (M, g') with M compact and g' defined by (1) have the same volume.*

Proof. The determinants in (1a) are divergences

$$\nabla^\lambda \begin{vmatrix} \nabla_\lambda\varphi & \nabla^\mu_\lambda\varphi & \cdots & \nabla^\nu_\lambda\varphi \\ \nabla_\mu\varphi & \nabla^\mu_\mu\varphi & & \\ \vdots & & \ddots & \\ \nabla_\nu\varphi & \nabla^\mu_\nu\varphi & & \nabla^\nu_\nu\varphi \end{vmatrix} = \begin{vmatrix} \nabla^\lambda_\lambda\varphi & \nabla^\mu_\lambda\varphi & \cdots & \nabla^\nu_\lambda\varphi \\ \nabla^\lambda_\mu\varphi & \nabla^\mu_\mu\varphi & & \\ \vdots & & \ddots & \\ \nabla^\lambda_\nu\varphi & & & \nabla^\nu_\nu\varphi \end{vmatrix}.$$

Indeed the differentiation of the other columns gives zero, because on a Kähler manifold $\nabla^\lambda\nabla^\mu_\lambda\varphi = \nabla^\mu\nabla^\lambda_\lambda\varphi$.

So integrating (1a) yields: $V' = \int_m dV' = \int_M \mathsf{M}(\varphi)\, dV = \int_M dV = V$. ∎

We can prove Lemma 7.5 by another method. Denote by ω^m (respectively, ω'^m) the m-fold tensor product of ω (respectively, ω'). $\omega' = \omega - (i/4\pi)\, dd^c\varphi$ and

$$\omega^m = \left(\frac{i}{2\pi}\right)^m m!(-2i)^m |g|\, dx^1 \wedge dy^1 \wedge dx^2 \wedge dy^2 \wedge \cdots \wedge dx^m \wedge dy^m.$$

Since $d\omega = 0$, then by Stokes' formula, $\int \omega'^m = \int \omega^m$. Hence

$$V' = \frac{\pi^m}{m!} \int \omega'^m = \frac{\pi^m}{m!} \int \omega^m = V.$$

§2. The Problems

2.1. Einstein–Kähler Metric

7.6 *Given a (compact) Kähler manifold M, does there exist an Einstein–Kähler metric on M?*

If $\tilde{g}$ is an Einstein–Kähler metric, there is a real number k such that $\tilde{R}_{\lambda\bar{\mu}} = k\tilde{g}_{\lambda\bar{\mu}}$. The Ricci form $\tilde{\Psi} = \frac{i}{2\pi}\tilde{R}_{\lambda\bar{\mu}} dz^\lambda \wedge dz^{\bar{\mu}}$ is equal to k times the first fundamental form $\tilde{\omega}$, so $k\tilde{\omega} \in C_1(M)$, the first Chern class and we have the following:

Proposition 7.6. *A necessary condition for a compact Kähler manifold to carry an Einstein–Kähler metric is that the first Chern class is positive, negative or zero.*

We say that $C_1(M)$ is positive (resp. zero or negative) if there is a positive (1-1) form ω in $C_1(M)$ (resp. $0 \in C_1(M)$ or a negative (1-1) form $\gamma \in C_1(M)$). It is easy to see that the three cases mutually exclude themselves.

2.2 Calabi's Conjecture

7.7 The *Calabi conjecture* ([73] and [74]), which is proved in 7.19, asserts that every form representing the first Chern class $\mathrm{C}_1(M)$ is the Ricci form Ψ' of some Kähler metric on a compact Kähler manifold (M, g).
Let $(i/2\pi)C_{\lambda\bar{\mu}}\, dz^\lambda \wedge dz^{\bar{\mu}}$ belong to $\mathrm{C}_1(M)$. According to Lemma 7.3, there exists an $f \in C^\infty$ such that $C_{\lambda\bar{\mu}} = R_{\lambda\bar{\mu}} - \partial_{\lambda\bar{\mu}} f$.
Consider a change of metric of type (3), the components of the corresponding Ricci tensor in a local chart are:

$$R'_{\lambda\bar{\mu}} = -\partial_{\lambda\bar{\mu}} \log |g'| = -\partial_{\lambda\bar{\mu}} \log \mathsf{M}(\varphi) + R_{\lambda\bar{\mu}}.$$

So we shall have $R'_{\lambda\bar{\mu}} = C_{\lambda\bar{\mu}}$, if there is an admissible function $\varphi \in C^\infty$ that satisfies

$$\log \mathsf{M}(\varphi) = f + k, \quad \text{with } k \text{ a constant.} \tag{2}$$

By Lemma 7.5, we can compute k, $k = \log V - \log \int e^f\, dV$.

§3. The Method

3.1. Reducing the Problem to Equations

7.8 If $C_1(M) > 0$, we consider as initial Kähler metric g some metric whose components $g_{\lambda\bar{\mu}}$ (in a complex chart) come from $\omega = \frac{i}{2\pi} g_{\lambda\bar{\mu}} dz^\lambda \wedge dz^{\bar{\mu}}$ with $\omega \in C_1(M)$ as above . If $C_1(M) < 0$, we choose g such that $\gamma = -\frac{i}{2\pi} g_{\lambda\bar{\mu}} dz^\lambda \wedge dz^{\bar{\mu}}$ belongs to $C_1(M)$.

If $C_1(M)$ is zero, we start with any Kähler metric. This case is a special case of the Calabi conjecture. We want to find a Kähler metric whose Ricci tensor vanishes, the zero-form belongs to $C_1(M)$.

Next we consider the new Kähler metric g' whose components are:

$$g'_{\lambda\bar{\mu}} = g_{\lambda\bar{\mu}} + \partial_{\lambda\bar{\mu}} \varphi,$$

where φ is a C^∞ admissible function (see definition below).

If $\lambda\omega \in C_1(M)$, since the Ricci form $\Psi = \frac{i}{2\pi} R_{\lambda\bar{\mu}} dz^\lambda \wedge dz^{\bar{\mu}} \in C_1(M)$, there exists, by Lemma 7.5, a C^∞ function f such that

$$R_{\lambda\bar{\mu}} = \lambda g_{\lambda\bar{\mu}} + \partial_{\lambda\bar{\mu}} f. \tag{3}$$

If g' is an Einstein–Kähler metric, $\lambda\omega \in C_1(M)$ and we can choose ω' homologous to ω. So according to Lemma 7.5, g' is of the form (1) and $R'_{\lambda\bar{\mu}} = \lambda g'_{\lambda\bar{\mu}}$ is equivalent to

$$\text{(4)} \qquad \lambda\partial_{\lambda\bar{\mu}}\varphi = R'_{\lambda\bar{\mu}} - R_{\lambda\mu} - \partial_{\lambda\bar{\mu}} f = -\partial_{\lambda\bar{\mu}} \log\big(|g'||g|^{-1}\big) - \partial_{\lambda\bar{\mu}} f,$$

since on a Kähler manifold, the components of the Ricci tensor are given by

$$\text{(5)} \qquad R'_{\lambda\bar{\mu}} = -\partial_{\lambda\bar{\mu}} \log|g'|.$$

Definition. *φ admissible means that g' is positive definite. $\mathcal{A}$ will be the set of the C^2 admissible functions.*

If $\tilde{g}$ is an Einstein–Kähler metric, it is proportional to a metric of the form (1), except in the null case when there are more than one positive (1-1) cohomology class. Then we proved that the problem is equivalent to solve the equation

$$\text{(6)} \qquad \log M(\varphi) = \varphi + f \qquad \text{if } C_1(M) < 0,$$

$$\log M(\varphi) = f + k \qquad \text{if } C_1(M) = 0,$$

$$\text{(7)} \qquad \log M(\varphi) = -\varphi + f \qquad \text{if } C_1(M) > 0,$$

where $M(\varphi) = |g' \circ g^{-1}| = |g'||g|^{-1}$ and f is some C^∞ function.

The proof is not difficult. Multiplying (4) by $g^{\lambda\bar{\mu}}$ and integrating yield

$$\Delta\big[\lambda\varphi + \log M(\varphi) + f\big] = 0.$$

Thus

$$\text{(8)} \qquad \lambda\varphi + \log M(\varphi) + f = \text{Const.},$$

which is nothing else than equation (2) when $\lambda = 0$, or equations (6) and (7), where the unknown function is φ − Const., when $\lambda = -1$ or +1.

3.2. The First Results

7.9 Equation (2) is the equation of the Calabi conjecture [*70]. T. Aubin [18], [20] and $S.T$. Yau [277] solved the two first equations (2) and (6), when $\lambda \leq 0$.

Theorem 7.9. *If $C_1(M) < 0$, there exists an Einstein–Kähler metric unique up to an homothety. If $C_1(M) = 0$, there exists a unique Einstein–Kähler metric (up to an homothety) in each positive* (1-1) *cohomology class.*

For the proof, it is possible to use the variational method as in the original proof (Aubin [20] and [18]), but here the continuity method is easier.

7.10 The continuity method. Let $E(\varphi) = 0$ be the equation to solve. We proceed in three steps:

a) We find a continuous family of equations E_τ, with $\tau \in [0, 1]$, such that $E_1 = E$ and $E_0(\varphi) = 0$ is a known equation which has one solution φ_0.

b) We prove that the set $\mathfrak{G} = \{\tau \in [0, 1]/E_\tau(\varphi) = 0$ has a solution$\}$ is open. For this, in general, we apply the inverse function theorem to the map $\Gamma : \varphi \to E_\tau(\varphi)$ in well chosen Banach spaces.

c) We prove that the set $\mathfrak{G}$ is closed. For this we have to establish a priori estimates.

§4. Complex Monge–Ampère Equation

7.11 More generally, we can consider an equation of the type

$$\mathsf{M}(\varphi) = \exp[F(\varphi, x)], \tag{9}$$

where $I \times M \ni (t, x) \to F(t, x)$ is a C^∞ function on $I \times M$ (or only C^3), with I an interval of $\mathbb{R}$.

(9) is called a Monge–Ampère Equation of complex type.

4.1. About Regularity

7.12 Proposition. *If F is in C^∞, then a C^2 solution of* (9) *is C^∞ admissible. If F is only $C^{r+\alpha}$ $r \geq 1$, $0 < \alpha < 1$, the solution is $C^{2+r+\alpha}$.*

Proof. At Q a point of M, where φ, a C^2 solution of (9), has a minimum, $\partial_{\lambda\bar{\lambda}}\varphi(Q) \geq 0$ for all directions λ. So at Q, g' is positive definite. By continuity, no eigenvalue of g' can be zero since $\mathsf{M}(\varphi) > 0$. Hence φ is admissible.

Consider the following mapping of the C^2 admissible functions to C^0:

$$\Gamma : \varphi \to F(\varphi, x) - \log \mathsf{M}(\varphi). \tag{10}$$

Γ is continuously differentiable. Let $d\Gamma_\varphi$ denote its differential at φ:

$$d\Gamma_\varphi(\psi) = F'_t(\varphi, x)\psi + \Delta'_\varphi\psi. \tag{11}$$

Δ'_φ is the Laplacian in the metric g'; $\Delta'_\varphi = -g'^{\nu\bar{\mu}}\,\partial_{\nu\bar{\mu}}$, where $g'^{\nu\bar{\mu}}$ denotes the components of the inverse matrix of $g'_{\nu\bar{\mu}}$. F'_t means $\partial F/\partial t$.
Since φ is admissible, Equation (9) is elliptic at φ. Hence, by Theorem 3.56, $\varphi \in C^2$ implies $\varphi \in C^\infty$. If F is only $C^{r+\alpha}(r \geq 1)$, φ belongs to $C^{2+r+\alpha}$. ■

4.2. About Uniqueness

7.13 Proposition. *Equation* (9) *has at most one* C^2 *solution, possibly up to a constant, if* $F_t'(t,x) \geq 0$ *for all* $(t,x) \in I \times M$.
In particular, Equation (8) has at most one C^2 *solution when* $\lambda < 0$*, while the solution is unique up to a constant if* $\lambda = 0$.

Proof. This follows from the maximum principle (Theorem 3.74). Let φ_1 and φ_2 be two solutions of (9). According to the mean value Theorem 3.6, there exists a function θ ($0 < \theta < 1$) such that $\psi = \varphi_2 - \varphi_1$ satisfies

(12) $$\Delta'_\gamma \psi + F_t'(\gamma, x)\psi = 0 \quad \text{with} \quad \gamma = \varphi_1 + \theta(\varphi_2 - \varphi_1).$$

Since $F_t' \geq 0$, Equation (12) has at most the constant solution.
If $F_t' > 0$, (12) has no solution except zero. ∎

§5. Theorem of Existence (the Negative Case)

7.14 *On a compact Kähler manifold, Equation* (8) *has a unique admissible* $C^{5+\alpha}$ *solution if* $\lambda < 0$ *and* $f \in C^{3+\alpha}$. *The solution is* C^∞ *if* $f \in C^\infty$.

Proof. We shall use the continuity method. For $t \geq 0$ a parameter, let us consider the equation:

(13) $$\log \mathsf{M}(\varphi) = -\lambda\varphi + tf,$$

with $f \in C^{3+\alpha}$. If for some t, Equation (13) has a C^2 solution φ_t, then φ_t is unique, admissible, and belongs to $C^{5+\alpha}$, by Proposition 7.12 and 7.13.

a) The set of functions f, for which Equation (8) has a $C^{5+\alpha}$ solution is open in $C^{3+\alpha}$.
To prove this, let us consider Γ, the mapping of the set Θ of the $C^{5+\alpha}$ admissible functions in $C^{3+\alpha}$ defined by:

$$C^{5+\alpha} \supset \Theta \ni \varphi \xrightarrow{\Gamma} -\lambda\varphi - \log \mathsf{M}(\varphi) \in C^{3+\alpha}.$$

$\log \mathsf{M}(\varphi) \in C^{3+\alpha}$ since φ is admissible and $\mathsf{M}(\varphi)$ involves only the second derivatives of φ.
Γ is continuously differentiable; its differential at φ is

$$d\Gamma_\varphi(\psi) = -\lambda\psi + \Delta'_\varphi \psi.$$

Indeed for φ given, $\|d\Gamma_\varphi(\psi)\|_{3+\alpha} \leq \text{Const} \times \|\psi\|_{5+\alpha}$ and $C^{5+\alpha} \supset \Theta \ni \varphi \to d\Gamma_\varphi \in \mathscr{L}(C^{5+\alpha}, C^{3+\alpha})$ is continuous since $C^{5+\alpha} \supset \Theta \ni \varphi \to g'^{\lambda\bar{\mu}} \in C^{3+\alpha}$ is continuous. By Theorem 4.18 the operator $d\Gamma_\varphi$ is invertible since $-\lambda > 0$. Indeed, we can write the equation $d\Gamma_\varphi(\Psi) = \tilde{f}$ in the form:

$$-\nabla_\nu\left[\mathsf{M}(\varphi)g'^{\nu\bar{\mu}}\nabla_{\bar{\mu}}\psi\right] + \lambda\mathsf{M}(\varphi)\psi = \tilde{f}\mathsf{M}(\varphi).$$

Since $C^{5+\alpha}$ and $C^{3+\alpha}$ are Banach spaces, we can use the inverse function Theorem 3.10. Thus if $\tilde{\varphi} \in C^{5+\alpha}$ satisfies

$$\log \mathsf{M}(\tilde{\varphi}) = \lambda\tilde{\varphi} + \tilde{f},$$

there exists $\mathscr{V}$, a $C^{3+\alpha}$ neighborhood of $\tilde{f}$, such that Equation (8) has a $C^{5+\alpha}$ solution when $f \in \mathscr{V}$.

Return to Equation (13), where f is given. Because $\varphi_0 = 0$ is the solution of (13) for $t = 0$, (13) has a solution for some interval $t \in [0, \tau[$, where $\tau > 0$. Let τ be the largest real number such that Equation (13) has a solution for all $t \in [0, \tau[$. If $\tau > 1$, then φ_1 is the solution of (8) and Theorem 7.14 is proved. So suppose $\tau \leq 1$, and come to a contradiction.

b) We claim that the set $\mathscr{B}$ of functions φ_t, $t \in [0, \tau[$, is bounded in $C^{2+\alpha}$, $(0 < \alpha < 1)$.
If φ_t has a maximum at P, then $\mathsf{M}(\varphi_t) \leq 1$. Indeed in a local chart for which $g_{\lambda\bar{\mu}}(P) = \delta^\mu_\lambda$ (δ^μ_λ the Kronecker tensor), and $\partial_{\lambda\bar{\mu}}\varphi_t = 0$ for $\lambda \neq \mu$, at P, we have $\mathsf{M}(\varphi_t) = \prod_{\lambda=1}^m (1 + \partial_{\lambda\bar{\lambda}}\varphi) \leq 1$, since all the terms are less than or equal to 1. Thus $\lambda\varphi_t(P) + tf(P) \leq 0$.
Similarly, we prove that if φ_t has a minimum at Q, then $\mathsf{M}(\varphi_t) \geq 1$ and

$$\lambda\varphi_t(Q) + tf(Q) \geq 0.$$

Hence $\sup|\varphi_t| \leq (\tau/\lambda)\sup|f|$. The set $\mathscr{B}$ is bounded in C^0. According to Proposition 7.23 below, $\mathscr{B}$ is bounded in $C^{2+\alpha}$.

c) We now show that (13) has a solution for $t = \tau$ and hence for some $t > \tau$ by a). This will give the desired contradiction.
According to Ascoli's theorem 3.15, the imbedding $C^{2+\alpha} \subset C^2$ is compact. Thus there exists $\varphi_\tau \in C^2$ and $t_i \to \tau$ an increasing sequence such that φ_{τ_i} converges to φ_τ in C^2.
Letting $i \to \infty$ in $\log \mathsf{M}(\varphi_{t_i}) = \lambda\varphi_{t_i} + t_i f$ we prove that φ_τ is the solution of (13) for $t = \tau$. According to the regularity theorem $\varphi_\tau \in C^{5+\alpha}$, and the contradiction follows from a), since (13) has a solution for t in a neighborhood of τ. ∎

§6. Existence of Einstein–Kähler Metric

7.15 Theorem (Aubin [18]). *A compact Kähler manifold with negative first Chern class has an Einstein–Kähler metric (all the Einstein–Kähler metrics are proportional).*

Proof. According to 7.8, finding an Einstein–Kähler metric when $\mathrm{C}_1(M) < 0$ is equivalent to solving Equation (8) with $\lambda > 0$. By Theorem 7.14, Equation (8) has a unique solution. Thus there exists a unique Einstein–Kähler metric whose Ricci curvature is equal to λ (we must choose g such that $\lambda\omega \in \mathsf{C}_1(M)$).

7.16 An application of the preceding theorem is the proof of the following, which is equivalent to the Poincaré conjecture in the case of a compact Kähler manifold of dimension 4:

Theorem. *A compact Kähler manifold homeomorphic to* $\mathsf{P}_2(\mathbb{C})$, *the complex projectif space of dimension* 2, *is biholomorphic to* $\mathsf{P}_2(\mathbb{C})$.

In their proof, Hirzebruch–Kodaira [143] supposed that the first Chern class is nonnegative. This extra hypothesis can be removed, as Yau [276] pointed out.
If $\mathsf{C}_1(M) < 0$, by Theorem 7.15 there exists an Einstein–Kähler metric. Some computations done with this metric (see Yau [276]) lead to a contradiction: the manifold would be covered by the ball and could not be simply connected.

§7. Theorem of Existence (the Null Case)

7.17 *On a compact Kähler manifold, Equation* (2) *has, up to a constant, a unique admissible* $C^{r+2+\alpha}$ *solution (respectively,* C^∞*) if* $f \in C^{r+\alpha}$, $r \geq 3$ *(respectively,* $f \in C^\infty$*).*

Proof. We shall use the continuity method. For $t \geq 0$ a parameter, let us consider the equation:

$$\mathsf{M}(\varphi) - 1 = t(e^f - 1) \tag{14}$$

with $f \in C^{3+\alpha}$ satisfying $\int e^f\, dV = \int dV$.

If for some t $(0 \leq t \leq 1)$, Equation (14) has a C^2 solution φ_t, then it is unique up to a constant, admissible, and belongs to $C^{5+\alpha}$. Indeed $\mathsf{M}(\varphi_t) = (1-t) + te^f$, so for t in a neighborhood of $[0,1]$, $\mathsf{M}(\varphi_t)$ is strictly positive and we can apply Propositions 7.12 and 7.13.
Set $\widetilde{C}^{r+\alpha} = \{f \in C^{r+\alpha} / \int f\, dV = 0\}$.

a) The set of the functions $h \in \widetilde{C}^{3+\alpha}$ for which the equation

$$\mathsf{M}(\varphi) - 1 = h$$

has a $C^{5+\alpha}$ admissible solution is open in $\widetilde{C}^{3+\alpha}$.
Let us consider the mapping Γ of the set Θ of the admissible functions belonging to $\widetilde{C}^{5+\alpha}$ in $\widetilde{C}^{3+\alpha}$ defined by

$$\widetilde{C}^{5+\alpha} \supset \Theta \ni \varphi \xrightarrow{\Gamma} \mathsf{M}(\varphi) - 1 \in \widetilde{C}^{3+\alpha}.$$

Γ is continuously differentiable; its differential at φ

$$d\Gamma_\varphi(\psi) = \mathsf{M}(\varphi)\Delta'_\varphi\psi,$$

is invertible. Indeed, $\int \mathsf{M}(\varphi)\Delta'_\varphi\psi\, dV = \int \Delta'_\varphi\psi\, dV' = 0$.
Since $\widetilde{C}^{5+\alpha}$ and $\widetilde{C}^{3+\alpha}$ are Banach spaces, we can use the inverse function Theorem 3.10. Thus if $\tilde{\varphi} \in \Theta$ satisfies $\mathsf{M}(\tilde{\varphi}) - 1 = \tilde{h}$ there exists $\mathscr{V}$, a $C^{3+\alpha}$ neighborhood of $\tilde{h}$ in $\widetilde{C}^{3+\alpha}$, such that equation $\mathsf{M}(\varphi) - 1 = h$ has a solution in Θ when $h \in \mathscr{V}$.

Return to Equation (14). Because $\varphi_0 = 0$ is the solution in Θ of (14) for $t = 0$, (14) has an admissible solution for $t \in [0, \tau[$, $\tau > 0$. Let τ be the largest real number such that Equation (14) has an admissible solution $\varphi_t \in \Theta$ for all $t \in [0, \tau[$. If $\tau > 1$, then φ_1 is the desired solution of (6) in Θ.
So we suppose $\tau \leq 1$ and come to a contradiction.

b) We claim that the set $\mathscr{B} \subset \widetilde{C}^{2+\alpha}$ of the functions φ_t, $t \in [0, \tau[$, is bounded in $C^{2+\alpha}$.
Let us prove that $\mathscr{B}$ is bounded in C^0. Then by Proposition 7.23 below, $\mathscr{B}$ is bounded in $C^{2+\alpha}$. Repeating the proof in 7.14c then establishes Theorem 7.17. The idea is to find a bound, uniform in t and $p \geq 2$, of $\|\varphi_t\|_p$ for $0 < t < \tau \leq 1$. Then $\|\varphi_t\|_p \leq \gamma$ and letting $p \to \infty$ will imply $\sup |\varphi_t| \leq \gamma$. For simplicity we drop the subscript t.

Setting $h(\varphi) = \varphi|\varphi|^{p-2}$ in Proposition 7.18 below yields:

$$(15) \qquad 4\frac{p-1}{mp^2} \int \nabla^\nu |\varphi|^{p/2} \nabla_\nu |\varphi|^{p/2}\, dV \leq \int [1 - \mathsf{M}(\varphi)]\varphi|\varphi|^{p-2}\, dV.$$

According to the Sobolev imbedding theorem, there is a constant, independent of p, such that

$$\|\, |\varphi|^p \|_{m/(m-1)} = \|\, |\varphi|^{p/2} \|^2_{2m/(m-1)} \leq \text{Const} \times \left(\|\nabla|\varphi|^{p/2}\|_2^2 + \|\varphi\|_p^p \right).$$

This inequality together with (15) leads to

$$(16) \qquad \|\, |\varphi|^p \|_{m/(m-1)} \leq \widetilde{C} \left(p \int |\varphi|^{p-1}\, dV + \int |\varphi|^p\, dV \right), \qquad (p > 1)$$

where $\widetilde{C}$ is a constant, since $\mathsf{M}(\varphi)$ is uniformly bounded; we pick $\widetilde{C} \geq 1$. The desired result, $\|\varphi\|_p \leq \gamma$ for all p and $\varphi \in \mathscr{B}$, will follow from:

Lemma. *There exists a constant γ such that for all real numbers $p \geq 1$ and all $\varphi \in \mathscr{B}$:*

$$(17) \qquad \|\varphi\|_p \leq \gamma(\alpha^{m-1} C p)^{-m/p}$$

with $\alpha = m/(m-1)$ and $C = \widetilde{C}(1 + V^{1/p})$, $\widetilde{C}$ the constant of (16).

Proof. Because φ is admissible, then $\Delta\varphi < m$. Thus $\|\Delta\varphi\|_1 < 2m \int dV$, since $\int \Delta\varphi\, dV = 0$. According to Theorem 4.13, as $\int \varphi\, dV = 0$, there exists a constant C_0 such that $\|\varphi\|_1 \leq C_0 \|\Delta\varphi\|_1 \leq 2mC_0 V$. Picking $p = 2$ in (15) gives

$\|\nabla\varphi\|_2 \leq$ Const, since $\mathsf{M}(\varphi)$ is uniformly bounded. Hence $\|\varphi\|_2 \leq$ Const, by Corollary 4.3 because $\int \varphi\, dV = 0$.
Choosing $p = 2$ in (16) yields $\|\varphi\|_{2m/(m-1)} \leq$ Const. By the interpolation inequality 3.69 there exists a constant k such that $\|\varphi\|_q \leq k$ for $1 \leq q \leq 2m/(m-1)$.
Set $\gamma = k\alpha^{m(m-1)}C^m e^{m/e}$. Then we can verify that inequality (17) is satisfied for $1 \leq p \leq 2m/(m-1)$. Either $\|\varphi\|_p$ is always smaller than 1, and there is nothing to prove (we pick $k \geq 1$ and (17) is satisfied); or else, for some p, $\|\varphi\|_p > 1$ and then by Hölder's inequality $\int |\varphi|^{p-1}\, dV \leq \|\varphi\|_p^{p-1}(\int dV)^{1/p} \leq V^{1/p}\int |\varphi|^p\, dV$.
Inequality (16) becomes

$$\left(\int |\varphi|^{p\alpha}\, dV\right)^{1/\alpha} \leq Cp \int |\varphi|^p\, dV$$

and inequality (17) follows by induction:

$$\int |\varphi|^{p\alpha}\, dV \leq (Cp)^\alpha \gamma^{p\alpha}(\alpha^{m-1}Cp)^{-m\alpha} = \gamma^{p\alpha}(\alpha^{m-1}Cp\alpha)^{-m}$$

since $\alpha(m-1) = m$, $(\alpha-1)(m-1) = 1$. ■

7.18 Proposition. *Let $h(t)$ be a C^1 increasing function on $\mathbb{R}$. Then all C^2 admissible functions φ satisfy:*

$$\int [1 - \mathsf{M}(\varphi)]h(\varphi)\, dV \geq \frac{1}{m}\int h'(\varphi)\nabla^\nu\varphi\nabla_\nu\varphi\, dV. \tag{18}$$

Proof. In the notation of 7.8,

$$\int [1 - \mathsf{M}(\varphi)]h(\varphi)\, dV = \frac{\pi^m}{m!}\int h(\varphi)(\omega^m - \omega'^m).$$

But $\omega^m - \omega'^m = (i/4\pi)\, dd^c\varphi \wedge (\omega^{m-1} + \omega^{m-2}\wedge\omega' + \cdots + \omega'^{m-1})$. Applying Stokes' formula leads to

$$\begin{aligned}\int & h(\varphi)(\omega^m - \omega'^m)\\ &= \frac{-i}{4\pi}\int h'(\varphi)\, d\varphi \wedge d^c\varphi \wedge (\omega^{m-1} + \omega^{m-2}\wedge\omega' + \cdots + \omega'^{m-1})\\ &\geq \frac{(m-1)!}{\pi}\left(\frac{i}{2\pi}\right)^{m-1}(-2i)^{m-1}\int h'(\varphi)\nabla^\nu\varphi\nabla_\nu\varphi\, dV,\end{aligned}$$

which gives (18). ■

§8. Proof of Calabi's Conjecture

7.19 Theorem. *On a compact Kähler manifold, every form representing* $\mathsf{C}_1(M)$ *is the Ricci form of some Kähler metric.*
To each positive cohomology class there corresponds one and only one metric. In particular, if $b_2(M) = 1$, *the solution is unique up to a homothetic change of metric.*

Proof. Let g be the Kähler metric, ω its first fundamental form, and $\gamma \in \mathsf{C}_1(M)$. According to 7.7, to find g', with $\omega' - \omega$ homologous to zero and $\Psi' = \gamma$, is equivalent to solving equation (2).
By Theorem 7.17, Equation (2) has a unique solution up to a constant. Thus in each positive cohomology class we find a unique ω' whose Ψ' equals γ. ∎

§9. The Positive Case

7.20 *According to* (7.8), *in the case* $\mathsf{C}_1(M) > 0$ *there exists a Einstein–Kähler metric, if and only if Equation* (8) *with* $\lambda > 0$ *has a* C^∞ *admissible solution.*

This problem is not yet solved. It is more difficult than the two preceding cases. First, since the linear map $d\Gamma_\varphi$ of 7.14 is not necessarily invertible, it is not obvious how to use the continuity method. Then we must find a C^0 estimate in order to use Proposition 7.23. On the other hand, Equation (8) with $\lambda > 0$ may have many solutions (see Aubin [20] pp. 85 and 86). For instance, on the complex projective space, $\log \mathsf{M}(\varphi) = -\lambda_1\varphi$ has many solutions; these solutions come from the infinitesimal holomorphic transformations which are not isometries. Worse, we know that some Equations (8) have no solution, since if we blow up one or two points of projective space, the manifold obtained cannot carry an Einstein–Kähler metric according to a theorem of Lichnerowicz [185] p. 156 (see Yau [275]).

We will see below that there has been great progress in the positive case (§13 and the ones following).

§10. A Priori Estimate for $\Delta\varphi$

7.21 Notations. On a compact Kähler manifold, let $\mathscr{B}$ be a set of $C^{5+\alpha}$ admissible functions and λ real numbers satisfying $|\lambda| \le \lambda_0$. We suppose that $\mathscr{C}$, the set of the corresponding functions $f = \log \mathsf{M}(\varphi) - \lambda\varphi$, is bounded in $C^{3+\alpha}$.
F_0 and F_1 are real numbers such that everywhere for all $\varphi \in \mathscr{B}$,

$$\Delta f \le F_1 \quad \text{and} \quad f \le F_0.$$

Proposition (Aubin [18]). *There exist two constants k and K depending only on λ_0, F_0, F_1, and the curvature, such that all $\varphi \in \mathscr{B}$ satisfy:*

$$0 < m - \Delta\varphi \leq K e^{k\varphi-(k-\lambda/m)\inf\varphi}.$$

Moreover, if $\mathscr{B}$ is bounded in C^0, all corresponding metrics g' are equivalent.

Proof. The first inequality is obvious. Since φ is admissible, for all directions μ, $g_{\mu\bar\mu} + \partial_{\mu\bar\mu}\varphi > 0$; thus by summing over μ we obtain $m - \Delta\varphi > 0$.
To prove the second inequality set

$$A = \log(m - \Delta\varphi) - k\varphi,$$

where k is a real number that we will choose later. Let us compute $\Delta'_\varphi A = -g'^{\lambda\bar\mu}_\varphi\, \partial_{\lambda\bar\mu} A$ (we will omit the index φ below).

(19) $$\Delta' A = -(m-\Delta\varphi)^{-1}\Delta'\Delta\varphi - k\Delta'\varphi + (m-\Delta\varphi)^{-2} g'^{\lambda\bar\mu}\nabla_\lambda\Delta\varphi\nabla_{\bar\mu}\Delta\varphi.$$

Recall $g'^{\lambda\bar\mu}$ are the components of the inverse matrix of $((g_{\lambda\bar\mu} + \partial_{\lambda\bar\mu}\varphi))$. Differentiating (8) yields:

$$\lambda\nabla_\nu\varphi + \nabla_\nu f = \nabla_\nu \log M(\varphi) = g'^{\alpha\bar\beta}\nabla_\nu\nabla_{\alpha\bar\beta}\varphi$$

(20) $$-\lambda\Delta\varphi - \Delta f = g'^{\alpha\bar\beta}\nabla^\nu\nabla_\nu\nabla_{\alpha\bar\beta}\varphi - g'^{\alpha\bar\mu}g'^{\gamma\bar\beta}\nabla^\nu\nabla_{\gamma\bar\mu}\varphi\nabla_\nu\nabla_{\alpha\bar\beta}\varphi.$$

But from 1.13,

(21) $$\Delta'\Delta\varphi - g'^{\alpha\bar\beta}\nabla^\nu\nabla_\nu\nabla_{\alpha\bar\beta}\varphi = R_{\alpha\bar\beta\lambda\bar\mu}\nabla^{\alpha\bar\beta}\varphi g'^{\lambda\bar\mu} - R_{\lambda\bar\mu}\nabla^\lambda_\nu\varphi g'^{\nu\bar\mu} = E$$

and there exists a constant C such that E satisfies

(22) $$|E| \leq C(m-\Delta\varphi)g'^{\lambda\bar\mu}g_{\lambda\bar\mu}.$$

Write $\Delta'\varphi = -g'^{\lambda\bar\mu}(g'_{\lambda\bar\mu} - g_{\lambda\bar\mu}) = g'^{\lambda\bar\mu}g_{\lambda\bar\mu} - m$ and observe that

(23) $$g'^{\alpha\bar\beta}g'^{\lambda\bar\mu}\nabla^\nu\nabla_{\alpha\bar\mu}\varphi\nabla_\nu\nabla_{\lambda\bar\beta}\varphi \geq (m-\Delta\varphi)^{-1}g'^{\lambda\bar\mu}\nabla_\lambda\Delta\varphi\nabla_{\bar\mu}\Delta\varphi.$$

To verify this inequality, we have only to expand

$$\left[(m-\Delta\varphi)\nabla_\nu\nabla_{\lambda\bar\beta}\varphi + \nabla_\lambda\Delta\varphi g'_{\nu\bar\beta}\right] \times \left[(m-\Delta\varphi)\nabla_{\bar\gamma}\nabla_{\alpha\bar\mu}\varphi + \nabla_{\bar\mu}\Delta\varphi g'_{\alpha\bar\gamma}\right]g'^{\alpha\bar\beta}g'^{\lambda\bar\mu}g^{\nu\bar\gamma} \geq 0.$$

(19)–(23) lead to

(24) $$\Delta' A \leq k(m - g'^{\lambda\bar\mu}g_{\lambda\bar\mu}) - (m-\Delta\varphi)^{-1}(E - \lambda\Delta\varphi - \Delta f).$$

At a point P where A has a maximum, $\Delta' A \geq 0$. We find, using (22),

(25) $$(k-C)g'^{\nu\bar\mu}g_{\nu\bar\mu} \leq (m-\Delta\varphi)^{-1}(\lambda\Delta\varphi + \Delta f) + mk.$$

Since the arithmetic mean is greater than or equal to the geometric mean,

$$1-\Delta\varphi/m \geq [\mathsf{M}(\varphi)]^{1/m} = e^{(\lambda\varphi+f)/m}.$$

Thus at P inequality (25) yields

$$(k-C)g'^{\nu\bar\mu}g_{\nu\bar\mu} \leq mk-\lambda+(\lambda+\Delta f/m)e^{-(\lambda\varphi+f)/m}. \tag{26}$$

However, $g'^{\nu\bar\mu}g_{\nu\bar\mu} \geq m[\mathsf{M}(\varphi)]^{-1/m}$, so that

$$m(k-C)-\lambda-\Delta f/m \leq (mk-\lambda)e^{(\lambda\varphi+f)/m}. \tag{27}$$

Pick k such that $m(k-C) \geq 1+\sup(\lambda,0)+\sup(\Delta f)/m$, expressions (26) and (27) lead to: There exists a constant K_0 such that at P

$$\left(g'^{\nu\bar\mu}g_{\nu\bar\mu}\right)_P \leq K_0. \tag{28}$$

K_0 depends on λ_0, F_1, and the curvature through C.
In an orthonormal chart at P for which $\partial_{\nu\bar\mu}\varphi=0$ if $\nu\neq\mu$

$$1+\partial_{\nu\bar\nu}\varphi \leq \mathsf{M}(\varphi)\prod_{\mu\neq\nu} g'^{\mu\bar\mu} \leq \mathsf{M}(\varphi)\left[\frac{1}{m-1}\sum_{\mu\neq\nu} g'^{\mu\bar\mu}\right]^{m-1}.$$

Taking the sum and using (28) yields at P

$$(m-\Delta\varphi)_P \leq m[K_0/(m-1)]^{m-1}e^{\lambda\varphi(P)+f(P)}.$$

Hence everywhere

$$(m-\Delta\varphi)e^{-k\varphi} \leq (m-\Delta\varphi)_P e^{-k\varphi(P)} \leq Ke^{-(k-\lambda/m)\varphi(P)}, \tag{29}$$

where K is a constant depending on K_0 and F_0.
The inequality of Proposition 7.21 now follows since $k\geq\lambda/m$.

If $\mathscr{B}$ is bounded in C^0 that is $|\varphi|\leq k_0$, using (29) for all $\varphi\in\mathscr{B}$ we have $\Delta\varphi$ uniformly bounded: $|\Delta\varphi|\leq k_1$. Therefore, in an orthonormal chart adapted to φ ($\partial_{\nu\bar\mu}=0$ if $\nu\neq\mu$),

$$\text{as } \partial_{\mu\bar\mu}\varphi > -1, \quad \partial_{\nu\bar\nu}\varphi < m-1+k_1,$$

and $(1+\partial_{\mu\bar\mu}\varphi)^{-1} \leq (m+k_1)^{m-1}[\mathsf{M}(\varphi)]^{-1} < (m+k_1)^{m-1}e^{k_0|\lambda|+f}$. Thus the metrics g'_φ, $\varphi\in\mathscr{B}$, are equivalent to g; for all directions μ

$$e^{-k_0|\lambda|-\sup f}(m+k_1)^{1-m}g_{\mu\bar\mu} \leq g'_{\mu\bar\mu} \leq (m+k_1)g_{\mu\bar\mu}. \qquad ■$$

§11. A Priori Estimate for the Third Derivatives of Mixed Type

7.22 Once we have uniform bounds for $|\varphi|$ and $|\Delta\varphi|$, to obtain estimates for the third derivatives of mixed type, consider

$$|\psi|^2 = g'^{\alpha\bar{\beta}} g'^{\lambda\bar{\mu}} g'^{\nu\bar{\gamma}} \nabla_\alpha \nabla_{\nu\bar{\mu}}\varphi \nabla_{\bar{\beta}} \nabla_{\lambda\bar{\gamma}}\varphi. \tag{30}$$

The choice of this norm instead of a simpler equivalent norm (in the metric g, for instance) imposes itself on those who make the computation. We now give the result; the reader can find the details of the calculation in Aubin (11) pp. 410 and 411.

Lemma.

$$\begin{aligned}
-\Delta'|\varphi|^2 = {} & g'^{\lambda\bar{\mu}} g'^{\alpha\bar{\beta}} g'^{a\bar{b}} g'^{c\bar{d}} [(\nabla_{\bar{\mu}\alpha\bar{b}c}\varphi - \nabla_{\bar{\mu}\gamma\bar{b}}\varphi \nabla_{\alpha\bar{\delta}c}\varphi g'^{\gamma\bar{\delta}}) (\text{conjugate expression}) \\
& + (\nabla_{\lambda\alpha\bar{b}c}\varphi - \nabla_{\lambda\bar{b}\rho}\varphi \nabla_{\alpha\bar{\nu}c}\varphi g'^{\rho\bar{\nu}} - \nabla_{\lambda\bar{\nu}c}\varphi \nabla_{\rho\bar{b}\alpha}\varphi g'^{\rho\bar{\nu}}) \\
& \times (\text{conjugate expression})] - g'^{c\bar{d}} (2 g'^{\alpha\bar{\delta}} g'^{\gamma\bar{\beta}} g'^{a\bar{b}} \\
& + g'^{\alpha\bar{\beta}} g'^{\alpha\bar{\delta}} g'^{\gamma\bar{b}}) \nabla_{\alpha\bar{b}c}\varphi \nabla_{\bar{\beta}a\bar{d}}\varphi [\nabla_{\gamma\bar{\delta}}(\lambda\varphi + f) - R_{\gamma\bar{\delta}}] \\
& + g'^{\alpha\bar{\beta}} g'^{a\bar{b}} g'^{c\bar{d}} [\nabla_{\bar{\beta}a\bar{d}}\varphi \nabla_{\alpha\bar{b}c}(\lambda\varphi + f) + \nabla_{\alpha\bar{b}c}\varphi \nabla_{\bar{\beta}a\bar{d}}(\lambda\varphi + f)] \\
& + g'^{\lambda\bar{\mu}} g'^{\alpha\bar{\beta}} g'^{a\bar{b}} g'^{c\bar{d}} [\nabla_{\bar{\beta}a\bar{d}}\varphi (R^{\nu}_{c\lambda\bar{b}} \nabla_{\alpha\bar{\mu}\nu}\varphi + R^{\bar{\rho}}_{\bar{b}\bar{\mu}\alpha} \nabla_{\lambda\bar{\rho}c}\varphi \\
& + R^{\nu}_{c\bar{\mu}\alpha} \nabla_{\lambda\bar{b}\nu}\varphi) + \text{conjugate expression} \\
& + g'^{\alpha\bar{\beta}} g'^{c\bar{d}} [\nabla_{\bar{\beta}a\bar{d}}\varphi (g'^{\lambda\bar{\mu}} \nabla_\lambda R^{a}_{c\bar{\mu}\alpha} - g'^{a\bar{b}} \nabla_\alpha R_{c\bar{b}}) \\
& + \text{conjugate expression}].
\end{aligned}$$

Hence there exists a constant k_2 which depends on λ_0, $\|\mathscr{B}\|_{C^0}$, $\|\mathscr{C}\|_{c^3}$, and the curvature such that

$$\Delta'|\psi|^2 \leq k_2(|\psi|^2 + |\psi|). \tag{31}$$

Proposition. *There exists a constant k_3, depending only on λ_0, $\|\mathscr{B}\|_{C^0}$, $\|\mathscr{C}\|_{C^3}$, and the curvature, such that $\nabla_{\lambda\bar{\mu}\nu}\varphi \nabla^{\lambda\bar{\mu}\nu}\varphi \leq k_3$, for all $\varphi \in \mathscr{B}$.*

Proof. Equations (20) and (21) give

$$\Delta'\Delta\varphi = g'^{\alpha\bar{\mu}} g'^{\gamma\bar{\beta}} \nabla^{\nu} \nabla_{\gamma\bar{\mu}}\varphi \nabla_\nu \nabla_{\alpha\bar{\beta}}\varphi - \lambda\Delta\varphi - \Delta f + E. \tag{32}$$

As all metrics g' are equivalent (Proposition 7.21), there exists a constant $B > 0$ such that

$$g'^{\alpha\bar{\mu}} g'^{\gamma\bar{\beta}} \nabla^{\nu} \nabla_{\gamma\bar{\mu}}\varphi \nabla_\nu \nabla_{\alpha\bar{\beta}}\varphi \geq B|\psi|^2.$$

Let $h > 0$ be a real number. According to (31),

$$\Delta'(|\psi|^2 - h\Delta\varphi) \leq k_2(|\psi|^2 + |\psi|) - hB|\psi|^2 + h(\lambda\Delta\varphi + \Delta f - E).$$

Picking $h = 2k_2B^{-1}$, we get

(33) $$\Delta'(|\psi|^2 - h\Delta\varphi) \leq -(k_2/2)|\psi|^2 + k_2/2 + 2k_2B^{-1}(\lambda\Delta\varphi + \Delta f - E).$$

At a point P where $|\psi|^2 - h\Delta\varphi$ has a maximum, the first member of (33) is nonnegative. Thus

$$|\psi(P)|^2 \leq 1 + 4B^{-1}(\lambda\Delta\varphi(P) + \Delta f(P) - E(P)).$$

So by Proposition 7.21, $|\psi(P)|^2 \leq$ Const. Hence everywhere, $|\psi|^2 \leq$ Const.

7.23 Proposition. *On a compact Kähler manifold, let $\mathscr{B}$ be a set of C^5 admissible functions, $\mathscr{B}$ bounded in C^0. Let $(\varphi, \lambda) \in \mathscr{B} \times [-\lambda_0, \lambda_0]$ with λ_0 a constant, and let*

$$f = \log \mathsf{M}(\varphi) - \lambda\varphi.$$

If the set $\mathscr{C}$ of corresponding functions f is bounded in C^3, then $\mathscr{B}$ is bounded in $C^{2+\alpha}$ for all $\alpha \in]0, 1[$.

Proof. According to Proposition 7.22, the third derivatives of mixed type of the functions $\varphi \in \mathscr{B}$ are uniformly bounded. Hence there exists a constant k such that for all $\varphi \in \mathscr{B}$ $|\nabla\Delta\varphi| \leq k$ since the gradient of $\Delta\varphi$ involves only third derivatives of mixed type. By the properties of Green's function (Theorem 4.13), for any $\alpha \in]0, 1[$, $\mathscr{B}$ is bounded in $C^{2+\alpha}$. ■

§12. The Method of Lower and Upper Solutions

7.24 Suppose we have to solve an elliptic differential equation $\mathscr{E}$. If there exist a lower solution u and an upper solution v satisfying $u \leq v$ we can hope to use the method of lower and upper solutions. But for this we also need to be able to solve an equation "close" to equation $\mathscr{E}$. Thus the method of lower and upper solutions requires an additional basic step. It is simpler to give an example.

7.25 Return to Equation (9),

(34) $$\log \mathsf{M}(\varphi) = F(\varphi, x),$$

and set $S(t) = \sup_{x \in M} F(t, x)$ and $P(t) = \inf_{x \in M} F(t, x)$. Recall that $F(t, x)$ is a C^∞ function on $I \times M$ where $I =]\alpha, \beta[$ is an interval of $\mathbb{R}$. We will prove:

Theorem 7.25. *Equation* (9) *has a C^∞ solution if there exist two real numbers a and b belonging to I, $(a \leq b)$ such that $S(a) = P(b) = 0$.*

In this problem a is a lower solution of Equation (9). Indeed, $\log \mathsf{M}(a) \geq F(a, x)$, since $F(a, x) \leq S(a) = 0$; and b is an upper solution, $\log \mathsf{M}(b) \leq F(b, x)$, since $F(b,x) \geq P(b) = 0$. Moreover, $a \leq b$. Before giving the proof of this theorem, let us establish the following:

Corollary 7.25. *The equation* $\log \mathsf{M}(\varphi) - F(\varphi, x) = \psi(x)$ *has a* C^∞ *solution for any function* $\psi \in C^\infty$ *if* $P(t) \to +\infty$ *as* $t \to \beta$, *and* $S(t) \to -\infty$ *as* $t \to \alpha$.

Proof. Corollary 7.25 follows from Theorem 7.25. Indeed set $P_0(t) = \inf_{x\in M}[F(t,x) + \psi(x)]$ and $S_0(t) = \sup_{x\in M}[F(t,x)+\psi(x)]$. These functions are continuous and obviously $P(t)+\inf\psi \leq P_0(t) \leq S_0(t) \leq S(t)+\sup\psi$. Thus, when $t \to \alpha$, $S_0(t)$ and $P_0(t)$ go to $-\infty$ and when $t \to \beta$, $S_0(t)$ and $P_0(t)$ go to $+\infty$. So there exists $a \in I$ such that $S_0(a) = 0$. But as $P_0(a) \leq 0$ there exists also $b \in [a, \beta[$ such that $P_0(b) = 0$. The hypotheses of Theorem 7.25 are satisfied. ■

If one solves the equation under the assumptions of Corollary 7.25, note that only the behavior of $F(t,x)$ as t goes to α and β is important.

Now we give the proof of Theorem 7.25.

α) By Theorem 7.14, the equation $\log \mathsf{M}(\varphi) - \lambda\varphi = f$ has a unique solution when $\lambda > 0$. With this result we will consider an increasing sequence of functions converging to a solution of Equation (9). Pick $\lambda > \sup[0, F'_t(t,x)]$ for all $(t,x) \in [a,b] \times M$. According to Theorem 7.14 we can define the sequence of functions φ_j by $\varphi_0 = a$ and

$$(35) \qquad \log \mathsf{M}(\varphi_j) - \lambda\varphi_j = F(\varphi_{j-1}, x) - \lambda\varphi_{j-1} \quad \text{for } j \geq 1.$$

This sequence of C^∞ functions is increasing and satisfies $a \leq \varphi_j \leq b$. The proof proceeds by induction by using the maximum principle. We suppose that for all $i \leq j$, $a \leq \varphi_{i-1} \leq \varphi_i \leq b$ and we write:

$$\begin{aligned} &\log \mathsf{M}(\varphi_{j+1}) - \log \mathsf{M}(\varphi_j) - \lambda(\varphi_{j+1} - \varphi_j) \\ &\qquad = [F(\varphi_j, x) - \lambda\varphi_j] - [F(\varphi_{j-1}, x) - \lambda\varphi_{j-1}] \leq 0. \end{aligned}$$

The last inequality is obtained by applying the mean value theorem. The maximum principle implies $\varphi_{j+1} - \varphi_j \geq 0$. Moreover we can start the induction because $\log \mathsf{M}(\varphi_1) - \lambda(\varphi_1 - a) = F(a,x) \leq S(a) = 0$. Similarly

$$\log \mathsf{M}(\varphi_{j+1}) - \lambda(\varphi_{j+1} - b) = [F(\varphi_j, x) - \lambda\varphi_j] + \lambda b \geq F(b,x) \geq P(b) = 0$$

shows that $\varphi_{j+1} - b \leq 0$. Therefore the sequence $\{\varphi_j\}$, which is increasing and bounded, converges pointwise to a function ψ which satisfies $a \leq \psi \leq b$. It remains to establish the regularity of ψ. For this we need estimates on the functions φ_j.

β) We will prove that the set of the functions φ_j is bounded in $C^{2+\alpha}$ ($0 < \alpha < 1$), because then, by Ascoli's theorem, there exists a subsequence $\{\varphi_i\}$ of the sequence $\{\varphi_j\}$ which converges in C^2 to a function which cannot be different from ψ. Then $\psi \in C^{2+\alpha}$ and by the regularity theorem 3.56 $\psi \in C^\infty$. We already have the C^0-estimate $a \leq \varphi_j \leq b$. To prove that $|\Delta\varphi_j|$ is smaller than a constant independent of j, we compute $\Delta'_j B$ where

$$B = \log(m - \Delta\varphi_j) - k\varphi_j + \sigma\varphi_{j-1} + \nu\varphi_{j-1}^2,$$

k, σ, and ν being real numbers which we will choose later. Recall that $\Delta'_j = -g_j'^{\lambda\bar{\mu}}\nabla_{\lambda\bar{\mu}}$ and $((g_j'^{\lambda\bar{\mu}}))$ is the inverse matrix of $((g_{\lambda\bar{\mu}} + \partial_{\lambda\bar{\mu}}\varphi_j))$. At a point P where B has a maximum, we note that $\Delta'_j B \geq 0$. First of all we choose ν so that the terms in $\Delta'_j B$ which involve the first derivatives of φ_{j-1} are less than some constant. Then we pick σ large enough to control the terms with the Laplacian of φ_{j-1}. Finally k is chosen sufficiently large (in particular $k > \sigma$) so that the inequality $\Delta'_j B \geq 0$ at P becomes $g_j'^{\alpha\bar{\beta}} g_{\alpha\bar{\beta}} \leq$ Const. (For the details of this computation see Aubin [20] p. 89). In 7.21 we saw that $g_j'^{\alpha\bar{\beta}} g_{\alpha\bar{\beta}} \leq$ Const at a point where B is maximum implies that the functions $|\Delta\varphi_j|$ are uniformly bounded, and consequently that the metrics g'_j are uniformly equivalent to g.

γ) Finally we prove that the set $\{\Delta\varphi_j\}$ is bounded in H_1^q for all $q \geq 1$ (see Aubin [20] p. 90) by using the following inequality instead of inequality (31):

$$|\Delta'_j|\psi_j|^2 + \Gamma_j^2| \leq k_2\left(|\psi_j|^2 + |\psi_j|\,|\psi_{j-1}| + |\psi_j|\right), \tag{36}$$

where $|\psi_j|$ is defined by (30) with $\varphi = \varphi_j$, and Γ_j^2 is a positive term which involves the fourth derivatives of φ_j.
Hence the set of the functions $\{\varphi_j\}$ is bounded in $C^{2+\alpha}$ for all α $(0 < \alpha < 1)$.

§13. A Method for the Positive Case

7.26 A priori it seemed that it was impossible to use the continuity method in this case, until Aubin [*7] showed how to proceed; indeed, the differential of: $\varphi \to \log M(\varphi) + \lambda\varphi$ is not necessarily invertible when $\lambda > 0$. Still, instead of (13), let us consider the following family of equations:

$$E_t : \log M(\varphi) = -t\varphi + f \qquad \text{for } \tau \in [\varepsilon, 1], \text{ with } \varepsilon > 0. \tag{37}$$

By this equation we control the Ricci curvature. If it exists, let φ_t be a solution of E_t for some t. According to (1) and (3), a computation gives:

$$R'_{t\lambda\bar{\mu}} = (1-t)g_{\lambda\bar{\mu}} + tg'_{t\lambda\bar{\mu}}, \tag{38}$$

where $R'_{t\lambda\bar{\mu}}$ are the components of the Ricci tensor corresponding to the metric g'_t of components $g'_{t\lambda\bar{\mu}} = g_{\lambda\bar{\mu}} + \partial_{\lambda\bar{\mu}}\varphi_t$. According to (8), the Ricci curvature of (M, g'_t) is greater than t for $t < 1$, and by Theorem 4.20 we know that the first eigenvalue λ_1^t of the Laplacian $\Delta'_t = -g_t'^{\lambda\bar{\mu}}\nabla_\lambda\nabla_{\bar{\mu}}$ satisfies $\lambda_1^t > t$; here $g_t'^{\lambda\bar{\mu}}$ are the components of the inverse matrix of $\left((g'_{t\lambda\bar{\mu}})\right)$.

7.27 Now we are in position to prove part *b*) of the continuity method. Define the map Γ :

$$\mathbb{R} \times \Theta \ni (t, \varphi) \xrightarrow{\Gamma} t\varphi + \log M(\varphi) \in C^{3+\alpha}.$$

Recall that $\Theta = \mathcal{A} \cap C^{5+\alpha}$. Γ is continuously differentiable and its partial differential with respect to φ is given by

$$(39) \qquad [D_\varphi \Gamma(t, \varphi)](\Psi) = t\Psi - \Delta'_\varphi \Psi ;$$

Γ is invertible for $\varepsilon \leq t < 1$ according to Theorem 4.20, since $\lambda_1^t > t$. Indeed (38) implies $R'_{t\lambda\bar{\mu}} - t g'_{t\lambda\bar{\mu}}$ is positive definite.

By the implicit function theorem, the map $(t, \varphi) \to [t, \Gamma(t, \varphi)]$ is a diffeomorphism of a neighbourhood of (τ, φ_τ) in $\mathbb{R} \times \Theta$ onto an open set of $\mathbb{R} \times C^{3+\alpha}$. So, if we can solve the equation $\Gamma(t, \varphi) = f$ at $t = \tau$, we can solve it when t is in a neighbourhood of τ.

7.28 Now, let us complete part a) of the continuity method. There is a difficulty: we cannot consider equation (37) at $t = 0$, even if f is chosen so that $\int e^f \, dV = \int dV$, because E_0 will have an infinity of solutions φ_0 (the solution is unique up to a constant) and, according to (39), the map Γ is not invertible with respect to φ at $(0, \varphi_0)$.

This is the reason why we consider E_t for $t \in [\varepsilon, 1]$ with $\varepsilon > 0$, but we have to prove the existence of φ_ε for some small ε. For this we consider the map

$\tilde{\Gamma} : \mathbb{R} \times \Theta \ni (t, \varphi) \to t\varphi + \log M(\varphi) + \beta \int \varphi \, dV \in C^{3+\alpha}$, where $\beta > 0$ is a given real number.

$\tilde{\Gamma}$ is continuously differentiable and its partial differential with respect to φ is

$$[D_\varphi \tilde{\Gamma}(t, \varphi)](\Psi) = t\Psi - \Delta'_\varphi \Psi + \beta \int \Psi \, dV.$$

$\tilde{\Gamma}$ is invertible even at $t = 0$. Since equation (2) has a unique solution up to a constant, the equation $\log M(\varphi) + \beta \int \varphi \, dV = f$ has a unique solution $\tilde{\varphi}_\circ$.

Now we apply the implicit function theorem to $\tilde{\Gamma}$ at $(0, \tilde{\varphi}_\circ)$, and deduce that, for some small $\varepsilon > 0$, the equation $\tilde{\Gamma}(\varepsilon, \varphi) = f$ has a solution $\varphi_\varepsilon \in \Theta$.

Thus $\varphi_\varepsilon = \tilde{\varphi}_\varepsilon + \frac{\beta}{2} \int \tilde{\varphi}_\varepsilon \, dV$ is a solution of E_ε.

7.29 *The estimates* (part *c* of the continuity method).

Set $\mathfrak{G} = \{t \in [\varepsilon, 1]/E_t$ has a solution$\}$.

Proposition. *If the set of* $\{\varphi_t\}(t \in \mathfrak{G})$ *is bounded in* C^0, *equation* (7) *has a* C^∞ *admissible solution.*

Since $\mathfrak{G}$ is open and non empty, if we prove that it is closed, $\mathfrak{G} = [0, 1]$ and equation (7) has a solution. If the set $\{\varphi_t\}$ $(t \in \mathfrak{G})$ is bounded in C^0, it is bounded in $C^{2+\alpha}$ by Proposition 7.23. Then $\mathfrak{G}$ is closed. Indeed let $\{t_i\} \subset \mathfrak{G}$ be a sequence which goes to $\tau (t_i \to \tau)$.

By Ascoli' s theorem, there exists a subsequence $\{t_j\}$ such that φ_{t_j} converges to a function $\Psi \in C^{2+\alpha}$ in C^2 when $j \to \infty$. Letting $j \to \infty$ in E_{t_j}, we prove that $\Psi = \varphi_\tau$, thus $\tau \in \mathfrak{G}$.

About the regularity, recall from proposition 7.12, that *a C^2 solution of* (5) *is C^∞ admissible.*

§14. The Obstructions when $C_1(M) > 0$

14.1. The First Obstruction

7.30 Let $G(M)$ be the group of automorphisms of M. By the Lichnerowicz–Matsushima theorem, we obtain the first known obstruction. This theorem ([185] p. 156) asserts that, if a compact Kähler manifold M has constant scalar curvature, then the group $G(M)$ is reductive. Thus we obtain

Proposition. *Any compact Kähler manifold, whose automorphism group is not reductive, does not admit a Kähler metric with constant scalar curvature.*

7.31 *Application to the projective space $P_m(\mathbb{C})$ blown up at one point.*

Let $(z_0, z_1, \dots, z_m)$ be homogeneous coordinates of $P_m(\mathbb{C})$.

Blowing up $P_m(\mathbb{C})$ at the point $Q = (1, 0, \dots 0, 0)$, we obtain a manifold M whose group $G(M)$ is not reductive (see below). So M cannot carry an Einstein–Kähler metric, although its first Chern class is positive.

We can visualize M as the set of the points of $P_m(\mathbb{C}) \times P_{m-1}(\mathbb{C})$ such that $z_1/\xi_1 = z_2/\xi_2 = \dots = z_m/\xi_m$ where $(\xi_1, \xi_2, \dots \xi_m)$ are homogeneous coordinates of $P_{m-1}(\mathbb{C})$. We get a holomorphic mapping π from M onto $P_m(\mathbb{C})$ such that $\pi^{-1}(Q) = D$ is isomorphic to $P_{m-1}(\mathbb{C})$ and $M - D$ is biholomorphic to $P_m(\mathbb{C}) - Q$ by π.

The (1-1) form $(i/2\pi)dd''\left[m \log(|z_0|^2 + r^2) + \log r^2\right]$, with $r^2 = \sum_{i=1}^m |z_i|^2$, belongs to $C_1(M)$ which is positive definite. $D = \pi^{-1}(Q)$ in M is an exceptional divisor which has a unique representative cycle.

Thus $G(M)$ consists of all automorphisms in $G(P_m(\mathbb{C}))$ preserving Q. GL $(m+1, \mathbb{C})$ acts on $P_m(\mathbb{C})$, its kernel is $K = \{\lambda I / \lambda \in \mathbb{C}\}$.

Let $\{e_j\}(j = 0, 1, \dots, m)$ be a natural basis of $\mathbb{C}^{m+1}$. $G(M)$ is isomorphic to S/K where $S = \{f \in GL(m+1)/f(e_0) = \lambda e_0 \text{ with } 0 \neq \lambda \in \mathbb{C}\}$.

Now a group is reductive if and only if any linear representation is completely reductible. This is not the case for S. In its natural representation $\mathbb{C}e_0$ is an invariant subspace which has no invariant supplementary subspace. Indeed S is represented by the matrices $((a_{i,j}))$(j for the column) with $a_{i,o} = 0$ for $1 \leq i \leq m$, and the group of the transposed matrices has no invariant subspace of dimension one.

The same argument proves that the manifolds, obtained by blowing up $P_m(\mathbb{C})$ at less than $m+1$ points in general position, have non-reductive automorphism groups. Conversely, the maximal connected group of automorphisms of $P_m(\mathbb{C})$ blown up at $m+1$ points in general position is reduced to the maximal connected group of automorphisms of $P_m(\mathbb{C})$ preserving each of the $m+1$ points. These automorphisms are represented by the diagonal matrices with $|a_{ii}| = 1$ for all i.

14.2. Futaki's Obstruction

7.32 If $C_1(M) > 0$, we can choose the Kähler metric such that the first fundamental form $\omega \in C_1(M)$. Then the Ricci form Ψ is homologous to ω, so that there exists a function F_ω such that $\Psi - \omega = (i/2\pi)dd''F_\omega$.

Denote by $h(M)$ the Lie algebra of holomorphic vector fields. Futaki considers the application of $h(M)$ in $\mathbb{C}$ defined by

$$h(M) \ni X \to f(X) = (i/2\pi)\int X(F_\omega)\omega^m.$$

Theorem (Futaki [*131]). *The linear function f does not depend on the choice of $\omega \in C_1(M)$. Therefore, if $h_0(M)$ is the kernel of f, the number $\delta_M = \dim\big[h(M)/h_0(M)\big]$ depends only on the complex structure of M. If M admits an Einstein–Kähler metric, then $\delta_M = 0$.*

In his article [*131] and his book [*132], we find examples of compact complex manifolds with $C_1(M) > 0$ and dimension $m > 2$ which are reductive but with number $\delta_M = 1$.

Futaki explains that his theorem is a complex version of the obstruction of Kazdan and Warner 6.66.

Remark. We can generalize Futaki's obstruction when $\omega \notin C_1(M)$.

Let $[\omega]$ be the cohomology class of ω and let F_ω be a function such that $\Delta F_\omega = R - V^{-1}\int R\,dV$. If there is a metric $\tilde{g}$ with $\tilde{\omega} \in [\omega]$ and $\tilde{R} = \text{Const.}$, then $\delta_M = 0$.

14.3. A Further Obstruction

7.33 If M is a compact Einstein–Kähler manifold, the tangent bundle TM satisfies the Einstein condition (trivial). So, by a theorem of Kobayashi [*201] (see also Lübke [*228]), TM is semi-stable. Thus we obtain the

Proposition. *Let M be a compact Kähler manifold. If TM is not semi-stable, M cannot carry an Einstein–Kähler metric.*

§15. The C^0-estimate

15.1. Definition of the Functionals $I(\varphi)$ and $J(\varphi)$

7.34 We set

$$I(\varphi) = \int \varphi[1 - M(\varphi)]\, dV = \int \varphi\, dV - \int \varphi\, dV'$$

and

$$J(\varphi) = (1/s) \int_0^1 I(s\varphi)\, ds.$$

Thus, if $t \to \varphi_t$ is a smooth map of an interval of $\mathbb{R}$ in the set of C^r admissible functions ($r > 2$), we have

$$\frac{d}{dt} J(\varphi_t) = \int \dot{\varphi}_t[1 - M(\varphi_t)]\, dV, \qquad \text{where } \dot{\varphi}_t = \tfrac{d}{dt}\varphi_t. \tag{40}$$

This comes from the fact that $1 - M(\varphi)$ is a divergence. Here, this is easy to verify since $M(\varphi)$ is the sum of m determinants but the result is true in general: see M.S. Berger [*41].

$I(\varphi)$ and $J(\varphi)$ satisfy the following inequalities (see Aubin [*7]):

$$J(\varphi) \le I(\varphi) \le (m + 1)J(\varphi); \tag{41}$$

in [*31] we find $(1 + 1/m)J(\varphi) \le I(\varphi)$.

For more details on these functionals see Bando–Mabuchi [*31]; these will be useful for the C^0-estimate. When $m = 1$,

$$I(\varphi) = \int |\nabla\varphi|^2\, dV = 2J(\varphi).$$

It is possible to prove the following

Proposition (Aubin [20]). *Let $h(t)$ be an increasing C^1 function on $\mathbb{R}$. Any C^2 admissible function φ satisfies*

$$\int [1 - M(\varphi)]h(\varphi)\, dV \ge (1/m) \int h'(\varphi)|\nabla\varphi|^2\, dV. \tag{42}$$

Choosing $h(t) = t$, we find $I(\varphi) \ge (1/m) \int |\nabla\varphi|^2\, dV$. Thus if $\varphi \not\equiv$ Const., $I(\varphi) > 0$.

15.2. Some Inequalities

7.35 Proposition. *If we have an estimate of $I(\varphi)$ (or $J(\varphi)$ according to (41)), we have the C^0-estimate.*

Proof. Recall that for $t \in \mathfrak{G}$, φ_t is a solution of E_t (37).

Since $\int M(\varphi_t)\,dV = V$ the volume of the manifold, we have: $V^{-1}\int(-t\varphi_t + f)\,dV \leq \log\int e^{-t\varphi_t+f}\,dV - \log V = 0$. Thus

$$\int \varphi_t\,dV \geq \inf\left[0, \varepsilon^{-1}\int f\,dV\right] = k_0. \tag{43}$$

Likewise, $V^{-1}\int\big(t\varphi_t - f\big)\,dV' \leq \log\int e^{t\varphi_t - f}\,dV' - \log V = 0$. Hence

$$\int \varphi_t\,dV' \leq \sup\big[0, \varepsilon^{-1}V\sup f\big] = k_1. \tag{44}$$

Multiplying $g'_{t\lambda\bar\mu} = g_{\lambda\bar\mu} + \partial_{\lambda\bar\mu}\varphi_t$ by its inverse matrix $g'^{\lambda\bar\mu}_t$, then by $g^{\lambda\bar\mu}$, we get:

$$m = g'^{\lambda\bar\mu}_t g_{\lambda\bar\mu} - \Delta'_t\varphi_t \quad\text{and}\quad 0 < g^{\lambda\bar\mu}g'_{t\lambda\bar\mu} = m - \Delta\varphi_t.$$

Thus,

$$\Delta\varphi_t < m \quad \text{and} \Delta'_t\varphi_t > -m. \tag{45}$$

Using the first inequality in the following equality (Theorem 4.13)

$$\varphi_t(P) = V^{-1}\int\varphi_t\,dV + \int G(P,Q)\Delta_t\varphi(Q)\,dV(Q) \tag{46}$$

where the Green function $G(P,Q)$ of the Laplacian Δ is chosen ≥ 0, we obtain:

$$\varphi_t(P) \leq V^{-1}\int\varphi_t\,dV + m\int G(P,Q)\,dV(Q) = V^{-1}\int\varphi_t\,dV + k, \tag{47}$$

with k a constant.

Since $t \in [\varepsilon, 1]$, the Ricci curvature of (M, g'_t) is greater than ε according to (38). By Myers' theorem 1.43, the diameter D_t of (M, g'_t) satisfies the inequality $D_t \leq \pi[(2m-1)/\varepsilon]^{1/2}$.

Consequently, Theorem 4.32 (or inequality 37 of 4.29) gives a uniform bound from below for the Green functions $G_t(P,Q)$ of the laplacian Δ'_t with integral zero ($\int G_t(P,Q)\,dV'_t(Q) = 0$):

$$G_t(P,Q) \geq -\,\text{Const.}\,D_t^2/V_t \geq -k_2 \quad \text{for } t \in \mathfrak{G};$$

since $V_t = \int dV'_t = \int M(\varphi_t)\,dV = \int dV = V$, k_2 is a positive real number which depends only on m and ε. Now, using the second inequality (45) in

$$\varphi_t(P) = V^{-1}\int \varphi_t\, dV'_t + \int \left[G_t(P,Q) + k_2\right]\Delta'_t\varphi_t(Q)\, dV'_t(Q),$$

we get:

$$\varphi_t(P) \geq V^{-1}\int \varphi_t\, dV'_t - mk_2 V. \tag{48}$$

Thus, since $I(\varphi) \leq K$ yields $\int \varphi_t\, dV \leq K + k_1$ by (44) and $\int \varphi_t\, dV' \geq k_0 - K$ by (43), using (47) and (48), we obtain:

$$V^{-1}(k_0 - K) - mk_2 V \leq \varphi_t(P) \leq V^{-1}(K + k_1) + k.$$

From (47) and (48) we deduce the following

7.36 Proposition. *On (M,g) a compact Kähler manifold, let us denote by $\mathcal{A}_\varepsilon(\varepsilon > 0)$ the set of the functions $\varphi \in \mathcal{A}$ such that $R'_{\lambda\bar\mu} - \varepsilon g'_{\lambda\bar\mu} \geq 0$ (g' is defined in* (1)*). There exists a constant k, depending on ε, such that any $\varphi \in \mathcal{A}_\varepsilon$ satisfies*

$$-V^{-1}I(\varphi) - k \leq \varphi - V^{-1}\int \varphi\, dV \leq k. \tag{49}$$

Remark. In [*291], Siu proves that the quantities $\sup(-\varphi)$, $\sup\varphi$, $\int \varphi\, dV$, $-\int \varphi\, dV'$, $\log\int e^{c\varphi}\, dV$ and $\log\int e^{-c\varphi}\, dV'$ $(c > 0)$ are comparable in the sense that any two such quantities Q, Q' satisfy $Q \leq AQ' + B$ for some a priori constants A and B.

Siu [*291] proves the following Harnack inequality: for each $\varepsilon > 0$, there exists $C(\varepsilon) > 0$ such that

$$\sup(-\varphi_t) \leq (m + \varepsilon)\sup\varphi_t + C(\varepsilon),$$

where φ_t satisfies (37) for $t \in [\varepsilon, t_0[$. The proof is by contradiction. This result was improved by the following:

Theorem 7.36 (Tian [*301]). *There exists a constant $C(t)$ such that, for any C^2 admissible function ψ satisfying $\int e^{f-t\psi}\, dV = V$, the solution φ_t of* (37) *satisfies*

$$\sup(\psi - \varphi_t) \leq m\sup(\varphi_t - \psi) + C(t).$$

Therefore, if the initial metric is Einstein–Kähler, any C^2 admissible function ψ with $\int e^{-\psi}\, dV = V$ satisfies

$$\sup\psi \leq -m\inf\psi + C.$$

Indeed, in this case, $\varphi_t \equiv 0$.

7.37 To proceed further, we need an inequality concerning the exponential function for admissible functions with integral zero.

For the dimension $m = 1$ (see Aubin [*9], p. 155), any function φ with $\int \varphi \, dV = 0$ satisfies

$$\int e^{-\varphi} \, dV \leq \text{Const.} \exp\left(\frac{\pi}{8} \int \nabla^\nu \varphi \nabla_\nu \varphi \, dV\right).$$

Recall that we are on a Kähler manifold, so $\int \nabla^\nu \varphi \nabla_\nu \varphi \, dV$ is one half of $\int \nabla^i \varphi \nabla_i \varphi \, dV$ on a Riemannian manifold (here the best constant is not $\pi/16$ but $\pi/8$).

By analogy we will suppose that any C^∞ admissible function with zero integral satisfies:

$$\text{(50)} \qquad \int e^{-\varphi} \, dV \leq C \exp\big[\xi I(\varphi)\big], \quad \text{with } C \text{ and } \xi \text{ two constants.}$$

With this inequality, we obtain the C^0-estimate (Aubin [*7]), see below.

Since we will apply (50) to the functions φ_t (solutions of (37)), it is necessary to prove (50) only for the functions $\varphi_t - V^{-1} \int \varphi_t \, dV$, or more generally for the functions in $\mathcal{A}_\varepsilon$ with zero integral ($\mathcal{A}_\varepsilon$ is defined in 7.36).

In our case $C_1 > 0$, $\omega \in C_1$, we can conjecture that the best constant ξ ($\xi_m = \inf \xi$ such that a constant C exists) is the one we found for the ball (see 8.30): $\xi_m = m^m m! \pi^{-m} (m+1)^{-2m-1}$.

15.3. The C^0-estimate (Aubin [*7])

7.38 Set $x(t) = \int \varphi_t \, dV$, $y(t) = J(\varphi_t)$ and $z(t) = I(\varphi_t)$ for $\varepsilon \leq t \in \mathfrak{G}$. Recall that φ_t is a solution of E_t (37).

Differentiating with respect to t the equality $\int e^{-t\varphi_t + f} \, dV = V$ gives

$$\int (-\varphi_t - t\dot{\varphi}_t) M(\varphi_t) \, dV = 0, \qquad \text{where } \dot{\varphi}_t = d\varphi_t/dt;$$

hence, according to (40),

$$\text{(51)} \qquad z(t) - x(t) + t(y' - x') = 0.$$

We have then:

$$V = \int e^{-t\varphi_t + f} \, dV \leq e^{\sup f - txV^{-1}} \int e^{-t(\varphi_t - xV^{-1})} \, dV.$$

Using Hölder's inequality and (50), we deduce:

$$V e^{\frac{tx}{V} - \sup f} \leq \left(\int e^{-(\varphi_t - \frac{x}{V})} \, dV\right)^t V^{1-t} \leq C^t V^{1-t} \exp\big[t\xi I(\varphi_t)\big];$$

thus

(52) $$x \leq k_3 + V\xi z, \qquad \text{with } k_3 \text{ a constant.}$$

From (51) and (52), we also have

(53) $$y' - x' = (x - z)/t \leq k_3(\varepsilon V\xi)^{-1} + x\big[1 - (V\xi)^{-1}\big]t^{-1}.$$

If $V\xi \leq 1$, since $x \geq k_0$ by (43), we have that $y' - x' \leq$ Const. and $y(t) - x(t) \leq k_4$. This inequality, together with (41) and (52), gives

$$y(t)\big[1 - (m+1)\xi V\big] \leq k_3 + k_4.$$

Hence if $(m+1)\xi V < 1$, we have $y(t) \leq$ Const. and an estimate of $z(t) = I(\varphi_t)$ since $z \leq (m+1)y$ by (41). Then, according to Proposition 7.35, we obtain the C^0-estimate and, by Proposition 7.29, the

7.39 Theorem. *If for some C and some $\xi < 1/(m+1)V$, the functions $\varphi_t - V^{-1}\int \varphi_t\, dV$ satisfy inequality* (50) *when $\omega \in C_1(M)$, then there exists an Einstein–Kähler metric. Recall that φ_t is a solution of E_t*(37).

15.4. Inequalities for the Dimension $m = 1$

7.40 Theorem (Aubin [*9]). *Let M be a C^∞ compact Riemannian manifold of real dimension $n = 2$. Define*

$$\mathcal{E}_\mu = \left\{\varphi \in C^2 \Big/ \int \varphi\, dV = 0 \quad \textit{and} \quad \int |\Delta\varphi|\, dV \leq 2\mu\right\}.$$

Given $\alpha < 4\pi/\mu$, there exists a constant C, depending on V, α and μ, such that any $\varphi \in \mathcal{E}_\mu$ satisfies

(54) $$\int e^{-\alpha\varphi}\, dV \leq C.$$

Proof. We can suppose that $\int |\Delta\varphi|\, dV = 2\mu$, from which the general case follows. Let $G(P,Q)$ be the Green function of the Laplacian Δ such that $G(P,Q) \geq 0$. Write $G(P,Q) = -(1/2\pi)\log f(r) + F(P,Q)$ with $f(r) = r = d(P,Q)$ in a neighbourhood of $r = 0$, $f(r)$ increasing and $f(r) = \delta/2$ for $r \geq \delta$ the injectivity radius. $\varphi(P) = \int G(P,Q)\Delta\varphi(Q)\, dV(Q)$ implies $-\varphi(P) \leq -\int_{\Delta\varphi<0} G(P,Q)\Delta\varphi(Q)\, dV(Q)$ and we have $-\int_{\Delta\varphi<0}\Delta\varphi\, dV \leq \mu$ since $-\int_{\Delta\varphi<0}\Delta\varphi\, dV = \int_{\Delta\varphi>0}\Delta\varphi\, dV$.

$(P,Q) \to F(P,Q)$ is a continuous function on $M \times M$, thus $|F(P,Q)| \leq a$, for some constant a. Hence, for any real number $\alpha > 0$, we have

$$e^{-\alpha\varphi} \leq e^{\alpha a\mu} \exp \int_{\Delta\varphi<0} (-\alpha\mu/2\pi)\big[\log f(r)\big](-\Delta\varphi/\mu)\, dV.$$

This yields

$$e^{-\alpha\varphi} \leq e^{\alpha a \mu} \int_{\Delta\varphi<0} [f(r)]^{-\alpha\mu/2\pi} (-\Delta\varphi/\mu)\, dV.$$

For $\alpha < 4\pi/\mu$, $e^{-\alpha\varphi}$ is integrable and

$$\int e^{-\alpha\varphi}\, dV \leq e^{\alpha a \mu} \sup_{P\in M} \int [f(r)]^{-\alpha\mu/2\pi}\, dV(Q) \leq \text{Const.}.$$

7.41 Corollary. *Let M and $\mathcal{E}_\mu$ be as in Theorem 7.40. Suppose there exists a group G of isometries such that each orbit admits at least $k > 1$ distinct points. Then, for $\beta < 4\pi k/\mu$, the G-invariant functions $\varphi \in \mathcal{E}_\mu$ satisfy*

$$\int e^{-\beta\varphi}\, dV \leq \text{Const.}.$$

7.42 *The case of the sphere S_2.* Let (S_2, g_0) be the sphere endowed with the canonical metric g_0. Its sectional curvature equals 1 and $V = 4\pi$. φ is admissible if its real Laplacian satisfies $\Delta\varphi < 2$.

Set $\tilde\varphi = \varphi - (1/4\pi)\int \varphi\, dV_0$. $\tilde\varphi$ satisfies (54) if $\alpha < 1/2$ (here $\mu = 8\pi$). More generally:

Proposition. *Let g' be a Riemannian metric on S_2, and V' be its volume. When $\beta < 2\pi/V'$, any function φ, admissible for g', satisfies*

$$\int e^{-\beta\tilde\varphi}\, dV \leq \text{Const.}.$$

15.5. Inequalities for the Exponential Function

7.43 We could think that inequality (54) is special for the dimension $m = 1$ and comes from the particularity of the Green function. In fact inequality (54) holds for the admissible functions when $m > 1$. This was shown by Hörmander [148] and Skoda [*292] for the plurisubharmonic functions.

7.44 Theorem (Hörmander [148]). *There is a constant C such that any plurisubharmonic function ψ in the unit ball in $\mathbb{C}^m$ with $\psi(z) \leq 1$ when $|z| \leq 1$ and $\psi(0) = 0$, satisfies*

$$\text{(55)} \qquad \int_{|z|<1/2} e^{-\psi(z)}\, dV \leq C.$$

Proof. When $m = 1$, the Green function for zero Dirichlet data, on the unit ball in $\mathbb{C}^m$ endowed with the euclidean metric, is $-\frac{1}{2\pi}\log\left(\frac{|z-\xi|}{|1-z\xi|}\right)$. According to (22) in 4.17 we have

$$2\pi\psi(z) = \int_{|\xi|<1} \log\left(\frac{|z-\xi|}{|1-z\bar{\xi}|}\right)|\Delta\psi|\, ds + \int_{|\xi|=1} \frac{(1-|z|^2)}{|z-\xi|^2}\psi(\xi)\, dl.$$

Choosing $z = 0$, we obtain

$$(56) \qquad 0 \le -\int_{|\xi|<1} (\log|\xi|)|\Delta\psi|\, ds = \int_{|\xi|=1} \psi(\xi)\, dl \le 2\pi.$$

Thus

$$\int_{|\xi|=1} |\psi(\xi)| dl \le 4\pi \qquad \text{and, when } |z| < 1/2,$$

$$(57) \qquad \left| \int_{|\xi|=1} (1-|z|^2)|z-\xi|^{-2}\psi(\xi)\, dl \right| \le 12\pi.$$

Choosing ρ so that $1/2 < \rho < e^{-1/2}$, we have according to (56):

$$(58) \qquad a = \int_{|\xi|<\rho} |\Delta\psi|\, ds/2\pi \le -1/\log\rho < 2.$$

Then, for some C, when $|z| < 1/2$

$$(59) \qquad \left| (1/2\pi) \int_{|\xi|>\rho} \log\big(|z-\xi|\,|1-z\bar{\xi}|^{-1}\big)|\Delta\psi|\, ds \right| \le C.$$

Using the inequality $\exp\big[\int_\Omega f\, dV / \int_\Omega dV\big] \le \int_\Omega e^f\, dV / \int_\Omega dV$, we obtain for $|z| < 1/2$, according to (57) and (59):

$$\begin{aligned} e^{-\psi(z)} &\le \exp\left[6 + C + \int_{|\xi|<\rho} -a\log\big(|z-\xi||1-z\bar{\xi}|^{-1}\big)|\Delta\psi| ds/2\pi a\right] \\ &\le e^{6+C} \int_{|\xi|<\rho} |z-\xi|^{-a}|1-z\bar{\xi}|^a |\Delta\psi|\, ds/2\pi a. \end{aligned}$$

Summing up, we have proved Theorem 7.44 for $m = 1$ since $a < 2$. When $m > 1$, we can apply the preceding result to any complex line through 0. Introducing polar coordinates (r, ζ) we have

$$\int_{|z|<\frac{1}{2}} e^{-\psi(z)}\, dV(z) = \int_{S_{2m-1}} d\sigma(\zeta) \int_{|w|<\frac{1}{2}} r^{2m-2} e^{-\psi(w\zeta)}\, ds(w)/2\pi$$

where $w \in \mathbb{C}$ and $d\sigma(\zeta)$ denotes the surface area on the unit sphere S_{2m-1}.

7.45 Corollary. *Let $B_R = \{z \in \mathbb{C}^m / |z| < R\}$ and $\lambda > 0$ be a real number. There exists a constant C which depends on m, λ and R such that any plurisubharmonic function ψ in B_R, with $\psi(0) \ge -1$ and $\psi(z) \le 0$ in B_R, satisfies $\int_{|z|<r_o} e^{-\lambda\psi(z)}\, dV \le C$, where $r_0 < Re^{-\lambda/2}$.*

Proof. By an homothety, we have to prove the inequality for $R = 1$ and $r_1 = r_0/R$. In the proof of Theorem 7.44, if we want an estimate of the integral of $e^{-\lambda\psi}$, we must have $\lambda a < 2$.

The inequality (58) is valid: $a \leq -1/\log\rho$, and we must choose $r_1 < \rho$. Thus we have $\lambda a < 2$ if $r_1 < e^{-\lambda/2}$.

7.46 Proposition (Tian [*300]). *Let (M, g) be a compact Kähler manifold. There exist two positive constants α, C depending only on (M, g) such that*

$$\int_M e^{-\alpha\varphi}\, dV \leq C \tag{60}$$

for each C^2 admissible function φ with $\sup\varphi = 0$ (or with $\int \varphi\, dV = 0$).

Proof. Choose $5r$ smaller than the injectivity radius of (M, g) and x_i points of M $(i = 1, 2, \ldots N)$ such that $M = \cup_{i=1}^{N} B_r(x_i)$.

Since $\sup\varphi = 0$, we have, according to (47):

$$\int \varphi_t\, dV \geq -kV.$$

Thus $\sup_{B_r(x_i)} \varphi(x) \geq -kV\left[\mathrm{Vol}\, B_r(x_i)\right]^{-1}$; let $y_i \in B_r(x_i)$ be such that

$$\varphi(y_i) \geq -kV[\mathrm{Vol}\, B_r(x_i)]^{-1} = -\nu.$$

Choose the Kähler potential ψ_i of (M, g) in $B_{5r}(x_i)$ such that $\psi_i(y_i) = 0$ and set $C = \sup_i \sup_{B_{5r}(x_i)} |\psi_i(x)|$.

We will apply Corollary 7.45 with $r_0 = 2r$, $R = 4r$ and $\lambda = 1$ to the function $\psi = \alpha(\varphi + \psi_i - C)$ with $\alpha = (\nu + C)^{-1}$. We verify that $\varphi + \psi_i - C \leq 0$ in $B_{4r}(y_i) \subset B_{5r}(x_i)$ and $\varphi(y_i) + \psi_i(y_i) - C \geq -(\nu + C)$. So

$$\int_{B_{2r}(y_i)} e^{-\alpha(\varphi+\psi_i-C)}\, dV \leq \text{Const.}.$$

Summing up, since $B_r(x_i) \subset B_{2r}(y_i)$, we obtain (60), when $\sup\varphi = 0$ and, in fact, for any C^2 admissible function positive somewhere. This is the case when $\int \varphi\, dV = 0$.

The converse is true; since $-\frac{1}{V}\int \varphi\, dV \leq k$ by (17), setting $\tilde\varphi = \varphi - \frac{1}{V}\int\varphi\, dV$, we have $\int e^{-\alpha\tilde\varphi}\, dV \leq \int e^{-\alpha\varphi}\, dV \leq e^{\alpha k}\int e^{-\alpha\tilde\varphi}\, dV$.

Remark. $\alpha(M, g) = \alpha(M, g')$ if g and g' belong to the same Kähler class. In the case $C_1(M) > 0$, we will write $\alpha(M)$ for $\alpha(M, g)$ with $\omega(g) \in C_1(M)$.

7.47 Theorem (Tian [*300], see also Ding [*116] and Aubin [*8]). *A compact Kähler manifold (M, g) of dimension m with $C_1(M) > 0$ admits an Einstein–Kähler metric if $\alpha(M) > m/(m+1)$, when $\omega \in C_1(M)$.*

Proof. Adding $-V^{-1}\int \varphi_t\, dV$ to both members of inequality (18) yields

$$(61) \qquad -\left[\varphi_t - V^{-1}\int \varphi_t\, dV\right] \leq V^{-1} I(\varphi_t) + mk_2 V.$$

If $\alpha(M) \geq 1$, (60) implies (50) with $\xi = 0$. Indeed

$$\int \exp\left[-\varphi_t + \frac{1}{V}\int \varphi_t\, dV\right] dV \leq C \exp\left[\frac{1}{V}\int \varphi_t\, dV - \sup \varphi_t\right] \leq C.$$

If $\alpha(M) < 1$, we apply to $\psi = \varphi_t - V^{-1}\int \varphi_t\, dV$ the inequality

$$\int e^{-\psi}\, dV \leq e^{(1-\alpha)\sup(-\psi)} \int e^{-\alpha\psi}\, dV.$$

By virtue of (60) and (61), we obtain (50) with $\xi = \left[1 - \alpha(M)\right]V^{-1}$. Moreover the hypothesis on $\alpha(M)$ implies $\xi < 1/(m+1)V$. The conditions of Theorems 7.39 are satisfied, hence there exists an Einstein–Kähler metric.

§16. Some Results

7.48 Let (M, g) be a compact Kähler manifold with $C_1(M) > 0$. How to know if (M, g) carries an Einstein–Kähler metric? At first, there may exist an obstruction, see §14. If there is none, we can compute $\alpha(M)$ to see if $\alpha(M)$ satisfies $\alpha(M) > m/(m+1)$ in order to apply Theorem 7.47. However, this procedure may not be viable: in fact, for the simplest Kähler manifold $P_m(\mathbb{C})$, which does carry Einstein–Kähler metrics, $\alpha\big(P_m(\mathbb{C})\big) = 1/(m+1)$ (see Aubin [*9] and Real [*275] for the proof). In dimension $m = 1$, on the sphere S_2, Moser (see 6.65) found the same difficulty. Here, if the Kähler manifold has some symmetries, we can hope to solve the problem, considering in (60) only functions φ having these symmetries.

7.49 Definition. Let G be a group of automorphisms of the compact Kähler manifold (M, g) with $\omega \in C_1(M) > 0$, ω G-invariant. We define $\alpha_G(M) = \sup \alpha$, for α such that any G-invariant admissible function φ with $\int \varphi\, dV = 0$ satisfies $\int e^{-\alpha\varphi}\, dV \leq C$ for some constant C which depends on α, G and M.

Suppose (M, g) has a non trivial group of automorphisms G. We can apply the continuity method in 7.10, considering instead of Θ, the set $\tilde{\Theta}$ of the G-invariant functions in $\Theta = \mathcal{A} \cap C^{5+\alpha}$, and instead of Γ, $\tilde{\Gamma}$ from $\mathbb{R} \times \tilde{\Theta}$ into $C_G^{3+\alpha}$ the set of G-invariant $C^{3+\alpha}$ functions.

$D_\varphi\tilde{\Gamma}(t, \varphi) \in \mathcal{L}\big(C_G^{5+\alpha}, C_G^{3+\alpha}\big)$ and it is inversible for $\varepsilon \leq t < 1$. Thus the functions φ_t belong to $\mathcal{A} \cap C_G^\infty$. For more details see Real [*274].

Thus, to obtain the C^0-estimate, we only have to verify that G-invariant admissible functions with $\int \varphi\, dV = 0$ satisfy (60). Proposition 7.29 then implies

7.50 Theorem. *If $\alpha_G(M) > m/(m+1)$, the compact Kähler manifold (M, g), with g G-invariant and $\omega \in C_1(M) > 0$, carries an Einstein–Kähler metric.*

7.51 Proposition (Real [*275]). *$\alpha_G(\mathbb{P}_m(\mathbb{C})) = 1$ where G is the compact subgroup of Aut $P_m(\mathbb{C})$ generated by the permutations $\sigma_{j,k}$ of the homogeneous coordinates together with the transformations $\gamma_{j,\theta}$, $j = 1, 2, \ldots m$ and $\theta \in [0, 2\pi]$*

$$\gamma_{j,\theta} : \left[z_0, \ldots z_j, \ldots z_m\right] \to \left[z_0, \ldots, z_j e^{i\theta}, \ldots z_m\right]$$

$$\sigma_{k,j} : \left[z_0, \ldots z_j, \ldots, z_k, \ldots z_m\right] \to \left[z_0, \ldots, z_k, \ldots, z_j, \ldots, z_m\right].$$

Proof. The Kähler potential is $K = (m+1)\log(1 + \sum_{i=1}^m x_i)$, where $x_i = |z_i]^2$, in U_0 defined by $z_0 \neq 0$, the usual metric is $g_{\lambda\bar{\mu}} = \partial_\lambda \partial_{\bar{\mu}} K$, ($\partial_\lambda = \partial/\partial z_\lambda$).

Since $idd''(K + \varphi)$ is positive definite,

$$(62) \qquad \partial_{\lambda\bar{\lambda}}(K+\varphi) = \frac{\partial}{\partial x_\lambda}\left(x_\lambda \frac{\partial(K+\varphi)}{\partial x_\lambda}\right) > 0\,;$$

φ is admissible and supposed to be a function of $x_1, x_2, \ldots, x_m$, moreover $\int \varphi \, dV = 0$. From (62) we obtain

$$(63) \qquad 0 \leq x_i \frac{\partial(K+\varphi)}{\partial x_i}(x_1, x_2, \ldots, x_m) \leq m+1$$

for $(x_1, x_2, \ldots, x_m) \in (\mathbb{R}_+^*)^m$. Indeed the expression of K gives

$$\left(x_i \frac{\partial K}{\partial x_i}\right)_{x_i=0} = 0 \quad \text{and} \quad \left(x_i \frac{\partial K}{\partial x_i}\right)_{x_i=+\infty} = m+1.$$

Moreover $\left(x_i \frac{\partial\varphi}{\partial x_i}\right)_{x_i=0} = \left(x_i \frac{\partial\varphi}{\partial x_i}\right)_{x_i=+\infty} = 0$ and $x_i \frac{\partial(K+\varphi)}{\partial x_i}$ is increasing in x_i.

Now, for $(x_1, x_2, \ldots, x_m) \in E =]0, 1]^m$,

$$(64) \qquad (K+\varphi)(x_1, \ldots, x_m) - (m+1)\log\left(\prod_{i=1}^m x_i\right) \geq (K+\varphi)(1, \ldots, 1)$$

since, according to (63), the partial derivatives of the left hand side of (64) are ≤ 0.

Since $\varphi \leq k$ (47), $-\int_{\varphi<0} \varphi \, dV = \int_{\varphi>0} \varphi \, dV \leq kV$. Consider $L = [1/2, 1]^m \subset P_m(\mathbb{C})$, $-kV \leq \int_L \varphi \, dV \leq V \sup_{x \in L} \varphi(x) = V\varphi(y)$ for some point $y \in L$.

Hence $(K+\varphi)(y) \geq -k$ and, since $K+\varphi$ is increasing in each of its variables (according to (63)), $(K+\varphi)(1, \ldots, 1) \geq -k$.

Thus, for all $(x_1, \ldots, x_m) \in E$:

$$(65) \qquad -(K+\varphi)(x_1, \ldots, x_m) \leq k - (m+1)\log\left(\prod_{i=1}^m x_i\right).$$

Henceforth, we suppose $\varphi \in \mathcal{A}_0$, the set of admissible G-invariant functions with zero integral. For $(x_1, \ldots, x_m) \in E$, Real proves that

$$-(K+\varphi)(x_1,\dots,x_m) \le -(K+\varphi)(\xi,\dots\xi) \quad \text{where } \xi = \left(\prod_{i=1}^{m} x_i\right)^{\frac{1}{m}}$$

and

$$-(K+\varphi)(\xi,\dots\xi) + m\log\xi \le -(K+\varphi)(1,\dots,1).$$

Thus, and this is the analogous of (65), for $(x_1,\dots,x_m) \in E$,

$$-(K+\varphi)(x_1,\dots,x_m) \le k - \log\left(\prod_{i=1}^{m} x_i\right). \tag{66}$$

It is now possible to complete the proof. For $\varphi \in \mathcal{A}_0$,

$$\int_{P_m(\mathbb{C})} e^{-\alpha\varphi}\,dV = (m+1)\int_E e^{-\alpha\varphi}\omega^m \qquad \left(\text{with } \omega = \frac{i}{2\pi}dd''K\right)$$
$$= (m+1)\int_0^1 \dots \int_0^1 e^{-(\alpha\varphi+K)(x_1,\dots,x_m)}\,dx_1\,dx_2\dots dx_m.$$

Choose $\alpha \in]0,1[$. According to (66) and since $K \ge 0$,

$$\int_{P_m(\mathbb{C})} e^{-\alpha\varphi}\,dV \le (m+1)e^{\alpha k}\left(\int_0^1 t^{-\alpha}\,dt\right)^m = (m+1)e^{\alpha k}(1-\alpha)^{-m}.$$

Thus $\alpha_G(P_m(\mathbb{C})) \ge 1$. For the details and the proof of $\alpha_G\big(P_m(\mathbb{C})\big) \le 1$, see Real [*274].

7.52 Proposition (Real [*275]). $\alpha_{G(p)}\big(P_m(\mathbb{C})\big) \ge \inf\{1, \frac{p}{m+1}\}$ *where* $G(p)$ *is the compact subgroup of Aut* $P_m(\mathbb{C})$ *generated by the permutations* $\sigma_{j,k}$ *and* $\gamma_{j,\theta}$ *with* $\theta = 2\pi/p$, $p \in \mathbb{N}^*$.

For the definition of $\sigma_{j,k}$ and $\gamma_{j,\theta}$ see Proposition 7.51. Picking $p = m+1$, we do not have an alternative proof of Proposition 7.51. Indeed, for the proof, Real uses the result of Proposition 7.51.

7.53 The dimension $m = 2$. The compact complex surfaces with $C_1(M) > 0$ are: $P_2(\mathbb{C})$, $S_2 \times S_2$ and $P_2(\mathbb{C})$ blown up at k generic points $(1 \le k \le 8)$.

We saw (7.31) that if $k = 1$ or 2 the corresponding manifolds have no Einstein–Kähler metric. Tian and Yau [*303] proved that for any k $(3 \le k \le 8)$ there is a compact complex surface of this type (with k exceptional divisors) which has an Einstein–Kähler metric.

Siu [*291] solved also the case $k = 3$. The following theorem solves entirely the case $m = 2$.

Theorem (Tian [*302]). *Any compact complex surface* M *with* $C_1(M) > 0$ *admits an Einstein–Kähler metric if its group of automorphisms is reductive.*

7.54 Conjecture (Calabi). *Any compact Kähler manifold with $C_1(M) > 0$ and without holomorphic vector field has an Einstein–Kähler metric.*

In [*71] and [*72] Calabi studied the functional $\int R^2\,dV$ when g belongs to a given cohomology class. Note that $\int R\,dV = \text{Const.}$ since $\int R\,dV = \pi^m \int \Psi \wedge \omega^{m-1}$, where Ψ is the Ricci form and ω the first fundamental form (see 7.1).

Let $[\omega]$ be a fixed class of Kähler metrics. The Euler–Lagrange equation of $S(g) = \int R^2\,dV$ when $g \in [\omega]$ is $\nabla_\alpha \nabla_\beta R = 0$ (or equivalently $\nabla_{\bar\alpha}\nabla_{\bar\beta} R = 0$). That is to say, the real vector field on M

$$X = g^{\alpha\bar\mu}\left(\nabla_\lambda R \frac{\partial}{\partial z^{\bar\mu}} + \nabla_{\bar\mu} R \frac{\partial}{\partial z^\lambda}\right)$$

generates a holomorphic flow (possibility trivial, if R is constant).

After this, Calabi proved that, if $\tilde g$ is a critical point of $S(g)$, then the second variation of $S(g)$ with respect to any infinitesimal deformation with $\delta g_{\alpha\bar\beta} = \partial_{\alpha\bar\beta} u$ is effectively positive definite (it is zero if and only if $\delta g_{\alpha\bar\beta}$ is induced by a holomorphic flow).

The problem of minimizing $\int R^2\,dV$ for all Kähler metrics in a given class is very hard. Solving it when $C_1(M) > 0$ and $[\omega] = C_1(M)$ would prove the conjecture. Indeed, if $R = \text{Const.}$ and $\tilde R_{\lambda\bar\mu} = \tilde g_{\lambda\bar\mu} + \partial_{\lambda\bar\mu} f$, we have $f = \text{Const.}$ and $\tilde g$ is an Einstein–Kähler metric.

To illustrate his study on $S(g)$, Calabi [*71] minimized $S(g)$ on $P_m(\mathbb{C})$ blown up at one point. This Calabi conjecture is proved for $m = 2$ (Theorem 7.53). In [*302] Tian discusses the problem when $m > 2$.

7.55 *Fermat hypersurfaces $X_{m,p}$.*

$$X_{m,p} = \left\{(z_0, \dots z_{m+1}) \in P_{m+1}(\mathbb{C}) / z_0^p + \dots + z_{m+1}^p = 0\right\}$$

where p is an integer satisfying $0 < p \leq m+1$. $C_1(X_{m,p}) > 0$, the restriction of $K = (m+2-p)\log\left(|z_0)|^2 + \dots + |z_{m+1}|^2\right)$ to $X_{m,p}$ is the potential of a Kählerian metric whose first fundamental form belongs to $C_1(X_{m,p})$.

Tian [*300] and Siu [*291] prove that $X_{m,m+1}$ and $X_{m,m}$ have an Einstein–Kähler metric. Tian proves that $\alpha_G(X_{m,p}) > m/(m+1)$ if $p = m$ or $m+1$. Here G is generated by $\sigma_{j,k}$ and $\gamma_{j,\theta}$ with $\theta \in [0, 2\pi]$ (see 7.51). Siu applies his method. φ being an admissible function, Siu [*291] considers restricting φ to algebraic curves in M. When $m = 1$ we saw (§15.4) that we can obtain the C^0-estimate by using the Green function. If the curves , considered by Siu, are invariant under a large group of automorphisms of M, the C^0-estimate obtained is sharp enough to infer the existence of an Einstein–Kähler metric (compare with 7.41 and 7.42, β is larger when the volume V' is smaller or when k is larger).

7.56 Theorem (Nadel [*248], Real [*274]). *The Fermat hypersurfaces $X_{m,p}$ with $1 + m/2 \leq p \leq m+1$ have an Einstein–Kähler metric.*

Real proves that $\alpha_G(X_{m,p}) \geq 1$ when $p \geq 1 + m/2$, by using Proposition 7.52; he then applies Theorem 7.50. For the proof of Theorem 7.56, Nadel uses the following:

7.57 Theorem (Nadel [*248]). *Let (M, g) be a compact Kähler manifold with $C_1(M) > 0$ and let G be a compact group of automorphisms of M. If M does not admit a G-invariant multiplier ideal sheaf, M admits an Einstein–Kähler metric.*

The proof proceed by contradiction. If M does not admit an Einstein–Kähler metric the C^0-estimate fails to hold. We saw that inequality (60) with $\alpha > m/(m+1)$ implies the C^0-estimate for the functions φ_t. φ_t solution of E_t (37) is G-invariant.

Hence for each $\alpha \in]m/(m+1), 1[$, there exists an increasing sequence $\{t_k\}$ $(t_k < 1)$ such that $\varphi_k = \varphi_{t_k} - \sup \varphi_{t_k}$ satisfies:

$$\int e^{-\alpha\varphi_k}\, dV \to \infty \qquad \text{when } k \to \infty.$$

After $S = \{\varphi_k\}$ is replaced by a suitable subsequence, we may find a nonempty open subset $U \subset M$ such that $\int_U e^{-\alpha\varphi_k}\, dV \leq$ Const..

Then, Nadel introduces the coherent sheaf of ideals I_s on M, called the multiplier ideal sheaf (in particular I_s is not equal to the zero sheaf of ideals and is not equal to all of θ_M). It is defined as follows: for each open subset $U \subset M$, $I_s(U)$ consists of the local holomorphic functions f such that $\int_U |f|^2 e^{-\alpha\varphi_k}\, dV \leq$ Const. for all k.

Various global algebro-geometric considerations lead to a contradiction.

7.58 Other results. Nadel [*248] uses his theorem to prove that the intersection of three quadrics in $P_6(\mathbb{C})$ or two quadrics in $P_5(\mathbb{C})$ or a cubic and a quadric in $\mathbb{P}_5(\mathbb{C})$ admit an Einstein–Kähler metric.

Ben Abdesselem and Cherrier [*33] proved that some manifolds carry Einstein–Kähler metrics. Among other things, they study manifolds obtained by blowing up $P_m(\mathbb{C})$ along l independent subprojective spaces $P_d(\mathbb{C})(ld = m+1)$. When $l = 2$ the manifold has an Einstein–Kähler metric.

§17. On Uniqueness

7.59 By the maximum principle, we prove that equations (2) and (6) have only one solution. Hence when $C_1(M) < 0$, there is a unique Einstein–Kähler metric if we fix the volume of the manifold, and when $C_1(M) = 0$, there is a unique Einstein–Kähler metric in each positive (1-1) cohomology class of the manifold with a given volume (Theorem 7.9).

When $C_1(M) > 0$, the identity component G of the group of holomorphic automorphisms of M is not necessarily a group of isometries. Suppose g is Einstein–Kähler; if $u \in G$, $u * g$ is Einstein–Kähler so the following result is the best possible.

7.60 Theorem (Bando, Mabuchi [*31]). *If (M, g) is a compact Einstein–Kähler manifold with $C_1(M) > 0$, g is unique up to G-action.*

The proof involves many steps. We will give some of them and a sketch of their proof. Let ω_0 be the first fundamental form of the initial Kähler metric g_0. Denote by $\omega_0(\varphi)$ the first fundamental form of the metric $g_0(\varphi)$, whose components are $g_{o\lambda\bar\mu} + \partial_{\lambda\bar\mu}\varphi$ (φ is supposed to be admissible for g_0). First we introduce the functionals

$$I(\varphi, \tilde\varphi) = V^{-1}\int(\tilde\varphi - \varphi)\left[\omega_0(\varphi)^m - \omega_0(\tilde\varphi)^m\right], \quad \text{and}$$

$$J(\varphi, \tilde\varphi) = -L(\varphi, \tilde\varphi) + V^{-1}\int(\tilde\varphi - \varphi)\omega_0(\varphi)^m,$$

with φ and $\tilde\varphi$ admissible functions for g_0, V the volume and

$$L(\varphi, \tilde\varphi) = V^{-1}\int_a^b\left[\int \dot\varphi_t\omega_0(\varphi_t)^m\right]dt$$

where $\dot\varphi_t = \partial\varphi_t/\partial t$, $(t, x) \to \varphi_t(x)$ being a smooth function satisfying $\varphi_a = \varphi$ and $\varphi_b = \tilde\varphi$.

We verify that $L(\varphi, \tilde\varphi)$ does not depend on the choice of the family φ_t, as $M(\varphi, \tilde\varphi)$ defined by

$$(67) \qquad M(\varphi, \tilde\varphi) = V^{-1}\int_a^b\left[\int(m - R_t)\dot\varphi_t\omega_0(\varphi_t)^m\right]dt,$$

where R_t is the scalar curvature of the metric $g_0(\varphi_t)$.

When $\tilde\varphi = 0$, we recognize Aubin's functionals $I(\varphi)$ and $J(\varphi)$ (see 7.34) in $I(\varphi, 0)$ and $J(\varphi, 0)$ respectively.

Bando and Mabuchi prove many properties of these functionals such as (41) and

$$(68) \qquad \frac{d}{dt}[I(0, \varphi_t) - J(0, \varphi_t)] = V^{-1}\int \dot\varphi_t\left[\Delta_{\omega_o(\varphi_t)}\varphi_t\right]\omega_0(\varphi_t)^m.$$

7.61 The family of generalized Aubin' s equations on (M, g_0) is defined by

$$(69) \qquad \log M(\varphi_t) = -t\varphi_t - L(0, \varphi_t) + f$$

where f is the function satisfying (3) and $\int e^f\omega_0^m = \int\omega_0^m = V$ (we suppose the manifold is positively oriented). (37) is the original family of equations.

For $t = 0$, equation (69) has a unique solution φ_0. φ_0 satisfies $L(0, \varphi_0) = 0$, and the Ricci form of $\omega_0(\varphi_0)$ is ω_0.

Lemma 7.61 (Bando–Mabuchi [*31]). *Let* $\{\varphi_t\}$ *be a* C^∞ *family of solutions of* (69) *on* $[a, b]$ $(0 \le a < b \le 1)$, *then*

$$\frac{d}{dt}\big[I(0,\varphi_t) - J(0,\varphi_t)\big] \ge 0. \tag{70}$$

Proof. A computation leads to

$$\frac{d}{dt}\big[I(0,\varphi_t) - J(0,\varphi_t)\big] = \frac{1}{V}\int \big[\Delta_{\omega_o(\varphi_t)}\dot\varphi_t - t\dot\varphi_t\big]\big[\Delta_{\omega_o(\varphi_t)}\dot\varphi_t\big]\omega_0(\varphi_t)^m.$$

According to Theorem 4.20, the right hand side is ≥ 0.

Theorem 7.61 (Bando–Mabuchi [*31]). *Let* $\{\varphi_t\}$ *be a* C^∞ *family of solutions of* (69) *on* $[a, b]$ $(0 \le a < b \le 1)$, *then*

$$\frac{d\mu(t)}{dt} = -(1-t)\frac{d}{dt}\big[I(0,\varphi_t) - J(0,\varphi_t)\big] \le 0 \tag{71}$$

where $\mu(t) = M(0,\varphi_t)$.

Proof. Multiplying (38) by the inverse of the metric $g_0(\varphi_t)$, we have $R_t = m + (1-t)\Delta_{\omega_o(\varphi_t)}\varphi_t$. (71) follows from (68), since

$$\frac{d\mu(t)}{dt} = \int (m - R_t)\dot\varphi_t\omega_0(\varphi_t)^m.$$

7.62 Theorem (Bando–Mabuchi [*31]). *Any solution* φ_τ *of* (69), $0 < \tau < 1$, *uniquely extends to a smooth family* $\{\varphi_t\}$ *of solutions of* (69), $0 \le t < \tau + \varepsilon$ *for some* $\varepsilon > 0$. *In particular* (69) *admits at most one solution at* $t = \tau$. *Moreover if* $\mu(t)$ *is bounded from below* $\tau + \varepsilon = 1$.

Proof. According to Aubin (see 7.27), the solution uniquely extends locally. We prove, by contradiction, that it extends until $t = 0$ (see [*31]). Moreover if we suppose that there are two smooth families $\{\varphi_t\}$ and $\{\tilde\varphi_t\}$ of solutions of (69) satisfying $\varphi_\tau = \tilde\varphi_\tau$, the set $\mathfrak{G}$ of the t, for which $\varphi_t = \tilde\varphi_t$, is open. But it is also closed since the families are smooth. Thus $\mathfrak{G} = [0, \tau + \varepsilon[$.

For the last part of the theorem, the hypothesis $\mu(t) \ge K$ implies that $I(0,\varphi_t) - J(0,\varphi_t)$ is bounded from above. The rest of the proof is similar to that of the first part.

7.63 *Sketch of the proof of Theorem* 7.60. Suppose (M, g_0) admits an Einstein–Kähler metric g. Then any $\tilde\omega$ in O, the orbit of ω under $\mathrm{Aut}(M)$, is Einstein–Kähler.

Now any $\tilde\omega \in O$ is of the form $\tilde\omega = \omega_0(\tilde\psi)$ for some C^∞ function $\tilde\psi$, since ω_0 and $\tilde\omega$ belong to $C_1(M)$.

If the first positive eigenvalue $\tilde{\lambda}_1$ of the Laplacian $\tilde{\Delta}$ on $(M, g_0(\tilde{\psi}))$ is equal to 1, there is a necessary condition to extend $\tilde{\psi} = \tilde{\psi}_1$ to a smooth family $\tilde{\psi}_t$ of solutions of (69). Indeed $v = \left(\frac{d}{dt}\tilde{\psi}_t\right)_{t=1}$ must satisfy $(\tilde{\Delta} - 1)v = \tilde{\psi}$. Thus $\int \tilde{\psi}\varphi\tilde{\omega}^m = 0$ for all φ in the first eigenspace.

Nevertheless, using a bifurcation technique, Bando and Mabuchi prove the existence of some $\theta \in O$, such that, for every sufficiently general $\bar{\omega} \in C_1(M)$ with positive definite Ricci tensor, there exists a smooth 1-parameter family of solutions Ψ_t of (69), $0 \leq t \leq 1$, satisfying

$$\omega_0(\psi_0) = \bar{\omega} \qquad \text{and} \qquad \omega_0(\psi_1) = \theta.$$

Now suppose there exists two distinct orbits θ and θ'. Consider the families of solutions ψ_t and ψ'_t of (69), ψ_t as before, and ψ'_t satisfying $\omega_0(\psi'_0) = \bar{\omega}$ and $\omega_0(\psi'_1) \in \theta'$.

According to Theorem 7.62, $\psi_t = \psi'_t$. Thus $\theta = \theta'$.

§18. On Noncompact Kähler Manifolds

7.64 Since problem 7.6 is now well studied when the Kähler manifold is compact, it is natural to seek complete Einstein–Kähler metrics on noncompact manifolds. Let us mention some references where the reader may find results on this topic.

In [*201] *R.* Kobayashi generalized Aubin's theorem 7.9 in the negative case, to the noncompact complex manifolds. The noncompact version of Calabi' *s* conjecture is studied on open manifold by Tian and Yau [*304], [*305] and solved on $\mathbb{C}^n$ by Jeune [*189]. Cheng and Yau [*93] constructed complete Einstein–Kähler metrics with negative Ricci curvature on some noncompact complex manifolds. Compactification of Kähler manifolds is studied by Nadel [*249], and Yeung [*321].

Chapter 8

Monge–Ampère Equations

§1. Monge–Ampère Equations on Bounded Domains of $\mathbb{R}^n$

8.1 In this chapter we study the *Dirichlet Problem* for real Monge–Ampère equations.

Let B be the ball of radius 1 in $\mathbb{R}^n$ and let I be a closed interval of $\mathbb{R}$. $f(x, t)$ will denote a C^∞ function on $\bar{B} \times I$ and g a C^∞ Riemannian metric on $\bar{B}$. Consider u(x) a C^∞ function on $S = \partial B$ with values in I, defined as the restriction to S of a C^∞ function γ on $\bar{B}$.

The problem is to prove the existence of a function $\varphi \in C^\infty(\bar{B})$ satisfying:

$$(1) \qquad \log \det((\nabla_{ij}\varphi + a_{ij})) = f(x, \varphi), \qquad \varphi/S = \mathrm{u},$$

where $a_{ij}(x) = a_{ji}(x)(1 \leq i, j \leq n)$ are $n(n + 1)/2$ C^∞ functions on $\bar{B}$. This problem is not yet solved, except for dimension two under some additional hypotheses. The reason for the difficulty is the following: for the present it is possible to obtain *a priori* estimates up to the second derivatives but not for the third derivatives in the general case. We need such estimates to exhibit a subsequence which converges in $C^2(\bar{B})$ to a C^2 function which will be a solution of (1). Then according to Nirenberg [217] the solution is C^∞. In the special case when $n = 2$, Nirenberg [216] found an estimate for the third derivatives in terms of a bound on the second derivatives. When $n \geq 3$ this estimate depends in addition on the modulus of continuity of the second derivatives.

1.1. The Fundamental Hypothesis

8.2 The hypothesis that B is convex in the metric g is fundamental: there exists $h \in C^\infty(\bar{B})$, $h/S = 0$ satisfying $\nabla_{ij}h(x)\xi^i\xi^j > 0$ for all vectors $\xi \neq 0$ and all points x in $\bar{B}$.

Proposition. *Under the hypothesis of convexity, there exists a lower solution of* (1): $\gamma_1 \in C^\infty(\bar{B})$ *if the right-hand side satisfies* $\lim_{t \to -\infty}[|t|^{-n} \exp f(x, t)] = 0$.

Proof. Consider the functions $\varphi_\alpha = \gamma + \alpha h$ for $\alpha > 0$.
They are equal to u on S and when $\alpha \to \infty$, $\det((\nabla_{ij}\varphi_\alpha + a_{ij}))$ converges to $\alpha^n \det((\nabla_{ij}h))$. Thus, for α large enough $\alpha^{-n}\det((\nabla_{ij}\varphi_\alpha + a_{ij})) \geq \text{Const} > 0$ and $\exp f(x, \varphi_\alpha) \leq \det((\nabla_{ij}\varphi_\alpha + a_{ij}))$.
Hence there exists $\gamma_1 \in C^\infty(\bar{B})$ satisfying

$$\text{(2)} \qquad \log\det((\nabla_{ij}\gamma_1 + a_{ij})) \geq f(x, \gamma_1), \qquad \gamma_1/S = \text{u}. \qquad \blacksquare$$

Remark. An open question: can one remove the hypothesis of convexity for some problems?

8.3 The problem. For simplicity we are going to consider the more usual Dirichlet problem for Monge–Ampère equations.

Let Ω be a bounded strictly convex domain in $\mathbb{R}^n$ ($n \geq 2$) defined by a C^∞ strictly convex function h on $\bar{\Omega}$ satisfying $h/\partial\Omega = 0$. Given u(x) *a* C^∞ function on $\partial\Omega$ which is the restriction to $\partial\Omega$ of a C^∞ function γ on $\bar{\Omega}$, we consider the equation:

$$\text{(3)} \qquad \log\det((\partial_{ij}\varphi)) = f(x, \varphi), \qquad \varphi/\partial\Omega = \text{u},$$

where $f(x, t) \in C^\infty(\bar{\Omega} \times \mathbb{R})$.

This equation was studied by Alexandrov [5], Pogorelov [235], and Cheng and Yau [89]. These authors all use the same method, that of Alexandrov, while the ideas for the estimates are due to Pogorelov. Under some hypotheses they prove the existence of a "generalized" solution of (3) (see 8.13 below) and then they try to establish its regularity. The result obtained is the following:

If $f'_t(x, t) \geq 0$ for $x \in \Omega$ and $t \leq \sup_{\partial\Omega} \text{u}$, then there exists $\varphi \in C^\infty(\Omega)$, a strictly convex solution of (3), which is Lipschitz continuous on $\bar{\Omega}$ (Ω is strictly convex).

Here we will use the continuity method advocated by Nirenberg [222]. The continuity method is simpler and allows us to prove the existence of a solution of (3) which belongs to $C^\infty(\bar{\Omega})$ if there exists a strictly convex upper solution of (3) (Theorem 8.5). Unfortunately the proof is complete only in dimension two. When $n \geq 3$, estimates for the third derivatives is still an open question.
We are going to show, among other things, how to obtain the estimates by using the continuity method. Pogorelov's estimates are different.

Notation. Henceforth we set $M(\varphi) = \det((\partial_{ij}\varphi))$.

1.2. Extra Hypothesis

8.4 For the continuity method we suppose that $f'_t(x, t) \geq 0$ on $\bar{\Omega} \times \mathbb{R}$. We will remove this hypothesis later by using the method of lower and upper solution. But we must suppose (the obviously necessary condition) that there exists a strictly convex upper solution of (3), $\gamma_0 \in C^2(\overline{\Omega})$, which satisfies:

$$(4) \qquad M(\gamma_0) = \det((\partial_{ij}\gamma_0) \leq \exp f(x, \gamma_0), \qquad \gamma_0/\partial\Omega = \mathrm{u}.$$

This hypothesis will be used to estimate the second derivatives on the boundary.
If there exists a convex function $\psi \in C^2(\overline{\Omega})$ satisfying:

$$M(\psi) = 0 \qquad \psi/\partial\Omega = \mathrm{u}$$

(ψ exists, in particular, if u is constant), then ψ satisfies (4) strictly for any function f and we can choose for γ_0 a function of the form $\gamma_0 = \psi + \beta h$ with $\beta > 0$.

1.3. Theorem of Existence

8.5 *The Dirichlet problem* (3) *has a unique strictly convex solution belonging to* $C^\infty(\overline{\Omega})$, *when* $n = 2$, *if there exists a strictly convex upper solution* $\gamma_0 \in C^2(\overline{\Omega})$ *satisfying* (4) *and if* $f'_t(x, t) \geq 0$ *for all* $x \in \Omega$ *and* $t \leq \sup_{\partial\Omega} \mathrm{u}$ (*we assume* Ω *is strictly convex*).

Proof. If $n > 2$ only inequality (23) is missing; otherwise the whole proof works for any dimension. This is why we give the proof for arbitrary n.
Let us consider the equations:

$$(5) \quad \log M(\varphi) = f(x, \varphi) + (1 - \sigma)[\log M(\gamma_1) - f(x, \gamma_1)], \qquad \varphi/\partial\Omega = \mathrm{u},$$

where $\sigma \geq 0$ is a real number and γ_1 is a lower solution, the existence of which was proved in Proposition 8.2. Thus γ_1 satisfies:

$$(6) \qquad \log M(\gamma_1) \geq f(x, \gamma_1), \qquad \gamma_1/\partial\Omega = \mathrm{u}.$$

Let $\mathscr{A}$ be the set of strictly convex functions belonging to $C^{2+\alpha}(\overline{\Omega})$ with $\alpha \in]0, 1[$ which are equal to u on $\partial\Omega$. The operator

$$\Gamma: \mathscr{A} \ni \varphi \to f(x, \varphi) - \log M(\varphi) \in C^\alpha$$

is continuously differentiable,

$$(d\Gamma_\varphi)(\psi) = f'_t(x, \varphi)\psi - g^{ij}_\varphi \partial_{ij}\psi,$$

$\mathscr{A} \ni \varphi \to d\Gamma_\varphi \in \mathscr{L}(C_0^{2+\alpha}(\overline{\Omega}), C^\alpha(\overline{\Omega}))$ is continuous, and $d\Gamma_\varphi$ is invertible because $f'_t \geq 0$ (Theorem 6.14, p. 101 of Gilbarg and Trüdinger [125]). $C_0^r(\overline{\Omega})$

denotes the functions of $C^r(\overline{\Omega})$ which vanish on the boundary $\partial\Omega$, and g_φ^{ij} are the components of the inverse matrix of $((\partial_{ij}\varphi))$.

Thus we can apply the inverse function theorem. If $\tilde{\varphi} \in \mathscr{A}$ satisfies $\log M(\tilde{\varphi}) = f(x, \tilde{\varphi}) + f_0(x)$, there exists $\mathscr{V}$, a neighborhood of f_0 in $C^\alpha(\overline{\Omega})$, such that the equation $\Gamma(\varphi) = -f_1(x)$ has a solution $\psi \in \mathscr{A}$ when $f_1 \in \mathscr{V}$. If $f_1 \in C^\infty(\overline{\Omega})$, $\psi \in C^\infty(\overline{\Omega})$, according to the regularity theorem 3.55. Moreover, ψ is strictly convex, because it is so at a minimum of ψ and remains so by continuity since $M(\psi) > 0$. Lastly, the solution is unique since $f'_t \geq 0$ (Theorem 3.74).

At $\sigma = 0$ Equation (5) has a solution $\varphi_0 = \gamma_1$; therefore there exists $\sigma_0 > 0$ such that (5) has a strictly convex solution $\varphi_\sigma \in C^\infty(\overline{\Omega})$ when $\sigma \in [0, \sigma_0[$. Let σ_0 be the largest real number having this property.

If $\sigma_0 > 1$, Equation (3) has a solution and Theorem 8.5 is proved. If $\sigma_0 \leq 1$, let us suppose for a moment the following, which we will prove shortly: the set of the functions φ_σ for $\sigma \in [0, \sigma_0[$ is bounded in $C^3(\overline{\Omega})$. Then there exist $\varphi_{\sigma_0} \in C^{2+\alpha}(\overline{\Omega})$ for some $\alpha \in]0, 1[$ and a sequence $\sigma_i \to \sigma_0$ such that $\varphi_{\sigma_i} \to \varphi_{\sigma_0}$ in $C^{2+\alpha}(\overline{\Omega})$.
Since φ_{σ_i} satisfies (5), letting $i \to \infty$, we see that φ_{σ_0} satisfies (5) with $\sigma = \sigma_0$. But now we can apply the inverse function theorem at φ_{σ_0} and find a neighborhood $\mathfrak{J}$ of σ_0 such that Equation (5) has a solution when $\sigma \in \mathfrak{J}$. This contradicts the definition of σ_0.

Now we have to establish the estimates, the hardest part of the proof.

§2. The Estimates

2.1. The First Estimates

8.6 C^0 and C^1 estimates. Henceforth, when no confusion is possible, we drop the subscript σ. Then $\varphi = \varphi_\sigma \in C^\infty(\overline{\Omega})$ is the solution of (5) with $\sigma \in [0, 1]$. We have

$$\log M(\varphi) - f(x, \varphi) \leq \log M(\gamma_1) - f(x, \gamma_1),$$

since the second term is positive. Thus by the maximum principle (Theorem 3.74), $\varphi - \gamma_1 \geq 0$ on Ω because $f'_t \geq 0$ and $\varphi - \gamma_1 = 0$ on $\partial\Omega$. This implies that on $\partial\Omega$: $\partial_\nu\varphi \leq \partial_\nu\gamma_1$, where ∂_ν denotes the exterior normal derivative.

We have thus proved the C^0 estimate:

$$\inf_{\Omega} \gamma_1 \leq \varphi \leq \sup_{\partial\Omega} u. \tag{7}$$

Since φ is convex, the gradient of φ attaines its maximum on the boundary. Let P be a point of $\partial\Omega$. The tangential derivatives of φ at P are bounded since $u \in C^2(\partial\Omega)$. On the other hand we previously saw that $\partial_\nu\varphi \leq \partial_\nu\gamma_1$. It remains to establish an inequality in the other direction. The normal at P intersects $\partial\Omega$ at one other point, which we call Q.

On the straight line $\mathfrak{D}$, through P and Q, let $\mathfrak{w}$ be the linear function equal to $\mathfrak{u}$ at P and Q. Let $\mu(P)$ be the gradient of $\mathfrak{w}$. Since φ is convex on $\mathfrak{D}$, $\varphi \le \mathfrak{w}$ on $\Omega \cap \mathfrak{D}$ and $(\partial_\nu \varphi)_P \ge \mu(P)$. $\mu(P)$ is a continuous function on $\partial\Omega$; let μ be its minimum on the compact set $\partial\Omega$. Hence $\partial_\nu \varphi \ge \mu$.

2.2. C^2-Estimate

8.7 C^2-estimate on the boundary. Let $P \in \partial\Omega$ and Y be a vector field on $\mathbb{R}^n$ which is tangent to $\partial\Omega$. Suppose $\| Y(P)\| = 1$ and choose on $\mathbb{R}^n$ orthonormal coördinates such that $\nu = (1, 0, 0, \dots, 0)$ is the unit exterior normal at P and $Y(P) = (0, 1, 0, \dots, 0)$. We will estimate $\partial^2_{x_2x_2}\varphi$, $\partial^2_{x_1x_2}\varphi$ and $\partial^2_{x_1x_1}\varphi$ at P.

a) Let R_2 be the radius of curvature of $\partial\Omega$ at P in the direction Y. Since $\partial\Omega$ is strictly convex, $R_2 \ge R_0 > 0$ (R_0 a real number independent of P and Y). At P:

$$\partial^2_{x_2x_2}\varphi = \frac{1}{R_2^2}\partial^2_{YY}\varphi + \frac{1}{R_2}\partial_\nu\varphi = \frac{1}{R_2^2}\partial^2_{YY}\mathfrak{u} + \frac{1}{R_2}\partial_\nu\varphi. \tag{8}$$

Therefore $\partial^2_{x_2x_2}\varphi$ is estimated

b) Let us consider a family $\mathfrak{g}$ of vector fields on $\mathbb{R}^n$ tangent to $\partial\Omega$ and bounded in $C^2(\overline{\Omega})$; thus the components $X^i(x)$ of the vector field $X \in \mathfrak{g}$ are uniformly bounded in C^2 on $\overline{\Omega}$.
Set $\psi = \varphi - \gamma$ and $L = X^i(x)\,\partial_{x_i}$ for $X \in \mathfrak{g}$.
Differentiating the equation

$$\log M(\varphi) = F(x, \varphi) \tag{9}$$

yields

$$LF = g^{ij}X^k\,\partial_{ijk}\varphi, \tag{10}$$

where $F(x, \varphi)$ is the right-hand side of (5) (recall that $((g^{ij}))$ is the inverse matrix of $((g_{ij}))$ with $g_{ij} = \partial_{ij}\varphi$. We will compute $B = g^{ij}\,\partial_{ij}(L\psi + \alpha h + \beta\psi)$, where α and β are two real numbers which we will choose later.

$$B = g^{ij}X^k\,\partial_{ijk}\psi + 2g^{ij}\,\partial_i X^k\,\partial_{jk}\psi + g^{ij}\,\partial_{ij}X^k\,\partial_k\psi$$
$$+ \alpha g^{ij}\,\partial_{ij}h + \beta(g^{ij}\,\partial_{ij}\varphi - g^{ij}\,\partial_{ij}\gamma).$$

Since $g^{ij}\,\partial_{ik}\varphi = \delta^j_k$, using (10) we obtain:

$$B = LF + \beta n + 2\,\partial_i X^i + g^{ij}(\mathfrak{m}_{ij} + \alpha\,\partial_{ij}h),$$

with

$$\mathfrak{m}_{ij} = -\beta\,\partial_{ij}\gamma - X^k\,\partial_{ijk}\gamma - 2\,\partial_i X^k\,\partial_{jk}\gamma + \partial_{ij}X^k\,\partial_k\psi.$$

At first we pick $\beta = \beta_0 \geq -(1/n) \inf(LF + 2\,\partial_i X^i)$, where the inf is taken for all $x \in \Omega$ and all functions φ_σ. Note that this inf is finite since the functions φ_σ are already estimated in C^1.

Then we choose $\alpha = \alpha_0$ large enough so that

$$(\mathfrak{m}_{ij} + \alpha_0\,\partial_{ij}h)g^{ij} \geq 0.$$

The real numbers α_0 and β_0 can be chosen independent of $X \in \mathfrak{g}$. This is possible by our hypothesis. Thus

$$g^{ij}\,\partial_{ij}(L\psi + \alpha_0 h + \beta_0 \psi) \geq 0.$$

Likewise, let $\beta = \beta_1 \leq -(1/n)\sup(LF + 2\,\partial_i X^i)$, where the sup is taken for all $x \in \Omega$ and all φ_σ, and let α_1 be such that $g^{ij}(\mathfrak{m}_{ij} + \alpha_1\,\partial_{ij}h) \leq 0$. β_1 and α_1 are chosen independent of $X \in \mathfrak{g}$. Thus

$$g^{ij}\,\partial_{ij}(L\psi + \alpha_1 h + \beta_1 \psi) \leq 0.$$

Since $L\psi$, h, and ψ vanish on $\partial\Omega$, by the maximum principle:

$$-(\alpha_1 h + \beta_1 \psi) \leq L\psi \leq -(\alpha_0 h + \beta_0 \psi)$$

and

$$-\partial_\nu(\alpha_0 h + \beta_0 \psi) \leq \partial_\nu L\psi \leq -\partial_\nu(\alpha_1 h + \beta_1 \psi).$$

These inequalities yield the estimate of $\partial_{\nu X}\psi$.

In order for the family $\mathfrak{g}$ to be large enough so that, for all pairs (P, Y) with $P \in \partial\Omega$, and $Y \in T_P(\partial\Omega)$ a unit vector, there exists $X \in \mathfrak{g}$ such that $X(P) = Y$, we define $\mathfrak{g}$ as follows.

Let B be the unit ball of $\mathbb{R}^n$ and Φ be a C^3-diffeomorphism from $\bar{B}$ to $\bar{\Omega}$. Then $\mathfrak{v}_{ij} = x_i\,\partial_j - x_j\,\partial_i$ are vector fields tangent to ∂B. Consider the family $\mathfrak{F} = \{\mathfrak{B} = \sum a_{ij}\mathfrak{v}_{ij}$, where a_{ij} are real numbers with $|a_{ij}| \leq \|(\Phi^{-1})_*\|\}$. Then $\mathfrak{g} = \Phi_*\mathfrak{F}$ has the desired property.

c) To estimate $\partial^2_{\nu\nu}\varphi$, we need to know a strictly convex upper solution γ_0 satisfying (4).

Since φ_σ satisfies (5), $\log M(\varphi_\sigma) \geq f(x, \varphi_\sigma)$. According to the maximum principle, since $f'_t \geq 0$, then as in 8.6, $\varphi_\sigma \leq \gamma_0$ and $\partial_\nu\varphi_\sigma \geq \partial_\nu\gamma_0$ on $\partial\Omega$.

Since γ_0 is strictly convex, there exists an $\varepsilon > 0$ such that for all $x \in \bar{\Omega}$ and all $i = 1, 2, \ldots, n$, $\partial^2_{x_i x_i}\gamma_0 \geq \varepsilon$. From (8) it follows for $i \geq 2$:

$$\partial^2_{x_i x_i}\varphi = \partial^2_{x_i x_i}\gamma_0 + \frac{1}{R_i}(\partial_\nu\varphi - \partial_\nu\gamma_0) \geq \varepsilon.$$

Suppose we choose the orthonormal frame in $T_P(\partial\Omega)$ such that for $j \geq i \geq 2$, $\partial^2_{x_i x_j}\varphi = 0$ if $i \neq j$. Then (9) implies

$$\partial^2_{x_1x_1}\varphi \prod_{p=2}^{n} \partial^2_{x_px_p}\varphi = \exp F(x, \varphi) + \sum_{p=2}^{n} (-1)^p \,\partial^2_{x_1x_p}\varphi\mu_{1p}, \tag{11}$$

where μ_{1p}, the minor of $\partial^2_{x_1x_p}\varphi$ in the determinant $M(\varphi)$, does not contain $\partial^2_{x_1x_1}\varphi$. Therefore μ_{1p} can be estimated by the inequalities in the preceding paragraphs.
Thus by (11) $\partial^2_{x_1x_1}\varphi$ is estimated on $\partial\Omega$ since $\partial^2_{x_px_p}\varphi \geq \varepsilon$.

8.8 C^2-estimate on $\bar{\Omega}$. Since φ is convex, $\partial_{yy}\varphi \geq 0$ for all directions y. Thus an upperbound for $\sum_{k=1}^{n} \partial_{kk}\varphi$ is enough to yield the C^2-estimate. Computing the Laplacian of (9) leads to:

$$\sum_{k=1}^{n} g^{ij}\,\partial_{ijkk}\varphi = \sum_{k=1}^{n} g^{im}g^{jl}\,\partial_{ijk}\varphi\,\partial_{mlk}\varphi + \sum_{k=1}^{n} \partial_{kk}F(x, \varphi). \tag{12}$$

Let $\beta_2 \leq 1/n \inf \sum_{k=1}^{n} \partial_{kk}F(x, \varphi_\sigma)$, where the inf is taken over Ω and for all functions φ_σ. It is finite since all of the terms have been estimated except the term which involves $\Delta\varphi$, and that term is positive: $F'_t(x, \varphi)\sum_{k=1}^{n} \partial_{kk}\varphi \geq 0$. β_2 can be chosen negative and independent of σ. Hence

$$g^{ij}\,\partial_{ij}\left(\sum_{k=1}^{n} \partial_{kk}\varphi - \beta_2\varphi\right) \geq -n\beta_2 + \sum_{k=1}^{n} \partial_{kk}F(x, \varphi) \geq 0.$$

By the maximum principle $\sum_{k=1}^{n} \partial_{kk}\varphi - \beta_2\varphi$ attains its maximum on $\partial\Omega$. But by (8.8) $\Delta\varphi$ is bounded on $\partial\Omega$. Hence the C^2-estimate follows:

$$0 \leq \partial_{yy}\varphi \leq 2|\beta_2| \sup_{\Omega} |\varphi| + \sup_{\partial\Omega} \sum_{k=1}^{n} \partial_{kk}\varphi.$$

Consequently the metrics $(g_\sigma)_{ij} = \partial_{ij}\varphi_\sigma$ are equivalent for $\sigma \in [0, 1]$. Indeed, according to the preceding inequality $(g_\sigma)_{yy} \leq C$, where C is a constant, and (5) implies

$$C^{n-1}(g_\sigma)_{xx} \geq B > 0,$$

where B is a constant. Thus for all vectors ξ and $\sigma \in [0, 1]$

$$C^{1-n}B\|\xi\|^2 \leq (g_\sigma)_{ij}\xi^i\xi^j \leq C\|\xi\|^2. \tag{13}$$

2.3. C^3-Estimate

8.9 Proposition. *$R = \frac{1}{4}g^{\alpha\beta}g^{ik}g^{j\ell}\,\partial_{\alpha ij}\varphi_{\beta k\ell}\varphi$ satisfies*

$$g^{ij}\nabla_{ij}R \geq \frac{2}{n-1}R^2 + CR^{1/2}, \tag{14}$$

where ∇ is the covariant derivative with respect to g and C is a constant which depends on the function $F(x, t)$ and on an upper bound of $\|\varphi\|_{C^2}$ (as do the constants introduced in the proof).

Proof. Calabi [75] p. 113 establishes the following inequality in the special case $F(x, \varphi) = 0$:

$$g^{ij}\nabla_{ij}R \geq \frac{2(n+1)}{n(n-1)}R^2.$$

He introduced $A_{ijk} = \Gamma_{ijk} = \frac{1}{2}\,\partial_{ijk}\varphi$. A_{ijk} is symmetric with respect to its subscripts and we can verify the following equalities:

$$g^{ij}A_{ijk} = \tfrac{1}{2}\,\partial_k F \quad \text{and} \quad \nabla_\ell A_{ijk} = \nabla_i A_{\ell jk} \tag{15}$$

where F is written in place of $F(x, \varphi)$ for simplicity.

A computation similar to that of Calabi (see Pogorelov [235] p. 39) leads to

$$g^{ij}\nabla_{ij}R \geq A^{ijk}\nabla_{jk}\nabla_i F + 2\frac{n+1}{n(n-1)}R^2 + C_1R^{3/2} + C_2R + 2\nabla^\ell A^{ijk}\nabla_\ell A_{ijk}.$$

C_1 and C_2 are two constants and the indices are raised using g^{ij}, for instance $\Gamma^j_{ik} = g^{j\ell}\Gamma_{i\ell k}$ are Christoffel's symbols of the Riemannian connection. Moreover:

$$\begin{aligned} A^{ijk}\nabla_{ijk}F = {} & A^{ijk}\,\partial_i(\partial_{jk}F - \Gamma^\ell_{jk}\,\partial_\ell F) \\ & - A^{ijk}\Gamma^\ell_{ij}(\partial_{\ell k}F - \Gamma^m_{\ell k}\,\partial_m F) \\ & - A^{ijk}\Gamma^\ell_{ik}[\partial_{j\ell}F - \Gamma^m_{j\ell}\,\partial_m F]. \end{aligned}$$

Thus

$$A^{ijk}(\nabla_{ijk}F + \nabla^\ell F\nabla_i A_{jk\ell}) \leq \text{Const} \times (1 + R)\sqrt{R}.$$

According to (15) for some constant C_3 we get:

$$g^{ij}\nabla_{ij}R \geq 2\frac{n+1}{n(n-1)}R^2 + C_3(1+R)\sqrt{R}$$
$$+ 2(\nabla^\ell A^{ijk} - \tfrac{1}{4}A^{ijk}\nabla^\ell F)(\nabla_\ell A_{ijk} - \tfrac{1}{4}A_{ijk}\nabla_\ell F),$$

and inequality (14) follows. ∎

8.10 Interior C^3-estimate. In this paragraph we assume that the derivatives of φ up to order two are estimated. A term which involves only derivatives of φ of order at most two is called a bounded term. First of all note that $\partial_i[g^{ij}M(\varphi)] = 0$. Indeed,

$$\partial_i[g^{ij}M(\varphi)] = M(\varphi)[g^{ij}g^{k\ell}\,\partial_{k\ell i}\varphi - g^{ik}g^{j\ell}\,\partial_{k\ell i}\varphi].$$

Interchanging ℓ and i in the last term, we obtain the result. Multiply (12) by $h^2M(\varphi)$ and integrate over Ω. Since $\partial_i[g^{ij}M(\varphi)] = 0$, integrating by parts twice leads to:

$$\int_\Omega h^2 \sum_{k=1}^{n} g^{im}g^{j\ell}\,\partial_{ijk}\varphi\,\partial_{m\ell k}\varphi M(\varphi)\,dx \leq \text{Const}.$$

Set $R = \frac{1}{4}g^{\alpha\beta}g^{ik}g^{j\ell}\,\partial_{\alpha ij}\varphi\,\partial_{\beta k\ell}\varphi$. Since the metrics g_σ are equivalent, the preceding inequality implies

(16) $$\int_\Omega h^2R\,dx \leq \text{Const}.$$

It is possible to show that $\int_\Omega R\,dx \leq \text{Const}$, but that yields nothing more here. Let us prove by induction that for all integer p:

(17) $$\int_\Omega h^{2p}R^p\,dx \leq \text{Const}.$$

Assume (17) holds for a given p. Multiplying (14) by $h^{2p+2}R^{p-1}\sqrt{M(\varphi)}$ and integrating by parts over Ω lead to:

$$(1-p)\int_\Omega h^{2p+2}g^{ij}R^{p-2}\,\partial_iR\,\partial_jR\sqrt{M(\varphi)}\,dx$$
$$-\int_\Omega g^{ij}\,\partial_ih^{2p+2}R^{p-1}\,\partial_jR\sqrt{M(\varphi)}\,dx$$
$$\geq \frac{2}{n-1}\int_\Omega h^{2p+2}R^{p+1}\sqrt{M(\varphi)}\,dx + C\int_\Omega h^{2p+2}R^{p-1/2}\sqrt{M(\varphi)}\,dx.$$

Integrating the second integral by parts again gives, by (15),

$$\frac{2}{n-1}\int_\Omega h^{2p+2}R^{p+1}\sqrt{M(\varphi)}\,dx \le \frac{1}{p}\int_\Omega R^p g^{ij}\nabla_{ij}h^{2p+2}\sqrt{M(\varphi)}\,dx + \text{Const}$$

$$\le \text{Const} \times \left[1 + \frac{1}{p}\int_\Omega h^{2p}R^p\,dx\right].$$

Thus, since (17) holds for $p = 1$ (inequality 16), (17) holds for all p. Accordingly, for any compact set $K \subset \Omega$ and any integer p, $\|\varphi_\sigma\|_{H_3^p(K)} \le$ Const for all $\sigma \in [0, 1]$.

By the regularity theorem (3.56), for all $r > 0$:

$$\|\varphi_\sigma\|_{C^r(K)} \le \text{Const}.$$

In particular the third derivatives of φ_σ are uniformly bounded on K.

8.11 C^3-estimate on the boundary.

a) Recall (8.7), where we defined $L = X^k\,\partial_k$ with $X \in \mathfrak{g}$. Differentiating (9) twice with respect to L gives: $L^2F(x,\varphi) = -g^{i\ell}g^{jk}L(\partial_{\ell k}\varphi)L(\partial_{ij}\varphi) + g^{ij}L^2(\partial_{ij}\varphi)$. Next we compute $g^{ij}\,\partial_{ij}L^2\varphi$. Since

$$\text{(18)}\qquad L^2\varphi = L(X^k\,\partial_k\varphi) = X^lX^k\,\partial_{lk}\varphi + X^l\,\partial_l X^k\,\partial_k\varphi,$$

then

$$g^{ij}(\partial_{ij}L^2\varphi) = g^{ij}L^2(\partial_{ij}\varphi) + 4g^{ij}(\partial_i X^\ell)X^k\,\partial_{j\ell k}\varphi + g^{ij}\,\partial_{ij}(X^\ell X^k)\,\partial_{\ell k}\varphi$$
$$+ g^{ij}\,\partial_{ij}(X^\ell\,\partial_\ell X^k)\,\partial_k\varphi + 2\,\partial_i(X^k\,\partial_k X^i).$$

Thus

$$\text{(19)}\qquad g^{ij}\,\partial_{ij}L^2\varphi = (X^k\,\partial_{j\ell k}\varphi g^{ij} + 2\,\partial_\ell X^i)(X^\lambda\,\partial_{\lambda\beta i}\varphi g^{\beta\ell} + 2\,\partial_i X^\ell)$$
$$+ \text{bounded terms}.$$

Consequently, there exists a constant α such that

$$g^{ij}\,\partial_{ij}(L^2\psi + \alpha h) \ge 0.$$

Since $L^2\psi$ and h vanish on the boundary $\partial\Omega$, by the maximum principle $L^2\psi + \alpha h \le 0$ on Ω and on the boundary

$$\text{(20)}\qquad \partial_\nu L^2\psi \ge -\alpha\,\partial_\nu h.$$

b) If we get an inequality in the opposite direction, the third derivatives will be estimated on the boundary. Indeed, then we have on the boundary:

$$(21) \qquad |\partial_\nu L^2 \varphi| < \text{Const.}$$

Consider $P \in \partial\Omega$ and use the coordinates in (8.7). Differentiating (18) with respect to L yields $L^3\varphi = X^i X^j X^k\, \partial_{ijk}\varphi$ + bounded terms.

On $\partial\Omega$, $L^3\varphi = L^3(\psi + \gamma) = L^3\gamma$, so $(\partial^3_{222}\varphi)_P$ is estimated. Likewise the third derivatives with respect to coordinates x_i with $i \geq 2$ are estimated. By (21), $(\partial_{1ij}\varphi)_P$ is estimated for i and $j \geq 2$.

To get an estimate for $(\partial_{11j}\varphi)_P, j > 1$, we differentiate (9) with respect to x_j. This yields:

$$g^{11}\, \partial_{11j}\varphi = \text{bounded terms.}$$

Because the metrics g_{ij} are equivalent, there exists a constant $\eta > 0$ such that $g^{11} \geq \eta > 0$ and the estimate of $(\partial_{11j}\varphi)_P$ follows.

Finally, differentiating (9) with respect to x_1 yields $g^{11}\, \partial^3_{111}\varphi$ = bounded terms. Hence all the third derivatives are estimated on the boundary.

c) It remains to find an upper bound for $\partial_\nu L^2\psi$ on the boundary. For the present such a bound is established only in the case $n = 2$.

From now on, $n = 2$; consequently the dimension of $\partial\Omega$ is equal to one. Consider a vector field X tangent to $\partial\Omega$ and of norm one on $\partial\Omega$. Since the second derivatives of φ are estimated, there exists M such that $|L^2\psi| \leq M$ on $\partial\Omega$. Recall that $\psi = \varphi - \gamma$.

Let p be an integer that we will choose later and set

$$\mathfrak{E} = (1 + M + L^2\psi)^{-p}.$$

Let $K \subset \Omega$ be a compact set such that $\|X\| > 1/2$ on $\Omega - K$. Compute $g^{\alpha\beta}\, \partial_{\alpha\beta}\mathfrak{E}$ on $\Omega - K$ to obtain

$$\begin{aligned} g^{\alpha\beta}\, \partial_{\alpha\beta}\mathfrak{E} = &-p(1 + M + L^2\psi)^{-p-1} g^{\alpha\beta}\, \partial_{\alpha\beta} L^2\psi \\ &+ p(p+1)(1 + M + L^2\psi)^{-p-2} g^{\alpha\beta}\, \partial_\alpha L^2\psi\, \partial_\beta L^2\psi. \end{aligned}$$

Using (19) gives

$$(21) \qquad \begin{aligned} g^{\alpha\beta}\, \partial_{\alpha\beta}\mathfrak{E} = p(1 + M + L^2\psi)^{-p-2}[&(p+1) g^{\alpha\beta}\, \partial_\alpha L^2\psi\, \partial_\beta L^2\psi \\ &- (1 + M + L^2\psi)(X^k\, \partial_{jlk}\varphi g^{ij} + 2\, \partial_l X^i) \\ &\times (X^\lambda\, \partial_{\lambda\beta i}\varphi g^{\beta l} + 2\, \partial_i X^l)] + \text{bounded terms.} \end{aligned}$$

At $Q \in \Omega - K$ suppose that X is in the direction of x_2. According to (10), there exists a constant k_0 such that

$$|\partial_{211}\varphi| \leq k_0(1 + |\partial_{221}\varphi| + |\partial_{222}\varphi|).$$

Thus there is a constant k_1 such that

$$(X^k\,\partial_{jlk}\varphi g^{ij} + 2\,\partial_l X^i)(X^\lambda\,\partial_{\lambda\beta i}\varphi g^{\beta l} + 2\,\partial_i X^l) \le k_1(1 + |\partial_{221}\varphi|^2 + |\partial_{222}\varphi|^2).$$

Moreover,

$$g^{\alpha\beta}\,\partial_\alpha L^2\psi\,\partial_\beta L^2\psi \ge k_2(|\partial_{221}\varphi|^2 + |\partial_{222}\varphi|^2 - 1),$$

where k_2 is a constant which depends on K and on the preceding estimates, much as did k_0 and k_1.

Pick p such that $(p+1)k_2 \ge (1+2M)k_1$. Since the third derivatives are bounded on K (by 8.9), (21) shows that $\mathfrak{E}$ satisfies the following inequality on Ω:

$$g^{\alpha\beta}\,\partial_{\alpha\beta}\mathfrak{E} \ge \text{Const}.$$

Hence there exists a constant $\mathfrak{k}$ such that

$$g^{\alpha\beta}\,\partial_{\alpha\beta}(\mathfrak{E} + \mathfrak{k}h) \ge 0 \tag{22}$$

since the metrics g_σ are equivalent. Because $\mathfrak{E}$ and $\mathfrak{h}$ are constant on $\partial\Omega$, according to the maximum principle.

$$\mathfrak{E} + \mathfrak{k}h \le (1+M)^{-p},$$

while on $\partial\Omega$: $\partial_\nu\mathfrak{E} + \mathfrak{k}\,\partial_\nu h \ge 0$. This gives

$$-p(1 + M + L^2\psi)^{-p-1}\,\partial_\nu L^2\psi \ge -\mathfrak{k}\,\partial_\nu h.$$

Hence we obtain the desired inequality

$$\partial_\nu L^2\psi \le \frac{\mathfrak{k}}{p}(1+2M)^{p+1}\sup_{\partial\Omega}\partial_\nu h. \tag{23}$$

C^3-estimate on $\bar{\Omega}$.

By inequality (14) there exist two positive constants a and α such that:

$$g^{ij}\nabla_{ij}R \ge \frac{2}{n-1}[(R-a)^2 - \alpha^2].$$

Two cases can occur: R attains its maximum on $\partial\Omega$ or in Ω. In the first case we saw that R is bounded, while if R attains its maximum at $P \in \Omega$, according to the preceding inequality:

$$R \le \sup_\Omega R \le a + \alpha$$

since $g^{ij}\nabla_{ij}R \le 0$ at P.

8.12 Using the method of lower and upper solutions (as in 7.25), and using Aubin [23] p. 374 for the estimates, we can prove the following.

Theorem. *Let $\Omega \subset \mathbb{R}^n$ be strictly convex. If there exist a strictly convex upper solution $\gamma_0 \in C^2(\overline{\Omega})$ satisfying (4) and a strictly convex lower solution $\gamma_1 \in C^\infty(\overline{\Omega})$ satisfying (6) with $\gamma_1 \leq \gamma_0$, the Dirichlet problem (3) has a strictly convex solution φ belonging to $C^\infty(\overline{\Omega})$ when $n = 2$, and φ satisfies $\gamma_1 \leq \varphi \leq \gamma_0$.*

Corollary. *If there exists γ_0 as above and if $\lim_{t\to-\infty}[|t|^{-n} \exp f(x, t)] = 0$, then the Dirichlet problem (3) has a strictly convex solution $\varphi \in C^\infty(\overline{\Omega})$ when $n = 2$.*

Proof. By Proposition (8.2), for α large enough $\gamma_1 = \gamma + \alpha h$ is a strictly convex lower solution satisfying (6). Thus we can choose α so that $\gamma_1 \leq \gamma_0$. Moreover $\gamma_1 \in C^\infty(\overline{\Omega})$.

The preceding theorem now implies the stated result.

§3. The Radon Measure $\mathscr{M}(\varphi)$

8.13 Definition. For a C^2 convex function on Ω, we set

$$\mathscr{M}(\varphi) = M(\varphi)\, dx^1 \wedge dx^2 \wedge \cdots \wedge dx^n$$

and we define the Radon measure:

$$C_0(\Omega) \ni \psi \to \int_\Omega \psi \mathscr{M}(\varphi),$$

where $C_0(\Omega)$ denote the set of continuous functions with compact support.

This definition extends to convex functions in general, according to Alexandrov. Here we will follow the analytic approach of Rauch and Taylor [242].

First let us remark that for a C^2 convex function

$$\mathscr{M}(\varphi) = d(\partial_1 \varphi) \wedge d(\partial_2 \varphi) \wedge \cdots \wedge d(\partial_n \varphi).$$

We are going to show by induction that $d(\partial_{i_1}\varphi) \wedge d(\partial_{i_2}\varphi) \wedge \cdots \wedge d(\partial_{i_m}\varphi)$ defines a current when φ is a convex function and $1 \leq i_1 < i_2 < \cdots < i_m \leq n$. Let us begin with $m = 1$. Let ω be a continuous $(n-1)$-form with compact support $K \subset \Omega$, and $\omega = \sum_{l=1}^n f_l(x)\, dx^1 \wedge \cdots \wedge \widehat{dx^l} \wedge \cdots \wedge dx^n$. For such ω we set $\|\omega\|_0 = \sup_{1\leq l\leq n} \sup_{x\in K} |f_l(x)|$. When φ is a convex function we will define

$$\int_\Omega d(\partial_i \varphi) \wedge \omega.$$

Choose a C^∞ positive function γ with compact support $\tilde{K} \subset \Omega$ which is equal to one on K.

Now consider a sequence of convex C^∞ functions φ_j which converges to φ uniformly on $\tilde{K}$. We have

$$\left|\int_\Omega f_1\, d(\partial_i \varphi_j) \wedge dx^2 \wedge \cdots \wedge dx^n\right| \le \sup|f_1| \int_K |\partial_{1i}\varphi_j|\, dx^1 \wedge dx^2 \wedge \cdots \wedge dx^n.$$

But since φ_j is convex,

$$(24) \qquad 2|\partial_{1i}\varphi_j| \le \partial_{11}\varphi_j + \partial_{ii}\varphi_j$$

and

$$2\int_K |\partial_{1i}\varphi_j|\, dx \le \int_\Omega \gamma(\partial_{11}\varphi_j + \partial_{ii}\varphi_j)\, dx = \int_\Omega \varphi_j(\partial_{11}\gamma + \partial_{ii}\gamma)\, dx.$$

Thus there exists a constant C_1 such that

$$(25) \qquad \left|\int_\Omega d(\partial_i \varphi_j) \wedge \omega\right| \le C_1 \|\omega\|_0 \sup|\varphi_j|.$$

The constants C_α ($\alpha \in \mathbb{N}$) introduced depend on K and γ only. In particular, they are independent of φ_j and ω.
Let ω_k be a sequence of C^∞ $(n-1)$-forms which converges uniformly to ω. We choose ω_k such that supp $\omega_k \subset K$.
We define $\int_\Omega d(\partial_i \varphi) \wedge \omega_k$ as a distribution and let $j \to \infty$,

$$\int_\Omega d(\partial_i \varphi_j) \wedge \omega_k \to \int_\Omega d(\partial_i \varphi) \wedge \omega_k.$$

Then by (25)

$$\left|\int_\Omega d(\partial_i \varphi) \wedge \omega_k\right| \le C_1 \|\omega_k\|_0 \sup|\varphi|,$$

so that $\int_\Omega d(\partial_i \varphi) \wedge \omega_k$ is a Cauchy sequence which converges to a limit independent of the sequence ω_k. We call this limit $\int_\Omega d(\partial_i \varphi) \wedge \omega$. It satisfies:

$$\left|\int_\Omega d(\partial_i \varphi) \wedge \omega\right| \le C_1 \|\omega\|_0 \sup|\varphi|.$$

Now suppose we have defined $\int_\Omega d(\partial_{i_1}\varphi) \wedge d(\partial_{i_2}\varphi) \wedge \cdots \wedge d(\partial_{i_m}\varphi) \wedge \check{\omega}$ for all families of integers $1 \le i_1 < i_2 < \cdots < i_m \le n$, $(m < n)$ and all continuous

$(n - m)$-forms $\tilde{\omega}$ with support contained in K. Suppose we have proved that there exists a constant C_2 such that:

$$(26) \quad \left| \int_\Omega d(\partial_{i_1}\varphi) \wedge d(\partial_{i_2}\varphi) \wedge \cdots \wedge d(\partial_{i_m}\varphi) \wedge \tilde{\omega} \right| \le C_2 \|\tilde{\omega}\|_0 [\sup|\varphi|]^m.$$

As before, $\|\tilde{\omega}\|_0$ is equal to the sup of the components of $\tilde{\omega}$.

Let $\tilde{\omega}$ be a continuous $(n - m - 1)$-form with support contained in K, and let $\tilde{\omega}_k$ be a sequence of C^∞ $(n - m - 1)$-forms which converge uniformly to $\tilde{\omega}$: $\|\tilde{\omega}_k - \tilde{\omega}\|_0 \to 0$ when $k \to \infty$. We pick $\tilde{\omega}_k$ such that $\operatorname{supp} \tilde{\omega}_k \subset K$. Consider $(m + 1)$ integers $1 \le i_1 < i_2 < \cdots < i_{m+1} \le n$. We want to define $\Gamma(\tilde{\omega}) = \int_\Omega d(\partial_{i_1}\varphi) \wedge \cdots \wedge d(\partial_{i_{m+1}}\varphi) \wedge \tilde{\omega}$. First we define $\int_\Omega d(\partial_{i_1}\varphi) \wedge \cdots \wedge d(\partial_{i_{m+1}}\varphi) \wedge \tilde{\omega}_k = \Gamma(\tilde{\omega}_k)$ by integrating by parts. For instance, if $f \in \mathscr{D}(\Omega)$, then by definition:

$$\begin{aligned}(m + 1) \int_\Omega & f(x)\, d(\partial_{i_1}\varphi) \wedge \cdots \wedge d(\partial_{i_{m+1}}\varphi) \wedge dx^{m+2} \wedge dx^{m+3} \wedge \cdots \wedge dx^n \\ &= \sum_{q=1}^{m+1} \int_\Omega \varphi\, d(\partial_{i_1}\varphi) \wedge \cdots \wedge d(\partial_{i_{q-1}}\varphi) \wedge d(\partial_{i_q} f) \wedge d(\partial_{i_{q+1}}\varphi) \wedge \cdots \wedge \\ &\quad \times d(\partial_{i_{m+1}}\varphi) \wedge dx^{m+2} \wedge \cdots \wedge dx^n.\end{aligned}$$

This equality holds if $\varphi \in C^\infty$. When φ is only convex the integrals of the second term make sense by our assumption, with the continuous $(n - m)$-forms

$$\tilde{\omega}_q = (-1)^{m+1-q} \varphi\, d(\partial_{i_q} f) \wedge dx^{m+2} \wedge \cdots \wedge dx^n.$$

Then letting $j \to \infty$,

$$(27) \quad \Gamma_j(\tilde{\omega}_k) = \int_\Omega d(\partial_{i_1}\varphi_j) \wedge d(\partial_{i_2}\varphi_j) \wedge \cdots \wedge d(\partial_{i_{m+1}}\varphi_j) \wedge \tilde{\omega}_k \to \Gamma(\tilde{\omega}_k)$$

On the other hand,

$$\begin{aligned}\left| \int_\Omega f\, d(\partial_{i_1}\varphi_j) \wedge \cdots \wedge d(\partial_{i_{m+1}}\varphi_j) \wedge dx^{m+2} \wedge \cdots \wedge dx^n \right| \\ \le \sup|f| \int_K |D(x)|\, dx,\end{aligned}$$

where $D(x) = \det((\partial_{l i_k} \varphi_j))$ is the $(m + 1)$-determinant with $1 \le l, k \le m + 1$. But because φ_j is convex, $2|D(x)| \le D_1(x) + D_2(x)$ where $D_1(x) = \det((\partial_{lk} \varphi_j))$ and $D_2 = \det((\partial_{i_k i_l} \varphi_j))$.

Indeed this inequality is the same as (24) on $\Lambda^{m+1}(\mathbb{R}^n)$. The extension to Λ^{m+1} of the quadratic form defined by $((\partial_{ih}\varphi_j))$ is a positive quadratic form whose components are the $(m+1)$-determinants extracted from $((\partial_{ih}\varphi_j))$ with $1 \le i, h \le n$.
Thus there exists a constant C_3 such that

$$2\int_K |D(x)|\, dx \le \int_\Omega \gamma[D_1(x) + D_2(x)]\, dx \le C_3[\sup|\varphi_j|]^{m+1}.$$

For instance, according to (26):

$$\int_\Omega \gamma D_1(x)\, dx$$

$$= \int_\Omega \gamma\, d(\partial_1\varphi_j) \wedge d(\partial_2\varphi_j) \wedge \cdots \wedge d(\partial_{m+1}\varphi_j) \wedge dx^{m+2} \wedge \cdots \wedge dx^n$$

$$= \int_\Omega \varphi_j\, d(\partial_1\varphi_j) \wedge d(\partial_2\varphi_j) \wedge \cdots \wedge d(\partial_m\varphi_j)$$

$$\wedge\, d(\partial_{m+1}\gamma) \wedge dx^{m+2} \wedge \cdots \wedge dx^n \le C_3[\sup|\varphi_j|]^{m+1}.$$

Consequently, there exists a constant C_4 such that

$$|\Gamma_j(\tilde{\omega}_k)| \le C_4\|\tilde{\omega}_k\|_0[\sup|\varphi_j|]^{m+1}.$$

Letting $j \to \infty$, by (27):

$$|\Gamma(\tilde{\omega}_k)| \le C_4\|\tilde{\omega}_k\|_0[\sup|\varphi|]^{m+1}.$$

Therefore $\Gamma(\tilde{\omega}_k)$ is a Cauchy sequence which converges to $\Gamma(\tilde{\omega})$ and this limit is independent of the sequence $\tilde{\omega}_k$.

By induction we have thus defined the Radon measure $\mathscr{M}(\varphi)$ when φ is a convex function. Moreover, for any compact set $K \subset \Omega$ there exists a constant C_5 such that

$$\left|\int_\Omega \psi\mathscr{M}(\varphi)\right| \le C_5 \sup|\psi|[\sup|\varphi|]^n \tag{28}$$

for all $\psi \in C(\Omega)$ with supp $\psi \subset K$.

8.14 Proposition. *Let $\{\varphi_p\}$ be a sequence of convex functions on Ω which converges uniformly to φ on Ω. Then $\mathscr{M}(\varphi_p) \to \mathscr{M}(\varphi)$ vaguely (i.e. for all $\psi \in C_0(\Omega)$, $\int_\Omega \psi\mathscr{M}(\varphi_p) \to \int_\Omega \psi\mathscr{M}(\varphi)$).*

Proof. As in the preceding paragraph, the proof proceeds by induction. We use the same notation.

Let ω be a continuous $(n-1)$-form and let $\{\omega_k\}$ be a sequence of C^∞ $(n-1)$-forms with support contained in K, and which converges uniformly to ω. Obviously

$$\int_\Omega d(\partial_i \varphi_p) \wedge \omega_k \to \int_\Omega d(\partial_i \varphi) \wedge \omega_k.$$

By definition

$$\int_\Omega d(\partial_i \varphi_p) \wedge \omega = \lim_{k\to\infty} \int_\Omega d(\partial_i \varphi_p) \wedge \omega_k.$$

But according to (25) this convergence is uniform in p:

$$\left| \int_\Omega d(\partial_i \varphi_p) \wedge (\omega - \omega_k) \right| \le C_1 \|\omega - \omega_k\|_0 \sup|\varphi_p| \le \tilde{C}_1 \|\omega_k - \omega\|_0$$

Thus we can interchange the limits in p and in k:

$$\left| \int_\Omega d(\partial_i \varphi) \wedge \omega - \int_\Omega d(\partial_i \varphi_p) \wedge \omega \right| \le \left| \int_\Omega d(\partial_i \varphi) \wedge (\omega - \omega_k) \right|$$
$$+ \left| \int_\Omega d(\partial_i \varphi) \wedge \omega_k - \int_\Omega d(\partial_i \varphi_p) \wedge \omega_k \right| + \left| \int_\Omega d(\partial_i \varphi_p) \wedge (\omega - \omega_k) \right|.$$

These three terms are smaller than $\varepsilon > 0$ if we choose k such that $\|\omega_k - \omega\|_0 < \varepsilon/\tilde{C}_1$, and then p large enough.

This proves the result for $m = 1$. We suppose now that it is true for some $m < n$, so we assume:

for all continuous $(n-m)$-form $\check{\omega}$ with support included in K,

$$\int_\Omega d(\partial_{i_1} \varphi_p) \wedge d(\partial_{i_2} \varphi_p) \wedge \cdots \wedge d(\partial_{i_m} \varphi_p) \wedge \check{\omega}$$
$$\to \int_\Omega d(\partial_{i_1} \varphi) \wedge d(\partial_{i_2} \varphi) \wedge \cdots \wedge d(\partial_{i_m} \varphi) \wedge \check{\omega}$$

We then prove the result for $m + 1$ in a similar way to the $m = 1$ case. ∎

§4. The Functional $\mathcal{I}(\varphi)$

8.15 Definition. For a C^2 convex function which is zero on the boundary $\partial\Omega$ we set:

$$\mathcal{I}(\varphi) = -n \int_\Omega \varphi M(\varphi)\,dx. \tag{29}$$

This definition makes sense if the convex function is not C^2. Indeed, for $a < 0$ set $\Omega_a = \{x \in \Omega/\varphi(x) \leq a\}$. According to the preceding paragraph

$$-\int_{\Omega_a} \varphi \mathcal{M}(\varphi) = \tau(a)$$

makes sense. Since $\mathcal{M}(\varphi)$ is a non-negative Radon measure and $\varphi \leq 0$, $\tau(a)$ is an increasing function. We define

$$\mathcal{I}(\varphi) = -n \lim_{a\to 0} \int_{\Omega_a} \varphi \mathcal{M}(\varphi).$$

Of course $\mathcal{I}(\varphi)$ may be infinite. The set of the convex functions for which $\mathcal{I}(\varphi)$ is finite will play surely an important role.

4.1. Properties of $\mathcal{I}(\varphi)$

8.16 Let us suppose that $\varphi \in C^2(\overline{\Omega})$ is a strictly convex function. As before, set $g_{ij} = \partial_{ij}\varphi$ and let g^{ij} be the components of the inverse matrix of $((g_{ij}))$. We will prove that

$$\mathcal{I}(\varphi) = \int_\Omega g^{ij}\,\partial_i\varphi\,\partial_j\varphi M(\varphi)\,dx. \tag{30}$$

$$M(\varphi) = \begin{vmatrix} \partial_{11}\varphi & \partial_{12}\varphi & \cdots & \partial_{1n}\varphi \\ \vdots & & & \vdots \\ \partial_{1n}\varphi & & & \partial_{nn}\varphi \end{vmatrix} = \frac{1}{n!}\sum \begin{vmatrix} \partial_{ii}\varphi & \partial_{ij}\varphi & \cdots & \partial_{ik}\varphi \\ \partial_{ij}\varphi & \partial_{jj}\varphi & \cdots & \\ \vdots & & & \\ \partial_{ik}\varphi & \cdots & & \partial_{kk}\varphi \end{vmatrix},$$

where the sum $\sum$ is extended to all $n \times n$ determinants obtained when the n subscripts $i, j, \ldots, k$ run from 1 to n. If two subscripts are equal the determinant is zero, otherwise it is equal to $M(\varphi)$. Thus we see that $M(\varphi)$ is a divergence:

$$M(\varphi) = \frac{1}{n!}\sum \partial_i \begin{vmatrix} \partial_i\varphi & \partial_{ij}\varphi & \cdots & \partial_{ik}\varphi \\ \partial_j\varphi & \partial_{jj}\varphi & & \\ \vdots & & & \vdots \\ \partial_k\varphi & \partial_{jk}\varphi & & \partial_{kk}\varphi \end{vmatrix},$$

$$nM(\varphi) = \partial_i(g^{ij}M(\varphi)\,\partial_j\varphi).$$

When we differentiate a column other than the first, we get a determinant which is zero. Thus integrating (29) by parts leads to (30).
For the preceding proof we supposed that $\varphi \in C^3$, but the result is true if φ is only C^2. In this case we approximate φ in C^2 by strictly convex C^∞ functions $\tilde{\varphi}$. We have

$$-n \int_\Omega \varphi M(\tilde{\varphi})\, dx = \int_\Omega \tilde{g}^{ij}\, \partial_i \varphi\, \partial_j \tilde{\varphi} M(\tilde{\varphi})\, dx.$$

By passing to the limit we get the result.

We will now show that the functional $\mathscr{I}(\varphi)$ is *convex* on the set of the strictly convex functions $\varphi \in C^2(\overline{\Omega})$. Let $\psi \in C^2(\overline{\Omega})$ be a function which vanishes on the boundary. $\varphi + t\psi$ is strictly convex for $|t|$ small enough and the second derivative of $\mathscr{I}(\varphi + t\psi)$ with respect to t at $t = 0$ is equal to

$$n(n+1) \int_\Omega g^{ij}\, \partial_i \psi\, \partial_j \psi M(\varphi)\, dx \geq 0.$$

Indeed, we have n terms of the form

$$\frac{1}{(n-1)!} \sum \int_\Omega \psi \begin{vmatrix} \partial_{ii}\psi & \partial_{ij}\varphi & \cdots & \partial_{ik}\varphi \\ \partial_{ij}\psi & \partial_{jj}\varphi & & \vdots \\ \vdots & & & \\ \partial_{ik}\psi & \cdots & & \partial_{kk}\varphi \end{vmatrix} dx = \int_\Omega \psi\, \partial_i [g^{ij} M(\varphi)\, \partial_j \psi]\, dx$$

and $n(n-1)/2$ terms of the form

$$\frac{1}{(n-1)!} \sum \int_\Omega \varphi \begin{vmatrix} \partial_{ii}\psi & \partial_{ij}\psi & \partial_{il}\varphi & \cdots & \partial_{ik}\varphi \\ \partial_{ij}\psi & \partial_{jj}\psi & \partial_{jl}\varphi & \cdots & \\ \vdots & & & & \\ \partial_{ik}\psi & \partial_{jk}\psi & \partial_{kl}\varphi & \cdots & \partial_{kk}\varphi \end{vmatrix} dx,$$

which are equal to

$$\frac{1}{(n-1)!} \sum \int_\Omega \partial_i \varphi \begin{vmatrix} \partial_i\psi & \partial_{ij}\psi & \partial_{il}\varphi & \cdots & \partial_{ik}\varphi \\ \partial_j\psi & \partial_{jj}\psi & \partial_{jl}\varphi & & \partial_{jk}\varphi \\ \vdots & \vdots & & & \vdots \\ \partial_k\psi & \partial_{jk}\psi & \partial_{kl}\varphi & & \partial_{kk}\varphi \end{vmatrix} dx.$$

This may be rewritten as

$$\frac{1}{(n-1)!} \sum \int_\Omega \partial_i \psi \begin{vmatrix} \partial_i\varphi & \partial_{ij}\psi & \partial_{il}\varphi & \cdots & \partial_{ik}\varphi \\ \partial_j\varphi & \partial_{jj}\psi & \partial_{jl}\varphi & & \vdots \\ \vdots & \vdots & \vdots & & \\ \partial_k\varphi & \partial_{jk}\psi & \partial_{kl}\varphi & \cdots & \partial_{kk}\varphi \end{vmatrix} dx,$$

which is equal to $\int_\Omega \psi\, \partial_i[g^{ij}M(\varphi)\, \partial_j\psi]\, dx$. Altogether we have $n(n+1)/2$ terms. Integrating by parts gives the result.

8.17 Remark. If Ω is a ball of radius τ and φ a C^2 radially symmetric function vanishing on the boundary: $\varphi(x) = g(\|x\|)$ with $g(\tau) = 0$.
The integrand in (30) is equal to $|g'|^{n+1}\rho^{1-n}$ with $\rho = \|x\|$. Thus in this case

$$\mathscr{I}(\varphi) = \omega_{n-1} \int_0^\tau |g'|^{n+1}\, d\rho.$$

Let f be a function belonging to $H_1^{n+1}([0, \tau])$. f is Hölder continuous. Indeed:

$$(31)\; |f(b) - f(a)| = \left| \int_a^b f'(s)\, ds \right| \le \left| \int_a^b |f'(s)|^{n+1}\, ds \right|^{1/(n+1)} |b - a|^{n/(n+1)}$$

Thus, $f(\tau) = 0$ makes sense, and if $\tilde{\varphi}(x) = f(\|x\|)$ we can define $\mathscr{I}(\tilde{\varphi})$ to be equal to $\omega_{n-1} \int_0^\tau |f'|^{n+1}\, d\rho$.

8.18 Proposition. *Let $\{\varphi_p\}_{p\in\mathbb{N}}$ be a sequence of convex functions on $\overline{\Omega}$, vanishing on $\partial\Omega$, which converges uniformly to φ on $\overline{\Omega}$. Then*

$$\mathscr{I}(\varphi) \le \liminf_{p\to\infty} \mathscr{I}(\varphi_p).$$

Thus, $\mathscr{I}(\varphi)$ is lower semi-continuous.

Proof. We use the notation of 8.14. For $a < 0$, according to (28):

$$\left| \int_{\Omega(a)} (\varphi - \varphi_p)\mathscr{M}(\varphi_p) \right| \le C(a) \sup|\varphi - \varphi_p| [\sup|\varphi_p|]^n$$

$$\le \text{Const} \times \|\varphi - \varphi_p\|_0:$$

thus $\int_{\Omega(a)} (\varphi - \varphi_p)\mathscr{M}(\varphi_p) \to 0$ when $p \to \infty$. But we can write

$$-n \int_{\Omega(a)} (\varphi_p - \varphi)\mathscr{M}(\varphi_p) - n \int_{\Omega(a)} \varphi\mathscr{M}(\varphi_p) = -n \int_{\Omega(a)} \varphi_p\mathscr{M}(\varphi_p) \le \mathscr{I}(\varphi_p).$$

Taking the lim inf when $p \to \infty$ leads to

$$-n \int_{\Omega(a)} \varphi\mathscr{M}(\varphi) \le \liminf_{p\to\infty} \mathscr{I}(\varphi_p),$$

by Proposition 8.14. Then letting $a \to 0$ we get the desired result. ∎

8.19 In this paragraph we will recall some properties of convex bodies which will be useful in 8.20.

As previously, Ω is a strictly convex bounded set of $\mathbb{R}^n$. If $Q \in \partial\Omega$, $\omega(Q)$ will denote the direction of the vector $\overrightarrow{PH}$ where P is the origin of the coordinates of $\mathbb{R}^n$ (which we assume is inside Ω) and $H(Q)$ the projection of P on the tangent plane of $\partial\Omega$ at Q.
We identify ω with a point on the sphere $\mathbb{S}_{n-1}(1)$. We let $\ell(Q)$ be the length of $\overrightarrow{PH}$.

For the proof of the next theorem, we are going to define a symmetrization procedure which is not the usual one of 2.17. It will be a consequence of the general inequality of Minkowski (see Buseman [71] p. 48) which applies to the convex bodies Ω and B and asserts that

$$\left(\frac{1}{n}\int_{\partial\Omega} \ell(Q)\, d\omega\right)^n \geq \mu(\Omega)[\mu(B)]^{n-1}, \tag{32}$$

where μ denote the Euclidean measure and $d\omega$ the element of measure on the unit sphere. Since equality holds for the ball, a consequence of (32) is:

Proposition 8.19. *Among the convex bodies with $\int_{\partial\Omega} \ell\, d\omega$ given, the ball has the greatest volume.*

8.20 Let d and D be the inradius and circumradius of Ω.

Theorem. *Let $\varphi \in C^2(\bar{\Omega})$ be a convex function which vanishes on $\partial\Omega$. There exists a radially symmetric function $\tilde{\varphi} \in C^1(\bar{B}_\tau)$ $(d \leq \tau \leq D)$, vanishing for $\|x\| = \tau$, with the following properties:*

$$\text{(33)}\qquad \begin{aligned} &\alpha)\ \ \tilde{\varphi} \text{ has the same extrema as } \varphi;\\ &\beta)\ \ \mathcal{I}(\tilde{\varphi}) \leq \mathcal{I}(\varphi);\\ &\gamma)\ \ \mu(\Omega_a) \leq \mu(\tilde{\Omega}_a) \text{ for all } a < 0, \end{aligned}$$

where $\tilde{\Omega}_a = \{x \in B_\tau \mid \tilde{\varphi}(x) \leq a\}$.

Proof. Let m be the minimum of φ. On $[m, 0]$ we define the function ρ by $\rho(0) = (1/\omega_{n-1}) \int_{\partial\Omega} \ell\, d\omega$ and $\rho(a) = (1/\omega_{n-1}) \int_{\partial\Omega_a} \ell\, d\omega$ for $m \leq a < 0$. ρ is strictly increasing and C^1 on $]m, 0[$. Indeed,

$$\rho'(a) = \frac{1}{\omega_{n-1}} \int_{\partial\Omega_a} \frac{d\omega}{|\nabla\varphi|} \quad \text{for } a \in]m, 0[. \tag{34}$$

$\rho'(a)$ is a continuous strictly positive function which goes to infinity as $a \to m$ because $|\nabla\varphi(x)| \to 0$ as $d(x, \Omega_m) \to 0$. Moreover, when $a \to 0$,

$$\rho'(a) \to \frac{1}{\omega_{n-1}} \int_{\partial\Omega} \frac{d\omega}{|\nabla\varphi|}.$$

Thus $a \to \rho(a)$ is invertible. Let g be its inverse function: $a = g(\rho)$.

On $\partial\Omega$ ℓ satisfies $d \le \ell \le D$; hence $d \le \tau = \rho(0) \le D$. If $\mu(\Omega_m) \ne 0$, $\rho(m) > 0$. In this case we set $g(\rho) = m$ for $0 \le \rho \le \rho(m)$. Thus $g \in C^1([0, \tau])$, $g'(0) = 0$, and $g' \ge 0$.

Consider now the radially symmetric function $\tilde{\varphi}(x) = g(\|x\|)$. Obviously α) is true and γ) holds by Proposition 8.19 since

$$\int_{\partial\Omega_a} \ell \, d\omega = \rho(a)\omega_{n-1} = \int_{\partial\tilde{\Omega}_a} \ell \, d\omega. \tag{35}$$

It remains to prove β). By (30)

$$\mathscr{I}(\varphi) = \int_m^0 da \int_{\partial\Omega_a} |\nabla\varphi| \prod_{i=1}^{n-1} \left(\frac{|\nabla\varphi|}{R_i}\right) d\sigma,$$

where $R_i(Q)$ are the principal radii of curvature of $\partial\Omega_a$ at $Q \in \partial\Omega_a$. Thus

$$\mathscr{I}(\varphi) = \int_m^0 da \int_{\partial\Omega_a} |\nabla\varphi|^n \, d\omega$$

because $d\sigma = \prod_{i=1}^{n-1} R_i \, d\omega$ and

$$\begin{aligned} \mathscr{I}(\tilde{\varphi}) &= \int_m^0 da \int_{\partial\tilde{\Omega}_a} |\nabla\tilde{\varphi}|^n \, d\omega = \omega_{n-1} \int_m^0 |g'[\rho(a)]|^n \, da \\ &= \omega_{n-1} \int_m^0 [\rho'(a)]^{-n} \, da. \end{aligned} \tag{36}$$

Applying Hölder's inequality yields

$$\omega_{n-1} = \int_{\partial\Omega_a} d\omega \le \left(\int_{\partial\Omega_a} |\nabla\varphi|^n \, d\omega\right)^{1/(n+1)} \left(\int_{\partial\Omega_a} \frac{d\omega}{|\nabla\varphi|}\right)^{n/(n+1)}.$$

Consequently by (34):

$$\int_{\partial\Omega_a} |\nabla\varphi|^n \, d\omega \ge \omega_{n-1}[\rho'(a)]^{-n}.$$

Integrating with respect to a over $[m, 0]$ gives β).

8.21 Theorem. *All convex functions φ, which are zero on $\partial\Omega$, satisfy:*

$$\left|\inf_{\Omega} \varphi\right|^{n+1} \le D^n \omega_{n-1}^{-1} \mathscr{I}(\varphi), \tag{37}$$

where D is the circumradius of Ω.

Proof. If $\varphi \in C^2(\overline{\Omega})$, Theorem 8.20 shows that it is enough to prove the result for radially symmetric function $\tilde{\varphi}$.
Set $\tilde{\varphi}(x) = f(\|x\|)$. We have $f(0) = \inf \tilde{\varphi}$ and $f(\tau) = 0$ with $\tau \leq D$. Using (31) gives

$$|f(0)|^{n+1} \leq \tau^n \int_0^\tau |f'(s)|^{n+1}\, ds = \tau^n \omega_{n-1}^{-1} \mathscr{I}(\tilde{\varphi}).$$

And we get the result from Theorem 8.20.

If φ is a convex function but not necessarily C^2, we consider a sequence of convex C^∞ functions ψ_i which converges uniformly to φ on $\overline{\Omega}_a$, where $a < 0$ is close to zero. We can suppose that on $\partial\Omega_a$, $\psi_i > 2a$. By Proposition 8.14, $\int_{\Omega_a} \psi_i \mathscr{M}(\psi_i) \to \int_{\Omega_a} \varphi \mathscr{M}(\varphi)$.
Applying (37) to the function $\inf(0, \psi_i - 2a)$ leads to

$$|\inf(\psi_i - 2a)|^{n+1} \leq D^n \omega_{n-1}^{-1}\left[-n \int_{\Omega_a} \psi_i \mathscr{M}(\psi_i)\right];$$

letting $i \to \infty$ and then $a \to 0$ yields (37). ■

§5. Variational Problem

8.22 Let $\tilde{f}(t) \in C^k(]-\infty, 0])$ $(k \geq 0)$ be a strictly positive function when $t \neq 0$ and greater than some $\varepsilon > 0$ for $t < t_0$, t_0 some real number. Set $F(t) = \int_{-|t|}^0 \tilde{f}(u)\, du$ and consider the functional Γ defined on the set of continuous functions on the unit ball $\bar{B}$ by:

$$\Gamma(\psi) = \int_B F(\psi(x))\, dx.$$

We are interested in the following problem:
Minimize $\mathscr{I}(\psi)$ over the set $\mathscr{A}$ of convex functions which are zero on ∂B and which satisfy $\Gamma(\psi) = \mathscr{k}$, for $\mathscr{k} > 0$ some given real number.

Theorem 8.22. *The* inf *of* $\mathscr{I}(\psi)$ *for all* $\psi \in \mathscr{A}$, *which we call* m, *is attained by a radially symmetric convex function* $\psi_0 \in C^{k+2}(\bar{B})$ *which vanishes on* ∂B *and which satisfies* $\mathscr{I}(\psi_0) = m$, $\Gamma(\psi_0) = \mathscr{k}$, *and for some* $v > 0$

$$M(\psi_0) = v\tilde{f}(\psi_0). \tag{38}$$

Proof. α) First of all $\mathscr{A}$ is not empty. More precisely if $\psi \leq 0$ $(\psi \not\equiv 0)$ is a continuous function on B there exists a unique real number $\mu_0 > 0$ for which $\Gamma(\mu_0 \psi) = \mathscr{k}$. Indeed for $\mu > 0$

$$\partial_\mu \Gamma(\mu\psi) = -\int_B \psi \tilde{f}(\mu\psi)\, dx > 0 \tag{39}$$

because the integrand is strictly negative somewhere. It is easy to verify that the hypotheses imply $\Gamma(\mu\psi) \to \infty$ as $\mu \to \infty$ and $\Gamma(\mu\psi) \to 0$ as $\mu \to 0$. Hence μ_0 exists and is unique.

β) One can show that the inf of $\mathscr{I}(\psi)$ for $\psi \in \mathscr{A}$ and that for $\psi \in \mathscr{A} \cap C^2(\bar{B})$ are equal. The reader will find the details in Aubin [23] p. 370.

γ) Now let $\psi \in \mathscr{A} \cap C^2(\bar{B})$ and $\tilde{\psi}$ be the corresponding radially symmetric functions introduced in 8.20. Then $\Gamma(\tilde{\psi}) \geq \Gamma(\psi)$. Indeed consider $\check{\psi}$, the radially symmetric function such that $\mu(\Omega_a) = \mu(\check{\Omega}_a)$ for all $a < 0$, where $\check{\Omega}_a = \{x \in B \,|\, \check{\psi}(x) \leq a\}$.
By Theorem 8.19, $\mu(\check{\Omega}_a) \leq \mu(\tilde{\Omega}_a)$. Thus $\tilde{\psi} \leq \check{\psi}$ on B and therefore $\Gamma(\tilde{\psi}) \geq \Gamma(\check{\psi})$ since Γ is decreasing in ψ. See (39). Moreover obviously, $\Gamma(\check{\psi}) = \Gamma(\psi)$.

Hence there exists $\mu_0 \leq 1$ such that $\Gamma(\mu_0\tilde{\psi}) = \Gamma(\psi) = k$. See α). But according to Theorem 8.20, $\mathscr{I}(\tilde{\psi}) \leq \mathscr{I}(\psi)$. Thus $\mathscr{I}(\mu_0\tilde{\psi}) = \mu_0^{n+1}\mathscr{I}(\tilde{\psi}) \leq \mathscr{I}(\psi)$. Therefore m is equal to the inf of $\mathscr{I}(\psi)$ for all radially symmetric functions $\psi \in \mathscr{A} \cap C^2(B)$.

δ) It remains for us to solve a variational problem in one dimension. That is the aim of the following.

8.23 Theorem. *The* inf *of* $\mathscr{I}(g) = \omega_{n-1}\int_0^1 |g'(r)|^{n+1}\,dr$ *for all nonpositive functions* $g \in H_1^{n+1}([0,1])$ *which vanish at* $r = 1$ *and which satisfy* $\Gamma(g) = \omega_{n-1}\int_0^1 F(g)r^{n-1}\,dr = k$ *is attained by a convex function* $g_0 \in C^{k+2}([0,1])$ *which is a solution of Equation* (38) *with* $v > 0$, g_0 *satisfying* $g_0'(0) = 0$ *and* $g_0(1) = 0$.

Proof. Since this is similar to several proofs done previously, we only sketch it. We already saw that this problem makes sense: g is Hölder continuous on $[0,1]$ (see 8.17), and there exist functions g satisfying $\Gamma(g) = k$ (see 8.22, α)).

Let $\{g_i\}$ be a minimizing sequence. These functions are equicontinuous by (31). Applying Ascoli's theorem 3.15 there exists a subsequence of the $\{g_i\}$ which converges uniformly to a continuous function g_0. Thus $\Gamma(g_0) = k$, $g_0 \leq 0$, and $g_0(1) = 0$. Moreover a subsequence converges weakly to g_0 in $H_1^{n+1}([0,1])$. Thus by 3.17 the inf of $I(g)$ is attained by g_0. Writing the Euler equation yields

$$\int_0^1 \psi'|g_0'|^{n-1}g_0'\,dr = -v\int_0^1 \psi\tilde{f}(g_0)r^{n-1}\,dr \tag{40}$$

for some real v and all function $\psi \in H_1^{n+1}([0,1])$ vanishing at $r = 1$. Picking $\psi = g_0$ we see that $v > 0$ ($v = 0$ is impossible because this would imply $\mathscr{I}(g_0) = 0$ and consequently $g_0 \equiv 0$).

We now prove that $g_0(r)$ is equal to $\tilde{g}(r) = \int_1^r [\nu \int_0^u \tilde{f}(g_0(t))t^{n-1}\, dt]^{1/n}\, du$. $\tilde{g}(1) = 0$, $\tilde{g} \in C^1([0, 1])$, and $(\tilde{g}'^n(r))' = \nu\tilde{f}(g_0(r))r^{n-1}$. Thus for all integrable functions γ on $[0, 1]$:

$$\int_0^1 \gamma(|g_0'|^{n-1}g_0' - \tilde{g}'^n)\, dr = 0.$$

This implies $|g_0'|^{n-1}g_0' = \tilde{g}'^n$, so $g_0' = \tilde{g}'$ since $\tilde{g}' \geq 0$. Hence $g_0 = \tilde{g}$. Considering the expression

$$g_0'(r) = \left[\nu \int_0^r \tilde{f}(g_0(t))t^{n-1}\, dt\right]^{1/n},$$

we see that $g_0'(1) > 0$ since $g_0 \not\equiv 0$ and therefore $g_0(r) < 0$ for $r < 1$. Thus $g_0'(r) \neq 0$ for $r > 0$ and g_0 is C^2 on $]0, 1]$ where $g_0'' > 0$. Moreover as $r \to 0$, $g_0'(r) \sim [(\nu/n)\tilde{f}(g_0(0))]^{1/n}r$. Thus $g_0 \in C^2([0, 1])$ and it is convex. If $\tilde{f} \in C^k$, $g_0 \in C^{2+k}$. ∎

8.24 Corollary. *Let $f(x, t)$ be a C^∞ function on $\bar{B} \times]-\infty, 0]$. There exists a real number $\nu_0 > 0$ such that the equation*

(41) $$M(\varphi) = \nu \exp f(x, \varphi), \qquad \varphi/\partial B = 0$$

has a strictly convex solution $\tilde{\varphi} \in C^\infty(\bar{B})$ when $n = 2$ and $0 < \nu \leq \nu_0$.

Proof. For some $\varepsilon > 0$ define $\tilde{f}(t) = \varepsilon + \sup_{x \in \bar{B}} \exp f(x, t)$. Consider $F(t) = \int_t^0 \tilde{f}(u)\, du$ and the functional $\Gamma(\psi) = \int_B F(\psi(x))\, dx$ as in 8.22.

By Theorem 8.22, there exist $\nu_0 > 0$ and a convex function $\psi_0 \in C^\infty(\bar{B})$ satisfying

$$M(\psi_0) = \nu_0 \tilde{f}(\psi_0), \qquad \psi_0/\partial B = 0.$$

Obviously ψ_0 is a strictly convex lower solution of (41) for $\nu \leq \nu_0$:

$$M(\psi_0) \geq \nu_0 \exp f(x, \psi_0(x))$$

and ψ_0 is strictly convex since $M(\psi_0) > 0$.

Then we choose $\beta > 0$ small enough so that $\gamma_0 = \beta(\|x\|^2 - 1)$ is an upper solution of (41) greater than ψ_0, where $\nu \leq \nu_0$ is given. Using Theorem 8.12 we obtain the stated result.

§6. The Complex Monge–Ampère Equation

8.25 The problem. We cannot end this paragraph without discussing the complex Monge–Ampère Equation.

Definition. A function φ with value in $[-\infty, +\infty[$, $(\varphi \not\equiv -\infty)$ is plurisubharmonic if it is lower semi-continuous and if the restriction of φ to any complex line is either a subharmonic function or else equal to $-\infty$. In case φ is C^2, φ is plurisubharmonic if the Hermitian form $\partial_{\lambda\bar{\mu}}\varphi\, dz^\lambda\, dz^{\bar{\mu}}$ is non-negative.

Henceforth Ω will be a strictly *pseudoconvex* bounded open set in $\mathbb{C}^m$ defined by a strictly plurisubharmonic function $h \in C^\infty(\bar{\Omega})$: $h/\partial\Omega = 0$ and let $\mathrm{u} \in C^k(\partial\Omega)$ $(k \geq 0)$.

We consider the *Dirichlet Problem*

$$\det((\partial_{\lambda\bar{\mu}}\varphi)) = f(x, \varphi), \qquad \varphi/\partial\Omega = \mathrm{u} \tag{42}$$

where $f(x, \varphi)$ is a non-negative function such that $f^{1/m}(x, \varphi) \in C^r(\bar{\Omega} \times \mathbb{R})$, $r \geq 0$. $\partial_{\lambda\bar{\mu}}\varphi$ denotes the second derivative of φ with respect to z^λ and $z^{\bar{\mu}}$ $(1 \leq \lambda, \mu \leq m)$.

This problem was studied by Bedford and Taylor [29] and [30], who use a very special method. They consider the upper envelope of the set of plurisubharmonic functions which are lower solutions of (42). They first prove that this upper envelope is a solution of (42) in a generalized sense and then try to prove that this solution is regular.

Remark 8.25. As boundary condition we can use $\varphi(x) \to +\infty$ when $x \to \partial\Omega$. Cheng and Yau [90] solve a problem of this kind:

They set $\varphi = u - \log(-h)$ and write the problem using the Kähler metric $\mathfrak{g}$ defined by $\mathfrak{g}_{\lambda\bar{\mu}} = -\partial_{\lambda\bar{\mu}} \log(-h)$.
They now solve a Monge–Ampère equation on a complete Kähler manifold, using the continuity method. The estimates are obtained by using the methods for compact Kähler manifolds (Chapter 7) thanks to their generalized maximum principle (Theorem 3.76).

6.1. Bedford and Taylor's Results

8.26 Bedford and Taylor [29] consider the case where the function f does not depend on φ and in Bedford and Taylor [30] the following is proved:

Theorem. *If $f^{1/m}(x, \varphi)$ is convex and nondecreasing in φ, then there exists a unique plurisubharmonic function $\varphi \in C(\bar{\Omega})$ which solves the Dirichlet problem (42) in a generalized sense. If $k \geq 2$ and $f^{1/m}(x, \varphi)$ is Lipschitz on $\bar{\Omega} \times \mathbb{R}$, then the solution φ is Lipschitz on $\bar{\Omega}$.*
In case Ω is the unit ball B, if in addition $r = k = 2$, then the function φ has second partial derivatives almost everywhere which are locally bounded.

6.2. The Measure $\mathfrak{M}(\varphi)$

8.27 Recall that Ω is a strictly pseudoconvex bounded open set in $\mathbb{C}^m$. For a continuous plurisubharmonic function φ on Ω, it is possible to define a measure $\mathfrak{M}(\varphi)$ which is equal to

$$\frac{i^m}{m!}\Lambda^m\, dd''\varphi = \det((\partial_{\lambda\bar{\mu}}\varphi))\, dz_1 \wedge d\bar{z}_1 \wedge \cdots \wedge dz_m \wedge d\bar{z}_m$$

in case φ is C^2. Recall Λ^m is m-times the exterior product and $d''\varphi = \partial_{\bar{\lambda}}\varphi\, dz^{\bar{\lambda}}$.

The method used in an earlier article by Chern, Levine, and Nirenberg [93] is similar to that of 8.13.
The main point is: for all compact $K \subset \Omega$ there is a constant $C(K)$ such that $\int_K \Lambda^m\, dd''\varphi \leq C(K)\sup_\Omega|\varphi|^m$ for all plurisubharmonic function $\varphi \in C^2(\Omega)$.

6.3. The Functional $\mathfrak{I}(\varphi)$

8.28 For a plurisubharmonic function $\varphi \in C(\bar{\Omega}) \cap C^2(\Omega)$ which is zero on the boundary $\partial\Omega$ we set:

$$\mathfrak{I}(\varphi) = -\frac{i^m}{(m-1)!}\int_\Omega \varphi\Lambda^m\, dd''\varphi. \tag{43}$$

If the plurisubharmonic function belongs only to $C(\bar{\Omega})$ we can extend this definition by the same procedure as in 8.15. For $a < 0$ set $\Omega_a = \{x \in \Omega/\varphi(x) \leq a\}$. Since $\mathfrak{M}(\varphi)$ is a non-negative Radon measure $-\int_{\Omega_a}\varphi\mathfrak{M}(\varphi) = \tau(a)$ makes sense and is an increasing function of a. We define

$$\mathfrak{I}(\varphi) = -m\lim_{a\to 0}\int_{\Omega_a}\varphi\mathfrak{M}(\varphi).$$

The set of the continuous plurisubharmonic functions on $\bar{\Omega}$ vanishing on $\partial\Omega$ for which $\mathfrak{I}(\varphi)$ is finite will play an important role.

6.4. Some Properties of $\mathfrak{I}(\varphi)$

8.29 Suppose $\varphi \in C^2(\bar{\Omega})$ is a strictly plurisubharmonic function vanishing on $\partial\Omega$. Set $g_{\lambda\bar{\mu}} = \partial_{\lambda\bar{\mu}}\varphi$ and let $g^{\lambda\bar{\mu}}$ be the components of the inverse matrix of $((g_{\lambda\bar{\mu}}))$. Integrating (43) by parts leads to:

$$\mathfrak{I}(\varphi) = i^m\int_\Omega g^{\lambda\bar{\mu}}\,\partial_\lambda\varphi\,\partial_{\bar{\mu}}\varphi\,\det((\partial_{\lambda\bar{\mu}}\varphi))\, dz^1 \wedge dz^{\bar{1}} \wedge \cdots \wedge dz^{\bar{m}}. \tag{44}$$

Thus $\mathfrak{I}(\varphi)$ is the integral on Ω of the square of the gradient of φ in the Kähler metric $g_{\lambda\bar{\mu}}$.
To carry out the integration by parts we need, in fact, $\varphi \in C^3$; however we obtain the result for $\varphi \in C^2$ by a density argument.

Proposition. *$\mathfrak{I}(\varphi)$ is convex on the set of the strictly plurisubharmonic functions $\varphi \in C^2(\bar{\Omega})$ vanishing on $\partial\Omega$.*

Proof. Let $\psi \in C^2(\Omega)$ be a function which is zero on the boundary. $\varphi + t\psi$ is strictly plurisubharmonic for $|t|$ small enough. Thus we have to verify that the second derivative with respect to t at $t = 0$ of $\mathfrak{I}(\varphi + t\psi)$ is non-negative. Integrating by parts enough times yields the following expression for this second derivative:

$$\left[\frac{\partial^2}{\partial t^2}\mathfrak{I}(\varphi + t\psi)\right]_{t=0}$$

$$= m(m-1)i^m \int_\Omega g^{\lambda\bar{\mu}}\, \partial_\lambda \psi\, \partial_{\bar{\mu}} \psi \det((\partial_{\lambda\bar{\mu}}\varphi))\, dz^1 \wedge dz^{\bar{1}} \wedge dz^2 \wedge \cdots \wedge dz^{\bar{m}}.$$

This is obviously non-negative. ∎

Theorem. *$\mathfrak{I}(\varphi)$ is lower semi-continuous: if $\{\varphi_p\}_{p\in\mathbb{N}}$ is a sequence of plurisubharmonic functions continuous on $\bar{\Omega}$ and vanishing on $\partial\Omega$ which converges uniformly to φ on $\bar{\Omega}$, then*

$$\mathfrak{I}(\varphi) \le \liminf_{p\to\infty} \mathfrak{I}(\varphi_p).$$

Proof. It is similar to that in 8.18. φ, being the uniform limit of the φ_p, is continuous on $\bar{\Omega}$, vanishes on $\partial\Omega$, and also is plurisubharmonic according to the definition. Thus $\mathfrak{I}(\varphi)$ makes sense. ∎

§7. The Case of Radially Symmetric Functions

8.30 If Ω is a ball of radius τ in $\mathbb{C}^m$ and φ a C^2 radially symmetric function plurisubharmonic on $\bar{\Omega}$ vanishing on the boundary, we can write $\varphi(z) = g(\|z\|)$ with $g(\tau) = 0$. In this paragraph we suppose that φ has these properties. In this case, by (44)

$$\mathfrak{I}(\varphi) = \omega_{2m-1}\int_0^\tau \frac{1}{4} g'^2 \left(\frac{g'}{2r}\right)^{m-1} 2^m r^{2m-1}\, dr$$

$$= \frac{1}{2}\omega_{2m-1}\int_0^\tau g'^{m+1} r^m\, dr$$

and we can apply Proposition 2.48 with $q = m + 1$.

Similarly we can associate to g the function ψ defined on the ball B_τ of $\mathbb{R}^{m+1}$ by $\psi(x) = g(\|x\|)$. Then $\mathfrak{I}(\varphi) = \frac{1}{2}\omega_{2m-1}\omega_m^{-1}\|\nabla\psi\|_{m+1}^{m+1}$ and we can apply all the results of 2.46–2.50 by noting that $r^{2m-1} \leq \tau^{m-1}r^m$. In particular, from Theorem 2.47 we get

Theorem 8.30. *If φ satisfies $\mathfrak{I}(\varphi) \leq 1$, then*

$$\int_\Omega \exp[v_m|\varphi|^{(m+1)/m}]\, dV \leq C \int_\Omega dV$$

where the constant C depends only on m and where

$$v_m = (m+1)2^{-1/m}\omega_{2m-1}^{1/m}.$$

From Corollary 2.49 we get

Corollary 8.30. *Set $\xi_m = 2m^m(m+1)^{-2m-1}\omega_{2m-1}^{-1}$. Then all φ satisfy*

$$\int_\Omega e^{-\varphi}\, dV \leq C\exp[\xi_m\mathfrak{I}(\varphi)] \int_\Omega dV$$

where C depends only on n.

Proposition 8.30. *Let $\mathscr{A}$ be a set of functions φ for which $\mathfrak{I}(\varphi) \leq$ Const. Then the set $\{e^\varphi\}_{\varphi\in\mathscr{A}}$ is precompact in L_1.*

The proof is similar to that of (2.46δ).
Using these results, under certain assumptions we can solve the Dirichlet Problem (42) on a ball of $\mathbb{C}^m$ with u = 0 on the boundary, assuming the dependence of $f(x, \varphi)$ on x only involves $\|x\|$.

By seeking radially symmetric solutions we actually obtain all solutions, provided f is decreasing in $r = \|x\|$, by a result of Gidas, Ni, and Nirenberg [124].

7.1. Variational Problem

8.31 Let Ω be a ball of radius τ in $\mathbb{C}^m$. We can consider the following problem: Find the Inf of $\mathfrak{I}(\varphi)$ for all plurisubharmonic function $\varphi \in C(\bar{\Omega})$ vanishing on the boundary such that $\|\varphi\|_q = 1$ for some $q \geq 1$ (or $\int_\Omega e^{-\varphi}\, dV = 1 + \int_\Omega dV$, for instance). We know that $\mathfrak{I}(\varphi)$ is convex. So for $q = 1$ we can consider only radially symmetric functions and it is possible to solve the problem in one dimension. The same result holds for $q \leq m$. For $q > m$ (or if the constraint is $\int_\Omega e^{-\varphi}\, dV = 1 + \int_\Omega dV$) it is conjectured that the inf of $\mathfrak{I}(\varphi)$ is unchanged if we consider only radially symmetric functions.

7.2. An Open Problem

8.32 The regularity of the solution of (42) obtained by Bedford and Taylor is an open problem.

If f is allowed to vanish, there is a counterexample of Bedford and Fornaess [27] showing that the solution may not be C^2.
Therefore, let us simply assume that $k = r = \infty$, $\Omega = B$, and $f < 0$ everywhere on $\bar{B}$. The problem of regularity is nevertheless open, not only on $\bar{B}$, of course (this case is not even solved for the real analog), but also on B.

One of the reasons we can get the interior regularity for the real Monge–Ampère equation, is that for each compact $\mathfrak{K} \subset \Omega$ it is possible to obtain a sequence of C^∞ convex functions φ_j converging uniformly to the generalized solution φ on $\mathfrak{K}$ while $M(\varphi_j) \to M(\varphi)$ on $\mathfrak{K}$ in C^r for large r. This result is obtained by geometrical considerations. At the present time a similar result has not been established in the complex case.[3]

§8. A New Method

8.33 In [190b] P. L. Lions presents a very interesting method for solving the Dirichlet problem for the real Monge–Ampère equation on a bounded strictly convex set Ω of $\mathbb{R}^n$:

$$\text{(45)} \qquad \log M(\varphi) = f(x), \qquad \varphi/\partial\Omega = 0$$

where f belongs to $C^\infty(\bar{\Omega})$. $\partial\Omega$ is supposed to be C^∞.

The method consists to exhibit a sequence of functions $\varphi_k \in C^\infty(\bar{\Omega})$ which are solutions of equation (45): $\log M(\varphi_k) = f(x)$, but with approximated boundary data: $\varphi_k/\partial\Omega = u_k$, u_k being an increasing sequence of functions converging uniformly to zero when $k \to \infty$.

Let $\tilde{f} \in \mathscr{D}(\mathbb{R}^n)$ be such that $\tilde{f}/\Omega = f$ and let $p(x) \in C^\infty(\mathbb{R}^n)$ be a function which are equal to zero on $\bar{\Omega}$ and to 1 outside a compact set. Moreover p satisfies: for all $\eta > 0$ there exists $\gamma > 0$ such that $p(x) > \gamma$ when $\operatorname{dist}(x, \bar{\Omega}) > \eta$.

The idea is to consider instead of (45) the following equation on $\mathbb{R}^n$:

$$\text{(46)} \qquad \log \det\left(\left(\partial_{ij}\varphi - \frac{p}{\varepsilon}\varphi\mathscr{E}_{ij}\right)\right) = \tilde{f}$$

with $\varepsilon > 0$ and $\mathscr{E}_{ij}$ the euclidean tensor.

8.34 We can prove that equation (46) has a unique solution belonging to $C_B^\infty(\mathbb{R}^n)$ for which the tensor $g_{ij} = \partial_{ij}\varphi - (p/\varepsilon)\varphi\mathscr{E}_{ij}$ defines a Riemannian metric.

[3] See 8.35 and 8.36 for new results.

The sketch of P. L. Lions' proof is the following. First he solves in $C_B^\infty(\mathbb{R}^n)$ an approximated equation of equation (46):

$$\log \det\left(\left(\partial_{ij}\varphi - \left(\frac{p}{\varepsilon} + \lambda\right)\varphi \mathscr{E}_{ij}\right)\right) = \tilde{f}$$

with $\lambda > 0$. For that he proves the existence of a solution in $H_2^\infty(\mathbb{R}^n)$ of an associated stochastic control problem; and for the regularity he uses the results of Evans (112b). These results yield uniform bounds with respect to λ for the second derivatives. So he obtains a solution $\varphi_\varepsilon \in C_B^\infty(\mathbb{R}^n)$ of equation (46), the regularity being given by Evans' results.

Instead of using the diffusion processes, it is possible by using the continuity method to solve directly equation (46).
In any case we must construct a subsolution $\varphi_0 \in C_B^\infty(\mathbb{R}^n)$ of (46) such that $(g_0)_{ij} = \partial_{ij}\varphi_0 - (p/\varepsilon)\varphi_0 \mathscr{E}_{ij}$ defines a Riemannian metric. Then the maximum principle implies that any solution of (46) satisfies $\varphi_0 \le \varphi < 0$.

For $t \ge 0$ a parameter we consider the equation:

$$\text{(47)} \quad \log \det\left(\left(\partial_{ij}\varphi - \frac{p}{\varepsilon}\varphi \mathscr{E}_{ij}\right)\right) = t\tilde{f} + (1-t)\log \det\left(\left(\partial_{ij}\varphi_0 - \frac{p}{\varepsilon}\varphi_0 \mathscr{E}_{ij}\right)\right)$$

and for $\alpha \in]0, 1[$ the operator Γ:

$$C_B^{2,\alpha}(\mathbb{R}^n) \supset \Theta \ni \varphi \to \log \det\left(\left(\partial_{ij}\varphi - \frac{p}{\varepsilon}\varphi \mathscr{E}_{ij}\right)\right) \in C_B^\alpha(\mathbb{R}^n)$$

where Θ is the subset of functions φ for which $((g_{ij}))$ is a positive definite bilinear form. Γ is continuously differentiable and its differential $d\Gamma_\varphi$ at φ is invertible. Indeed the equation

$$\text{(48)} \qquad g^{ij}\partial_{ij}\psi - \frac{p}{\varepsilon}\psi g^{ij}\mathscr{E}_{ij} = F \in C_B^\alpha(\mathbb{R}^n)$$

has a unique solution belonging to $C_B^{2,\alpha}(\mathbb{R}^n)$.
To establish this result we consider for instance the solution $\psi_k \in C^{2,\alpha}(\bar{B}_k)$ of (48) on $\bar{B}_k$ for Dirichlet's data equal to zero on the boundary. It is easy to prove that the set of functions $\{\psi_k\}_{k\in\mathbb{N}}$ is uniformly bounded, then we use the Schauder Interior Estimates 3.61. At $x \in \mathbb{R}^n$ we have for $k > \|x\| + 2$:

$$\|\psi_k\|_{C^{2,\alpha}(K)} \le C[\|\psi_k\|_{C^0(\mathbb{R}^n)} + \|F\|_{C^\alpha(\mathbb{R}^n)}] \le \text{Constant}$$

with $K = \bar{B}_x(1)$ and the constant C independant on x.
It follows that a subsequence of ψ_k converges to a solution of (48) which belongs to $C_B^{2,\alpha}(\mathbb{R}^n)$. The generalized maximum principle 3.76 implies the uniqueness assertion. By Theorem 3.56 the solution belongs to $C_B^\infty(\mathbb{R}^n)$.

The inverse function theorem 3.10 establishes that the set $\mathscr{C}$ of the $t \in [0, 1]$, for which (47) has a solution, is open in $[0, 1]$. To prove that $\mathscr{C}$ is closed we need uniform estimates of the solutions of (47) in $C^3(\mathbb{R}^n)$. To get them we do similar computations to those done at the beginning of this chapter, but here we use the generalized maximum principle 3.76.

8.35 Pogorelov, Cheng and Yau get C^∞ approximated solutions of (45) by geometrical considerations. Here we get the functions $\varphi_k = \psi_k/\overline{\Omega}$ by analysis. It is the distinction between both proofs, because we proceed as Pogorelov [235] to have uniform estimates in $C^3(K)$ with the compact $K \subset \Omega$. That is why Lions' result is not an improvement for equation (45).

But we can apply the method to the Dirichlet problem for the complex Monge–Ampère equation 8.25. By a similar approach, we get C^∞ approximated solutions of the complex equation.
Then we need estimates. The C^0 and C^1 estimates are not difficult to obtain. But if there exists Aubin's estimate 7.22 for the third derivatives of mixed type (and so for the gradient of the laplacian) when we assume the estimate for the laplacian, there is no complex equivalent of Pogorelov's C^2-estimate [235 pp. 73–75]. This is still an open problem.

8.36 Note added in proofs. In June 1982, Cafarelli L., Nirenberg L. and Spruck J. proved [*66] that in the general case they have obtained the estimate of the third derivatives of the functions φ_σ on the boundary (see 8.11). They claim that there exists a modulus of continuity for the second derivatives of the function φ_σ:

$$|\partial_{ij}\varphi_\sigma(x) - \partial_{ij}\varphi_\sigma(y)| \leq C[1 + |\log \|x - y\| |]^{-1}$$

with C a constant. It is also possible to obtain this modulus of continuity for the complex equation 8.25 (see [*67] and [*68]).

Chapter 9

The Ricci Curvature

§1. About the Different Types of Curvature

9.1 In this chapter we deal with problems concerning Ricci Curvature mainly:

- Prescribing the Ricci curvature
- Ricci curvature with a given sign
- Existence of Einstein metrics.

This latter problem: to decide if a Riemannian manifold carries an Einstein metric, will be one of the important questions in Riemannian geometry for the next decades. Indeed, in spite of recent results that we will talk about, Ricci curvature is not yet well understood. Ricci curvature lies between sectional and scalar curvatures. We saw that scalar curvature is now well-known and we recall below some results concerning sectional curvature. In this chapter (except in §1.4) we suppose that the dimension of the manifold is greater than 2.

1.1 The Sectional Curvature

9.2 We will mention some well-known results which prove that it is a strong property for a manifold to have its sectional curvature of a given sign. We see that it is impossible for a manifold to carry two metrics, the sectional curvatures of which are of opposite sign.

Theorem 9.2. *A complete connected Riemaniann manifold (M_n, g) has constant sectional curvature if and only if it is isometric to S_n, $\mathbb{R}^n$ or H_n the hyperbolic space, or one of their quotients by a group Γ of isometries which acts freely and properly. S_n, $\mathbb{R}^n$ and H_n are endowed with their canonical metrics.*

9.3 Theorem (Synge). *A compact connected orientable Riemannian manifold of even dimension with strictly positive sectional curvature is simply connected.*

The proof is by contradiction. If the manifold is not simply connected there is a shortest closed geodesic Γ in any nontrivial homotopy class. As the manifold is orientable and of even dimension, there exists a unit parallel vector field along Γ orthogonal to Γ. Then we can consider the second variation of the length integral in the direction of this vector field (related to a family Γ_λ of closed curves near $\Gamma_0 = \Gamma$), as we do for the proof of Myers' Theorem 1.43. The hypothesis on the sign of the sectional curvature implies that this second variation is negative, which is a contradiction, since Γ would not be the shortest curve in its homotopy class.

9.4 Theorem. *A complete simply-connected Riemannian manifold* (M, g) *with nonpositive sectional curvature is diffeomorphic to* $\mathbb{R}^n$.

Proof. Let P be any point of M. P has no conjugate point (Theorem 1.48), so $\exp_P$ is a diffeomorphism from $\mathbb{R}^n$ to M (Theorem 1.46).

Corollary 9.4. *A compact Riemannian manifold* (M, g) *with non-positive sectional curvature cannot carry a metric* $\tilde{g}$ *with positive Ricci curvature.*

Indeed by Myers' Theorem 9.6, if $\tilde{g}$ exists, $(M, \tilde{g})$ has a compact universal covering space $(\tilde{M}, \pi^*\tilde{g})$, $\pi : \tilde{M} \to M$. This is in contradiction with Theorem 9.4 which asserts that the universal covering space of (M, g) is diffeomorphic to $\mathbb{R}^n$.

1.2. The Scalar Curvature

9.5 In this section we suppose $n \geq 3$. We saw that there are obstructions for a manifold to carry a metric with positive scalar curvature. But in Aubin [21] it is proven that we can locally decrease the average of the scalar curvature by some local change of metrics. Thus we get a metric with negative scalar curvature on any manifold. So there are three types of compact manifolds. Those which carry a metric with positive constant scalar curvature, those which carry a metric with zero scalar curvature and no metric with positive scalar curvature, and those which carry no metric with non-negative scalar curvature. For complete non-compact manifolds Aviles–Mc Owen [*18] proved that there always exists a complete metric the scalar curvature of which is constant and negative.

1.3. The Ricci Curvature

9.6 Between the scalar curvature which has little significance, and the sectional curvature which has strong meaning, there is the Ricci curvature. Recently Lohkamp [*226] (see 9.44) proved that any Riemannian manifold of dimension $n \geq 3$ carries a complete metric with negative Ricci curvature. On the other hand to carry a metric with positive Ricci curvature implies strong result such as

Theorem 9.6 (Myers [*247]). *A connected complete Riemannian manifold* M_n *with Ricci curvature* $\geq (n-1)k^2 > 0$ *is compact and its diameter is* $\leq \pi/k$. *Its fundamental group is finite.*

For the proof see [*247] or 1.43.

1.4. Two Dimension

9.7 The two-dimension case is very particular. If the local chart is chosen so that at $P, g_{ij}(P) = \delta_i^j$, we have at $P : R_{11} = R_{22} = R_{1212} = R/2$. Moreover, if (M, g) is compact, the Gauss–Bonnet theorem asserts that $\int_M R\,dV = 4\pi\chi(M)$, where χ is the Euler–Poincaré characteristic. For complete manifolds Cohn–Vossen proved the following inequality $\int_M R\,dV \leq 4\pi\chi(M)$. So it is obvious that a compact manifold cannot carry two metrics whose curvatures are of different sign. It is well known that there exists on a compact manifold a metric with constant sectional curvature $R/2$. The problem of prescribing the scalar curvature R is discussed in chapter 3, and the problem of prescribing the Ricci curvature is considered in this chapter . For details on the Gaussian curvature see Kazdan [*194].

§2. Prescribing the Ricci Curvature

2.1. DeTurck's Result

9.8 Theorem (DeTurck [*109]). *If* $T = \{T_{ij}\}$ *is a* $C^{m+\alpha}$ (resp. C^∞, analytic) *symmetric tensor field* ($m > 2$) *in a neighbourhood of a point* P *on a manifold of dimension* $n \geq 3$, *and if the matrix* $((T_{ij}))$ *is invertible at* P, *then there is a* $C^{m+\alpha}$ (C^∞, analytic) *Riemaniann metric* g *such that* $R_{ij} = T_{ij}$ *in a neighbourhood of* P.

Recall the expression for the curvature tensor in local coordinates (see 1.13), in terms of the Christoffel symbols of the metric:

(1) $$R^\ell_{kij} = \partial_i\Gamma^\ell_{jk} - \partial_j\Gamma^\ell_{ik} + \Gamma^\ell_{im}\Gamma^m_{jk} - \Gamma^\ell_{jm}\Gamma^m_{ik}.$$

Because the Ricci tensor is R^l_{klj}, we calculate that

(2) $$E(g) \equiv \tfrac{1}{2}g^{il}\big(\partial_{ik}g_{jl} + \partial_{jl}g_{ik} - \partial_{il}g_{jk} - \partial_{kj}g_{il}\big) + Q_{jk}(g) = T_{jk}$$

where $Q_{jk}(g)$ is quadratic in the first derivatives $\partial_i g_{jk}$.

(2) is a non linear second-order differential equation, its linearization gives at P:

(3) $$\big[DE(g)\big](h) = \tfrac{1}{2}g_P^{i\ell}\big(\partial_{ik}h_{jl} + \partial_{jl}h_{ik} - \partial_{il}h_{jk} - \partial_{kj}h_{il}\big) - R_{klij}(P)g_P^{ia}g_P^{\ell b}h_{ab}$$

if we suppose that the coordinate system is normal at P. Let $\xi = \{\xi_i\}$ be a unit vector of $T_P(M)$, the symbol σ of DE (g) at P is

(4) $$\big[\sigma(\xi)h\big]_{ij} = h_{ij} - \sum_{k=1}^{n}\big(\xi_i\xi_k h_{jk} + \xi_j\xi_k h_{ik} - \xi_i\xi_j h_{kk}\big)$$

To see how acts $\sigma(\xi)$ on the symmetric tensor h, we can suppose without loss of generality that $\xi_1 = 1$ and $\xi_i = 0$ for $i > 1$. We find $[\sigma(\xi)h]_{ij} = h_{ij}$ if $i \neq 1$ and $j \neq 1$, $[\sigma(\xi)h]_{1i} = 0$ if $i \neq 1$ and $[\sigma(\xi)h]_{11} = \sum_{k=2}^n h_{kk}$. Since there are zero eigenvalues, the symbol $\sigma(\xi)$ is not an isomorphism. Thus equation (2) is not strictly elliptic. The presence of zero eigenvalues is not surprising since under a diffeomorphism φ

$$\varphi^* \operatorname{Ricci}(g) = \operatorname{Ricci}(\varphi^* g).$$

We can verify that for any ξ the kernel of $\sigma(\xi)$ consists of all tensors h of the form $h_{ij} = v_i\eta_j + \eta_i v_j$ where v_i and η_j are the components of two 1-forms. For different points of view of this fact, see DeTurck [*109] p. 181, Hamilton [*151] p. 261 and Besse [*44] p. 139. To overcome this difficulty DeTurck considers the "gravitation operator" G.

2.2. Some Computations

9.9 Definition. We define $(Gh)_{ij} = h_{ij} - \frac{1}{2}\left(g^{kl}h_{kl}\right)g_{ij}$ and $(\delta h)_i = -\nabla^j h_{ij}$ on a symmetric tensor h and on a 1-form $v = \{v_i\}$,$(\delta^* v)_{ij} = \frac{1}{2}\left(\nabla_i v_j + \nabla_j v_i\right)$. Moreover set $B(g,T) = -\delta GT$ and for any tensor field S, $\Delta S = -g^{ij}\nabla_{ij}S$.

The second Bianchi identity (see 1.20) gives

$$B(g, \operatorname{Ricci}) = g^{jk}\left(\nabla_k R_{ij} - \tfrac{1}{2}\nabla_i R_{jk}\right) = 0 \tag{5}$$

$$B(g,T)_j = g^{ik}\left(\partial_k T_{ij} - \tfrac{1}{2}\partial_j T_{ik}\right) + \left(\tfrac{1}{2}\partial_k g_{il} - \delta_l g_{ik}\right)T_j^k g^{il}.$$

As we suppose that T is invertible, differentiating $T^{-1}B(g,T)$ with respect to g yields

$$D_g\left[T^{-1}B(g,T)\right](h) = \delta \mathrm{Gh} + \text{terms in h}. \tag{6}$$

Thus the leading part of $D_g\left[\delta^* T^{-1}B(g,T)\right](h)$ is

$$\tfrac{1}{2}\left(\partial_{jk}h_{il} - \partial_{jl}h_{ik} - \partial_{ik}h_{jl}\right)g^{il}. \tag{7}$$

Comparing (3) and (7) we find

$$D_g\left[\operatorname{Ricci}(g) + \delta^* T^{-1}B(g,T)\right](h) = \tfrac{1}{2}\Delta h + \text{lower order terms}. \tag{8}$$

2.3. DeTurck's Equations

9.10 We saw that (2) $E(g) = T$ is not elliptic. But we know that if g is solution of (2): $\operatorname{Ricci}(g) = T$, when T will satisfy $B(g,T) = 0$. Moreover we saw that (8) is elliptic. This is the reason for which DeTurck considered the new system:

$$R_{ij} + [\delta^* T^{-1}B(g,T)]_{ij} = T_{ij}. \tag{9}$$

$$B(g,T) = 0 \tag{10}$$

He proved that this system is elliptic and it is equivalent to the original one (2). As there exists a metric g_0 which satisfies (2) at P, g_0 satisfies (9) and (10) at P and the local theory of elliptic systems can be used. For the proof of Theorem 9.8 DeTurck considered an iteration scheme and showed that it converges. In [*111] DeTurck gives an alternative proof of Theorem 9.8 which is not so hard as the original one. The new idea is to find a metric g and a diffeomorphism φ such that Ricci $(g) = \varphi^* T$

9.11 Remark. We cannot drop the hypothesis $\big((T_{ij})\big)$ invertible at P. Consider, as DeTurck did, the tensor field $T_{ij} = x^i + x^j + 2\delta_i^j \sum_{k=1}^n x^k$, which vanishes at P. We verify that it cannot satisfy the Bianchi identity (5) at P for any Riemaniann metric. Indeed it gives at P $\sum_{j=1}^n g^{ij} = 0$ for $i = 1, \ldots, n$.

2.4. Global Results

9.12 On a compact kählerian manifold $(M, \tilde{g})$, we completely solved the problem of prescribed Ricci curvature (see 7.19). This problem was known as the Calabi conjecture. Recall the answer : Let $R_{\lambda\bar{\mu}}$ be a 1-1 covariant tensor field. The necessary and sufficient condition for which there exists a kählerian metric with Ricci tensor $R_{\lambda\bar{\mu}}$ is that the Ricci form $\frac{i}{2\pi} R_{\lambda\bar{\mu}} dz^\lambda \wedge dz^{\bar{\mu}}$ belongs to $C_1(M)$ the first Chern class. Moreover, in each positive cohomology class there is a solution g, which is unique up to a homothetic change of metric.

9.13 Myers' theorem 9.6 gives obstructions for a compact manifold to carry a metric with positive Ricci curvature. On the other hand, there is no obstruction for a manifold to carry a metric with negative Ricci curvature (see Lohkamp' s result in 9.44). In Kazdan [*194] we find other cases of non existence, such as:

Theorem 9.13 (DeTurck–Koiso [*112]). *On a compact manifold (M, g), if the Ricci curvature is positive, the tensor cR_{ij} is not the Ricci tensor of any metric for c large enough. We may take $c > 1$ if R_{ij} is the Ricci tensor of an Einstein metric, or if the sectional curvature of R_{ij} considered as a metric on M, is $\leq 1/(n-1)$.*

When $0 \leq c < 1$, we can conjecture that there is no metric with Ricci $(g) = cR_{ij}$, and Cao–DeTurck [*75] proved that there is no conformally flat metric with this property. DeTurck–Koiso [*112] also established some results of uniqueness for Ricci curvature.

§3. The Hamilton Evolution Equation

3.1. The Equation

9.14 One of the most famous problems in geometry is:

The Poincaré conjecture.

A compact simply-connected Riemannian manifold (M, g) *of dimension* $n = 3$ *is diffeomorphic to* S_3.

To attack this problem we can think of trying to deform the initial metric to an Einstein metric. If we succeeded we would get a metric of constant curvature since the Weyl tensor vanishes identically when $n = 3$. And we know (9.2) that a compact simply connected Riemannian manifold with constant curvature is isometric to the sphere. In his theorem (9.37) Hamilton supposes that the Ricci curvature of the initial metric is positive. Of course a hypothesis of this type is necessary since $S_2 \times C$ has non-negative Ricci curvature. (C is the circle). Actually we don't know how to express the hypothesis "simply connected" of the Poincaré conjecture, by means of Riemannian invariants.

9.15 To carry out this idea, R. Hamilton [*151] introduced the following evolution equation:

$$\frac{\partial}{\partial t} g_{ij} = (2r/n) g_{ij} - 2R_{ij} \tag{11}$$

where g_{ij} and R_{ij} are the components of the metric g_t and the Ricci tensor of g_t in a local chart.(To simplify we drop the subscript t when there is no ambiguity). The solution g_t of this equation will be a smooth family of metrics on the compact manifold M, and r is the average of the scalar curvature $R : r = \int R\,dV / \int dV$. Because $\frac{\partial}{\partial t}\sqrt{|g|} = \frac{1}{2}\sqrt{|g|} g^{ij} \frac{\partial g_{ij}}{\partial t} = (r - R)\sqrt{|g|}$ the volume of (M, g_t) is constant. In order to make the computations easier, R. Hamilton [*151] considered the evolution equation

$$\frac{\partial}{\partial t} g_{ij} = -2R_{ij} \tag{12}$$

Proposition 9.15. *Suppose* g_t *is a solution of* (12). *We define the function* $m(t)$ *so that* $(M, \tilde{g}_t)$ *has volume* 1 *with* $\tilde{g}_t = m(t) g_t$. *Set* $\tilde{t} = \int_0^t m(s)\,ds$, *then* $\tilde{g}$ *satisfies equation* (11) *with* $\tilde{t}$ *instead of* t.

Proof. First of all, in a homothetic change of metric, the Ricci curvature remains unchanged: $\tilde{R}_{ij} = R_{ij}$.

So $\tilde{r} = \int \tilde{R} d\tilde{V} = [m(t)]^{n/2-1} \int R\,dV$. But by hypothesis

$$1 = \int d\tilde{V} = \left[m(t)\right]^{n/2} \int dV,$$

hence

$$\frac{n}{2}\frac{m'(t)}{m(t)} = -\frac{1}{2}\big[m(t)\big]^{n/2}\int g^{ij}\frac{\partial g_{ij}}{\partial t}\,dV = \big[m(t)\big]^{n/2}\int R\,dV$$

using (12). Thus $\tilde{r} = \frac{n}{2}m'(t)/m^2(t)$. Now we verify that $\tilde{g}_{\tilde{t}}$ satisfies equation (11).

$$\frac{\partial}{\partial \tilde{t}}\tilde{g}_{ij} = \frac{1}{m(t)}\frac{\partial}{\partial t}\tilde{g}_{ij} = \frac{m'(t)}{m^2(t)}\tilde{g}_{ij} - 2R_{ij}.$$

9.16 Let $\{x^i\}$ be a normal coordinate system at $P \in M$ (see 1.25). We will write equation (12) at P in this local chart. According to the expression of the components of the curvature tensor (1):

(13) $$R_{ij}(P) = \tfrac{1}{2}g_P^{kl}\big(\partial_{ik}g_{jl} + \partial_{jl}g_{ik} - \partial_{ij}g_{kl} - \partial_{kl}g_{ij}\big)_P$$

If the coordinate system would not normal at P, there would be, in the expression of R_{ij}, additional terms involving only g_{ij}, g^{kl} and quadratic in the first derivatives $\partial_i g_{jk}$. So from (13) we get the linearization $D\tilde{E}(g)$ of the right hand side of (12): $\tilde{E}(g) = -2\,\text{Ricci}(g)$. We have $D\tilde{E}(g) = -2DE(g)$ where $E(g)$ was defined in (2). Equation (12) is not strictly parabolic, as (2) is not strictly elliptic (See 9.8).

3.2. Solution for a Short Time

9.17 Theorem (Hamilton [*151], DeTurck [*111]). *On any compact Riemaniann manifold* (M, g_0)*, the evolution equation* (12) *has a unique solution for a short time with initial metric* g_0 *at* $t = 0$.

For the proof Hamilton used the Nash–Moser inverse function theorem [*150], some special technique is required because equation (12) is not strictly parabolic. When this proof appeared, DeTurck [*109] had already solved the local existence of metrics with prescribed Ricci curvature (that we saw above §2), and then he gave a proof of Theorem 9.17 which uses Theorem 4.51 for parabolic equations.

DeTurck's idea is to show that (12) is equivalent to a strictly parabolic equation (15) when $n = 3$ or (16) for $n > 3$.

Let c be any constant such that $L_{ij} = R_{ij} + cg_{ij}$ is positive definite at any point of (M, g_0). So L^{-1} exists. Recall Definition 9.9: For a symmetric tensor $h = \{h_{ij}\}$,

$$(Gh)_{ij} = h_{ij} - \tfrac{1}{2}g^{kl}h_{kl}g_{ij}$$

and for a 1-form $v = \{v_i\}$, $(\delta^* v)_{ij} = \nabla_i v_j + \nabla_j v_i$.

The second Bianchi identity implies (see 1.20):

(14) $$(\delta GL)_j = -\nabla^i(GL)_{ij} = -\nabla^i R_{ij} + \tfrac{1}{2}\nabla_j R = 0.$$

9.18 When $n = 3$, DeTurck [*111] considers the following parabolic equation:

$$(15)\quad \begin{cases} \dfrac{\partial}{\partial t} g_{ij} = -2\left[R_{ij} - \left(\delta^*\left[L^{-1}\delta GL\right]\right)_{,ij}\right] \\ \dfrac{\partial}{\partial t} L_{ij} = -\Delta L_{ij} - 2c\left(L_{ij} - cg_{ij}\right) - \left[Q(L-cg)\right]_{,ij} \\ g(x,0) = g_0(x), L(x,0) = \mathrm{Ricci}(g_0)(x) + cg_0(x) \end{cases}$$

where the unknown is the pair $[g_{ij}(x,t),\ L_{ij}(x,t)]$. $Q(S)$ is some quadratic expression in S using the metric. This system is strictly parabolic. Indeed by (8) we have that the symbol of the right hand side of the first equation with respect to g is the symbol of minus the laplacian.

Hence from Theorem 4.51 (15) has a unique solution for a short time. We have to show that this solution solves (12).

For this DeTurck considers the quantities

$$u_i = \left[L^{-1}\delta G(L)\right]_i \quad \text{and} \quad P_{ij} = L_{ij} - (R_{ij} + cg_{ij}).$$

A computation gives the evolution equations for u and P. It is a parabolic system which admits the solution $u \equiv 0$ and $P \equiv 0$. As the initial conditions are $P(x,0) = 0$ and $\delta GL(x,0) = 0$, we have indeed $P \equiv 0$ since the solution is unique. Since any solution of (12) is a solution of (16), the resulting solution of (12) is unique.

When $n \geq 4$, the Weyl tensor does not vanish identically, and equation (15) involves the curvature tensor. We must introduce a new unknown T_{ijkl}. The parabolic equation to consider is of the form.

$$(16)\quad \begin{cases} \dfrac{\partial}{\partial t} g_{ij} = -2\left[R_{ij} - (\delta^*[L^{-1}\delta GL])_{ij}\right] \\ \dfrac{\partial}{\partial t} L_{ij} = -\Delta L_{ij} - 2c(L_{ij} - cg_{ij}) + 2g^{pr}g^{qs}T_{ipqj}L_{rs} - 2g^{pq}L_{pi}L_{jq} \\ \dfrac{\partial}{\partial t} T_{ijkl} = -\Delta T_{ijkl} + \text{quadratic expression in } T_{ijkl} \text{ using the metric} \\ g(x,0) = g_0(x),\ L(x,0) = \mathrm{Ricci}(g_0)(x) + cg_0(x), T(x,0) = \mathrm{Riem}(g_0) \end{cases}$$

where $\mathrm{Riem}(g_0)$ is the curvature tensor of g_0.

Thanks to (8) it is obvious that this system is strictly parabolic. Hence (16) has a unique solution for a short time. We prove that this solution satisfies (12) by the same way as above for the dimension 3. The evolution equation for u, P and $S = T - \mathrm{Riem}(g)$ is strictly parabolic and admits the solution $u \equiv 0$, $P \equiv 0$ and $S \equiv 0$.

So (12) has a solution. This solution is unique since any solution of (12) is a solution of (16).

9.19 DeTurck found a simpler proof of the existence, for a short time, of solutions for the evolution equation (12).

Since his proof is unpublished, we reproduce it now. As before, DeTurck replace (12) with a strictly parabolic equation. Let T_{ij} be any symmetric tensor field on M which has the property that T_{ij} is invertible $\big($ as a map from $T_P(M)$ to $T_P^*(M)\big)$ at every point P of M. One could, for instance, take T equal to g_0. Then the equation

$$(17) \qquad \frac{\partial}{\partial t} g_{ij} = -2R_{ij} - 2\big[\delta^* T^{-1} B(g,T)\big]_{ij}, g(0) = g_0$$

has a unique solution for small time by the parabolique existence Theorem 4.51

For the notations δ^*, B, see Definition 9.9. The proof that the right side of (17) is elliptic appears in [*109], see also (8) in 9.9.

The introduction of T breaks the diffeomorphism-invariance of (12) and renders (17) parabolic. To show how to get solutions of (12) from those of (17), we need the following two results.

Proposition 9.19. *Let $v(y,t)(y \in M,\ t \in \mathbb{R}^+)$ be a time-varying vector field on M. Then for small t, there exists a unique family of diffeomorphisms $\varphi_t : M \to M$ such that $\frac{\partial \varphi_t(x)}{\partial t} = v\big(\varphi_t(x,),t\big)$ for all $x \in M$, and with $\varphi_0 =$ identity.*

Proof. The standard proof when v does not depend on t still applies, via the existence and uniqueness theorem for ordinary differential equations (see for instance Warner [*313]).

Lemma 9.19. *Let $g_{ij}(y,t)$ $(y \in M,\ t \in \mathbb{R}^+)$ be a time-varying Riemannian metric on M, and φ_t the family of diffeomorphisms from Lemma* 9.19. *Then*

$$\frac{\partial \varphi_t^*(g)}{\partial t}(x) = \varphi_t^*\left[\frac{\partial g}{\partial t}(\varphi_t(x))\right] + 2\varphi_t^*\left[\delta^* w(\varphi_t(x))\right]$$

where w is the one-form $w_i = g_{ik} v^k$.

Proof. Compute

$$\varphi^*(g)_{ij} = \frac{\partial \varphi^\alpha}{\partial x^i}\frac{\partial \varphi^\beta}{\partial x^j} g_{\alpha\beta}\big(\varphi(x),t\big)$$

so

$$\begin{aligned}
\left(\frac{\partial \varphi^*(g)}{\partial t}\right) &= \frac{\partial v^\alpha}{\partial x^i}\frac{\partial \varphi^\beta}{\partial x^j} g_{\alpha\beta} + \frac{\partial \varphi^\alpha}{\partial x^i}\frac{\partial v^\beta}{\partial x^j} g_{\alpha\beta} + \frac{\partial \varphi^\alpha}{\partial x^i}\frac{\partial \varphi^\beta}{\partial x^j}\frac{\partial}{\partial t} g_{\alpha\beta} \\
&\quad + \frac{\partial \varphi^\alpha}{\partial x^i}\frac{\partial \varphi^\beta}{\partial x^j}\frac{\partial g_{\alpha\beta}}{\partial y^k} v^k \\
&= \varphi^*\left(\frac{\partial g}{\partial t}\right)_{ij} + \frac{\partial \varphi^\alpha}{\partial x^i}\frac{\partial \varphi^\beta}{\partial x^j}\left[\frac{\partial v^k}{\partial y^\alpha} g_{k\beta} + \frac{\partial v^k}{\partial y^\beta} g_{k\alpha} + \frac{\partial g_{\alpha\beta}}{\partial y^k} v^k\right] \\
&= \varphi^*\left(\frac{\partial g}{\partial t}\right)_{ij} + 2\varphi^*\left(\delta^* w\right)_{ij}.
\end{aligned}$$

Proof of Theorem 9.17. Let w be the one-form $w = T^{-1}B(g,T)$ obtained using T and the solution g of (17) above, and let φ_t be the family of diffeomorphisms obtained by integrating v using Proposition 9.19 ($v^k = g^{ki}w_i$). Then according to Lemma 9.19

$$\begin{aligned}\frac{\partial\varphi_t^*(g)}{\partial t} &= \varphi_t^*\frac{\partial g}{\partial t} + 2\varphi_t^*(\delta^* w)\\ &= -2\varphi_t^*\left[\mathrm{Ricci}(g) + \delta^* T^{-1}B(g,T)\right] + 2\varphi_t^*\left[\delta^* T^{-1}B(g,T)\right]\\ &= -2\,\mathrm{Ricci}(\varphi_t^* g).\end{aligned}$$

Thus $\varphi_t^*(g)$ satisfies (12).

3.3. Some Useful Results

9.20 Generalized maximum principle. Let F be a vector bundle over a compact manifold M. To a doubly covariant symmetric tensor field T on F we associate an other two covariant symmetric tensor field on F, $N = p(T,g)$ which is a polynomial in T formed by contracting products of T with itself using the metric g. We say $T \geq 0$ if $\langle T(v), v\rangle \geq 0$ for any $v \in F$.

Theorem 9.20 (Hamilton [*151] p. 279, Margerin [*233] p. 311). *Let T_t and g_t be smooth families on $0 \leq t \leq \tau$ which satisfy*

$$\frac{\partial}{\partial t}T_t = -\Delta T_t + u^k\nabla_k T_t + N_t \tag{18}$$

where $N_t = p(T_t, g_t)$. We suppose $N = p(T,g)$ has the following property: $T(v) = 0$ implies $\langle N(v), v\rangle \geq 0$. Then $T_t \geq 0$ for any $t \in [0,\tau]$ if $T_0 \geq 0$.

Proof. Set $T'_t = T_t + \varepsilon(\delta + t)\,\mathrm{Id}$, where $\varepsilon > 0$ and $\delta > 0$ are small and $\mathrm{Id} = g$ if $F = T(M)$ and where $(\mathrm{Id})_{ijkl} = \frac{1}{2}(g_{ik}g_{jl} - g_{il}g_{jk})$ if $F = \Lambda^2(M)$. These are the bundle F for which we will use Theorem 9.20.

We assert that, for some $\delta > 0$, $T'_t > 0$ on $[0,\delta]$ and for every $\varepsilon > 0$. Then letting $\varepsilon \to 0$ yields $T_t \geq 0$ on $[0,\delta]$, hence on $[0,\tau]$. If not there is a first time θ $(0 < \theta \leq \delta)$ and a unit vector $v \in F_{x_o}$ for some ($x_0 \in M$) such that $T'_\theta(v) = 0$. Thus $\langle p(T'_\theta, g_\theta)(v), v\rangle \geq 0$.

As N is a polynomial, $\|p(T',g) - p(T,g)\| \leq C_1\|T' - T\|$ for some constant C_1 which depends only on $\max(\|T'\|, \|T\|)$.

Then

$$\langle N_\theta(v), v\rangle \geq -C_2\varepsilon\delta \tag{19}$$

We extend v in a neighbourhood of x_0 to a vector field denoted $\tilde{v}$, in such a way that $\tilde{v}$ is independent of t and such that $\nabla\tilde{v}(x_0) = 0$.

Set $f(t,x) = \langle T'_t\tilde{v}, \tilde{v}\rangle$. Then $f \geq 0$ on $[0,\theta] \times M$ and at (θ, x_0), $f = 0$, $\frac{\partial f}{\partial t} \leq 0$, df = 0 and $\Delta f \leq 0$.

This implies among other things $\langle \nabla T_\theta(\tilde{v}), \tilde{v}\rangle_{x_o} = 0$ and $\langle \Delta T_\theta(\tilde{v}), \tilde{v}\rangle_{x_o} \leq 0$. But (18) gives

$$\frac{\partial f}{\partial t} = -\langle \Delta T(\tilde{v}), \tilde{v}\rangle + \langle u^k \nabla_k T(\tilde{v}), \tilde{v}\rangle + \langle N(\tilde{v}), \tilde{v}\rangle + \varepsilon(\delta + t)\frac{\partial}{\partial t}\langle \tilde{v}, \tilde{v}\rangle + \varepsilon\langle \tilde{v}, \tilde{v}\rangle.$$

At (θ, x_0) we get

$$\langle N(\tilde{v}), \tilde{v}\rangle \leq -\varepsilon\big[1 - 2C\delta\big] \tag{20}$$

where $C \geq C_2$ is chosen so that $\frac{\partial}{\partial t}\langle v, v\rangle \leq C$.

If we choose $\delta < 1/3C$, (20) is in contradiction with (19).

9.21 Theorem (Hamilton [*151]). *Interpolation inequality for tensors.*

Let (M_n, g) a compact Riemannian manifold and let p, q, r be real numbers ≥ 1. If $1/r = 1/p + 1/q$, any tensor field T on M satisfies

$$\begin{aligned}(21)\quad &\left[\int |\nabla T|^{2r}\, dV\right]^{1/r} \\ &\leq \left[\sqrt{n} + 2(r-1)\right]\left[\int |\nabla^2 T|^p\, dV\right]^{1/p}\left[\int |T|^q\, dV\right]^{1/q}.\end{aligned}$$

If $p = r \geq 1$ and $q = \infty$, then T satisfies

$$\left[\int |\nabla T|^{2p}\, dV\right]^{\frac{1}{p}} \leq \left[\sqrt{n} + 2(p-1)\right]\sup_M |T|\left[\int |\nabla^2 T|^p\, dV\right]^{\frac{1}{p}}. \tag{22}$$

Proof. Set $T = (T_\alpha)$, α multi-index. Integrating by parts yields:

$$\begin{aligned}\int |\nabla T|^{2r}\, dV &= \int \nabla_i T_\alpha \nabla^i T^\alpha |\nabla T|^{2(r-1)}\, dV \\ (23)\quad &= -\int T^\alpha \nabla^i \nabla_i T_\alpha |\nabla T|^{2(r-1)}\, dV \\ &\quad - 2(r-1)\int T^\alpha \nabla^i \nabla^j T^\beta \nabla_i T_\alpha \nabla_j T_\beta |\nabla T|^{2(r-2)}\, dV.\end{aligned}$$

Now

$$|T^\alpha \nabla^i \nabla_i T_\alpha|^2 \leq n|T|^2|\nabla^2 T|^2 \tag{24}$$

$$T^\alpha \nabla^i \nabla^j T^\beta \nabla_i T_\alpha \nabla_j T_\beta \leq |T||\nabla^2 T||\nabla T|^2. \tag{25}$$

In fact, expanding $|T_\beta \nabla_i \nabla_j T_\alpha - \lambda \nabla_i T_\alpha \nabla_j T_\beta|^2 \geq 0$ yields a polynomial in λ of order 2. The nonpositivity of the discriminant is of this polynomial gives (25). To verify (24), we write

$$|T^\alpha \nabla_i \nabla_j T_\alpha - T^\alpha \nabla^k \nabla_k T_\alpha g_{ij}/n|^2 \geq 0$$

and

$$|T_\beta \nabla_i \nabla_j T_\alpha - T_\alpha \nabla_i \nabla_j T_\beta|^2 \geq 0.$$

The first inequality is $|T^\alpha \nabla^k \nabla_k T_\alpha|^2 \leq n |T^\alpha \nabla_i \nabla_j T_\alpha|^2$ and the second $|T^\alpha \nabla_i \nabla_j T_\alpha|^2 \leq |T|^2 |\nabla_i \nabla_j T_\alpha|^2$.

Putting (24) and (25) in (23) implies

$$\text{(26)} \qquad \int |\nabla T|^{2r} \, dV \leq \left[2(r-1) + \sqrt{n}\right] \int |T| |\nabla^2 T| |\nabla T|^{2(r-1)} \, dV.$$

As $1/p + 1/q + (r-1)/r = 1$ the Hölder inequality then implies

$$\int |\nabla T|^{2r} \, dV \leq \left[2(r-1) + \sqrt{n}\right] \left[\int |\nabla^2 T|^p \, dV\right]^{1/p} \times \left[\int |T|^q| \, dV\right]^{1/q} \left[\int |\nabla T|^{2r} \, dV\right]^{1-1/r}$$

which is (21). Similarly (26) implies (22) when $q = \infty$.

9.22 Corollary (Hamilton [*151]). *Let (M_n, g) be a compact Riemannian manifold and let $m \in \mathbb{N}$. There exists a constant $C(n,m)$ independent of g such that any tensor field T satisfies*

$$\text{(27)} \qquad \int |\nabla^k T|^{2m/k} \, dV \leq C(n,m) \sup_M |T|^{2(m/k-1)} \int |\nabla^m T|^2 \, dV$$

for all integers k with $1 \leq k \leq m-1$.

Proof. Set $f(0) = \sup_M |T|$ and $f(k) = \left[\int |\nabla^k T|^{2m/k} \, dV\right]^{k/2m}$.

Applying (21) to the tensor field $(\nabla_{i_1 i_2 .. i_{k-1}} T_\alpha)$ with $p = \frac{2m}{k+1}$, $q = \frac{2m}{k-1}$ and $r = m/k$ yields

$$\text{(28)} \qquad f^2(k) \leq C^2 f(k+1) f(k-1)$$

where we can choose C depending only on n and m. But (28) implies, as we will see below,

$$\text{(29)} \qquad f(k) \leq C^{m(m-k)} \left[f(0)\right]^{1-k/m} \left[f(m)\right]^{k/m}$$

which is (27). Let us now prove (29). Set $g(k) = \log f(k)$ and a $= \log C$. By (28) we have $2g(k) - g(k+1) - g(k-1) \leq 2a$.

$$\sum_{j=1}^{k} j\left[2g(j) - g(j+1) - g(j-1)\right] \leq k(k+1)a.$$

As $(k+1)g(k) - kg(k+1) - g(0) = \sum_{j=1}^{k} j[2g(j) - g(j+1) - g(j-1)]$ we find

$$\text{(30)} \qquad (1 + 1/k)g(k) - g(k+1) \leq (k+1)a + g(0)/k.$$

And since $(m-k)g(k+1)-(m-k-1)g(k)-g(m) = \sum_{j=k+1}^{m-1}(m-j)[2g(j)-g(j+1)-g(j-1)] \leq (m-k)(m-k-1)a$, we can sum this inequality with $(m-k)$ times (30) to get

$$\frac{m}{k}g(k) \leq g(m) + m(m-k)a + \frac{m-k}{k}g(0)$$

which implies (29).

3.4. Hamilton's Evolution Equations

9.23 Theorem (Hamilton [*151] p. 274). *If $g(t)$ satisfies the evolution equation (12) on $[0,\tau[$, then the curvature tensor $R_{ijkl}(t)$, the Ricci tensor $R_{ij}(t)$ and the scalar curvature $R(t)$ satisfy the following equations:*

$$\text{(31)}\quad \frac{\partial}{\partial t}R_{ijkl} = -\Delta R_{ijkl} + 2\big(B_{ijkl} - B_{ijlk} - B_{iljk} + B_{ikjl}\big) - g^{pq}\big(R_{pjkl}R_{qi} + R_{ipkl}R_{qj} + R_{ijpl}R_{qk} + R_{ijkp}R_{ql}\big)$$

$$\text{(32)}\quad \frac{\partial}{\partial t}R_{ik} = -\Delta R_{ik} + 2g^{pa}g^{qb}R_{piqk}R_{ab} - 2g^{pq}R_{pi}R_{qk}$$

$$\text{(33)}\quad \frac{\partial}{\partial t}R = -\Delta R + 2g^{ij}g^{kl}R_{ik}R_{jl}$$

where $B_{ijkl} = g^{pa}g^{qb}R_{piqj}R_{akbl}$. Recall $\Delta R_{ijkl} = -\nabla^{\nu}\nabla_{\nu}R_{ijkl}$.

Proof. As $\frac{\partial}{\partial t}R_{ik} = g^{jl}\frac{\partial}{\partial t}R_{ijkl} - R_{ijkl}g^{jp}g^{lq}\frac{\partial}{\partial t}g_{pq}$, contracting (31) by g^{jl} gives (32). Similarly contracting (32) yields (33). In normal coordinates for $g(t)$ at P:

$$\frac{\partial}{\partial t}\Gamma^m_{jl} = g^{mk}\big(\partial_k R_{jl} - \partial_j R_{kl} - \partial_l R_{jk}\big).$$

As $\frac{\partial}{\partial t}\Gamma^m_{jl}$ is a tensor field in the local chart

$$\text{(34)}\quad \frac{\partial}{\partial t}\Gamma^m_{jl} = g^{mk}\big(\nabla_k R_{jl} - \nabla_j R_{kl} - \nabla_l R_{jk}\big).$$

According to (1) we have

$$\frac{\partial}{\partial t}R_{ij}{}^m{}_l = \nabla_i\left(\frac{\partial}{\partial t}\Gamma^m_{jl}\right) - \nabla_j\left(\frac{\partial}{\partial t}\Gamma^m_{il}\right).$$

Thus we get

$$\text{(35)}\quad \frac{\partial}{\partial t}R_{ijkl} = \nabla_{ik}R_{jl} - \nabla_{il}R_{jk} - \nabla_{jk}R_{il} + \nabla_{jl}R_{ik} - R^m_k R_{ijml} - R^m_l R_{ijmk}$$

since $\nabla_{ji}R_{kl} - \nabla_{ij}R_{kl} - 2R_{km}R_{ij}{}^m{}_l = R_{ij}{}^m{}_k R_{ml} - R_{km}R_{ij}{}^m{}_l$.

Differentiating the second Bianchi identity we obtain

$$\Delta R_{ijkl} = \nabla^m\nabla_i R_{jmkl} + \nabla^m\nabla_j R_{mikl}.$$

Permuting the covariant derivatives and using the contracted second Bianchi identity we obtain (31) from (35).

9.24 Theorem (Hamilton [*151]). *If $R(0) \geq C \geq 0$ then $R(t) \geq C$ for $0 \leq t < \tau$. In dimension $n = 3$, if $R_{ij}(0) \geq 0$ then $R_{ij}(t) \geq 0$ for $0 \leq t < \tau$. Moreover, when $n = 3$, if $R_{ij}(0) - aR(0)g_{ij}(0) \geq 0$ with $R(0) > 0$ then for $0 \leq t < \tau$, $R_{ij}(t) - aR(t)g_{ij}(t) \geq 0$.*

Proof. The maximum principle for the heat equation implies the first result since according to (33)

$$\frac{\partial}{\partial t}R + \Delta R \geq 0.$$

When $n = 3$ the curvarture tensor expresses itself in terms of the Ricci curvature and R since the Weyl tensor W_{ijkl} vanishes. Thus (32) becomes

$$\text{(36)} \qquad \frac{\partial}{\partial t}R_{ij} + \Delta R_{ij} = N_{ij}$$

with $N_{ij} = -6g^{kl}R_{il}R_{jk} + 3RR_{ij} - (R^2 - 2R_{kl}R^{kl})g_{ij}$.
We remark now that if $R_{ij}u^i = 0$ with $|u| \neq 0$, $N_{ij}u^iu^j = (2R_{ij}R^{ij} - R^2)|u|^2 \geq 0$.

Indeed if the eigenvalues of $((R_{ij}))$ are λ, μ and zero $2R_{ij}R^{ij} - R^2 = (\lambda - \mu)^2$, zero is an eigenvalue since we suppose $R_{ij}u^i = 0$.

Theorem 9.20 then implies the second assertion.

For the third we apply the same theorem to the tensor field $T_{ij} = R_{ij}/R - ag_{ij}$. Indeed we verify that

$$\frac{\partial}{\partial t}T_{ij} + \Delta T_{ij} = \frac{2}{R}g^{kl}\nabla_k R \nabla_l T_{ij} + \tilde{N}_{ij}$$

with $\tilde{N}_{ij} = 2aR_{ij} + N_{ij} - 2R^{-2}R_{kl}R^{kl}R_{ij}$.

As before $R_{ij}u^i = 0$ implies $\tilde{N}_{ij}u^iu^j \geq 0$.

9.25 The curvature tensor defines a linear operator on the space $\Lambda^2(M)$ of two differential forms (ω_{ij}):

$$\text{Riem}(g)(\omega) = R_{ijkl}\omega^{kl}\, dx^i \wedge dx^j.$$

Theorem 9.25 (Margerin [*233]). *If* Riem(g_0) *is positive,* Riem(g_t) *remains positive for all $0 \leq t < \tau$. The smallest eigenvalue λ_t of* Riem(g_t) *satisfies $\lambda_t \geq \lambda_0$.*

Proof. One more time we apply Theorem 9.20.

This time $F = \Lambda^2(M)$, $T_t = \text{Riem}(g_t)$, $u = 0$ and N_t is given by (31). We verify that $\text{Riem}(g)(\omega) = 0$ implies $\langle N_t(\omega), \omega \rangle \geq 0$.

We prove the second part of the theorem by using Theorem 9.20 with T_t defined by

$$T_t(\omega) = \text{Riem}(g_t)(\omega) - \lambda_0 \omega.$$

9.26 The solution g_t of (12) exists on $[0, \tau[$ for some $\tau > 0$ according to Proposition 9.17.

Theorem 9.26 (Hamilton [*151]). *If $R_0 \geq c > 0$, then $\tau \leq n/2c$.*

Proof. Set $f(x, t) = nc/(n - 2ct)$. $\frac{\partial f}{\partial t} = 2f^2/n$ thus

$$\frac{\partial}{\partial t}(R - f) \geq -\Delta(R - f) + \frac{2}{n}(R + f)(R - f)$$

since $|R_{ij} - \frac{R}{n} g_{ij}|^2 \geq 0$ implies $R_{ij}R^{ij} \geq R^2/n$.

As $R - f \geq 0$ at $t = 0$, $R - f$ remains ≥ 0 on $[o, \tau[$. But $f(x, t) \to \infty$ when $t \to n/2c$, so $\tau \leq n/2c$.

9.27 Let us return to the normalized equation (11). In 9.15 we have written $\tilde{g}(\tilde{t})$ for the solution of (11).

The key point is to prove that the solution $\tilde{g}(\tilde{t})$ exists for all $\tilde{t} \geq 0$ and converges to a smooth metric when $\tilde{t} \to \infty$.

E. Hebey pointed out to me that, altough the following theorem is not explicitly stated in Hamilton, all the ingredients needed for its proof were proved in Hamilton.

This theorem is basic in the works of Hamilton [*151], [*179], Huisken [*179], Margerin [*233], [*234] and Nishikawa [*259].

Let Z be the concircular curvature tensor

$$Z_{ijkl} = R_{ijkl} - \frac{R}{n(n-1)}\big(g_{ik}g_{jl} - g_{il}g_{jk}\big). \tag{37}$$

(M, g) has a constant sectional curvature if and only if $Z(g) = 0$.

Theorem 9.27. *Let (M_n, g_0) be a compact Riemannian manifold of dimension $n \geq 3$ and scalar curvature $R_0 > 0$. If there exist positive constants α, β, γ, independent of t such that on M and for all $t \in [0, \tau[$*

$$R_{ij}(t) - \alpha R_t g_{ij}(t) \geq 0 \tag{38}$$

$$|Z(g_t)| \leq \beta R_t^{1-\gamma} \tag{39}$$

then the solution $\tilde{g}(\tilde{t})$ of equation (11) *exists for all $\tilde{t} > 0$, and $\tilde{g}(\tilde{t})$ converges to a metric with constant positive sectional curvature when $\tilde{t} \to \infty$.*

The proof is given in 9.36 and uses many results that we give now, following Hebey (private communication).

9.28 Proposition (Analogous to theorem 11.1 of Hamilton [*151]). *Under the hypothesis of Theorem 9.27, for any $\eta \in]0, 1/n[$, there exists a constant $c(\eta)$ independent of t such that*

$$|\nabla R_t|^2 \leq \eta R_t^3 + c(\eta) \qquad \textit{for } 0 \leq t < \tau. \tag{40}$$

Proof. $R_0 > 0$ implies there exists $\varepsilon > 0$ such that $R_t \geq \varepsilon$ according to Theorem 9.24. Thus $|Z(g_t)| \leq C_1 R_t$ with $C_1 = \beta\varepsilon^{-\gamma}$.

Set A $(t) = |\nabla R|^2 R^{-1} - \eta R^2$. Using (33) a computation gives:

$$\frac{\partial}{\partial t} A = -\Delta A + \frac{4}{R}\nabla_\alpha R \nabla^\alpha |R_{ij}|^2 + K$$

where

$$K = -\frac{2}{R^3}|R\nabla_i\nabla_j R - \nabla_i R\nabla_j R|^2 - \frac{2}{R^2}|R_{ij}|^2|\nabla R|^2$$
$$+ 2\eta|\nabla R|^2 - 4\eta R|R_{ij}|^2.$$

We verify that $K \leq 2R^2(\eta - |R_{ij}|^2 R^{-2}) - 4\eta R|R_{ij}|^2$ and since $|R_{ij}|^2 \geq R^2/n$ we obtain $K \leq -4\eta R^3/n$ if $0 \leq \eta \leq 1/n$.

Now $|Z|^2 = |R_{ijkl}|^2 - \frac{2}{n(n-1)}R^2 \geq \frac{2}{n-1}|R_{ij}|^2 - \frac{2}{n(n-1)}R^2$.
So by (40) there exists a constant C_2 such that $|R_{ij}|^2 \leq C_2 R^2$.

Since $|\nabla_i R_{jk}|^2 \geq |\nabla R|^2/n$ (obtained by developing $|\nabla_i E_{jk}|^2 \geq 0$ with $E_{jk} = R_{jk} - Rg_{jk}/n$) we get

$$\frac{4}{R}\nabla_\alpha R\nabla^\alpha |R_{ij}|^2 \leq \frac{8}{R}|R_{ij}||\nabla R||\nabla_i R_{jk}| \leq C_3|\nabla_i R_{jk}|^2$$

where $C_3 = 8\sqrt{nC_2}$. Hence

$$\frac{\partial A}{\partial t} \leq -\Delta A + C_3|\nabla_i R_{jk}|^2 - 4\eta R^3/n. \tag{41}$$

Moreover using (31) a computation leads to

$$\frac{\partial}{\partial t}|Z|^2 = -\Delta|Z|^2 - 2|\nabla Z|^2 + \frac{16}{n(n-1)}R|E_{ij}|^2$$
$$- 8\left(Z_{ijkl}Z^{pklq}Z_p{}^{ij}{}_q + Z_{ijkl}Z^{pikq}Z_p{}^{jl}{}_q\right).$$

As $|Z|^2 = |W|^2 + \frac{4}{n-2}|E_{ij}|^2 \geq \frac{4}{n-2}|E_{ij}|^2$ (see 37)

$$\frac{\partial}{\partial t}|Z|^2 = -\Delta|Z|^2 - 2|\nabla Z|^2 + \frac{4(n-2)}{n(n-1)}R|Z|^2 + 16|Z|^3. \tag{42}$$

According to Lemma 9.29 and using 37 leads to

$$|\nabla Z|^2 = |\nabla_i R_{jklm}|^2 - \frac{2}{n(n-1)}|\nabla R|^2$$
$$\geq \frac{2}{n-1}\left[1 - \frac{2(n-1)(n+2)}{n(3n-2)}\right]|\nabla_i R_{jk}|^2$$

$$|\nabla Z|^2 \geq \frac{2(n-2)^2}{n(n-1)(3n-2)}|\nabla_i R_{jk}|^2. \tag{43}$$

Now set $F = A + \alpha|Z|^2$ where $\alpha > 0$. For $0 \leq \eta \leq 1/n$ (41), (42) and (43) give $\frac{\partial}{\partial t}F \leq -\Delta F + C_4 R|Z|^2 + C_5|Z|^3 - 4\eta R^3/n$ if α is chosen so that $C_3 \leq \frac{4\alpha(n-2)^2}{n(n-1)(3n-2)}$.

The hypothesis $|Z| \leq \beta R^{1-\gamma}$ then implies the existence of some constant $\tilde{C}(\eta)$ such that

$$C_4 R|Z|^2 + C_5|Z|^3 - \frac{4\eta}{n} R^3 \leq \tilde{C}(\eta).$$

Hence $\frac{\partial}{\partial t} F \leq -\Delta F + \tilde{C}(\eta)$.

By virtue of the maximum principle 4.46, $F \leq C(\eta)$ for $0 \leq t < \tau$.

The result follows.

9.29 Lemma. *On any Riemannian manifold*

$$|\nabla_i R_{jk}|^2 \geq \frac{3n-2}{2(n-1)(n+2)} |\nabla R|^2.$$

Proof. We set $F_{ijk} = \nabla_i R_{jk} - \alpha \nabla_i R g_{jk} - \beta\big(\nabla_j R g_{ik} + \nabla_k R g_{ij}\big)$, and seek α and β such that $g^{jk} F_{ijk} = g^{ij} F_{ijk} = g^{ik} F_{ijk} = 0$.

By virtue of the second Bianchi identity α and β are the solutions of $\alpha + \beta(1+n) = 1/2$ and $n\alpha + 2\beta = 1$.

These are $\alpha = \frac{n}{(n-1)(n+2)}$ and $\beta = \frac{n-2}{2(n-1)(n+2)}$.

So we find $|\nabla_i R_{jk}|^2 = |F_{ijk}|^2 + \lambda |\nabla_i R|^2$ with

$$\lambda = n\alpha^2 + 2\beta^2(1+n) + 4\alpha\beta = \frac{3n-2}{2(n-1)(n+2)}.$$

9.30 Proposition (Hamilton [*151] p. 296). *For any $m \in N$ there exists a constant $C(n,m)$ independent of the metric and t such that*

$$\frac{d}{dt} \int |\nabla^m \operatorname{Riem}|^2 \, dV + 2 \int |\nabla^{m+1} \operatorname{Riem}|^2 \, dV \leq C(n,m) \sup |\operatorname{Riem}| \int |\nabla^m \operatorname{Riem}|^2 \, dV. \tag{44}$$

Here we set $|\nabla^m \operatorname{Riem}| = |\nabla_{\alpha_1 \alpha_2 \ldots \alpha_m} R_{ijkl}|$.

Proof. For any tensor A and B, we write $A * B$ to denote any bilinear combination of these tensors formed by contraction using the metric.

Differentiating (31) gives

$$\frac{\partial}{\partial t}(\nabla^m \operatorname{Riem}) = -\Delta(\nabla^m \operatorname{Riem}) + \sum_{p+q=m} (\nabla^p \operatorname{Riem}) * (\nabla^q \operatorname{Riem}).$$

Then

$$\frac{\partial}{\partial t} |\nabla^m \operatorname{Riem}|^2 = -\Delta |\nabla^m \operatorname{Riem}|^2 - 2|\nabla^{m+1} \operatorname{Riem}|^2 + \sum_{p+q=m} (\nabla^p \operatorname{Riem}) * (\nabla^q \operatorname{Riem}) * (\nabla^m \operatorname{Riem}).$$

For any C^1 function $f(x,t)$

$$\frac{d}{dt}\int f(x,t)\,dV = \int \frac{\partial}{\partial t} f(x,t)\,dV + \int f(x,t)\frac{\partial}{\partial t}\,dV$$

this last term being smaller than $\sup_M |\,\mathrm{Riem}\,| \int |f(x,t)|\,dV$. Therefore for $p+q=m$

$$\text{(45)} \qquad \int |\nabla^p \mathrm{Riem}\,|\,|\nabla^q \mathrm{Riem}\,|\,|\nabla^m \mathrm{Riem}\,|\,dV \leq \mathrm{Const.}\Big(\sup_M |\,\mathrm{Riem}\,|\Big)\int |\nabla^m \mathrm{Riem}\,|^2\,dV.$$

Now we verify this inequality. According to the Hölder inequality the left hand side is smaller than

$$\|\nabla^p \mathrm{Riem}\,\|_{2m/p}\|\nabla^q \mathrm{Riem}\,\|_{2m/q}\|\nabla^m \mathrm{Riem}\,\|_2.$$

and by (27)

$$\|\nabla^k \mathrm{Riem}\,\|_{2m/k} \leq \mathrm{Const.}\sup_M \mathrm{Riem}\,|^{1-k/m}\|\nabla^m \mathrm{Riem}\,\|_2^{k/m}.$$

Thus (45) follows.

9.30 Lemma. *If the maximal interval* $[0,\tau[$ *where the solution* g_t *exists is finite,* $R_0 > 0$ *implies* $\limsup_{t\to\tau}(\sup_M |\,\mathrm{Riem}\,|) = \infty$.

The proof is by contradiction. Suppose $|\,\mathrm{Riem}\,| \leq C$ for any $x \in M$ and $t \in [0,\tau[$. Then g_t converges, when $t \to \tau$, to a smooth metric g_τ and $[0,\tau[$ would not be the maximal interval. Indeed any derivative of g is bounded. We have

$$\frac{\partial}{\partial t} g_{ij} = -2R_{ij}$$

and we can prove that $|\,\mathrm{Riem}\,| \leq C$ implies $|\nabla^m \mathrm{Riem}\,| \leq C(m)$.

For the complete proof see Hamilton [*151] p. 298.

9.31 Lemma. *Under the hypothesis of theorem* 9.27

$$\lim_{t\to\tau}\inf_M R_t = +\infty \quad \textit{and} \quad \lim_{t\to\tau}\big[\sup_M R_t/\inf_M R_t\big] = 1.$$

Proof. According to Lemma 9.30, there exists a sequence $t_j \to \tau$ such that

$$\lim_{j\to\infty}\sup_M |\,\mathrm{Riem}(g_{tj})| = \infty.$$

As

$$|Z|^2 = |\,\mathrm{Riem}\,|^2 - \frac{2}{n(n-1)}R^2 \quad \text{and} \quad |Z| \leq \beta R^{1-\gamma}$$

we have $\lim_{j\to\infty}\sup_M R_{t_j} = \infty$.

Now consider $x_j \in M$ a point where R_{t_j} is maximum. Integrating $|\nabla R_{t_j}|$ on any geodesic γ emanating from x_j yields a bound from below of R_{t_j} as we will see.

We rewrite (40) in the form $|\nabla R_t| \leq \frac{\eta^2}{2} R_t^{3/2} + C(\eta)$. For any $\eta > 0$ there exists $C(\eta)$ independent of t. Fix η small. For j large enough $\big(j > j(\eta)\big)$, $C(\eta) \leq \frac{\eta^2}{2}\big(\sup_M R_{t_j}\big)^{3/2}$ and $|\nabla R_t| \leq \eta^2 (\sup{}_M R_{t_j})^{3/2}$. So at $y \in \gamma$ with

$$d(x_j, y) \leq S = \frac{1}{\eta}\big(\sup_M R_{t_j}\big)^{-1/2}, \qquad R_{t_j}(y) \geq (1-\eta)\sup_M R_{t_j}.$$

But in fact, for η small enough, we get $\inf_M R_{t_j} \geq \big(1-\eta\big)\sup_M R_{t_j}$ when $j > j(\eta)$, since the diameter of M is smaller than s. Indeed by the proof of Myers' theorem 1.43, the hypothesis (38) implies that the Cut-locus of x_j is reached along γ at a point z satisfying $d(x_j, z) \leq \pi\sqrt{n-1}[\alpha(1-\eta)\sup_M R_{t_j}]^{-1/2}$. The assertion is valid if η satisfies $(n-1)\pi^2\eta^2 \leq \alpha(1-\eta)$. Hence $\lim_{j\to\infty}[\sup_M R_{t_j}/\inf_M R_{t_j}] = 1$.

Now, by the maximum principle, for $t \geq t_j$, $\inf{}_M R_t \geq \inf_M R_{t_j}$. Thus $\lim_{t\to\tau}\inf_M R_t = \infty$.

By the same reasoning as above, this implies $\lim_{t\to\tau}[\sup_M R_t/\inf_M R_t] = 1$.

9.32 Lemma. *Under the hypothesis of Theorem* 9.27

$$\int_0^\tau \sup_M R_t dt = \infty, \qquad \int_0^\tau r_t\, dt = \infty$$

where $r_t = \int R_t\, dV_t / \int dV_t$ *and* $\lim_{t\to\tau} |Z_t|/R_t = 0$.

Proof. Let $f(t)$ be the solution of $f'(t) = 2C\sup_M R_t f(t)$ such that $f(0) = \sup_M R_0$ where C is some constant for which $|R_{ij}|^2 \leq CR^2$ (C exists according to (42)).

(33) then implies

$$\frac{\partial}{\partial t}(R-f) \leq -\Delta(R-f) + 2C(R-f)\sup_M R.$$

By virtue of the maximum principle $R_t \leq f(t)$ for $t \in [0,\tau[$ and Lemma 9.31 yields $\lim_{t\to\tau} f(t) = +\infty$.

As

$$\log f(t) = \log f(0) + 2C\int_0^t \sup_M R_t\, dt,$$

we have $\int_0^\tau \sup_M R_t\, dt = +\infty$. Hence $\int_0^\tau r_t\, dt = \infty$ according to Lemma 9.31. (39) and Lemma 9.31 then give $\lim_{t\to\tau}|Z_t|/R_t = 0$.

9.33 Let us return to the normalized equation (11).

$\tilde{g}(\tilde{t})$ denotes the solution of (11) obtained from $g(t)$ as shown in Proposition 9.15

$\tilde{g}(\tilde{t})$ exists on the maximal interval $[0, \tilde{\tau}[$.

Lemma 9.33. *Under the hypothesis of Theorem* 9.27

$$\lim_{\tilde{t}\to\tilde{\tau}} \Big[\sup_M \tilde{R}(\tilde{t}) / \inf_M \tilde{R}(\tilde{t})\Big] = 1, \qquad \tilde{R}_{ij} \geq \alpha \tilde{R} \tilde{g}_{ij}, \tag{48}$$

$$\lim_{\tilde{t}\to\tilde{\tau}} |\tilde{Z}(\tilde{t})| / \tilde{R}(\tilde{t}) = 0 \quad \textit{and}$$

$$\sup_M \tilde{R}(\tilde{t}) \leq C < \infty \quad \textit{for any } \tilde{t} \in [0, \tilde{\tau}[\textit{ and } \tilde{\tau} = +\infty. \tag{49}$$

Proof. Under dilations inequality (38) and $|Z|/R$ are unchanged. Hence (48) comes from (38), Lemma 9.31 and 9.32.

As $\tilde{R}_{ij} \geq \alpha \tilde{R} \tilde{g}_{ij} > 0$, we have $\tilde{V} \leq \sigma_{n-1} \tilde{d}^n / n$ where $\tilde{V}$ and $\tilde{d}$ denote the volume and the diameter of $(M, \tilde{g})$. Now by definition $\tilde{V} = 1$ and according to Myers' theorem $\tilde{d}^2 \leq (n-1)\pi^2 / \alpha \inf_M \tilde{R}$, thus $\inf_M \tilde{R} \leq$ constant. (48) then implies $\sup_M \tilde{R} \leq C < \infty$ and consequently $\tilde{r} \leq C$.

Moreover $d\tilde{t} = m(\tilde{\tau})\, dt$ and $m(t)\tilde{r}(\tilde{t}) = r_t$ (see the proof of Proposition 9.15) yield $\int_0^{\tilde{\tau}} \tilde{r}(\tilde{t})\, d(\tilde{t}) = \int_0^{\tau} r_t\, dt$ which is equal to $+\infty$, according to Lemma 9.32.Thus $\tilde{\tau} = +\infty$.

9.34 Lemma. *Under the hypothesis of Theorem* 9.27, *there exists a constant* $C > 0$ *such that* $\inf_M \tilde{R}(\tilde{t}) \geq C$ *for all* $\tilde{t}$.

Proof. According to Lemma 9.33, $|\tilde{Z}(\tilde{t})| < \varepsilon \tilde{R}(\tilde{t})$ for some $\varepsilon < \frac{1}{2n(n-1)}$ when $\tilde{t}$ is large enough ($\tilde{t} > T_1$). The sectional curvature $\tilde{K}$ of $(M, \tilde{g})$ then satisfies

$$\left(\frac{1}{n(n-1)} - \varepsilon\right) \tilde{R}(\tilde{t}) \leq \tilde{K}(\tilde{t}) \leq \left(\frac{1}{n(n-1)} + \varepsilon\right) \tilde{R}(\tilde{t}) \qquad \text{for } \tilde{t} > T_1.$$

By (48) there exists T_2 such that for $\tilde{t} > T_2$

$$\frac{1}{4}\left[\frac{2}{n(n-1)} \inf_M \tilde{R}(\tilde{t})\right] \leq \tilde{R}(\tilde{t}) < \frac{2}{n(n-1)} \inf_M \tilde{R}(\tilde{t}). \tag{50}$$

Now let us consider the Riemannian universal cover $(\hat{M}, \hat{g})$ of $(M, \tilde{g})$. According to the Klingenberg Theorem, the injectivity radius $\hat{\delta}$ of $(\hat{M}, \hat{g})$ satisfies $\hat{\delta} \geq \pi \big(\frac{2}{n(n-1)} \inf_M \tilde{R}\big)^{-1/2}$. So there exists a constant C such that $\mathrm{Vol}(\hat{M}) \geq C \big(\inf_M R\big)^{-n/2}$ since the sectional curvature is bounded, and we get $\inf_M \tilde{R} \geq [C/\nu]^{2/n}$ where ν is the number of elements in the fundamental group of $(M, \tilde{g})$. Indeed $\mathrm{Vol}(\hat{M}, \hat{g}) = \nu \,\mathrm{Vol}(M, \tilde{g}) = \nu$.

Moreover, by Myers' Theorem ν is finite since $\tilde{R}_{ij} > 0$ (48), and of course ν does not depend on the metric.

9.35 Lemma. *Under the assumptions of Theorem 9.27, there exist two positive real numbers C and δ such that for all $\tilde{t}$:*

$$|\tilde{Z}(\tilde{t})| \leq Ce^{-\delta\tilde{t}} \quad \textit{and} \quad \sup_M \tilde{R}(\tilde{t}) - \inf_M \tilde{R}(\tilde{t}) \leq Ce^{-\delta\tilde{t}}.$$

Proof. From (33) and (44), we get $\frac{\partial}{\partial t}R \geq -\Delta R + \frac{2R^2}{n}$ and

$$\frac{\partial}{\partial t}|Z|^2 \leq -\Delta|Z|^2 - 2|\nabla Z|^2 + \frac{4(n-2)}{n(n-1)}R|Z|^2 + 16|Z|^3 - \lambda|\nabla R|^2 \tag{51}$$

for some constant $\lambda > 0$. Set A $= |Z|^2R^{-2}$. As A is homogeneous (unchanged under dilations), A $= \tilde{A} = |\tilde{Z}|^2\tilde{R}^{-2}$.

We compute $B = \frac{\partial A}{\partial t} + \Delta A - \frac{4}{R}\nabla_i R \nabla^i A$.

$$B = R^{-2}\frac{\partial}{\partial t}|Z|^2 - 2|Z|^2R^{-3}\frac{\partial R}{\partial t} - 2|\nabla Z|^2R^{-2} + R^{-2}\Delta|Z|^2 + 2R^{-4}|Z|^2\big(|\nabla R|^2 - R\Delta R\big),$$

$$B \leq AR\left[\frac{4(n-2)}{n(n-1)} - \frac{4}{n} + 16A^{1/2}\right] + R^{-2}|\nabla R|^2(2A-\lambda) - 2R^{-2}|\nabla Z|^2. \tag{52}$$

Since $\tilde{g} = m(t)g$ and $d\tilde{t} = m(t)\,dt$, we get

$$\tilde{B} = \frac{\partial\tilde{A}}{\partial\tilde{t}} + \tilde{\Delta}\tilde{A} - \frac{4}{\tilde{R}}\tilde{\nabla}_i\tilde{R}\tilde{\nabla}^i\tilde{A} \leq \tilde{A}\tilde{R}\left(16\tilde{A}^{1/2} - \frac{4}{n(n-1)}\right) + \tilde{R}^{-2}|\tilde{\nabla}\tilde{R}|^2(2\tilde{A}-\lambda). \tag{53}$$

By (48), there exists $s > 0$ such that for $\tilde{t} \geq s$

$$\tilde{A}^{1/2}(\tilde{t}) = |\tilde{Z}(\tilde{t})|/\tilde{R}(\tilde{t}) \leq \inf\big(\sqrt{\lambda/2}, 1/8n(n-1)\big).$$

Set $\delta = \frac{\inf \tilde{R}_1}{n(n-1)}$, (53) yields $\tilde{B} \leq -2\delta\tilde{A}$ and by the maximum principle $e^{2\delta\tilde{t}}\tilde{A}(\tilde{t}) \leq e^{2\delta s}\sup\tilde{A}_s$ for $\tilde{t} \geq s$. Hence for all $\tilde{t}$, $|\tilde{Z}_t|e^{2\delta\tilde{t}} \leq C$ some constant since $\tilde{R}$ is bounded by (49).

The proof of the second part of Lemma 9.35 is similar. By virtue of (43) we have

$$\frac{\partial}{\partial t}\left(\frac{|\nabla R|^2}{R^3}\right) \le -R^{-2}\Delta\left(\frac{|\nabla R|^2}{R}\right) + C_3R^{-2}|\nabla_i R_{jk}|^2$$
$$- 2R^{-4}|\nabla R|^2(-\Delta R + 2|R_{ij}|^2)$$

$$(54)\qquad \frac{\partial}{\partial t}(R^{-3}|\nabla R|^2) \le -\Delta\left(\frac{|\nabla R|^2}{R^3}\right) + \frac{4}{R}\nabla^i\left(\frac{|\nabla R|^2}{R^3}\right)\nabla_i R$$
$$+ C_3R^{-2}|\nabla_i R_{jk}|^2 - 2\frac{|\nabla R|^4}{R^5} - 4R^{-4}|\nabla R|^2|R_{ij}|^2.$$

Set $f = R^{-3}|\nabla R|^2 + kR^{-2}|Z|^2$ for some constant $k > 0$. (52) and (54) yield

$$\frac{\partial}{\partial t}f + \Delta f - \frac{4}{R}\nabla^i f\nabla_i R \le \left[C_3|\nabla_i R_{jk}|^2 - 2k|\nabla Z|^2\right]R^{-2}$$
$$- \frac{4}{nR^2}|\nabla R|^2 - 4k\frac{|Z|^2}{R}\left(\frac{1}{n(n-1)} - 4\frac{|Z|}{R}\right)$$
$$- R^{-2}|\nabla R|^2\left(\lambda - 2\frac{|Z|^2}{R^2}\right).$$

Pick k large enough, $k > \frac{n(n-1)(3n-2)}{4(n-2)^2}C_3$.

As $R^{-3}|\nabla R|^2$ is homogeneous $\tilde f = f$ and for $\tilde t \ge s$ (s defined above) we have by (45):

$$\frac{\partial}{\partial \tilde t}\tilde f + \tilde\Delta\tilde f - \frac{4}{\tilde R}\tilde\nabla^i\tilde f\tilde\nabla_i\tilde R \le -\frac{2}{n(n-1)}\tilde R\tilde f \le -2\delta\tilde f.$$

Thus $\tilde f e^{2\delta\tilde t} \le C_1^2$ some constant. Hence

$$|\tilde\Delta\tilde R| \le C_1\tilde R^{3/2}e^{-\delta\tilde t} \le C_2e^{-\delta\tilde t}$$

and $\sup_M \tilde R(\tilde t) - \inf_M \tilde R(\tilde t) \le C_2\tilde d(\tilde t)e^{2\delta\tilde t}$. But we saw in 9.33 that the diameter $\tilde d(\tilde t)$ of $\big(M, \tilde g(\tilde t)\big)$ is uniformly bounded, thus the result follows.

9.36 *Proof of Theorem* 9.27. We have

$$\int_{t_1}^{t_2}\sup_M\left|\frac{\partial}{\partial\tilde t}\tilde g_{ij} biggr\right| d\tilde t = 2\int_{t_1}^{t_2}\sup_M\left|\tilde R_{ij}(\tilde t) - \frac{1}{n}r(\tilde t)\tilde g_{ij}(\tilde t)\right| d\tilde t$$
$$\le 2\int_{t_1}^{t_2}\sup_M|\tilde E_{ij}(\tilde t)|\,d\tilde t + \frac{2}{\sqrt n}\int_{t_1}^{t_2}\sup_M|\tilde R(\tilde t) - r(\tilde t)|\,d\tilde t$$
$$\le \text{Const}\big(e^{-\delta t_1} - e^{-\delta t_2}\big)$$

according to Lemma 9.35.

The metrics $\tilde g(\tilde t)$ are all uniformly equivalent and converge to some metric $\tilde g_\infty$ as $\tilde t \to \infty$ in C^0. Using (21), (22) and (27), we see that all the derivatives of $\tilde g(\tilde t)$ are uniformly bounded and $\tilde g(\tilde t)$ converge to $\tilde g_\infty$ in the C^∞ topology when $\tilde t \to \infty$. (48) together with lemma 9.34 then implies $\tilde Z(\tilde g_\infty) = 0$. Thus $\tilde g_\infty$ has constant positive sectional curvature.

§4. The Consequences of Hamilton's Work

4.1. Hamilton's Theorems

9.37 Theorem (Hamilton [*151]). *A compact Riemannian manifold of dimension 3, which has strictly positive Ricci curvature, carries a metric of constant positive sectional curvature. It is thus diffeomorphic to a quotient of S_3.*

Proof. As Ricci $(g_0) > 0$, we have $R_0 > 0$ and there exists α such that Ricci $(g_0) \geq \alpha R_0 g_0$. By Theorem 9.24, $R_{ij}(t) \geq \alpha R(t) g_{ij}(t)$ for $0 \leq t < \tau$. In order to apply the main Theorem 9.27, which implies the announced result, we have only to show that (39) $|Z(g_t)| \leq \beta R_t^{1-\gamma}$ for some positive constants β, γ, this for all $t \in [0, \tau[$. In dimension 3, inequality (39) is equivalent to $|E_{ij}(t)| \leq \beta R_t^{1-\gamma}$ for some positive constants β, γ, since $|Z|^2 = |W_{ijkl}|^2 + \frac{4}{n-2}|E_{ij}|^2$ with the Weyl tensor $W_{ijkl} \equiv 0$.

Set A $= R_t^{-a}|R_{ij}(t)|^2 - R_t^{2-a}/3 = R_t^{-a}|E_{ij}(t)|^2$ with $1 < a < 2$.

A computation, using (32), (33) and the expression of the Weyl tensor, leads to (see Hamilton [*152] p. 285):

$$(55)\quad \frac{\partial}{\partial t}A + \Delta A \leq 2(a-1)R^{-1}\nabla_i R \nabla^i A + 2R^{-1-a}\big[(2-a)|R_{ij}|^2|E_{ij}|^2 - 2Q\big]$$

where $Q = |R_{ij}|^4 + R\big[R\big(R^2 - 5|R_{ij}|^2\big)/2 + 2R_{ij}R^{ki}R^j_k\big]$.
According to Lemma 9.38 below $Q \geq \alpha^2 |R_{ij}|^2|E_{ij}|^2$. Pick a such that $2 - a \leq 2\alpha^2$, we get

$$\frac{\partial}{\partial t}A + \Delta A \leq 2(a-1)R^{-1}\nabla_i R \nabla^i A.$$

By the maximum principle $A_t \leq A_0$ for all $t \in [0, \tau[$. This is the inequality we need.

9.38 Lemma. $Q \geq \alpha^2 |R_{ij}|^2 |E_{ij}|^2$.

Pick normal coordinates at $x \in M$ such that $R_{ij}(x)$ is diagonal. Let $\lambda \geq \mu \geq \nu \geq 0$ be the eigenvalues of $R_{ij}(x)$. We have

$$R(x) = \lambda + \mu + \nu, \qquad \big|R_{ij}(x)\big|^2 = \lambda^2 + \mu^2 + \nu^2$$

and

$$Q(x) = (\lambda^2 + \mu^2 + \nu^2)^2 + (\lambda + \mu + \nu)$$
$$\times \big[(\lambda + \mu + \nu)(\lambda\mu + \lambda\nu + \nu\mu - 2\lambda^2 - 2\mu^2 - 2\nu^2) + 2\lambda^3 + 2\mu^3 + 2\nu^3\big]$$

$$Q(x) = (\lambda - \mu)^2\big[\lambda^2 + (\lambda + \mu)(\mu - \nu)\big] + \nu^2(\lambda - \nu)(\mu - \nu).$$

Since both sides of the inequality that we wish to prove, are homogeneous of degree 4 in λ, μ, ν, we can suppose $\lambda^2 + \mu^2 + \nu^2 = 1$.

This implies $R^2 = (\lambda + \mu + \nu)^2 \geq 1$, and since $R_{ij} \geq \alpha R g_{ij}$, $\nu \geq \alpha$.

Now $Q(x) \geq \lambda^2(\lambda - \mu)^2 + \nu^2(\mu - \nu)^2 \geq \alpha^2\big[(\lambda - \mu)^2 + (\mu - \nu)^2\big]$ and $|E_{ij}|^2 = \frac{1}{3}\big[(\lambda - \mu)^2 + (\lambda - \nu)^2 + (\mu - \nu)^2\big] \leq (\lambda - \mu)^2 + (\mu - \nu)^2$.

Thus the inequality is proved.

9.39 Theorem. *A compact manifold of dimension* 3, *for which the Ricci curvature is non-negative and strictly positive at some point, is diffeomorphic to a quotient of* S_3.

The proof comes at once from Theorem 9.37 together with the following result (Aubin [21]): If the Ricci curvature of a compact Riemannian manifold is non-negative and positive somewhere, then the manifold carries a metric with positive Ricci curvature.

9.40 Theorem (Hamilton [*152]). *A compact Riemannian manifold of dimension* 4, *whose curvature tensor is strictly positive, carries a metric of constant positive sectional curvature. It is therefore diffeomorphic to* S_4 *or* $\mathbb{P}^4(\mathbb{R})$.

Curvature tensor strictly positive means that the bilinear form on the two-forms, defined by $(\varphi, \Psi) \to \text{Riem}(\varphi, \Psi) = R^{ijkl}\varphi_{ij}\Psi_{kl}$ is positive: $\text{Riem}(\varphi, \varphi) > 0$ if $\varphi \neq 0$.

The proof uses Theorem 9.27 after proving (39) holds.

4.2. Pinched Theorems on the Concircular Curvature

9.41 In order to use Theorem 9.27, the condition (39) suggests that a good hypothesis on the initial metric g_0 would be

$$|Z(g_0)|^2 < C(n)R_0^2. \tag{56}$$

But $|Z|^2 = 4R^2/n(n-1)(n-2)$ for $S_1 \times S_{n-1}$ endowed with the canonical product metric. Consequently if the condition (56) is sufficient to apply Theorem 9.27 when $R_0 > 0$, it must be therefore that $C(n) \leq 4/n(n-1)(n-2)$.

Lemma 9.41. *If* $C(n) \leq 4/n(n-1)(n-2)$, $R_0 > 0$ *together with* (56) *imply* $\text{Ricci}(g_0) > 0$. *More precisely*

$$|E_{ij}(g_0)|^2 < \frac{n-2}{4}C(n)R_0^2.$$

Proof. λ_i $(i = 1, 2, \ldots, n)$ the eigenvalues of E_{ij} satisfy $\sum_{i=1}^n \lambda_i = 0$. This implies $\sup_{1\leq i\leq n} |\lambda_i|^2 \leq \frac{n-1}{n}|E_{ij}|^2$ and if $g_{ij}v^iv^j = 1$,

$$E_{ij}v^iv^j \geq -(1-1/n)^{1/2}|E_{ij}| > -\big[(n-1)(n-2)c/4n\big]^{1/2}R.$$

Thus $R_{ij}v^iv^j > \big[\frac{1}{n} - \big(\frac{(n-1)(n-2)c}{4n}\big)^{1/2}\big]R \geq 0$ if $C \leq 4/n(n-1)(n-2)$.

Remark. $R > 0$ and (56) with $C = 4/n(n-1)(n-2)$ do not imply $\text{Riem}(g_0) > 0$. We get $\text{Riem}(g_0) > 0$ when $C \leq 4/n(n-1)(n-2)(n+1)$.

9.42 With a hypothesis of the type 56, Huisken [*179], Margerin [*233], [*234] and Nishikawa [*259] succeeded in using Theorem 9.27. For Nishikawa, $C(n) = 1/16n^2(n-1)^2$.

For Huisken, $C(n) = 4/n(n-1)(n-2)(n+1)$ if $n \geq 6$, $C(5) = 1/100$ and $C(4) = 1/30$. In these cases Riem(g_0) is positive.

For Margerin, $C(n) = 1/2n(n-1)(n-2)$ if $n \geq 6$, $C(5) = 4/625$ and $C(4) = 1/6$. The constant $C(4)$ of Margerin is optimal. When $n = 4$ and $n > 6$, the best value for $C(n)$ is those announced by Margerin. In particular Riem(g_0) is not necessarily positive.

Theorem 9.42. *A compact Riemannian manifold* (M_n, g_0) *of dimension* $n \geq 4$, *with positive scalar curvature* ($R_0 > 0$), *satisfying*

$$|Z(g_0)|^2 < C(n)R_0^2$$

where $C(4) = 1/6$, $C(5) = 1/100$, $C(6) = 1/210$ *and* $C(n) = 1/2n(n-1)(n-2)$ *when* $n > 6$, *carries a metric with constant sectional curvature. The manifold is diffeomorphic to a quotient of* S_n.

Proof. First we prove that, if (56) holds at time $t = 0$, it remains so on $0 \leq t < \tau$. This gives (38), (see Lemma 9.41). Then we prove the existence of some β and γ such that (39) holds. According to Lemma 9.41 the hypotheses of Theorem 9.27 are then satisfied, it implies Theorem 9.42. The evolution equations of Theorem 9.23 yield

$$\frac{\partial}{\partial t}\left(|Z|^2R^{-\alpha}\right) = -\Delta\left(|Z|^2R^{-\alpha}\right) + 2(\alpha-1)R^{-1}\nabla^i\left(|Z|^2R^{-\alpha}\right)\nabla_i R + A_\alpha$$

where (see for instance Huisken [*179] p. 52):

$$\begin{aligned} A_\alpha = &-2R^{-(2+\alpha)}|R\nabla_i R_{jklm} - R_{jklm}\nabla_i R|^2 \\ &- (2-\alpha)(\alpha-1)R^{-(\alpha+2)}|Z|^2|\nabla R|^2 + 4R^{-(\alpha+1)}\big[(1-\alpha/2)|Z|^2|R_{ij}|^2 \\ &+ 2RR_{ijkl}R^{imkn}R_m{}^j{}_n{}^l + (1/2)R_{ijkl}R^{klmn}R_{mn}{}^{ij} - |R_{ijkl}|^2|R_{ij}|^2\big]. \end{aligned}$$

The problem is to find a constant K such that $|Z| \leq KR$ implies $A_2 \leq 0$ and the existence of some $\alpha < 2$ such that A $_\alpha \leq 0$. Then we can apply the maximum principle and the result will follow. The constant K is found by algebraic computations.

Remark. Recently Margerin [*234] proved that we can take for n large, $C(n) = 4/n(n-1)(n-2)$ in Theorem 9.42.

§5. Recent Results

5.1. On the Ricci Curvature

9.43 We could hope that Hamilton's equation (11) or (12) would yield results on the Ricci curvature, especially after his first article [*151] in dimension 3. But in dimension 3, the curvature tensor expresses itself in terms of the Ricci tensor, and Hamilton's theorem (9.37) was a result on the sectional curvature.

In higher dimension, the method yields, under some hypotheses, a metric with constant sectional curvature.

The most general result on Ricci curvature is the following:

9.44 Theorem (Lohkamp [*226]). *Every Riemannian manifold M_n of dimension $n \geq 3$ carries a complete metric g whose Ricci curvature satisfies*

$$-a(n) < \text{Ricci}(g) < -b(n) \tag{57}$$

where $a(n) > b(n) > 0$ are two constants depending only on n.

So, as for the scalar curvature, when $n \geq 3$, the negative sign for the Ricci curvature has no topological meaning. Previously Gao and Yau [*136] proved that any compact Riemannian manifold of dimension 3 has a metric with negative Ricci curvature.

But the proof of Lohkamp is quite different and begins with the existence on $\mathbb{R}^3$, then on $\mathbb{R}^n$, of a metric g_n which satisfies Ricci $(g_n) < 0$ on a ball B, and $g_n = \mathcal{E}$ the euclidean metric outside B. Surgical techniques are used.

Then, using some deformation techniques, Lohkamp exhibits from g_n a metric g which satisfies (57).

Lohkamp ([*226], [*227]) studied the space of all metrics with negative Ricci curvature. He also proved the following results.

9.45 Theorem (Lohkamp [*226]). *A Riemannian manifold M_n of dimension $n \geq 3$ carries a complete metric g with negative Ricci curvature and finite volume.*

9.46 Theorem (Lohkamp [*226]). *A subgroup G of the group of diffeomorphisms of a compact manifold $M_n (n \geq 3)$ is the isometry group of (M_n, g) for some metric g with negative Ricci curvature, if and only if G is finite.*

Bochner's result asserts that the isometry group of a compact manifold with negative Ricci curvature is finite, Lohkamp proved the converse.

5.2. On the Concircular Curvature

9.47 The concircular curvature tensor Z is defined in 9.27. We saw (see Theorem 9.42) that under an hypothesis of the type (56): $|Z|^2 < C(n)R$ with $R > 0$, we can prove the existence of a metric with positive constant sectional curvature on a compact manifold.

Instead of to have (56) satisfied at each point of M_n, we can ask the following question:

Can we get similar results with only integral assumptions on $|Z|$?

The components Z_{ijkl} of the tensor Z express itself in terms of W_{ijkl} (W the Weyl tensor) and of $E_{ij} = R_{ij} - Rg_{ij}/n$. If one of the two orthogonal components of Z vanishes, Theorem 9.48 gives a first answer to this question.

On a compact Riemannian manifold (M_n, g_0) $n \geq 3$, a Yamabe metric is a metric g such that $\int dV(g) = 1$ and such that $\int R(g)\, dV(g) \leq \int \tilde{R}(\tilde{g}) d\tilde{V}(\tilde{g})$ for all metric $\tilde{g} \in [g]$ (the conformal class of g) with $\int d\tilde{V}(\tilde{g}) = 1$. We know that there always exists at least one Yamabe metric in each conformal class and that the scalar curvature $R(g)$ is constant. If g is Einstein, g is unique in $[g]$.

9.48 Theorem (Hebey-Vaugon [*170]). *Let (M_n, g_0) be a compact Riemannian manifold with $n \geq 3$ and conformal invariant $\mu\big([g_0]\big) > 0$ (see 5.8). We suppose either $[g_0]$ has an Einstein metric or g_0 is locally conformally flat. Then there exists a positive constant $C(n)$, with depends only on n, so that if for some Yamabe metric $g \in [g_0]$, $\|Z(g)\|^2_{g,n/2} < C(n)R^2(g)$, then (M_n, g) is isometric to a quotient of S_n endowed with the standard metric.*

Here $\|Z(g)\|_{g,n/2} = \left[\int |Z(g)|^{n/2}\, dV(g)\right]^{2/n}$. If $[g_0]$ has an Einstein metric, we can pick $C(n) = [(n-2)/20(n-1)]^2$ when $3 \leq n \leq 9$ and $C(n) = (2/5n)^2$ when $n \geq 10$. If g_0 is locally conformally flat $C(3) = 25/6^3$, $C(4) = 6/64$ and $C(n) = 4/n(n-1)(n-2)$ when $n \geq 5$ suffices.

The last constant is optimal. Indeed on $(S_{n-1} \times C, g)$, g the product metric with volume 1, $|Z(g)|^2 = 4R^2/n(n-1)(n-2)$, and g is a Yamabe metric when the radius of C is small enough.

Corollary 9.48 (Hebey–Vaugon [*170]). *$P_4(\mathbb{R})$ and S_4, with their standard metrics, are the only locally conformally flat manifold of dimension 4, which have positive scalar curvature and positive Euler–Poincaré caracteristic. In particular if (M_4, g) is not diffeomorphic to $P_4(\mathbb{R})$ or S_4, M_4 does not carry an Einstein metric if g is locally conformally flat with $R(g) > 0$.*

Chapter 10

Harmonic Maps

§1. Definitions and First Results

10.1 Let (M,g) and $(\tilde M,\tilde g)$ be two C^∞ riemannian manifolds, M of dimension n and $\tilde M$ of dimension m. M will be compact with boundary or without and $\{x^i\}(1\le i\le n)$ will denote local coordinates of x in a neighbourhood of a point $P\in M$ and y^α $(1\le\alpha\le m)$ local coordinates of y in a neighbourhood of $f(P)\in\tilde M$.

We consider $f\in C^2(M,\tilde M)$ the set of the maps of class C^2 of M into $\tilde M$.

Definition 10.1. The first fundamental form of f is $h=f^*\tilde g$. Its components are $h_{ij}=\partial_i f^\alpha\partial_j f^\beta\tilde g_{\alpha\beta}$ where $\partial_i f^\alpha=\frac{\partial f^\alpha}{\partial x^i}$. The energy density of f at x is $e(f)_x=\frac12\left(h_{ij}g^{ij}\right)_x$ and the energy of the map f is defined by $E(f)=\int_M e(f)\,dV$.

As $\tilde g$ is positive definite, the eigenvalues of h are non negative and $E(f)=0$ if and only if f is a constant map.

10.2 Definition. The tension field $\tau(f)$ of the map f is a mapping of M into $T(\tilde M)$ defined as follows. $\tau(f)_x\in T_{f(x)}(\tilde M)$ and its components are:

(1) $$\tau^\gamma(f)_x=-\Delta f^\gamma(x)+g^{ij}(x)\tilde\Gamma^\gamma_{\alpha\beta}\big(f(x)\big)\partial_i f^\alpha(x)\partial_j f^\beta(x).$$

Proposition 10.2 (Eells–Sampson [*124]). *The Euler equation for E is $\tau(f)=0$. For any $v\in C\big(M,T(\tilde M)\big)$ satisfying $v(x)\in T_{f(x)}(\tilde M)$ and $v(x)=0$ for $x\in\partial M$ in case $\partial M\neq\emptyset$:*

(2) $$E'(f)\cdot v=-\int_M\tilde g_{\alpha\beta}\big(f(x)\big)\tau^\alpha(f)_x v^\beta(x)\,dV.$$

Proof.

$$E'(f)\cdot v=\frac12\int_M\partial_\gamma\tilde g_{\alpha\beta}\big(f(x)\big)v^\gamma(x)g^{ij}(x)\partial_i f^\alpha\partial_j f^\beta\,dV$$
$$+\int_M\tilde g_{\alpha\beta}\big(f(x)\big)g^{ij}(x)\partial_i v^\alpha(x)\partial_j f^\beta(x)\,dV.$$

Integrating by parts the second integral in the right hand side, we get

$$E'(f)\cdot v=\int_M \tilde g_{\alpha\beta}\big(f(x)\big)v^\alpha(x)\Delta f^\beta(x)\,dV$$
$$+\frac{1}{2}\int_M\big(\tilde\Gamma^\lambda_{\gamma\beta}\tilde g_{\lambda\alpha}+\tilde\Gamma^\lambda_{\gamma\alpha}\tilde g_{\lambda\beta}\big)_{f(x)}v^\gamma(x)g^{ij}(x)\partial_i f^\alpha(x)\partial_j f^\beta(x)\,dV$$
$$-\int v^\gamma(x)g^{ij}(x)\partial_j f^\beta(x)\partial_\alpha\tilde g_{\gamma\beta}\big(f(x)\big)\partial_i f^\alpha(x)\,dV.$$

Since $\partial_\alpha\tilde g_{\gamma\beta}=\tilde\Gamma^\lambda_{\alpha\beta}\tilde g_{\lambda\gamma}+\tilde\Gamma^\lambda_{\gamma\alpha}\tilde g_{\lambda\beta}$, the symmetry beetwen α and β induced by g^{ij} gives the result (2).

10.3 Definition. A harmonic map $f\in C^2(M,\tilde M)$ is a critical point of E (Definition 10.1). That is to say, f satisfies $\tau(f)=0$.

We can introduce the harmonic maps in another way. Suppose f is an immersion; f is injective on Ω a neighbourhood of P. Let Y be a vector field on Ω; $\tilde Y=f_*Y$ can be extended to a neighbourhood of $f(P)$. For X belonging to $T_x(\Omega)$, we set $\tilde X=f_*X$. We verify that $\tilde\nabla_{\tilde X}\tilde Y$ is well defined and that $\tilde\nabla_{\tilde X}\tilde Y-f_*(\nabla_X Y)=\alpha_x(X,Y)$ is bilinear in X and Y. Indeed

$$\alpha_x(X,Y)=\big[\partial^2_{ij}f^\gamma(x)-\Gamma^k_{ij}\partial_k f^\gamma(x)+\tilde\Gamma^\gamma_{\alpha\beta}\big(f(x)\big)\partial_i f^\alpha(x)\partial_j f^\beta(x)\big]X^iY^j\frac{\partial}{\partial y^\gamma}. \tag{3}$$

We call α_x the *second fundamental form* of f at x. It is a bilinear form on $T_x(M)$ with values in $T_{f(x)}(\tilde M)$. The *tension field* $\tau(f)$ is the trace of α_x for g. f is *totally geodesic* if $\alpha_x=0$ for all $x\in M$ and f is *harmonic* if $\tau(f)=0$.

10.4 Proposition (Ishiara [*183]). **(i)** *f is totally geodesic if and only if for any C^2 convex function φ defined on an open set $\theta\subset\tilde M$, $\varphi\circ f$ is convex on $f^{-1}(\theta)$.*

(ii) *f is harmonic if and only if for any φ as above, $\varphi o f$ is subharmonic.*

Proof. We suppose that the coordinates $\{x^i\}$ are normal at x and that the coordinates $\{y^\alpha\}$ are normal at $f(x)$.

$$\partial_{ij}(\varphi\circ f)_x=(\partial_{\alpha\beta}\varphi)_{f(x)}\partial_i f^\alpha(x)\partial_j f^\beta(x)+(\partial_\alpha\varphi)_{f(x)}\partial_{ij}f^\alpha(x). \tag{4}$$

If $\partial_{ij}f^\alpha(x)=0$, (Hess $\varphi)_{f(x)}\geq 0$ implies Hess $(\varphi\circ f)_x\geq 0$.

Conversly if $\partial_{ij}f^\alpha(x)\neq 0$, we can exhibit a convex function φ such that $\varphi\circ f$ is not convex.

From (4), we get if f is harmonic

$$-\Delta(\varphi\circ f)_x=g^{ij}(x)(\partial_{\alpha\beta}\varphi)_{f(x)}\partial_i f^\alpha(x)\partial_j f^\beta(x)$$

φ convex implies $\Delta(\varphi\circ f)_x\leq 0$. Conversly if $\tau(f)_x\neq 0$, we can exhibit a convex function φ such that $\Delta(\varphi\circ f)_x>0$.

10.5 Proposition. *A C^2 harmonic map f is C^∞.*

f satisfies $\tau(f) = 0$ which is in local coordinates an elliptic equation. By the standard theorems of regularity $f \in C^\infty$.

Examples 10.5. If $(\tilde{M}, \tilde{g})$ is $(\mathbb{R}^m, \mathcal{E})$, we can choose the coordinates $\{y^\alpha\}$ such that $\tilde{\Gamma}^\gamma_{\alpha\beta} \equiv 0$. Thus f is harmonic if and only if $\Delta f^\alpha = 0$ for $1 \leq \alpha \leq m$. In case $\partial M = \emptyset$, f is a constant map.

Suppose $g = f^*\tilde{g}$ for $f \in C^2(M, \tilde{M})$, then $e(f) = n/2$.

10.6 Examples. *The case* $n = 1$. *Suppose* M *is the unit circle* C *and* $f \in C^2(C, \tilde{M})$ *harmonic, then* $f(C)$ *is a closed geodesic on* $\tilde{M}$.

Choose t, the central angle of C, as coordinate.

$$e(f) = \frac{1}{2}\tilde{g}_{\alpha\beta}\frac{df^\alpha}{dt}\frac{df^\beta}{dt}$$

and the tension field is

$$\tau^\gamma(f) = \frac{d^2 f^\gamma}{dt^2} + \tilde{\Gamma}^\gamma_{\alpha\beta}\frac{df^\alpha}{dt}\frac{df^\beta}{dt}.$$

This is the equation of the geodesics. Conversly if $f(C)$ is a closed geodesic on $\tilde{M}$, f is harmonic.

When $n = 2$, there are some relations between the *Plateau problem* and the problem of harmonic maps (see Eells–Sampson [*124]).

The case $m = 1$. *In every homotopy class of maps* $M \to C$, *there is an harmonic map.*

For other examples see Eells–Sampson [*124].

10.7 Proposition. *Consider a third* C^∞ *Riemannian manifold* (M', g') *and* $\tilde{f} \in C^2(\tilde{M}, M')$. *If* f *and* $\tilde{f}$ *are totally geodesic maps, then* $\tilde{f} \circ f$ *is totally geodesic. If* f *is harmonic and* $\tilde{f}$ *totally geodesic, then* $\tilde{f} \circ f$ *is harmonic.*

The composition of harmonic maps is not harmonic in general (see Eells–Sampson [*124]).

10.8 By definition, E is defined if $f \in H_1(M, \tilde{M})$, *i.e.* for each α, f^α belongs locally to $H_1(M)$.

Definition 10.8. A map $f \in H_1(M, \tilde{M})$ is weakly harmonic if it is a critical point of E.

Thanks to the Nash Theorem, if $\tilde{M}$ is compact without boundary, there is an isometric imbedding of $\tilde{M}$ in $\mathbb{R}^k$ for k large enough. We can view $\tilde{M}$ as a submanifold of $\mathbb{R}^k$, $\tilde{M} \subset \mathbb{R}^k$ and $i^*\mathcal{E} = \tilde{g}$, i being the inclusion map. The second fundamental form A of $\tilde{M}$ is given by the $\tilde{\Gamma}^\alpha_{\beta\gamma}$ in a suitable coordinates system $\{z^a\}(1 \leq a \leq k)$ of $\mathbb{R}^k$. For $f \in C^2(M, \tilde{M})$, set $\varphi = i \circ f$. Then f is harmonic if and only if

(5) $$\Delta\varphi^a(x) = -g^{ij}(x)A^a_{f(x)}\Big(\frac{\partial\varphi}{\partial x^i}, \frac{\partial\varphi}{\partial x^j}\Big) \qquad \text{for all } a.$$

When $f \in H_1(M, \tilde{M})$, $\varphi \in H_1(M, \mathbb{R}^k)$ and f is weakly harmonic if and only if for any $\Psi \in C^\infty(M, \mathbb{R}^k)$:

(6) $$\sum_{a=1}^{k} \int_M g^{ij}(x)\Big(\frac{\partial\varphi^a}{\partial x^i}\frac{\partial\Psi^a}{\partial x^j} + A^a_{f(x)}\frac{\partial\varphi}{\partial x^i}, \frac{\partial\varphi}{\partial x^j}\Big)\Psi^a(x)\,dV = 0.$$

To see this, we introduce for instance $\varphi_t(x) = \pi \circ [\varphi(x) + t\Psi(x)]$, where π is the orthogonal projection of $\mathbb{R}^k$ on $\tilde{M}$ which is well defined for t small (see Eells–Lemaire [*121] p. 397).

10.9 Theorem. *If $f \in C^0(M, \tilde{M}) \cap H_1(M, \tilde{M})$ is weakly harmonic, then $f \in C^\infty(M, \tilde{M})$. Thus f is harmonic.*

For the proof see Ladyzenskaya–Ural'ceva [*206].

When $n \geq 3$, there exist weakly harmonic maps which are not C^0 and so not harmonic.

Example 10.9. Consider the case $\tilde{M} = S_m \subset \mathbb{R}^{m+1}$. We can view the maps $f \in H_1(M, S_m)$ as maps $f \in H_1(M, \mathbb{R}^{m+1})$ such that $\sum_{a=1}^{m+1}(f^a)^2 = 1$, $\{\xi^a\}$ $1 \leq a \leq m+1$ being coordinates on $\mathbb{R}^{m+1}$.

Set $|\nabla f|^2 = \sum_{a=1}^{m+1} g^{ij}\partial_i f^a \partial_j f^a$.

Then f is weakly harmonic, if it satisfies in the distributional sense $\Delta f^a = f^a|\nabla f|^2$.

Indeed the second fundamental form of S_m is given at $\xi \in S_m \subset \mathbb{R}^{m+1}$ (see Kobayashi–Nomizu [*202]) by $A^a_\xi(X, Y) = \xi^a \mathcal{E}(X, Y)$.

§2. Existence Problems

2.1. The Problem

10.10 Let (M_n, g) and $(\tilde{M}_m, \tilde{g})$ be two C^∞ compact Riemannian manifolds. Given $f_0 \in C^1(M, \tilde{M})$ does there exist a deformation of f_0 to a harmonic map?

This question was asked by Eells–Sampson who gave the first results in [*124]. They approach the problem through the gradient-line technique. Instead of solving the equation $\tau(f) = 0$ (see (1) for the definition), they consider the parabolic equation

(7) $$\frac{\partial f}{\partial t} = \tau(f_t) \qquad \text{with } f_0 \text{ as initial value.}$$

If f_t satisfies (7), from (2) we have

$$(8)\qquad \frac{dE(f_t)}{dt} = -\int_M \tilde{g}_{\alpha\beta}\big(f_t(x)\big)\tau^\alpha(f_t)_x \frac{\partial f_t^\beta(x)}{\partial t}\,dV = -\int_M \tilde{g}_{\alpha\beta}\big(f_t(x)\big)\frac{\partial f_t^\alpha(x)}{\partial t}\frac{\partial f_t^\beta(x)}{\partial t}\,dV.$$

So $E(f_t)$ is a strictly decreasing function, except for the t for which $\tau(f_t) = 0$, *i.e.* when f_t is harmonic.

The basic result is Theorem 10.16; its proof is of independent interest. We will give a sketch of it, but first we need some computations.

2.2. Some Basic Results

10.11 Lemma. *If f is harmonic*

$$(9)\qquad -\Delta e(f) = |\alpha|^2 + Q(f)$$

with $|\alpha|^2(x) = \tilde{g}_{\beta\gamma}\big(f(x)\big)g^{ik}(x)g^{jl}(x)(\alpha_x)^\beta_{ij}(\alpha_x)^\gamma_{kl}$ *and*

$$Q(f) = -\tilde{R}_{\alpha\beta\gamma\delta}\big(f(x)\big)g^{ik}(x)g^{jl}(x)\partial_i f^\alpha(x)\partial_j f^\beta(x)\partial_k f^\gamma(x)\partial_l f^\delta(x) + R^{ij}(x)\tilde{g}_{\alpha\beta}\big(f(x)\big)\partial_i f^\alpha(x)\partial_j f^\beta(x).$$

Recall α_x is the second fundamental form of f at x (see 10.3).

Proof. We suppose the coordinates $\{x^i\}$ normal at x and the coordinates $\{y^\alpha\}$ normal at $f(x)$.

$$(10)\qquad -\Delta e(f) = \frac{1}{2}\sum_{\alpha=1}^{m}\sum_{k=1}^{n}\partial^2_{kk}g^{ij}(x)\partial_i f^\alpha(x)\partial_j f^\alpha(x) + \sum_{\alpha=1}^{m}\sum_{i,j=1}^{n}\big[\partial^2_{ij}f^\alpha(x)\big]^2 + \sum_{\mu=1}^{m}\sum_{i,k=1}^{n}\partial^3_{ikk}f^\mu(x)\partial_i f^\mu(x) + \frac{1}{2}\sum_{i,j=1}^{n}\partial^2_{\alpha\beta}\tilde{g}_{\lambda\mu}\big(f(x)\big)\partial_i f^\alpha(x)\partial_i f^\beta(x)\partial_j f^\lambda(x)\partial_j f^\mu(x).$$

In normal coordinates, according to (3) $\big(\alpha_x\big)^\gamma_{ij} = \partial^2_{ij}f^\gamma(x)$, thus

$$(11)\qquad \sum_{\gamma=1}^{m}\sum_{i,j=1}^{n}\big[\partial^2_{ij}f^\gamma(x)\big]^2 = |\alpha|^2(x).$$

Since f is harmonic $\tau(f) = 0$. Differentiating (1), we get

$$(12)\qquad \sum_{k=1}^{n}\partial^3_{ikk}f^\mu(x) = \sum_{k=1}^{n}\Big[\partial_i\Gamma^j_{kk}(x)\partial_j f^\mu(x) - \partial_\lambda\tilde{\Gamma}^\mu_{ab}\big(f(x)\big)\partial_i f^\lambda(x)\partial_k f^\alpha(x)\partial_k f^\beta(x)\Big].$$

Now at x, since the coordinates $\{x^i\}$ are normal

$$(13) \qquad R_{ij} = \frac{1}{2}\sum_{k=1}^{n}\left(\partial_j\Gamma^i_{kk} + \partial_i\Gamma^j_{kk} - \partial^2_{kk}g_{ij}\right)$$
$$= \frac{1}{2}\sum_{k=1}^{n}\left(\partial_j\Gamma^i_{kk} + \partial_i\Gamma^j_{kk} - \partial^2_{kk}g^{ij}\right)$$

and at $f(x)$, since the coordinates $\{y^\alpha\}$ are normal

$$(14) \qquad \tilde{R}_{\alpha\mu\beta\lambda} + \tilde{R}_{\beta\mu\alpha\lambda} = \partial_\mu\tilde{\Gamma}^\lambda_{\alpha\beta} + \partial_\lambda\tilde{\Gamma}^\mu_{\alpha\beta} - \partial_{\alpha\beta}\tilde{g}_{\lambda\mu}.$$

Putting (11) and (12) in (10), then using (13) and (14), we obtain (9).

10.12 Proposition (Eells–Sampson [*124]). *If f is a harmonic map, then $\int_M Q(f)\,dV \leq 0$ and equality holds if and only if f is totally geodesic. Furthermore if $Q(f) \geq 0$ on M, then f is totally geodesic and has constant energy density $e(f)$.*

Integrating (9) yields the result.

Corollary 10.12 (Eells–Sampson [*124]). *Suppose that the Ricci curvature of M is non-negative and that the sectional curvature of $\tilde{M}$ is non-positive. Then a map f is harmonic if and only if it is totally geodesic. If in addition there is at least one point of M at which the Ricci curvature is positive, then every harmonic map is constant.*

If the Ricci curvature of M is nonnegative and the sectional curvature of $\tilde{M}$ everywhere negative, then every harmonic map is either constant or maps M onto a closed geodesic of $\tilde{M}$.

Proof. The assumption implies $Q(f) \geq 0$ and $e(f)$ is constant. Thus

$$R^{ij}(x)\tilde{g}_{\alpha\beta}\big(f(x)\big)\partial_i f^\alpha(x)\partial_j f^\beta(x) = 0 \qquad \text{for any } x \in M.$$

If at x_0 the Ricci curvature is negative, $\partial_i f^\alpha(x_0) = 0$ for all i and α, thus $e(f)_x = e(f)_{x_o} = 0$ and f is a constant map.

$Q(f) \geq 0$ implies also

$$(15) \qquad \tilde{R}_{\alpha\beta\gamma\delta}\big(f(x)\big)g^{ik}(x)g^{jl}(x)\partial_i f^\alpha(x)\partial_j f^\beta(x)\partial_k f^\gamma(x)\partial_l f^\delta(x) = 0.$$

If the sectional curvature of $\tilde{M}$ is negative, (15) holds when and only when dim $f_*\big(T_x(M)\big) \leq 1$. The result follows, since $e(f)$ constant implies dim $f_*\big(T_x(M)\big)$ constant in that case.

10.13 Corollary. *Assume there exist two positive constants k and C such that $R_{ij} - kg_{ij}$ is non negative on M and*

$$\left[\tilde{R}_{\alpha\beta\gamma\delta} - C(\tilde{g}_{\alpha\gamma}\tilde{g}_{\beta\delta} - \tilde{g}_{\alpha\delta}\tilde{g}_{\beta\gamma})\right]X^\alpha X^\gamma Y^\beta Y^\delta \leq 0$$

for any $y \in \tilde{M}$ and all X, Y in $T_y\tilde{M}$. Then $e(f)$ is sub-harmonic if f is an harmonic map which satisfies $e(f) \leq k/2C$.

If in addition $\Delta e(f) = 0$, f *is a constant map.*

Thus in example 10.9, *if* f *is harmonic and satisfies* $2e(f) \leq \lambda_1$, *where* λ_1 *is the first positive eigenvalue of* S_n, *then* f *is a constant map.*

10.14 If $f \in C^2(M, \tilde{M})$, we define the *stress-energy tensor* of f: $S(f) = e(f)g - f^*\tilde{g}$. We can prove that

$$\operatorname{div} S(f) \cdot X = -\tilde{g}_{\alpha\beta}\tau(f)^\alpha (f_* X)^\beta;$$

see Baird–Eells [*29] and Pluzhnikov [*264] or Eells–Lemaire [*121].

In particular if f is harmonic, then $\operatorname{div} S(f) = 0$ and conversly, if f is a submersion which satisfies $\operatorname{div} S(f) = 0$, then f is harmonic.

Application. *If* (M, g) *has strictly positive Ricci curvature, then* $\mathrm{Id}\colon (M, g) \to (M, \mathrm{Ricci}(g))$ *is harmonic.*

In fact the condition $\operatorname{div} S(f) = 0$ is only the contracted second Bianchi's identity. It is not difficult to show that

$$\left[\operatorname{div} S(Id)\right]_i = \tfrac{1}{2}\nabla_i R - \nabla^j R_{ij}.$$

10.15 Theorem. *On the unique continuation* (*see* Eells–Lemaire [*120] p. 13).

Let f *and* φ *be two* C^∞ *harmonic maps* $M \to \tilde{M}$. *If at a point* $x_0 \in M$, f *and* φ *are equal and have all their derivatives of any order equal, then* $f \equiv \varphi$.

According to Hartman [*61], given two homotopic harmonic maps f_0 and f_1, there is a C^∞ family f_t, $t \in [0, 1]$ of harmonic maps.

2.3. Existence Results

10.16 Theorem (Eells–Sampson [*124]). *Let* (M, g) *and* $(\tilde{M}, \tilde{g})$ *be two compact Riemannian manifolds. If* $(\tilde{M}, \tilde{g})$ *has nonpositive sectional curvature, then any* $f \in C^1(M, \tilde{M})$ *is homotopic to a harmonic map* φ *for which the energy* $E(\varphi)$ *is minimum in the homotopy class of* f.

Eells–Sampson consider the nonlinear parabolic equation

$$\frac{\partial f_t}{\partial t} = \tau(f_t), \quad f_0 = f. \tag{16}$$

There exists a family of maps f_t, $t \in [0, \tau[$ for some $\tau > 0$, which is continuous at $t = 0$ along with their first-order space derivatives and which satisfies (16) on $]0, \tau[$ with $f_0 = f$.

Such f_t is unique and C^∞ on $\left]0, \tau\right[$. By (8) $\frac{dE(f_t)}{dt} < 0$ except for those values of t for which $\tau(f_t) = 0$.

As $\tilde{M}$ has nonpositive sectional curvature, $e(f_t)$ is bounded on $M \times [0, \tau[$. Moreover Eells–Sampson proved that there is $\varepsilon > 0$ independent of t such that any f_t can be continued as a solution of (16) on the interval $]t, t + \varepsilon[$. Thus $\tau = +\infty$.

Then as f_t, along with their first order space derivatives, form equicontinuous families, there is a sequence $\alpha = \{t_k\}$, $k \in N$, such that the maps f_{t_k} converge uniformly to a harmonic map f_α. A subsequence of these f_α converges uniformly to a harmonic map φ which has the desired property.

K. Uhlenbeck [*307] gave a proof of Theorem 10.16 by using the calculus of variation. See the definition of Sobolev spaces in 10.20. For any $\alpha \in \left[0, \frac{1}{2}\right]$, $H_1^{2n}(M, \tilde{M}) \subset C^\alpha(M, \tilde{M})$ and the inclusion is compact. We define for $\varepsilon > 0$ a map E_ε of $\mathrm{H}_1^{2n}(M, \tilde{M})$ into $\mathbb{R}$ by

$$E_\varepsilon(f) = E(f) + \varepsilon \int_M \left[e(f)\right]^n dV.$$

E_ε is C^∞ and satisfies the Palais–Smale condition. It follows that there exists a minimum of E_ε in each connected component $\mathcal{H}$ of $H_1^{2n}(M, \tilde{M})$.

If ε is small enough, these minima are C^∞. When $\tilde{M}$ has nonpositive sectional curvature, it is possible to show that for all a > 0, $\exists \delta > 0$ such that the set $\left\{f \in \mathcal{H}/f \text{ is a critical point of } E_\varepsilon \text{ for some } \varepsilon < \delta \text{ with } E_\varepsilon(f) \leq a\right\}$ has a compact closure in $\mathcal{H}$. Then there is in $\mathcal{H}$ a harmonic map.

The homotopy comes from the fact, that $\mathcal{H}$ is connected by arcs.

Remark 10.16. According to the Nash theorem, there is a Riemannian imbedding of $\tilde{M}$ in R^k for k large enough. In Theorem 10.16, we can drop the hypothesis $\tilde{M}$ compact if $\tilde{M}$ is complete and if the imbedding $\tilde{M} \to R^k$ satisfies some boundedness conditions (see Eells–Sampson [*124]).

If M is complete and $\tilde{M}$ compact with nonpositive sectional curvature, a map $f \in C^1(M, \tilde{M})$ with finite energy is homotopic to a harmonic map on every compact set of M (Schoen–Yau [*287]).

There are other particular results in Eells [*119], White [*316], Lemaire [177] and Sacks–Uhlenbeck [244].

10.17 Remarks. The uniqueness of the Eells–Sampson flow was studied by Coron [*101] when M has a boundary.

The existence of a global flow was studied by Struwe [*295] and Chen–Struwe [*91] (with a flow in the weak sense), as well as Naito [*250] under some conditions on the initial data. Blow-up phenomenon at finite time was studied by Coron and Ghidaglia [*102].

10.18 Corollary. *On a compact Riemannian manifold with nonpositive sectional curvature, there is no metric whose Ricci curvature is non negative and not identically zero.*

Proof. Without loss of generality, we can suppose M orientable. The proof is by contradiction. Let us suppose that M is endowed with the metrics g and h, the sectional curvature of h being non-positive and the Ricci curvature of g being nonnegative and not identically zero.

Then the identity of (M, g) into (M, h), which is of degree 1, would be homotopic to a harmonic map of degree one. But this is impossible since a harmonic map of (M, g) into (M, h) is a constant map (see Corollary 10.12).

10.19 Theorem (Eells–Lemaire [*122]). *Let f be a C^∞ harmonic map of (M, g) into $(\tilde{M}, \tilde{g})$ such that $\nabla^2 E(f)$ is nondegenerate. For any integer $k \geq 1$ and all $r \in N$, there exists a neighbourhood $\mathcal{V}$ of $(g, \tilde{g})$ in $\mathcal{M}^{r+1+\alpha} \times \tilde{\mathcal{M}}^{r+k+\alpha}$ and a unique C^k map S of $\mathcal{V}$ into $C^{r+1+\alpha}(M, \tilde{M})$ satisfying $S(g, \tilde{g}) = f$ and $S(h, \tilde{h})$ harmonic for $(h, \tilde{h}) \in \mathcal{V}$.*

Here $S(h, \tilde{h})$ is a map of (M, h) into $(\tilde{M}, \tilde{h})$ and $\mathcal{M}^{r+\alpha}$ denotes the set of the C^r Riemannian metrics on M whose derivatives of order r are C^α, $0 < \alpha < 1$.

§3. Problems of Regularity

3.1. Sobolev Spaces

10.20 According to the Nash theorem, there is a Riemannian imbedding i of $\tilde{M}$ in $\mathbb{R}^k$ for some $k \in N$. $H_1^2(M, \mathbb{R}^k)$ is the completion of $\mathcal{D}(M, \mathbb{R}^k)$ with respect to the norm

$$\|f\|^2 = \int_M (|\nabla f|^2 + |f|^2)\, dV. \tag{17}$$

Remember that M is compact. We consider $\tilde{M}$ as a submanifold of $\mathbb{R}^k (\tilde{M} \subset \mathbb{R}^k)$ and we identify f and $\varphi = i \circ f$.

Definition 10.20. $H_1^2(M, \tilde{M})$ is the set of $f \in H_1^2(M, \mathbb{R}^k)$, such that $f(x) \in \tilde{M}$ for all $x \in M$.

$H_1^2(M, \tilde{M})$, which does not depend on k and on the Riemannian imbedding, has a structure of a C^∞ manifold. The tangent space at f is defined by

$$T_f\left[H_1^2(M, \tilde{M})\right] = \left[\psi \in H_1^2(M, \mathbb{R}^k) / \psi(x) \in T_{f(x)} \quad \tilde{M} \text{for all } x \in M\right]. \tag{18}$$

10.21 Theorem. *$C^\infty(M, \tilde{M})$ is dense in $C^0(M, \tilde{M}) \cap H_1^2(M, \tilde{M})$. If dim $M = 2$, $C^k(M, \tilde{M})$ is dense in $H_1^2(M, \tilde{M})$ for all $k \geq 1$. When $n \geq 3$, $C^k(M, \tilde{M})$ is not dense in general in $H_1^2(M, \tilde{M})$.*

$C^\infty(M, \tilde{M})$ is dense in $H_1^2(M, \tilde{M})$ if and only if the homotopy group $\pi_2(\tilde{M})$ is trivial.

These different results where proved by Bethuel [*45], Bethuel–Zang [*51] and Schoen–Uhlenbeck [*285].

For more details see Eells–Lemaire [*121] and Coron [*100].

3.2. The Results

10.22 The first important result is Theorem 10.9: A continuous weakly harmonic map is harmonic.

Theorem 10.22 (Helein [*174]). *When $n = 2$, the weakly harmonic maps are harmonic.*

The other results deal with the maps which minimize the energy E and the subset of M where they are singular.

When $n = 1$, $H_1^2(M, \tilde{M}) \subset C^0(M, \tilde{M})$, and when $n = 2$ we knew for a long time by Morrey [*242] that the minimizers of E were regular. When $n \geq 3$ we define S_f.

10.23 Definition. Let f be a map $M \to \tilde{M}$. The singular set S_f of f is defined by:

$$S_f = M - \text{the open set where } f \text{ is continuous.} \tag{19}$$

We recall the definition of *Hausdorff dimension.*

Let X be a metric space and let $p \geq 0$ be a real number. We set $m_p(X) = \lim_{\varepsilon \to 0} m_{p,\varepsilon}(X)$, $\varepsilon > 0$, where $m_{p,\varepsilon}(X) = \inf \sum_{i=1}^{\infty} (\operatorname{diam} A_i)^p$ for all denumerable partitions $\{A_i\}_{i \in N}$ of X such that $\operatorname{diam} A_i < \varepsilon$, $i \in N$.

The Hausdorff dimension of X, $\dim_H X$ is defined by

$$\dim_H X = \sup\{p / m_p(X) > 0\}. \tag{20}$$

Note that $m_p(X) < \infty$ implies $m_k(X) = 0$ for any $k > p$.

10.24 Theorem (Schoen–Uhlenbeck [*284]). *Let $f \in H_1^2(M, \tilde{M})$ be a weakly harmonic map which minimizes E. Then $\dim_H S_f \leq n - 3$. When $n = 3$, S_f is finite. If $x \in S_f$, there exists a sequence ε_i of positive real numbers, satisfying $\lim_{i \to \infty} \varepsilon_i = 0$, such that the sequence of maps $h_i \in H_1^2(B, \tilde{M})$, defined by $h_i(z) = f \exp_x(\varepsilon_i z)$, converges to a map $u \in H_1^2(B, \tilde{M})$ which is a "minimizing tangent map".*

Here $B = B_n$ is the unit ball in $\mathbb{R}^n$.

Definition 10.24. A homogeneous map $u \in H_1^2(B, \tilde{M})$ (i.e. satisfying $\partial u / \partial r = 0$) is called a minimizing tangent map (MTM) if $E(u) \leq E(v)$, for all $v \in H_1^2(B, \tilde{M})$ such that $v = u$ on ∂B.

The maps MTM characterize the behaviour, near their singularities, of the weak harmonic maps which minimize E. One proves that a MTM is of the form $u(x) = w(x/|x|)$, where $w : S^{n-1} \to \tilde{M}$ is a (weak) harmonic map.

If $x \in S_f$ is isolated, $0 \in \mathbb{R}^n$ is then an isolated singularity of u. In this case, if $\tilde{M}$ is real analytic, Simon [*290] proved that u is unique (see also Gulliver–White). Recently White [*317] gave the first examples for which u is not unique.

10.25 Theorem (Schoen–Uhlenbeck [*284]). *Assume there exists an integer $p \geq 2$, such that u is trivial as soon as $u \in H_1^2(B_q, \tilde{M})$ is a MTM of isolated singularity at 0, for any $q \leq p$. Then $\dim_H S_f \leq n - p - 1$ for all weak harmonic maps $f \in H_1^2(M, \tilde{M})$ of minimizing energy. If $n = p + 1$, S_f is finite. If $n < p + 1$, S_f is empty.*

This theorem is a generalisation of Theorem 10.24 and we have

Corollary 10.25. *If $\tilde{M}$ has non positive sectional curvature, any weak harmonic map $M \to \tilde{M}$ of minimizing energy is harmonic.*

Proof. If $u \in H_1^2(B_q, \tilde{M})$, $q \geq 3$, is a MTM with an isolated singularity at 0, then there exists $w : S_{q-1} \to \tilde{M}$ a harmonic map (smooth) such that $u(x) = w(x/|x|)$. According to Corollary 10.12, w is a constant map and the hypothesis of Theorem 10.25 is satisfied with $p > n - 1$.

10.26 Proposition (Schoen–Uhlenbeck [*284]). *A MTM $u \in H_1^2(B_n, S_m)$ whose singularity at 0 is isolated, is trivial if $n \leq d(m)$ where $d(2) = 2$, $d(3) = 3$ and $d(m) = 1 + \inf\big([m/2], 5\big)$ for $m \geq 4$.*

This result together with Theorem 10.24 implies

Theorem 10.26. *If $n \leq d(m)$, any weak harmonic map f of minimizing energy of M into S_m is harmonic (i.e. smooth). If $n = 1 + d(m)$, S_f is finite and if $n > 1 + d(m)$, $dim_H S_f < n - d(m) - 1$.*

There are very few examples of MTM. Let us mention some of them.

10.27 Proposition (see Lin [*222]). *The map of B_n into S_{n-1} ($n \geq 3$) defined by $x \to x/|x|$ is a MTM.*

Proof. First we establish the following inequality for any $u \in C^1(\mathbb{R}^n, \mathbb{R}^n)$ with $|u| = 1$:

$$|\nabla u|^2 + \frac{1}{n-2}\big[tr(\nabla u)^2 - (\operatorname{div} u)^2\big] \geq 0. \tag{21}$$

Then we verify that $u \in H_1^2(B_n, S_{n-1})$ with $u(x) = x$ on ∂B_n satisfies

$$\int_B \big[(\operatorname{div} u)^2 - \operatorname{tr}(\nabla u)^2\big]^2 dx = (n-1)\omega_{n-1}. \tag{22}$$

Set $u_0(x) = x/|x|$. (21) and (22) imply that any $u \in H_1^2(B_n, S_{n-1})$ such that $u = u_0$ on ∂B_n satisfies

$$\int_B |\nabla u|^2 \geq \frac{n-1}{n-2}\omega_{n-1}. \tag{23}$$

We have only to remark now that $\int_B |\nabla u_0|^2 = \frac{n-1}{n-2}\omega_{n-1}$.

10.28 Proposition. a) (Jäger–Kaul [*186]) *The map of B_n into $S_n \subset \mathbb{R}^{n+1}$ defined by $x \to (x/|x|, 0)$ is a MTM if and only if $n \geq 7$.*

b) (Brezis–Coron–Lieb [*59]) $u \in H_1^2(B_3, S_2)$ *is a MTM if and only if* $u(x) = \pm A(x/|x|)$ *with* $A \in SO(3)$.

c) (Coron–Gulliver [*103]) *For $2 \leq n \leq m-1$, the map of $B_{n+m} \subset \mathbb{R}^{n+1} \times \mathbb{R}^{2m-n-1}$ into S_m defined by $(x, y) \to x/|x|$ is a MTM.*

§4. The Case of $\partial M \neq \emptyset$

4.1. General Results

10.29 From now on M is a compact C^∞ manifold with boundary ($\partial M \neq \emptyset$) and $\tilde{M}$ is a compact C^∞ manifold without boundary. We deal with the Dirichlet problem (see Eells–Lemaire [*120] and [*121] for other boundary conditions, such as Neumann conditions).

For the existence problem, the equivalent of Theorem 10.16 for manifolds with boundary was proved by Hamilton [*149].

If ψ is a C^∞ map of ∂M into $\tilde{M}$, we consider $\mathcal{M}_\psi(M, \tilde{M})$ the set of the map f of M into $\tilde{M}$ such that $f/_{\partial M} = \psi$.

Theorem 10.29 (Hamilton [*149]). *If $\tilde{M}$ has non-positive sectional curvature, there exists, in each connected component of $\mathcal{M}_\psi(M, \tilde{M})$, a unique harmonic map which is a minimizer of E on the component.*

10.30 The results on the regularity of weak harmonic maps which minimizes E, obtained by Schoen and Uhlenbeck (Theorem 10.24), are valid when $\partial M \neq \emptyset$.

Recall $u \in H_1^2(M, \tilde{M}) \cap \mathcal{M}_\psi(M, \tilde{M})$ is a minimizer of E if $E(u) \leq E(v)$ for all $v \in H_1^2(M, \tilde{M}) \cap \mathcal{M}_\psi(M, \tilde{M})$.

Theorem 10.30 (Schoen–Uhlenbeck [*285]). *If $f \in H_1^2(M, \tilde{M})$ is a weak harmonic extension, with minimizing energy, of $\psi \in C^{2,\alpha}(\partial M, \tilde{M})$, then*

a) *S_f, the singular set of f, is compact and $S_f \cap \partial M = \emptyset$.*

b) *f is $C^{2,\alpha}$ in a neighbourhood of ∂M*

c) *The results on* $\dim_H S_f$ *mentioned in Theorems* 10.24 *and* 10.25 *are valid. In particular* $\dim_H S_f \leq n-3$.

The same goes for Theorem 10.26. Let f be a weak harmonic map of minimizing energy of M into S_m. If $n \leq d(m)$, then f is regular.

10.31 When f is no longer minimizing, S_f may be strange.

Theorem 10.31 (Rivière [*278]). *Let $\psi \in C^\infty(\partial B_3, S_2)$ be a non constant map. Then there exists a weak harmonic map f of B_3 into S_2, satisfying $f/_{\partial B_3} = \psi$, such that $S_f = B_3$.*

4.2. Relaxed Energies

10.32 The *relaxed energies* were introduced by Brezis, Coron and Lieb [*59], see also Bethuel–Brezis–Coron [*48].

The problem comes from a fact discoverd by Hardt and Lin [*160B]: *There exist maps $\psi \in C^\infty(\partial B_3, S_2)$ of degree zero, such that*

$$\begin{aligned}(24)\qquad &\inf\{E(u)/u \in H_1^2(\Omega, S_2), u = \psi \text{ on } \partial\Omega\}\\ &< \inf\{E(u)/u \in C^1(\bar{\Omega}, S_2), u = \psi \text{ on } \partial\Omega\},\end{aligned}$$

where $\Omega = B_3$.

It is not difficult to prove the left hand side of (24) is attained. More generally, if $H_1^2(M, \tilde{M}) \cap \mathcal{M}_\psi(M, \tilde{M})$ is not empty, the inf of $E(f)$ on this set is attained by a weak harmonic map φ. Moreover if $\psi \in C^{2,\alpha}$, then the minimizer φ is $C^{2,\alpha}$ on a neighbourhood of ∂M and $\dim_H S_\varphi \leq n-3$ (Theorem 10.30).

To prove the existence of φ, let us consider $\{f_i\} \subset H_1^2(M, \tilde{M}) \cap \mathcal{M}_\psi(M, \tilde{M})$ a minimizing sequence. Since $E(f_i) \leq$ Const., there exists a subsequence which converges weakly in $H_1^2(M, \tilde{M})$ to some minimizer $\varphi \in H_1^2(M, \tilde{M}) \cap \mathcal{M}_\psi(M, \tilde{M})$.

10.33 We are interested now in the Hardt–Lin problem:

Is $\inf\{E(f)/f \in C^1(\bar{\Omega}, S_2),\ f = \psi \text{ on } \partial\Omega\}$ attained?

Let Ω be a bounded open set of $\mathbb{R}^3$ such that $\bar{\Omega}$ is a manifold with C^∞ boundary, and $\psi \in C^\infty(\partial\Omega, S^2)$ a given map of degree zero.

We set

$$H_{1,\psi}^2(\Omega, S_2) = \{f \in H_1^2(\Omega, S_2)/f = \psi \text{ on } \partial\Omega\},$$

$$C_\psi^1(\bar{\Omega}, S_2) = \{f \in C^1(\bar{\Omega}, S_2)/f = \psi \text{ on } \partial\Omega\},$$

$$\begin{aligned}R_\psi(\Omega, S_2) = \{f \in H_{1,\psi}^2(\Omega, S_2) &\text{ which are } C^1 \text{ on } \bar{\Omega}\\ &\text{except at a finite number of points of } \Omega\}.\end{aligned}$$

Let f be a map of $R_\psi(\Omega, S_2)$ and let $\{a_1, a_2, \dots, a_k\}$ be the points of Ω where f is not C^1. We define $d_i = \deg(f, a_i)$ as equal to the degree of the restriction of f to a sphere centered at a_i of small radius $r(B_{ai}(r) \cap \{a_j\} = \emptyset$ if $i \neq j)$. As $\deg \psi = 0, \sum_{i=1}^k d_i = 0$.

Now we denote by $\{P_i\}(1 \leq i \leq p)$ the family of the a_i for which $d_i > 0$, each a_i being repeated d_i times, and by $\{Q_j\}(1 \leq j \leq q)$ the family of the a_i for which $d_i < 0$, each a_i being repeated $|d_i|$ times. Of course $p = q$.

Definition 10.33. $L(f) = \inf$ of $\sum_{i=1}^p d(P_i, Q_{\sigma(i)})$ on all permutations σ of $\{1, 2, \dots, p\}$. Here d is the geodesic distance on Ω.

10.34 Lemma (Brezis–Coron–Lieb [*59]). *If $D(f)$ is the vector field whose components are $D^1(f) = \det(f, \partial f/\partial y, \partial f/\partial z)$, $D^2 f = \det(\partial f/\partial x, f, \partial f/\partial z)$ and $D^3 f = \det(\partial f/\partial x, \partial f/\partial y, f)$,*

$$(25) \qquad L(f) = \frac{1}{4\pi} \sup\left\{ \int_\Omega D(f).\nabla\xi - \int_{\partial\Omega} \xi D(f).\nu \right\}$$

for all $\xi \in C^1(\bar\Omega)$ with $\|\nabla\xi\|_\infty \leq 1$. Here ν is the outside normal.

Thus we can extend the definition of $L(f)$ when $f \in H^2_{1,\psi}(\Omega, S_2)$.

10.35 Proposition (Bethuel–Brezis–Coron [*48]).
Define for $f \in H^2_{1,\psi}(\Omega, S_2)$, $E_1(f) = E(f) + 8\pi L(f)$.

a) *E_1 is l.s.c. on $H^2_{1,\psi}(\Omega, S_2)$ for the weak topology of H^2_1. In particular, $\inf E_1(f)$ for $f \in H^2_{1,\psi}(\Omega, S_2)$ is attained.*
b) $\inf\{E_1(f)/f \in H^2_{1,\psi}(\Omega, S_2)\} = \inf\{E(f)/f \in C^1_\psi(\Omega, S_2)\}$.

With this last equality, the Hardt–Lin problem is reduced to proving the regularity of the minimizers of E_1. In this direction there are some results.

Giaquinta, Modica and Soucek [139] proved that if φ is a minimizer of E_1, $\dim_H S_\varphi \leq 1$.

Bethuel and Brezis [46] proved that the minimizers of the functionals $E_\lambda, \lambda \in [0, 1[$, are in $R_\psi(\Omega, S_2)$ (i.e. regular except at a finite number of points). E_λ is defined by $E_\lambda(f) = E(f) + 8\pi\lambda L(f)$.

Let us mention to finish this section the following

10.36 Theorem (Bethuel, Brezis and Coron [*48]). *If, for some ψ, inequality (24) is strict, then there exists an infinity of weak harmonic maps of Ω into S_2 which are equal to ψ on $\partial\Omega$.*

4.3. The Ginzburg–Landau Functional

10.37 The functional. Let $\Omega \subset \mathbb{R}^2$ be a smooth bounded domain in $\mathbb{R}^2$. For maps $u: \Omega \to \mathbb{C}$ and $\varepsilon > 0$, we consider the functional

$$(26) \qquad E_\varepsilon(u) = \frac{1}{2}\int_\Omega |\nabla u|^2\,dx + \frac{1}{4\varepsilon^2}\int_\Omega \left(|u|^2 - 1\right)^2 dx$$

where dx is the Euclidean measure on R^2.

Bethuel, Brezis and Helein [*50] considered the minimization problem of $E_\varepsilon(u)$ for $u \in H^1_g = \{u \in H_1(\Omega, \mathbb{C})/u = g \text{ on } \partial\Omega\}$ where $g : \partial\Omega \to \mathbb{C}$ is a fixed boundary condition. Here g is assumed to be smooth with values in C the unit circle ($|g| = 1$ on $\partial\Omega$).

The problem consists in studying the behavior of minimizers u_ε of (26) when $\varepsilon \to 0$. It depends on the degree d of g $d = \deg(g, \partial\Omega)$. Obviously such minimizers exist.

10.38 The case $d = 0$. There exists a unique harmonic map $u_0 \in C^\infty(\Omega, C)$ such that $u_0 = g$ on $\partial\Omega$. u_0 satisfies in Ω (see Example 10.9) the equation

$$\Delta u_0 = u_0|\nabla u_0|^2 \quad \text{and} \quad |u_0| = 1 \tag{27}$$

Indeed set $u_0 = e^{i\varphi_o}$; then (27) is equivalent to $\Delta\varphi_o = 0$ in Ω. Now we know that the equation

$$\Delta\varphi_0 = 0 \quad \text{in} \quad \Omega, \quad \varphi_0 = \psi_0 \text{ on } \partial\Omega \tag{28}$$

has a unique solution.

When $d = 0$ there exists $\psi_0 \in C^\infty(\partial\Omega, \mathbb{R})$, defined up to a multiple of 2π, such that $g = e^{i\psi_o}$.

This gives the existence and uniqueness of a solution of (27) satisfying $u_0 = g$ on $\partial\Omega$.

Theorem 10.38 (Bethuel, Brezis and Helein [*50]). *As $\varepsilon \to 0$, $u_\varepsilon \to u_0$ in $C^{1,\alpha}(\bar{\Omega})$ for every $\alpha < 1$. The u_ε satisfies the equation*

$$\Delta u_\varepsilon = \varepsilon^2 u_\varepsilon\left(1 - |u_\varepsilon|^2\right). \tag{29}$$

Thus

$$\tfrac{1}{2}\Delta|u_\varepsilon|^2 = \varepsilon^{-2}|u_\varepsilon|^2\left(1 - |u_\varepsilon|^2\right) - |\nabla u_\varepsilon|^2. \tag{30}$$

Hence $|u_\varepsilon|$ cannot achieve a maximum greater than one. At such point the right side of (30) would be negative and he left side nonnegative.

So u_ε satisfies $|u_\varepsilon| \le 1$ on Ω.

Moreover, $E_\varepsilon(u_\varepsilon) \le E_\varepsilon(u_0) = \frac{1}{2}\int_\Omega |\nabla u_0|^2$ dx; thus $u_\varepsilon \to u_0$ in H_1 since u_0 is unique.

Using a uniform bound for Δu_ε in L_∞, Bethuel, Brezis and Helein deduce Theorem 10.38.

10.39 The case $d \neq 0$. This case is very different since $E_\varepsilon(u_\varepsilon) \to +\infty$ as $\varepsilon \to 0$. Without loss of generality we may assume that $d > 0$.

Theorem 10.39 (Bethuel, Brezis and Helein [*50]). *Suppose Ω is star-shaped. There is a sequence $\varepsilon_n \to 0$ and there exist exactly d points a_i in Ω $(i = 1, 2, \ldots, d)$ and a smooth harmonic map u_0 from $\Omega - K$ into C with $u_0 = g$ on $\partial\Omega$ $\left(K = \bigcup_{i=1}^d \{a_i\}\right)$, such that $u_{\varepsilon_n} \to u_0$ uniformly on compact subsets of $\bar{\Omega} - K$.*

The energy of u_0, $\int |\nabla u_0|^2\, dx$, is infinite and each singularity a_i has degree $+1$.

More precisely $u_0(z) \sim \alpha_i(z - a_i)/|z - a_i|$ in a neighourhood of a_i, with $|\alpha_i| = 1$.

Let (r, θ) be radial coordinates centered at $a_i \in \Omega$. Consider the map v_α defined by $v_\alpha = \frac{r}{\alpha} e^{im_j\theta}$ for $r \leq \alpha$ and $v_\alpha = e^{im_j\theta}$ for $r > \alpha$.

Here α is a positive real number and m a positive integer. If we compute $E_\varepsilon(v_\alpha)$, we see that the leading part in ε will be smaller if we choose $\alpha = \varepsilon$, and $E_\varepsilon(v_\varepsilon) \sim -\pi m_j^2 \log \varepsilon$ when $\varepsilon \to 0$.

Furthermore the degree of $v_\alpha/\partial\Omega$ is m_j. Let u be the sum of some functions of this type centered at different points $a_j \in \Omega$. In order to have the degree of $u/\partial\Omega$ equal to d, the m_j must satisfy $\sum m_j = d$. But $E_\varepsilon(u) \sim -\pi\left(\sum m_j^2\right)\log \varepsilon$.

So to make $E_\varepsilon(u)$ as small as possible, we must choose each m_j equal to +1. This prove the first part of the

10.40 Lemma. *There exists a constant C such that*

$$E_\varepsilon(u_\varepsilon) \leq -\pi d \log \varepsilon + C. \tag{31}$$

Moreover

$$\varepsilon^{-2} \int_\Omega \left(|u_\varepsilon|^2 - 1\right)^2 dx < C. \tag{32}$$

We drope the subscript ε for simplicity. Integrating the scalar product of (29) with $x^j \partial_j u$, we get after integrating by parts

$$\int_\Omega \left[\nabla^j u_i \nabla_j u^i + \frac{1}{2} x^j \partial_j \left(\nabla^k u_i \nabla_k u^i\right)\right] dx - \int_{\partial\Omega} x^j \partial_j u_i \partial_\nu u^i \, ds \tag{33}$$
$$= \frac{1}{2\varepsilon^2} \int_\Omega x^j (1 - |u|^2) \partial_j |u|^2 dx.$$

∂_ν is the outside normal derivative, ds the measure on $\partial\Omega$ and $\{x^j\}$ a coordinate system centered at a point of Ω. We set $r^2 = \sum (x^i)^2$. Integrating by parts the second and the last terms of (33) gives

$$\int_{\partial\Omega} \left[\frac{1}{4} \partial_\nu r^2 (\nabla^k u_i \nabla_k u^i) - x^j \partial_j u_i \partial_\nu u^i\right] ds = \frac{1}{2\varepsilon^2} \int_\Omega \left(1 - |u|^2\right)^2 dx.$$

This can be rewritten as (∂_s is the derivative on $\partial\Omega$)

$$\frac{1}{\varepsilon^2} \int_\Omega \left(1 - |u|^2\right)^2 dx \tag{34}$$
$$= \int_{\partial\Omega} \sum \left\{\frac{1}{2} \partial_\nu r^2 [(\partial_s u^i)^2 - (\partial_\nu u^i)^2] - \partial_s r^2 \partial_s u^i \partial_\nu u^i\right\} ds$$

$\partial_s u^i$ is given and smooth on $\partial\Omega$, so it is bounded. If $\frac{1}{2}\partial_\nu r^2 \geq a > 0$, and this is the case when Ω is star-shaped, (34) yields

$$\frac{1}{\varepsilon^2} \int_\Omega \left(1 - |u|^2\right)^2 dx \leq -a \int_{\partial\Omega} \left(\sum |\partial_\nu u^i| - b\right)^2 ds + C \leq c$$

for some constants b and c.

10.41 *Proof of Theorem* **10.39** *(continuation).* Blowing up u_ε at a point $y \in \Omega$, we find that $v_\varepsilon(\xi) = u_\varepsilon(y + \varepsilon\xi)$ satisfies in some ball in $\mathbb{R}^2$

$$\Delta v_\varepsilon = v_\varepsilon\left(1 - |v_\varepsilon|^2\right) \quad \text{and} \quad |v_\varepsilon| \le 1.$$

A subsequence, noted always v_ε, converges to v which satisfies

$$\Delta v = v\left(1 - |v^2|\right) \quad \text{and} \quad |v| \le 1 \text{ on } \mathbb{R}^2. \tag{35}$$

Proposition 10.41. *If a function v satisfies* (35), *then*

$$\int \left(|v|^2 - 1\right)^2 dx = 2\pi n^2,$$

$n = 0, 1, 2, \ldots \infty$.

For the proof see Brezis, Merle and Rivière [*62]. As v depends on y, we denote this limit by v_y. So if v_y is not a constant of modulus 1, $\int \left(|v_y|^2 - 1\right)^2$ dx $\ge 2\pi$.

According to (32), only for a finite subset K of Ω, v_y is not constant. Now we can write with $\tilde{K} = \{b_1, b_2, \ldots, b_d\}$,

$$\int \left(|\nabla u_\varepsilon|^2\right) dx = \pi d |\log \varepsilon| + W(\tilde{K}) + O(\varepsilon).$$

$W(\tilde{K})$ can be expressed in terms of the Green function of the Laplacian on Ω with some Neumann condition depending on g.

There is K which minimizes $W(\tilde{K})$. u_0 satisfies (27) in $\Omega - K$ with $u_0 = g$ on $\partial\Omega$ and for each i $(1 \le i \le d)$, there exists $\alpha_i \in \mathbb{C}$ with $|\alpha_i| = 1$ such that

$$|u_0(z) - \alpha_i(z - a_i)|z - a_i|^{-1}| \le \text{Const.}\, |z - a_i|^2.$$

Remark. This problem in dimension 2 is very different than in the other dimensions. If B_n is the unit ball in $\mathbb{R}^n$, the map: $B_2 \ni x \to x/|x| \in C$ does not belong to $H_1(B_2, C)$. For $n > 2$ the map $x \to x/|x|$ belongs to $H_1(B_n, S_{n-1})$. See for instance Bethuel, Brezis [*46] and Brezis [*58].

Bibliography

[1] Adams R.— *Sobolev Spaces.* Academic Press, New York, 1975.
[2] Agmon S.— The L_p approach to the Dirichlet Problem. *Ann. Scuola Norm. Sup. Pisa* **13**, (1959) 405–448.
[3] Agmon S.— *Lectures on elliptic boundary value problems.* Van Nostrand, New York, 1965.
[4] Agmon S., Douglis A., and Nirenberg L.— Estimates near the boundary for solutions of elliptic partial differential equations satisfying general boundary conditions I. *Comm. Pure Appl. Math.* **12**, (1959) 623–727.
[5] Alexandrov A.D.— Dirichlet's problem for the equation

$$\text{Det } \|z_{ij}\| = \varphi(z_1, \ldots, z_n, z, x_1, \ldots, x_n).$$

Vestinik Leningrad *Univ. Ser. Math. Meh. Astr.* **13**, (1958) 5–24. MR 20 ♯ 3385.
[6] Amann H. and Crandall M.— On some existence theorems for semi-linear elliptic equations. Math. Research Center, Uni. Wisconsin, Madison 1977.
[7] Atiyah M. and Bott R.— On the Yang–Mills Equations over Riemannian Surfaces. (to appear).
[8] Atiyah M. F. and F. Hirzebruch— Riemann–Roch theorems for differentiable manifolds. *Bull. Amer. Math. Soc.* **65**, (1959) 276–281.
[9] Atiyah M., Hitchin N., and Singer I.— Self-duality in four-dimensional Riemannian geometry. *Proc. Roy. Soc. Lond.* **362**, (1978) 425–461.
[10] Aubin T.— Sur la fonction exponentielle. *C.R. Acad. Sci. Paris* **270A**, (1970) 1514.
[11] Aubin T.— Métrique riemanniennes et courbure. *J. Diff. Geo.* **4**, (1970) 383–424.
[12] Aubin T.— Fonction de Green et valeurs propres du Laplacien. *J. Math. pures et appl.* **53**, (1974) 347–371.
[13] Aubin T.— Problèmes isopérimétriques et espaces de Sobolev. *J. Diff. Géo.* **11**, (1976) 573–598. *Abstract in C.R. Acad. Sc. Paris* **280A**, (1975) 279.
[14] Aubin T.— Equations différentielles non linéaires et Problème de Yamabe concernant la courbure scalaire. *J. Math. Pures et appl.* **55**, (1976) 269–296.
[15] Aubin T.— Inégalités concernant la lère valeur propre non nulle du laplacien *C.R.A.S. Paris* **281A**, (1975) 979.
[16] Aubin T.— Equations différentielles non linéaires. *Bull. Sc. Math.* **99**, (1975) 201–210.
[17] Aubin T.— Espaces de Sobolev sur les variétés Riemanniennes. *Bull. Sc. Math.* **100**, (1976) 149–173.
[18] Aubin T.— Equations du type Monge–Ampère sur les variétés kählériennes compactes. *C.R.A.S. Paris* **283A**, (1976) 119.
[19] Aubin T.— The Scalar Curvature. *Differential Geometry and Relativity*, edited by M. Cahen and M. Flato, Reidel Publ. Co. Dordrecht, Holland (1976).
[20] Aubin T.— Equations du type Monge–Ampère sur les variétés kählériennes compactes. *Bull. Sc. Math.* **102**, (1978) 63–95.
[21] Aubin T.— Meilleures constantes dans le théorème d'inclusion de Sobolev et un théorème de Fredholm non linéaire pour la transformation conforme de la courbure scalaire. *J. Funct. Ana.* **32**, (1979) 148–174.
[22] Aubin T.— Inégalités d'Interpolation. *Bull. Sc. Math.* **105**, (1981) 229–234.
[23] Aubin T.— Equations de Monge–Ampère réelles. *J. Funct. Ana.* **41**, (1981) 354–377.
[24] Avez A.— Characteristic Classes and Weyl Tensor: Applications to general Relativity. *Proc. Nat. Acad. Sci. U.S.A.* **66**, (1970) 265–268.

[25] Avez A.— Le laplacien de Lichnerowicz sur les tenseurs *C.R. Acad. Sc. Paris* **284A**, (1977) 1219.

[26] Avez A.— Le laplacien de Lichnerowicz. *Rend. Sem. Mat. Univ. Poli. Torino* **33**, (1976) 123–127.

[27] Bedford E. and Fornaess J.— Counterexamples to regularity for the complex Monge–Ampère Equation. *Invent. Math.* **50**, (1979) 129–134.

[28] Bedford E. and Gaveau B.— Hypersurfaces with bounded Levi Form. *Indiana Uni. Math. J.* **27**, (1978) 867–873.

[29] Bedford E. and Taylor B.— The Dirichlet Problem for a complex Monge–Ampère equation. *Invent. Math.* **37**, (1976) 1–44.

[30] Bedford E. and Taylor B.— The Dirichlet Problem for an equation of complex Monge–Ampère Type. Partial differential equations and Geometry. *Proc. Conf. Park City Utah* (1977), 39–50. *Lectures Notes in Pure and Appl. Math.* **48**, Dekker, New-York (1979).

[31] Bedford E. and Taylor B.— Variational Properties of the complex Monge–Ampère equation. I Dirichlet Principle. *Duke Math. J.* **45**, (1978) 375–403.

[32] Bedford E. and Taylor B.— Variational Properties of the complex Monge–Ampère equation. II Intrinsic Norms. *Am. J. Math.* **101**, (1979) 1131–1166.

[33] Berard P. et Meyer D.— Inégalités isopérimétriques et applications. *Ann. Sci. Ec. Norm. Super.* **15** (1982) 513–542.

[34] Berard Bergery L.— La courbure scalaire des variétés riemanniennes. Séminaire N. Bourbaki (1980) **556**.

[35] Berger Marcel—Blaschke's Conjecture for Spheres. Appendix D in the book below by A. Besse.

[36] Berger M.— Lectures on geodesics in Riemannian geometry. *Lecture Notes in Math.* **33**, Tata Inst. Bombay (1965).

[37] Berger M., Gauduchon P. et Mazet E.— Le spectre d'une variété riemannienne. *Lecture Notes in Math.* **194** Springer-Verlag, Berlin, Heidelberg, New York, 1971.

[38] Berger M. et B. Gostiaux— *Géométrie Différentielle.* Armand Colin, Paris, 1972.

[39] Berger Melvyn S.— An eigenvalue problem for nonlinear elliptic partial differential equations. *Trans. Amer. Math. Soc.* **120**, (1965) 145–184.

[40] Berger M. S.— On Riemannian structures of prescribed Gaussian curvature for compact 2-manifolds. *J. Diff. Geo.* **5**, (1971) 325–332.

[41] Berger M. S.— Constant scalar curvature metrics for complex manifold. *Proc. Amer. Math. Soc. Inst. Diff. Geo.* **27**, (1975) 153–170.

[42] Berger M. S.— *Nonlinearity and Functional Analysis.* Academic Press, New York, 1977.

[43] Berger M. S.— Nonlinear Problems with exactly three Solutions. *Ind. Univ. Math. J.* **28**, (1979) 689–698.

[44] Berger M. S.— Two minimax problems in the Calculus of variations. In Partial differential equations. *Proc. Sympos. Pure Math.* **23**, Berkeley (1971) 261–267.

[45] Berger M. S. and Bombieri E.— On Poincaré's isoperimetric problem for simple closed geodesic. *J. Funct. Ana.* **42**, (1981) 274–298.

[46] Berger M. S. and Church P.— On the complete integrability of nonlinear partial differential equations. *Bull. Am. Math. Soc.* **1**, (1979) 698–701.

[47] Berger M. S. and Fraenkel L.— Nonlinear desingularization in certain free-boundary Problems. *Com. Math. Phys.* **77**, (1980) 149–172.

[48] Berger M. S. and Plastock R.— On the singularities of nonlinear Fredholm Operators of positive index. *Proc. Am. Math. Soc.* **79**, (1980) 217–221.

[49] Berger M. S. and Podolak E.—On the solutions of a nonlinear Dirichlet Problem. *Indiana J. Math.* **24**, (1975) 837–846.

[50] Bers L., John F., and Schechter M.— *Partial Differential Equations.* Wiley, New York (1964).

[51] Besse A.—Manifolds all of whose Geodesics are closed. *Ergebnisse der Mathematik* **93**, Springer-Verlag, Berlin 1978.

[52] Bidal P. et De Rham G.— Les formes différentielles harmoniques. *Comm. Math. Helv.* **19**, (1946) 1–49.

[53] Bishop R. and Crittenden R.— *Geometry of Manifolds.* Academic Press, New York (1964).

[54] Blaschke W.— *Vorlesungen über Differential Geometrie*, Vol. 1. Springer-Verlag, 1924.

[55] Bliss G.— An integral inequality. *J. London. Math. Soc.* **5**, (1930) 40–46.

[56] Bombieri E.— Variational problems and elliptic equations. *Proceedings of the international Congress of Math., Vancouver* **1**, (1974) 53–63.

[57] Bombieri E. and Giusti E.— Harnack's inequality for elliptic differential equations on minimal surfaces. *Invent. Math.* **15**, (1972) 24–46.

[58] Bombieri E. and Giusti E.— Local estimates for the gradient of nonparametric surfaces of prescribed mean curvature. *Comm. Pure Appl. Math.* **26**, (1973) 381–394.

[59] Bourguignon J. P.— Séminaire Palaiseau. Première classe de Chern et courbure de Ricci: preuve de la conjecture de Calabi. *Astérisque* **58**.

[60] Bourguignon J. P.— Premières formes de Chern des variétés kählériennes compactes. Séminaire N. Bourbaki (1977) **507**.

[61] Bourguignon J. P.— Deformations des métriques d'Einstein. *Asterisque* **80**, (1980) 21–32.

[62] Bourguignon J. P., Lawson H. B., and Simons J.— Stability and gapphenomena in Yang–Mills fields. *Proc. Nat. Acad. Sci. USA* **76**, (1979) 1550–1553.

[63] Brezis H. et Browder F.— Existence theorems for nonlinear integral equations of Hammerstein type. *Bull. Am. Math. Soc.* **81**, (1975) 73–78.

[64] Brezis H. and Lions P. L.— Boundary regularity for some nonlinear elliptic degenerate equation. *Comm. Math. Phy.* **70**, (1979) 181–185.

[65] Brezis H. and Nirenberg L.— Characterizations of the ranges of some nonlinear operators and applications to boundary value problems. *Ann. Scuola Norm. Sup. Pisa Cl. Sci.* **5**, (1978) 225–326.

[66] Brezis H. and Veron L.— Removable singularities of some nonlinear elliptic equations. *Arch. Rat. Mech. and Analysis* **75**, (1980) 1–6.

[67] Brezis H. and Waigner S.— A note on limiting cases of Sobolev Imbeddings and Convolution Inequalities. *Comm. Part. Diff. Eq.* **5**, (1980) 773–789.

[68] Browder F.— Nonlinear operators in Banach spaces. *Proc. Symp. P.M. 18 pt. 2—Amer. Math. Soc.* Providence, Rhode Island.

[69] Browder F.— Problèmes non-linéaires. *Presses de l'Université de Montreal*, Montreal, 1966.

[70] Browder F.— Functional Analysis and related fields. *Proceedings of a Conference in honor of Professor Marshall Stone, University of Chicago 1968*. Springer-Verlag, New York, 1970.

[71] Buseman H.— Convex surfaces. *Interscience Tracts in Pure and Applied Mathematics* **6** (1958).

[72] Buser P.— On Cheeger inequality $\lambda_1 \geq h^2/4$. *Proc. Symp. Pure Math. A.M.S.* **36**, (1980) 29–77.

[73] Calabi E.— The space of Kähler metrics, *Proc. Internat. Congress Math. Amsterdam* **2**, (1954) 206–207.

[74] Calabi E.— On Kähler manifolds with vanishing canonical class. Algebraic geometry and topology. A symposium in Honor of S. Lefschetz, Princeton University Press, Princeton (1955) 78–79.

[75] Calabi E.— Improper affine hyperspheres and a generalization of a theorem of K. Jörgens. *Mich. Math. J.* **5**, (1958) 105–126.

[76] Calabi E.— On Ricci curvature and geodesics. *Duke Math. J.* **34**, (1967) 667–676.

[77] Calabi E.— Examples of Bernstein Problems for some nonlinear equations. *Proceedings of symposia in pure Mathematics. Amer. Math. Soc.* **15**, (1970) 223–230.

[78] Calabi E.— Complete hyperspheres I, Ist. *Naz. Alta Mat. Symp. Math.* **10**, (1972), 19–38.

[79] Calabi E.— On manifolds with non-negative Ricci curvature II. *Notices Amer. Math. Soc.* **22**, (1975) A 205.

[80] Calabi E.— Métriques Kählériennes et fibrés holomorphes. *Ann. Sc. Ec. Norm. Sup. Paris* **12**, (1979) 269–294.

[81] Chaljub-Simon A. et Choquet-Bruhat Y.— Solutions asymptotiquement euclidiennes de l'équation de Lichnerowicz. *C.R. Acad. Sc. Paris* **286**, (1978) 917.

[82] Cheeger J.— A lower bound for the smallest eigenvalue of the Laplacian. Problems in Analysis. *Symposium in Honor of S. Bochner*. Princeton University Press, Princeton (1970) 195–199.

[83] Cheeger J. and Ebin D.— *Comparison Theorems in Riemannian Geometry*. North Holland, Amsterdam 1975.

[84] Cheeger J. and Gromoll D.— The splitting theorem for manifolds of non-negative Ricci curvature. *J. Diff. Geo.* **6**, (1971) 119–128.

[85] Cheng S. Y.— Eigenfunctions and eigenvalues of Laplacian. *Proc. Symp. Pure Math. Am. Math. Soc.* **27**, (1975) 185–193.

[86] Cheng S. Y.— Eigenvalue comparison theorems and its geometric applications. *Math. Z.* **143**, (1975) 289–297.

[87] Cheng S. Y.— On the Hayman–Osserman–Taylor Inequality. (to appear).

[88] Cheng S. Y. and Yau S. T.— On the regularity of the solution of the *n*-dimensional Minkowski problem. *Comm. Pure Appl. Math.* **29**, (1976) 495–515.

[89] Cheng S. Y. and Yau S. T.— On the regularity of the Monge–Ampère equation det $((\partial^2 u/\partial x^i \partial x^j)) = F(x, u)$. *Comm. Pure Appl. Math.* **30**, (1977) 41–68.

[90] Cheng S. Y. and Yau S. T.— On the Existence of a complete Kähler metric on non-compact complex Manifolds and the regularity of Fefferman's equation. *Comm. Pure Appl. Math.* **33**, (1980) 507–544.

[91] Chern S. S.— Characteristic classes of Hermitian Manifold. *Ann. of Math.* **47**, (1946) 85–121.

[92] Chen S. S.— On curvature and characteristic classes of a Riemannian manifold. *Abh. Math. Sun. Univ. Hamburg* **20**, (1956) 117–126.

[93] Chern S. S., Levine and Nirenberg L.— Intrinsic norms on a complex manifold. In "Global Analysis" Papers in Honor of Kodaira K. ed. Spencer D. and Iyanaga S. University of Tokyo Press and Princeton University Press 1969.

[94] Chern S. S. and Moser J.— Real hypersurfaces in complex manifolds. *Acta. Math.* **133**, (1974) 219–271.

[95] Cherrier P.— Une inégalité de Sobolev sur les variétés Riemanniennes. *Bull. Sc. Math.* **103**, (1979) 353–374.

[96] Cherrier P.— Cas d'exception du théorème d'inclusion de Sobolev sur les variétés Riemanniennes et applications. *Bull. Sc. Math.* **105**, (1981) 235–288.

[97] Cherrier P.— Problèmes de Neumann non linéaires sur les Variétés Riemanniennes. *J. Diff. Geo.* (to appear).

[98] Cherrier P.—Etude d'une équation du type Monge–Ampère sur les variétés kählériennes compactes. *J. Funct. Ana.* (to appear).

[99] Choquet-Bruhat Y.— *Geométrie Différentielle et systèmes extérieurs.* Dunod, Paris (1968).

[100] Choquet-Bruhat Y. et J. Leray— Sur le problème de Dirichlet quasi-linéaire d'ordre 2. *Compt. Rendu Acad. Sc. Paris* **274**, (1972) 81–85.

[101] Choquet-Bruhat Y. et Marsden J.— Sur la positivité de la masse. *Compt. Rendu Acad. Sc. Paris* **282**, (1976) 609.

[102] Choquet-Bruhat Y. and Marsden J.—Solution of the local mass problem in general relativity. *Comm. Physics* **51**, (1976) 283–296.

[103] Croke C.— Some isoperimetric inequalities and eigenvalue estimates. *Ann. Sc. Ec. Norm. Sup.* **13**, (1980) 419–435.

[104] Delanoë P.— Equations du type de Monge–Ampère sur les variétés Riemanniennes compactes. *J. Funct. Anal.* Part I, **40**, (1980) 358–386. Part II, **41**, (1981) 341–353. Part III, **45**, (1982) 403–430.

[105] De Rham G.— Sur la théorie des formes différentielles harmoniques. *Ann. Univ. de Grenoble* **22**, (1946) 135–152.

[106] De Rham G.— *Variétés différentielles.* Hermann, Paris (1955).

[107] Deturck D. and Kazdan J.— Some regularity theorems in Riemannian Geometry. *Ann. Sc. Ec. Norm. Sup.* **14**, (1981).

[108] Dieudonné J.— *Panorama des Mathématiques Pures* Gauthier Villars, Paris (1979).

[109] Dieudonne J.— *Treatise on Analysis.* Academic Press, New York.

[110] Dinghas A.— Einfacher Beweis der isoperimetrischen Eigenschaft der Kugel in Riemannschen Räumen konstanter Krümmung. *Math. Nachr.* **2**, (1949) 148–162.

[111] Dunford N. and Schwartz J.— *Linear Operators.* Wiley, New York. Part I (1958), Part II (1963), Part III (1971).

[112a] Eells J. Jr. and Sampson J.— Harmonic mappings of Riemannian manifolds. *Amer. J. Math.* **86**, (1964) 109–160.

[112b] Evans L. C.— Classical solutions of fully non linear, convex second order, elliptic equations (to appear).

[113] Federer H.— *Geometric Measure Theory*. Springer-Verlag, New York (1969).
[114] Fefferman C.— Monge–Ampère equations, the Bergman kernel, and geometry of pseudoconvex domains. *Ann. of Math.* **103**, (1976) 395–416.
[115] Fischer A. and Marsden J.— Deformations of the scalar curvature. *Duke Math. J.* **42**, (1975) 519–547.
[116] Friedman A.— *Partial Differential Equations*. Holt, New York (1969).
[117] Friedman A.— *Partial differential equations of parabolic type*. Prentice-Hall, Englewood Cliffs, New Jersey (1964).
[118] Gagliardo E.— Properietà di alcune classi di funzioni in piu variabili. *Ric. Mat.* **7**, (1958) 102–137.
[119] Gallot S.— Un theorème de pincement et une estimation sur la première valeur propre du laplacien d'une variété Riemannienne. *C.R. Acad. Sc. Paris* **289A**, (1979) 441.
[120] Gallot S.— Estimées de Sobolev quantitatives sur les variétés Riemanniennes et applications. *C.R. Acad. Sc. Paris* **292A**, (1981) 375.
[121] Gallot S.— Minorations sur le λ_1 des variétés Riemanniennes. Séminaire N. Bourbaki (1981) **569**.
[122] Gaveau B.— Méthode de contrôle optimal en analyse complexe, I: Résolution d'équations de Monge–Ampère, *J. Funct. Anal.* **25**, (1977) 391–411, abstract in *C.R. Acad. Sc. Paris* **284A**, (1977) 593–596.
II: Equation de Monge–Ampère dans certains domaines faiblement pseudoconvexes. *Bull. Sc. Math.* **102**, (1978) 101–128.
III: Diffusion des fonctions convexes, plurisousharmoniques et des courants positifs fermés. *J. Funct. Anal.* **32**, (1979) 228–253. Abstract in *C.R. Acad. Sc. Paris* **287A**, (1978) 413.
[123] Giaquinta M. and Giusti E.— Nonlinear elliptic systems with quadratic growth. *Manuscripta Math.* **24**, (1978) 323–349.
[124] Gidas B., Ni W-M. and Nirenberg L.— Symmetry and related properties via the Maximum Principle. *Comm. Math. Phys.* **68**, (1979), 209–243.
[125] Gilbarg D., Trüdinger N. S.— *Elliptic Partial Differential Equations of Second Order*. Springer-Verlag, New York (1977).
[126] Giraud G.— Sur la problème de Dirichlet généralisé. *Ann. Sc. Ecole Norm. Sup.* **46**, (1929) 131–245.
[127] Giraud G.— Sur différentes questions relatives aux équations du type elliptique. *Ann. Sc. Ecole Norm. Sup.* **47**, (1930) 197–266.
[128] Glimm J. and Jaffe A.— Multiple meron solutions of the Classical Yang–Mills equation. *Physics Letters* **73B**, (1978) 167–170.
[129] Glimm J. and Jaffe A.— Droplet model for quark confinement. *Phys. Rev. D* **18**, (1978) 463–467.
[130] Glimm J. and Jaffe A.— Meron Paris and quark confinement. *Phy. Rev. Letters* **40**, (1978) 277–278.
[131] Golberg S.— Mappings of nonpositively curved manifolds. *Nagoya Math. J.* **61**, (1976) 73–84.
[132] Gray A.— Chern numbers and curvature. *Am. J. Math.* **100**, (1978) 463–476.
[133] Gray A. and Vanhecke L.— Riemannian geometry as determined by the volumes of small geodesic balls. *Acta Math.* **142**, (1979) 157–198.
[134] Gromov M.— Homotopical effects of dilatation. *J. Diff. Geo.* **13**, (1978) 303–310.
[135] Gromov M.— Paul Levy's isoperimetric inequality. *Publ. Math. Inst. Hautes Etud. Sci.* (1980).
[136] Gromov M., Lawson H. B. Jr.— Spin and scalar curvature in the presence of a fundamental group I. *Ann. of Math.* **111**, (1980) 209–230.
[137] Gromov M., Lawson H. B. Jr.— The classification of a simply connected manifolds of positive scalar curvature. *Ann. of Math.* **111**, (1980) 423–434.
[138] Hardy G., Littlewood J. et Polya G.— *Inequalities*. Cambridge Univ. Press, London and New York 1943.
[139] Helgason S.— *Differential Geometry and Symmetric Spaces*. Academic Press, New York, 1962.
[140] Hilbert D.— *Grundzüge einer allgemeinen Theorie der linearen Integralgleichungen*. Teubner (1912) Chapter XVIII.
[141] Hilbert Problems. Edited by F. Browder. *Proc. Symp. Pure Math. A.M.S. V.28* Providence, Rhode Island 1976.

[142] Hirzebruch F.— *Topological Methods in Algebraic Geometry*, Springer-Verlag, 1962.
[143] Hirzebruch F. and Kodaira K.— On the complex projective spaces. *J. Math. Pures Appl.* **36**, (1957) 201–216.
[144] Hitchin N.— Compact four-dimensional Einstein manifolds. *J. Diff. Geom.* **9**, (1974) 435–441.
[145] Hitchin N.— The space of harmonic spinors. *Advances in Math.* **14**, (1974) 1–55.
[146] Hitchin N.— On the curvature of rational surfaces. In Differential Geometry (*Proc. Sympos. Pure Math.* **27**, Stanford 1973) 65–80.
[147] Hopf E.— Uber den functionalen insbesondere den analytischen Charakter der Lösung elliptischer Differentialgleichungen zweiter Ordnung. *Math. Zeit.* **34**, (1931) 194–233.
[148] Hörmander L.— *An introduction to several complex variables.* Van Nostrand-Reinhold, Princeton.
[149] Hsiung C-C.— Some integral formulas for closed hypersurfaces. *Math. Scand.* **2**, (1954) 286–294.

[150] Inoue A.— On Yamabe's problem—by a modified direct method.
[151] Itô S.— The Fundamental solution of the Parabolic Equation in a differentiable Manifold II. *Osaka Math. J.* **6**, (1954) 167–185.
[152] Itô S.— Fundamental solutions of parabolic differential equations and boundary value problems. *Japan J. Math.* **27**, (1957) 55–102.

[153] Jaffe A.— Introduction to gauge theories. *Proceedings of the International Congress of Mathematicians*, Helsinki (1978) 905–916.
[154] Jonsson T., McBryan O., Zirilli F., and Hubbard J.— An existence theorem for multimeron solution to classical Yang–Mills field equations. *Comm. Math. Phys.* **68**, (1979) 259–273.

[155] Kazdan J.— An isoperimetric inequality and wiedersehen manifolds. Seminar on Differential Geometry, ed. S. T. Yau *Annals of Math. Studies* **102**.
[156] Kazdan J. and Kramer R.— Invariant criteria for existence of solutions to second order quasilinear elliptic equations. *Comm. Pure Appl. Math.* **31**, (1978) 619–645.
[157] Kazdan J. and Warner F.— Curvature functions for compact 2-manifolds. *Ann. of Math.* **99**, (1974) 14–47.
[158] Kazdan J. and Warner F.— Curvature functions for open 2-manifolds. *Ann. of Math.* **99**, (1974) 203–219.
[159] Kazdan J. and Warner F.— Existence and conformal deformation of metrics with prescribed Gaussian and scalar curvatures. *Ann. of Math.* **101**, (1975) 317–331.
[160] Kazdan J. and Warner F.— Scalar Curvature and conformal deformation of Riemannian structure. *J. Diff. Geom.* **10**, (1975) 113–134.
[161] Kazdan J. and Warner F.— Prescribing curvatures. *Proc. Symposia Pure Math.* **27**. *Am. Math. Soc.* (1975).
[162] Kazdan J. and Warner F.— A direct approach to the determination of Gaussian and scalar curvature functions. *Inv. Math.* **28**, (1975) 227–230.
[163] Kazdan J. and Warner F.— Remarks on some quasilinear elliptic equations. *Comm. Pure Appl. Math.* **38**, (1975) 567–579.
[164] Kinderlehrer D., Nirenberg L., and Spruck J.— Regularity in elliptic free boundary Problems. I, *J. d'Analyse Math.* **34**, (1978) 86–119. II, *Ann. Sc. Norm. Sup. Pisa* **6**, (1979) 637–683.
[165] Kobayashi S.— The First Chern class and holomorphic symmetric tensor fields. *J. Math. Soc. Japan* **32**, (1980) 325–329.
[166] Kobayashi S.— First Chern class and holomorphic tensor fields. *Nagoya Math. J.* **77**, (1980) 5–11.
[167] Kobayashi S. and Nomizu K.— *Foundations of Differential Geometry I and II.* Interscience Publishers, New York, 1963.
[168] Kodaira K.— *Collected Works I–III.* Princeton University Press 1975.
[169] Kodaira K. and Morrow J.— *Complex manifolds.* Holt, New York (1971).
[170] Kramer R.— Sub- and Super-Solutions of quasilinear elliptic boundary value problems. *J. Diff. Equat.* **28**, (1978) 278–283.
[171] Krasnoselki M.— Topological Methods in the theory of nonlinear integral equations. *Pergamon*, Oxford.

[172] Kuiper N. H. On conformally flat spaces in the large. *Ann. of Math.* **50**, (1949) 916–924.

[173] Ladyzenskaya O. and Uralsteva N.— *Linear and Quasilinear Elliptic Partial Differential Equations*. Academic Press, New York, 1968.

[174] Lelong P.— *Fonctions plurisousharmoniques et formes différentielles positives*. Gordon et Breach, Paris.

[175] Lelong-Ferrand J.— Transformations conformes et quasi conformes des variétés riemanniennes compactes (démonstration de la conjecture de Lichnerowicz), *Mem. Acad. Roy. Belg. Cl. Sc. Mem. Coll.*, in *8°*, **39** (1971).

[176] Lelong-Ferrand J.— Construction de modules de continuité dans le cas limite de Sobolev et applications à la Géométrie Différentielle. *Arch. Rat. Mech. Ana.* **52**, (1973) 297–311.

[177] Lemaire L.— Applications harmoniques de surfaces Riemanniennes. *J. Diff. Geo.* **13**, (1978) 51–78.

[178] Leray J.— Etude de diverses équations intégrales non linéaires. *J. Math.* **12**, (1933) 1–82.

[179] Leray J.— La théorie des points fixes et ses applications en analyse. (1952) 202–208. *Amer. Math. Soc.*, Providence, Rhode Island.

[180] Leray J. and Schauder J.— Topologie et équations fonctionnelles. *Ann. Sci. Ecole Norm. Sup.* **51**, (1934) 45–78.

[181] Li P.— A lower bound for the first eigenvalue of the Laplacian on a compact Riemannian manifold. *Ind. J. Math.* **28**, (1979) 1013–1019.

[182] Li P.— On the Sobolev constant and the p-spectrum of a compact Riemannian manifold. *Ann. Ec. Norm. Sup.* **13A**, (1980) 451–469.

[183] Li P. and Yau S-T.— Estimates of eigenvalues of a compact Riemannian manifold. *Proc. Symp. Pure Math. A.M.S.* **36**, (1980) 205–239.

[184] Lichnerowicz A.— *Theorie globale des connexions et des groupes d'holonomie*, Edizioni Cremonese, Roma (1962).

[185] Lichnerowicz A.— *Geométrie des groupes de transformations*. Dunod, Paris (1958).

[186] Lichnerowicz A.— Spineurs harmoniques. *C.R. Acad. Sc. Paris* **257A**, (1963) 7–9.

[187] Lichnerowicz A.— Propagateurs et commutateurs en relativité générale. *Publications Math. Inst. H.E.S.* **10**, (1961).

[188] Lions J. L.— *Equations différentielles opérationnelles et problèmes aux limites*. Springer-Verlag, Berlin (1961).

[189] Lions J. L.— *Quelques methodes de résolution des problèmes aux limites non-linéaires*. Dunod, Paris (1969).

[190a] Lions P. L.—Résolution analytique des problèmes de Bellman-Dirichlet. *Acta Math.* **146**, (1981) 151–166.

[190b] Lions P. L.— Une méthode nouvelle pour l'existence de solutions régulières de l'équation de Monge-Ampère. *C.R. Acad. Sc. Paris* **293**, (1981) 589.⁺

[191a] Lions J. L. and Magenes F.— *Problèmes aux limites non homogènes*, Vols. 1, 2. Dunod, Paris 1968.

[191b] Lions J. L. and Magenes F.— On a mixed problem for □ with a discontinuous boundary condition II. An example of moving boundary. *J. Math. Soc. Japan* **30**, (1978) 633–651.

[192] Ljusternik L.— Topology of the Calculus of Variations in the Large. *Amer. Math. Soc. Transl.* **16**. *Amer. Math. Soc.*, Providence, Rhode Island (1966).

[193] Ljusternik L. and Schnirelmann L.— Methode topologique dans les Problèmes Variationnelles. *Actualités Sci. Indust.* **188**. Hermann, Paris (1930).

[194] Loewner C. and Nirenberg L.—Partial differential equations invariant under conformal or projective transformations. *Contributions to Analysis* 245–272, Academic Press, New York, 1974.

[195] Malliavin P.— *Géométrie différentielle intrinsèque*. Hermann, Paris (1972).

[196] Malliavin P.— Formules de la moyenne, calcul des perturbations et theorèmes d'annulation pour les formes harmoniques. *J. Funct. Anal.* **17**, (1974) 274–291.

[197] Malliavin P.— Annulation de cohomologie et calcul des perturbations dans L_2. *Bull. Sc. Math.* **100**, (1976) 331–336.

[198] Meyer D.— Courbure des sommes de métriques. *Bull. Soc. Math. France* **109**, (1981) 123–142.

[199] Michael J. and Simon L.— Sobolev and mean-value inequalities on generalized submanifolds of $\mathbb{R}^n$. *Comm. Pure Appl. Math.* **26**, (1973) 361–379.

⁺The details will appear *Manuscripta Math.* and *Arch. Rat. Mech. Anal.*

[200] Milnor J.— *Morse Theory*. Princeton Univ. Press, Princeton, New Jersey (1963).

[201] Milnor J.— *Topology from the differentiable viewpoint*. Univ. of Virginia Press, Charlottesville, Virginia (1965).

[202] Minakshisundaram S.— Eigenfunctions on Riemannian manifolds. *J. Indian Math. Soc.* **17**, (1953) 158–165.

[203] Minakshisundaram S. and Pleijel A.— Some properties of the eigenfunctions of the Laplace operator on Riemannian manifolds. *Canad. J. Math.* **1**, (1949) 242–256.

[204] Morrey C. Jr.— *Multiple Integrals in the Calculus of Variations*. Springer-Verlag, New York (1966).

[205] Morse M.— The Calculus of Variations in the Large (Colloq. Publ. 18), *Amer. Math. Soc.*, Providence, Rhode Island (1934).

[206] Morse M.— Global variational Analysis: Weierstrass Integrals on a Riemannian manifold. *Princeton Math. Notes*, Princeton University Press, Princeton (1976).

[207] Morse M. and Cairns S.— *Critical Point Theory in Global Analysis*. Academic Press, New York (1969).

[208] Moser J.— A rapidly convergent iteration method and nonlinear partial differential equations, I, *Ann. Scuola Norm. Sup. Pisa* **20**, (1966) 226–315, II, 449–535.

[209] Moser J.— A sharp form of an inequality of N. Trudinger. *Indiana Univ. Math. J.* **20**, (1971) 1077–1092.

[210] Moser J.— On a nonlinear problem in differential geometry. In *Dynamical Systems* edited by M. Peixoto. Academic Press, New York (1973).

[211] Myers S.— Riemannian manifolds with positive mean curvature. *Duke Math. J.* **8**, (1941) 401–404.

[212] Narasimhan R.—*Analysis on real and complex manifolds*. Masson, New York (1971).

[213] Nash J.— The imbedding problem for Riemannian manifolds. *Ann. of Math.* **63**, (1956) 20–63.

[214] Ni W-M.— Conformal metrics with zero scalar curvature and a symmetrization process via Maximum Principle. Seminar on Differential Geometry, ed. S-T. Yau. *Annals of Math. Studies* **102**.

[215] Ni W-M.— Some Minimax principles and their applications in nonlinear elliptic equations. *J. d'Analyse Math.* **37**, (1980) 248–275.

[216] Nirenberg L.— On nonlinear elliptic differential equations and Hölder continuity. *Comm. Pure Appl. Math.*, **6**, (1953) 103–156.

[217] Nirenberg L.— Contributions to the theory of partial differential equations. *Ann. Math. Studies* **33**, Princeton (1954) 95–100.

[218] Nirenberg L.— Remarks on strongly elliptic partial differential equations. *Comm. Pure Appl. Math.* **8**, (1955) 649–675.

[219] Nirenberg L.— Estimates and existence of solutions of elliptic Equations. *Comm. Pure Appl. Math.* **9**, (1956) 509–530.

[220] Nirenberg L.—On elliptic partial differential equations. *Ann. Scuola Norm. Sup. Pisa Sci. Fis. Mat.* **13**, (1959) 116–162.

[221] Nirenberg L.— Topics in nonlinear functional analysis (Lecture Notes), New York Univ. (1974).

[222] Nirenberg L.— Monge–Ampère equations and some associated problems in geometry. *Proceedings of the International Congress of Mathematicians*, Vancouver (1974) 275–279.

[223] Nirenberg L.— Variational and topological methods in nonlinear problems. *Bull. Amer. Math. Soc.* **4**, (1981) 267–302.

[224] Obata M.— Certain conditions for a Riemannian manifold to be isometric with a sphere. *J. Math. Soc. Japan* **14**, (1962) 333–340.

[225] Obata M.— The Conjectures on conformal transformations of Riemannian Manifolds. *J. Diff. Geom.* **6**, (1971) 247–258.

[226] Osserman R.— A note on Hayman's theorem on the bass note of a drum. *Comm. Math. Helvetici* **52**, (1977) 545–555.

[227] Osserman R.— The isoperimetric inequality. *Bull. Am. Math. Soc.* **84**, (1978) 1182–1238.

[228] Osserman R.— Bonnesen-style isoperimetric inequalities. *Am. Math. Month.* **86**, (1979) 1–29.

[229] Palais R.— *Foundations of Global nonlinear analysis*. Benjamin, Reading, Mass. (1967).

[230] Palais R.— The principle of symmetric criticality. *Comm. Math. Phy.* **69**, (1979) 19–30.

[231] Pham Mau Quan— *Introduction a la géométrie des variétés différentiables*—Dunod, Paris, 1969.

[232] Pinsky M.— The spectrum of the laplacian on a manifold of negative curvature I. *J. Diff. Geo.* **13**, (1978) 87–91.

[233] Pogorelov A. V.— *Monge–Ampère Equations of Elliptic Type*. Noordhoff, Groningen (1964).

[234] Pogorelov A. V.—Extrinsic geometry of convex surfaces. *Translations of Mathematical Monographs V.* **35**, A.M.S. Providence (1973).

[235] Pogorelov A. V.— *The Minkowski Multidimensional Problem*, translated by V. Oliker, Winston et Sons (1978).

[236] Pohozaev S.— Eigenfunctions of the equation $\Delta u + \lambda f(u) = 0$. *Doklady* **165**, (1965) 1408–1410.

[237] Pohozaev S.— The set of critical values of a functional. *Math. USSR* **4**, (1968) 93–98.

[238] Polyakov A.— Compact Gauge fields and the infrared catastrophe. *Phys. Lett.* **59B**, (1975) 82–84.

[239] Protter M. and Weinberger H.— *Maximum Principles in Partial Differential Equations*. Prentice-Hall, Englewood Cliffs, New Jersey (1967).

[240] Rabinowitz P. H.— Some global results for nonlinear eigenvalue problems. *J. Funct. Ana.* **7**, (1971) 487–513.

[241] Rabinowitz P. H.—Some aspects of nonlinear eigenvalue problems. *Rocky Mountain J. Math.* **3**, (1973) 161–202.

[242] Rauch J. and Taylor B. A.— The Dirichlet Problem for the multidimensional Monge–Ampère Equation. *Rocky Mountain J. of Math.* **7**, (1977) 345–363.

[243] Rudin W.— *Functional analysis*. McGraw-Hill, New York (1973).

[244] Sacks J. and Uhlenbeck K.— The existence of minimal immersions of 2-spheres. *Ann. of Math.* **113**, (1981) 1–24.

[245] Scheffer V.— The Navier–Stokes equations on a bounded domain. *Comm. Math. Phy.* **73**, (1980) 1–42.

[246] Schoen R. and Yau S-T.— Existence of incompressible minimal surfaces and the topology of manifolds with non-negative scalar curvature. *Ann. of Math.* **110**, (1979) 127–142.

[247] Schoen R. and Yau S-T.— On the structure of manifolds with positive scalar curvature. *Man. Math.* **28**, (1979) 159–183.

[248] Schoen R. and Yau S-T.— Proof of the positive mass theorem II. *Comm. Math. Phy.* **79**, (1981) 231–260.

[249] Schwartz J.— *Nonlinear functional analysis*. Gordon and Breach, New York (1969).

[250] Schwartz L.— *Théorie des distributions*. Hermann, Paris (1966).

[251] Schwarz A.— On regular solutions of Euclidean Yang–Mills equations. *Phys. Lett.* **67B**, (1977) 172–174.

[252] Seminar on Differential Geometry, ed. S-T. Yau. *Annals of Math. Studies* ♯ **102**.

[253] Sin Y-T.— Pseudo-convexity and the Problem of Levi. *Bull. Amer. Math. Soc.* **84**, (1978) 481–512.

[254] Sin Y-T. and Yau S-T.— Compact Kähler manifolds of positive bisectional curvature. *Invent. Math.* **59**, (1980) 189–204.

[255] Sobolev S. L.— Sur un théorème d'analyse fonctionnelle. *Math. Sb. (N.S)* **46**, (1938) 471–496.

[256] Sternberg S.— *Lectures on Differential Geometry*. Prentice-Hall, Englewood Cliffs, N.J. (1964).

[257] Talenti G.— Best constant in Sobolev inequality. *Ann. Mat. Pura Appl.* **110**, (1976) 353–372.

[258] Talenti G.— Elliptic equations and rearrangements. *Ann. Scuola Norm. Sup. Pisa* **3**, (1976) 697–718.

[259] Talenti G.— Nonlinear elliptic equations, rearrangement of functions and Orlitz spaces. *Ann. Mat. Pura Appl.* **120**, (1979) 159–184.

[260] Talenti G.— Some estimates of solutions to Monge–Ampère type equations in dimension two. *Ann. Scuola Norm. Sup. Pisa* **7**, (1981) 183–230.

[261] Trüdinger N.— On imbedding into Orlitz spaces and some applications. *J. Math. Phys.* **17**, (1967) 473–484.

[262] Trüdinger N.— Remarks concerning the conformal deformation of Riemannian structures on compact manifolds. *Ann. Scuola Norm. Sup. Pisa* **22**, (1968) 265–274.
[263] Trüdinger N.— Maximum principles for linear, non-uniformly elliptic operators with measurable coefficients. *Math. Z.* **156**, (1977) 299–301.

[264] Uhlenbeck K.— Removable singularities in Yang–Mills fields. *Bull. Amer. Math. Soc.* **1**, (1979) 579–581.

[265] Vaugon M.— Equations différentielles non linéaires sur les variétés Riemanniennes compactes. *Bull. Sc. Math.* (to appear).

[266] Ward R.— On self-dual gauge fields. *Phys. Lett.* **61A**, (1977) 81–82.
[267] Warner F.— *Foundations of differentiable manifolds and Lie Groups*. Academic Press, New York (1971).
[268] Wells R. O.— *Differential Analysis on Complex Manifolds*. Springer-Verlag, New York (1973).

[269] Yamabe H.— On the deformation of Riemannian structures on compact manifolds. *Osaka Math. J.* **12**, (1960) 21–37.
[270] Yang P. and S-T. Yau— Eigenvalues of the Laplacian of compact Riemannian surfaces and minimal submanifolds. *Ann. Scuola Norm. Sup. Pisa* **7**, (1980) 55–63.
[271] Yano K.— *Differential geometry on complex and almost complex spaces*. Pergamon Press, Oxford (1965).
[272] Yano K. and Bochner S.— Curvature and Betti numbers. *Annals of Math. Studies* **32**. Princeton University Press, 1953.
[273] Yano K. and Ishihara S.— *Tangent and Cotangent Bundles*. Marcel Dekker, New-York 1973.
[274] Yau S-T.— Isoperimetric inequalities and the first eigenvalue of a compact Riemannian manifold. *Ann. Sci. Ecole Norm. Sup. Paris* **8**, (1975) 487–507.
[275] Yau S-T.— On the curvature of compact hermitian manifolds. *Inv. Math.* **25**, (1974) 213–239.
[276] Yau S-T.— Calabi's conjecture and some new results in algebraic geometry. *Proc. Nat. Acad. Sc. U.S.A.* **74**, (1977) 1798–1799.
[277] Yau S-T.— On the Ricci curvature of a compact Kähler manifold and the complex Monge–Ampère equation I. *Comm. Pure Appl. Math.* **31**, (1978) 339–411.
[278] Yosida K.— *Functional Analysis*, Springer-Verlag, Berlin (1965).

Bibliography*

[*1] Alvino A., Lions P.L., and Trombetti G.— Comparison results for elliptic and parabolic equations via Schwarz symmetrization. *Ann. Inst. Henri Poincaré. Analyse non lineaire* **7** (1990) 37–65.

[*2] Ambrosetti A., Rabinowitz P.— Dual variational methods in critical point theory and applications. *J. Funct. Anal.* **14** (1973) 349–381.

[*3] Anderson M.— Ricci curvature bounds and Einstein metrics on compact manifolds. *J. Am. Math. Soc.* **2** (1989) 455–490.

[*4] Anderson M.— Convergence and rigidity of manifolds under Ricci curvature bounds. *Invent. Math.* **102** (1990) 429–445.

[*5] Anderson M. and Cheeger J.— C^α-compactness for manifolds with Ricci curvature and injectivity radius bounded below. *J. Differ. Geom.* **35** (1992) 265–281.

[*6] Aubin T.— *Nonlinear Analysis on Manifolds. Monge-Ampère Equations.* Springer, New York 1982.

[*7] Aubin T.— Réduction du cas positif de l'équation de Monge-Ampère sur les variétés kählériennes compactes à la démonstration d'une inégalité. *J. Funct. Anal.* **57** (1984) 143–153.

[*8] Aubin T.— Métriques d'Einstein-Kähler. Differential Geometry. Proceedings of the 3rd international Symposium held at Peniscola 1988. *Lectures Notes* **1410**. Springer (1989).

[*9] Aubin T.— Métriques d'Einstein-Kähler et exponentiel des fonctions admissibles. *J. Funct. Anal.* **88** (1990) 385–394.

[*10] Aubin T.— *Prescribed scalar curvature and the method of isometry concentration Partial Differential Equations of Elliptic Type* (A. Alvino, E. Fabes and G. Talenti ed.). Symposia Mathematica 35. Cambridge University Press (1994).

[*11] Aubin T. et Bahri A.— Méthodes de topologie algébrique pour le problème de la courbure scalaire prescrite. *J. Math. Pures Appl.* **76** (1997) 525–549.

[*12] Aubin T. et Bahri A.— Une hypothèse topologique pour le problème de la courbure scalaire prescrite. *J. Math. Pures Appl.* **76** (1997) 843–850.

[*13] Aubin T. et Bismuth S.— Courbure scalaire prescrite sur les variétés Riemanniennes compactes dans le cas négatif. *J. Funct. Anal.* **143** (1997) 529–541.

[*14] Aubin T. et Coulouvrat F.— Ondes acoustiques non linaires dans un fluide avec relaxation. *J. Math. Pures et Appli.* (to appear)

[*15] Aubin T. et Hebey E.— Courbure scalaire prescrite. *Bull. Sci. Math.* **115** (1991) 125–132.

[*16] Aviles P.— A study of the singularities of solutions of a class of nonlinear elliptic partial differential equations. *Commun. Partial Differ. Equations* **7** (1982) 609–643.

[*17] Aviles P. and Escobar J.— On the Sobolev quotient of an Einstein manifold. *Indiana Univ. Math. J.* **41** (1992) 435–438.

[*18] Aviles P. and Mc Owen R.— Conformal deformation to constant negative scalar curvature on noncompact riemannian manifolds. *J. Differ. Geom.* **27** (1988) 225–239.

[*19] Bahri A.— Critical points at infinity in some variational problems. *Research Notes in Mathematics*, vol. **182**. Longman Pitman, London (1989).

[*20] Bahri A.— *Proof of the Yamabe conjecture, without the positive mass theorem, for locally conformally flat manifolds. In Einstein metrics and Yang-Mills connections* (Mabuchi T., Mukai S. ed.). Marcel Dekker, New York 1993.

[*21] Bahri A.— An invariant for Yamabe-type flows with applications to the scalar curvature problem in high dimensions. *Duke Math. J.*, a book in honor of Nash (to appear)

[*22] Bahri A. et Brezis H.— Equations elliptiques non linéaires sur des variétés avec exposant de Sobolev critique. *C.R. Acad. Sci. Paris* **307** (1988) 537–576.

[*23] Bahri A. et Brezis H.— Elliptic Differential Equations involving the Sobolev critical exponent on Manifolds. (to appear)

[*24] Bahri A. et Coron J.M.— Une théorie des points critiques à l'infini pour l'équation de Yamabe et le problème de Kazdan-Warner. *C.R. Acad. Sci. Paris* **300** (1985) 513.

[*25] Bahri A. and Coron J.M.— On a nonlinear elliptic equations involving the critical Sobolev exponent: The effect of the topology of the domain. *Commun. Pure Appl. Math.* **41** (1988) 253–294.

[*26] Bahri A. and Coron J.M.— The scalar curvature problem on the standard three dimensional sphere. *J. Funct Anal.* **95** (1991) 106–172.

[*27] Bahri A., Li Y-Y. and Rey O.— On a variational problem with lack of compactness: the topological effect of the critical points at infinity. *Calc. Var. Partial Differ. Equ.* **3** (1995) 67–93.

[*28] Bahri A. and Lions P.L.— On the existence of a positive solution of semilinear elliptic equations in unbunded domains. (to appear)

[*29] Baird P. et Eells J.— A conservation law for harmonic maps. *Geometry Symp. Utrecht 1980. Lecture notes in Math.* **894**. Springer (1981) 1–25.

[*30] Bando S.— The K-energy map, almost Einstein-Kähler metrics and an inequality of the Miyaoka-Yau type. *Tôhoku Math. J.* **39** (1987) 231–235.

[*31] Bando S. and Mabuchi T.— Uniqueness of Einstein-Kähler metrics modulo connected group actions. Algebraic geometry. Proc. Symp. Sendai/Japan 1985. *Adv. Stud. Pure Math.* **10** (1987) 11–40. Kinokuniya, Tokyo and North-Holland.

[*32] Bartnik R.— The mass of an asymptotically flat manifold. *Commun. Pure Appl. Math.* **34** (1986), 661–693.

[*33] Ben Abdesselem A. et Cherrier P.— Métriques d'Einstein-Kähler sur certaines variétés kählériennes à première classe de Chern positive. (to appear)

[*34] Ben Ayed M., Chen Y., Chtioui H., and Hammami M.— On the prescribed scalar curvature. Problem on four-manifolds. (to appear)

[*35] Bérard P.— Spectral Geometry. Direct and inverse problems. *Lecture Notes in Math.* **1207**. Springer (1986).

[*36] Bérard P., Besson G. et Gallot S.— Sur une inégalité isopérimétrique qui généralise celle de Paul Lévy Gromov. *Invent. Math.* **80** (1985) 295–308.

[*37] Bérard P. et Gallot S.— Remarques sur quelques estimées géométriques explicites. *C.R. Acad. Sci. Paris* **297** (1983) 185–188.

[*38] Bérard P. et Meyer D.— Inégalités isopérimétriques et applications. *Ann. Sci. Ec. Norm. Super.* **15** (1982) 513–542.

[*39] Berestycki H. and Nirenberg L.— On the method of moving planes and the sliding method. *Bol. Soc. Bras. Mat.* **22** (1991) 1–37.

[*40] Berestycki H., Nirenberg L., and Varadhan S.— The principal eigenvalue and maximum principle for second order elliptic operators in general domains. *Commun. Pure Appl. Math.* **47** (1994) 47–92.

[*41] Berger M.S.— *Perspectives in Nonlinearity.* Benjamin, New York (1968).

[*42] Berger M.S.— *Nonlinearity and Functional Analysis.* Academic Press, New York (1977).

[*43] Berger M.S.— *Mathematical Structures of Nonlinear Science.* Kluwer Academic Publishers (1990).

[*44] Besse A.— *Einstein Manifolds.* Springer, New York (1987).

[*45] Bethuel F.— *Approximation in Sobolev spaces between two manifolds and homothopy groups. Variational problems.* (H. Berestycki, J.M. Coron, I. Ekeland eds). Birkhäuser (1990).

[*46] Bethuel F. and Brezis H.— Regularity of minimizers of relaxed problems for harmonic maps. *J. Funct. Anal.* **101** (1991) 145–161.

[*47] Bethuel F. and Brezis H.— Minimisation de $\int |\nabla(u - x/|x|)|^2$ et divers phénomènes de gap. *C.R. Acad. Sci. Paris* **310** (1990) 859–864.

[*48] Bethuel F., Brezis H. et Coron J.M.— Relaxed energies for harmonic maps. Variational problems (H. Berestycki, J.M. Coron, I. Ekelande eds.). Birkhäuser (1990).

[*49] Bethuel F., Brezis H. et Helein F.— Limite singulière pour la minimisation de fonctionelles du type Ginzburg Landau. *C.R. Acad. Sci. Paris* **314** (1992) 891–895.

[*50] Bethuel F., Brézis H. et Helein F.— *Ginzburg Landau Vortices.* Birkhäuser (1994).

[*51] Bethuel F. and Zang X.— Density of smooth functions between two manifolds in Sobolev spaces. *J. Funct. Anal.* **80** (1988) 60–75.

[*52A] Bianchi G.— Non-existence and symmetry of solutions to the scalar curvature equation. (to appear)

[*52B] Bianchi G.— The scalar curvature equation on $\mathbb{R}^n$ and S^n. (to appear)

[*53A] Bianchi G. and Egnell H.—An ODE approach to the equation $\Delta u + Ku^{(n+2)/(n-2)} = 0$ in $\mathbb{R}^n$. *Math. Z.* **210** (1992) 137–166.

[*53B] Bianchi G. and Egnell H.— A variational approach to the equation $\Delta u + Ku^{\frac{n+2}{n-2}} = 0$, in $\mathbb{R}^n$. *Arch. Rational. Mech. Anal.* **122** (1993) 159–182.

[*54A] Bidaut, Veron M.F., and Veron L.— Nonlinear elliptic equations on compact Riemann manifolds and asymptotics of Emden equations. *Invent. Math.* **106** (1991) 489–539.

[*54B] Bismuth S.— Courbure scalaire prescrite sur une variété Riemannienne compacte à courbure scalaire nulle. *C.R. Acad. Sci. Paris* **321** (1995) 1035.

[*55] Bony J.M.— Principe du maximum dans les espaces de Sobolev. *C.R. Acad. Sci. Paris* **265** (1967) 333–336.

[*56] Bourguignon J.P. and Ezin J.P.— Scalar curvature functions in a conformal class of metrics and conformal transformations. *Transl. Am. Math. Soc.* **301** (1987) 723–736.

[*57] Brezis H.— Elliptic equations with limiting Sobolev exponents. The impact of topology. *Commun. Pure Appl. Math.* **39** (1986) S17–S39.

[*58] Brezis H.— *Liquid crystals and energy estimates for S^2-valued maps. Theory and Applications of Liquid Crystals* (J. Erickson and D. Kinderlehrer ed.). Springer, Berlin, New York (1987).

[*59] Brezis H., Coron J.M., and Lieb E.H.— Harmonic maps with defects. *Commun. Math. Phys.* **107** (1986) 649–705.

[*60] Brezis H., Li Y-Y., and Shafrir I.— A sup + inf inequality for some nonlinear elliptic equations involving exponential nonlinearities. *J. Funct. Anal.* **115** (1993) 344–358.

[*61] Brezis H. and Merle F.— Uniform estimates and blow-up behavior for solutions of $-\Delta u = Ve^u$ in two dimension. *Commun Partial Differ. Equations* **16** (1991) 1223–1253.

[*62] Brezis H., Merle F., and Rivière T.— Quantization effects for solutions of the equation $-\Delta u = (1 - |u|^2)u$ in $\mathbb{R}^2$. *Arch. Ration, Mech. Anal.* **126** (1994) 35–58.

[*63] Brezis H. and Nirenberg L.— Positive solutions of nonlinear elliptic equations involving critical Sobolev exponents. *Commun. Pure Appl. Math.* **36** (1983) 437–477.

[*64] Brezis H. and Peletier L.— *Asymptotics for elliptic equations involving critical growth. Partial Differential Equations and the Calculus of Variations I* (F. Colombini ed.). Birkhaüser (1989).

[*65] Caffarelli L., Gidas B., and Spruck J.— Asymptotic symmetry and local behavior of semilinear elliptic equations with critical Sobolev growth. *Commun. Pure Appl. Math.* **42** (1989) 271–297.

[*66] Caffarelli L., Nirenberg L., and Spruck J.— The Dirichlet problem for nonlinear second-order elliptic equations I. Monge-Ampère equation. *Commun. Pure Appl. Math.* **37** (1984) 369–402.

[*67] Caffarelli L., Kohn J.J., Nirenberg L., and Spruck J.— The Dirichlet problem for nonlinear second-order elliptic equations II. Complex Monge-Ampère and uniformly elliptic, equations. *Commun. Pure Appl. Math* **38** (1985) 209–252.

[*68] Caffarelli L., Nirenberg L., and Spruck J.— Nonlinear second order elliptic equations V. The Dirichlet problem for Weingarten hypersurfaces. *Commun. Pure Appl. Math.* **41** (1988), 47–70.

[*69] Caffarelli L. and Spruck J.— Variational problems with critical Sobolev growth and positive Dirichlet data. *Indiana Univ. Math. J.* **39** (1990) 1–18.

[*70] Calabi E.— The space of Kähler metrics. *Proc. Internat. Congress Math. Amsterdam* **2** (1954) 206–207.

[*71] Calabi E.— *Extremal Kähler metrics.* Seminars on Differential Geometry (S-T. Yau ed.). Princeton Univ. Press (1982) 259–290.

[*72] Calabi E.— *Extremal kähler metrics II.* Differential Geometry and Complex Analysis, (Chavel I., Farkas H.M. ed.). Springer, Berlin (1985) 95–114.

[*73] Cao J.— The existence of generalized isothermal coordinates for higher dimensional Riemannian manifolds. *Trans. Am. Math. Soc.* **324** (1991) 901–920.

[*74] Cao J. and DeTurck D.— Prescribing Ricci curvature on open surfaces. *Hokkaido Math. J.* **20** (1991) 265–278.

[*75] Cao J. and DeTurck D.— The Ricci curvature equation with rotationaly symmetry. *Am. J. Math.* **116** (1994) 219–241.

[*76] Carleson L. and Chang S-Y. A.— On the existence of an extremal function for an inequality of J. Moser. *Bull. Sci. Math.* **110** (1986) 113–127.

[*77] Chaljub-Simon A.— Quelques résultats d'existence pour le problème de la courbure dans $\mathbb{R}^n$. *Rend. Circ. Mat. Palermo II, Ser., Suppl.* **8** (1985) 337–344.

[*78] Chang S-Y. A., Gursky M., and Yang P.— The scalar curvature equation on 2- and 3-spheres. *Calculus of Variations and Partial Differ. Equations* **1** (1993) 205–229.

[*79] Chang K.C. and Liu J.Q.— On Nirenberg's problem. *Int. J. Math.* **4** (1993) 35–58.

[*80] Chang S-Y. A. and Yang P.— Prescribing Gaussian curvature on S^2. *Acta Math.* **159** (1987) 215–259.

[*81] Chang S-Y. A. and Yang P.— Conformal deformation of metrics on S^2. *J. Differ. Geom.* **23** (1988) 259–296.

[*82] Chang S-Y. A. and Yang P.— The conformal deformation equation and isospectral set of conformal metrics. *Contemp. Math.* **101** (1989) 165–178.

[*83] Chang S-Y. A. and Yang P.— Compactness of isospectral conformal metrics on S^3. *Comment. Math. Helv.* **64** (1989) 363–374.

[*84] Chang S-Y. A. and Yang P.— Isospectral conformal metrics on 3-manifolds. *J. Am. Math. Soc.* **3** (1990) 117–145.

[*85] Chang S-Y. A. and Yang P.— A perturbation result in prescribing scalar curvature on S^n. *Duke Math. J.* **64** (1991) 27–69.

[*86] Cheeger J.— Finiteness theorems for Riemannian manifolds. *Am. J. Math.* **92** (1970) 61–74.

[*87] Chen W.— A note on the Kazdan Warner type conditions. *J. Differ. Geom.* (to appear)

[*88] Chen W. and Ding W.— Scalar curvatures on S^2. *Transl. Am. Math. Soc.* **303** (1987) 365–382.

[*89] Chen W. and Li C.— Scalar curvature on S^n. *Math. Ann.* **283** (1989) 353–365.

[*90] Chen W. and Li C.— A necessary and sufficient condition for the Nirenberg problem. *Commun. Pure Appl. Math.* **48** (1995) 657–667.

[*91] Chen Y. and Struwe M.— Existence and partial regularity results for the heat flow for harmonic maps. *Math. Z.* **201** (1989) 83–103.

[*92] Cheng K. and Smoller J.— Conformal metrics with prescribed Gaussian curvature on S^2. *Trans. Am. Math. Soc.* **336** (1993) 219–255.

[*93] Cheng S.Y. and Yau S-T.— On the existence of complete Kähler metric on noncompact complex manifolds and the regularity of Fefferman's equation. *Commun. Pure Appl. Math.* **33** (1980) 507–544.

[*94] Chern S.S.— An elementary proof of the existence of isothermal parameters on a surface. *Proc. Am. Math. Soc.* **6** (1955) 771–782.

[*95] Chern S.S., Levine and Nirenberg L.— *Intrinsic norms on a complex manifold.* In "Global Analysis" Papers in Honor of Kodaira K. (Spencer D. and Iyanaga S. ed.). University of Tokyo Press and Princeton University Press (1969).

[*96] Cherrier P.— Etude d'une équation de Monge-Ampère sur les variétés kählériennes compactes. *J. Funct. Anal.* **53** (1983) 231–245.

[*97] Cherrier P.— Problèmes de Neumann non linéaires sur les variétés riemanniennes. *J. Funct. Anal.* **57** (1984) 154–206.

[*98] Cherrier P.— Equations de Monge-Ampère sur les variétés hermitiennes compactes. *Bull. Sci. Math.* **111** (1987) 343–385.

[*99] Cherrier P. and Milani A.— Equations von Karman's type on compact Kähler manifolds. *Bull. Sci. Math.* **116** (1992) 325–352.

[*100] Coron J.M.— *Harmonic maps with values into spheres.* Proceedings of the International Congress of Mathematicians, Kyoto, Japan, 1990. Springer (1991).

[*101] Coron J.M.— Nonuniqueness for the heat flow of harmonic maps. *Ann. Inst. Henri Poincaré* **7** (1990) 335–344.

[*102] Coron J.M. et Ghidaglia J.M.— Explosion en temps fini pour le flot des applications harmoniques. *C.R. Acad. Sci. Paris* **308** (1989) 399–344.

[*103] Coron J.M. and Gulliver R.— Minimizing p-harmonic maps into spheres. *J. Reine Angew. Math.* **401** (1989) 82–100.

[*104] Crandall M. and Rabinowitz P.— Some continuation and variational methods for positive solutions of nonlinear elliptic eigenvalue problems. *Arch. Rat. Mech. Anal.* **58** (1975) 208–218.

[*105] Delanoe P.— Generalized stereographic projections with prescribed scalar curvature. *Contemp. Math.* **127** (1992) 17–25.

[*106] Demailly J.P.— Regularization of closed positive currents and intersection theory. *J. Algebra Geom.* **1** (1992) 361–409.

[*107] Demailly J.P.— A numerical criterion for very ample line bundles. *J. Differ. Geom.* **37** (1993) 323–374.

[*108] Demailly J.P., Peternell T., and Schneider M.— Kähler manifolds with numerical effective Ricci class. *Compos. Math.* **89** (1993) 217–240.

[*109] DeTurck D.— Existence of metrics with prescribed Ricci curvature: local theory. *Invent. Math.* **65** (1981) 179–207.

[*110] DeTurck D.— Nonlinear Douglis-Nirenberg systems in geometry. *Lecture notes.* Universita di Roma (1982).

[*111] DeTurck D.— Deforming metrics in the direction of their Ricci tensors. *J. Differ. Geom.* **18** (1983) 157–162.

[*112] DeTurck D. and Koiso N.— Uniqueness and non-existence of metrics with prescribed Ricci curvature. *Ann. Sci. Ec. Norm. Supér.* **14** (1984) 249–260.

[*113] Dieudonné J.— *Eléments d'analyse* n°8. Gauthier-Villars (1978).

[*114] Ding W.— Positive solutions of $\Delta u = u^{(n+2)/(n-2)}$ on contractible domains. *J. Partial Differ. Equations* **2** (1989) 83–88.

[*115] Ding W.— On a conformally invariant elliptic equation on $\mathbb{R}^n$. *Commun. Math. Phys.* **107** (1986) 331–335.

[*116] Ding W.— Remarks on the existence problem of positive Kähler-Einstein metrics. *Math. Am.* **282** (1988) 463–471.

[*117] Ding W. and Ni W-M.— On the elliptic equation $\Delta u + Ku^{(n+2)/(n-2)} = 0$ and related topics. *Duke Math. J.* **52** (1985) 485–506.

[*118] Dong R.— Another approach to Yamabe's problem. *Scientia Sinica* **29** (1986) 125–137.

[*119] Eells J.— *Regularity of certain harmonic maps. Global Riemannian geometry.* Durham (1982) 137–147.

[*120] Eells J. and Lemaire L.— A report on harmonic maps. *Bull. Lond. Math. Soc.* **10** (1978) 1–68.

[*121] Eells J. and Lemaire L.— Another report on harmonic maps. *Bull. London Math. Soc.* **20** (1988) 385–524.

[*122] Eells J. et Lemaire L.— Deformations of metrics and associated harmonic maps. Patodi memorial vol., Geometry and Analysis. Proc. Indian Acad. Math. Sci. **90**. Tata Inst. (1981) 33–45.

[*123] Eells J. and Ratto A.— Harmonic maps and minimal immersions with symmetries. *Ann. Math. Stud.* **130**. Princeton University Press (1993).

[*124] Eells J. and Sampson J.H.— Harmonic mappings of Riemannian manifolds. *Am. J. Math.* **86** (1964) 109–160.

[*125] Eidelman S.D.— *Parabolic systems.* North-Holland publ. Comp. Amsterdam (1969). J. Wiley

[*126] Escobar J.— Sharp constant in a Sobolev trace inequality. *Indiana Univ. Math. J.* **37** (1988) 687–698.

[*127] Escobar J.— Conformal deformation of a Riemannian metric to a scalar flat metric with constant mean curvature on the boundary. *Ann. Math.* **136** (1992) 1–50 and **139** (1994) 749–750.

[*128] Escobar J.— The Yamabe problem on manifolds with boundary. *J. Differ. Geom.* **35** (1992) 21–84.

[*129] Escobar J. and Schoen R.— Conformal metrics with prescribed scalar curvature. *Invent. Math.* **86** (1986) 243–254.

[*130] Friedman A.— *Partial Differential Equations.* Holt, New York (1969).

[*131] Futaki A.— An obstruction to the existence of Einstein-Kähler metrics. *Invent. Math.* **73** (1983) 437–443.

[*132] Futaki A.— Kähler-Einstein metrics and integral invariants. *Lectures Notes in Math.* **1314**. Springer, Berlin (1988).

[*133] Gallot S.— Inégalités isopérimétriques et analytiques sur les variétés Riemanniennes. *Astérisque* **163-164** (1988) 31–91.

[*134] Gallot S.— Isoperimetric inequalities based on integral norms of Ricci curvature. *Astérisque* **157-158** (1988) 191–216.

[*135] Gallot S., Hulin D. et Lafontaine J.— *Riemannian Geometry.* Springer, Berlin (1987).

[*136] Gao L.Z. and Yau S-T.— The existence of negatively Ricci curved metrics on three manifolds. *Invent. Math.* **85** (1986) 637–652.

[*137] Gaveau B. et Mazet E.— Méthode de plus grande descente pour la construction du problème de Plateau. *Bull. Sci. Math.* **111** (1987) 111–187.

[*138] Giaquinta M., Modica G. et Soucek J.— The Dirichlet energy of mapping with values into the sphere. *Manuscr. Math.* **65** (1989) 489–507.

[*139] Giaquinta M., Modica G. and Soucek J.— Singular perturbations associated to the Dirichlet energy of maps with values in S^2. *J. Funct. Anal.* **118** (1993) 188–197.

[*140] Gidas B., Ni W-M. and Nirenberg L.— Symmetry and related properties via the maximum principle. *Commun. Math. Phys.* **68** (1979) 209–243.

[*141] Gidas B. and Spruck J.— Global and local behavior of positive solutions of nonlinear elliptic equations. *Commun. Pure Appl. Math.* **34** (1981) 525–598.

[*142] Gil-Medrano O.— On the Yamabe problem concerning the compact locally conformally flat manifolds. *J. Funct. Anal.* **66** (1986) 42–53.

[*143] Gilbarg D. and Trudinger N.— *Elliptic partial differential equations of second order.* Springer, New York (1983).

[*144] Gray A.— Comparison theorems for the volumes of tubes as generalizations of the Weyl tube formula. *Topology* **21** (1982) 201–228.

[*145] Greene R.E. and Wu H.— Lipschitz convergence of Riemannian manifolds. *Pacific J. Math.* **131** (1988) 119–141.

[*146] Griffiths P. and Harris J.— Principles of Algebraic Geometry. *Pure and Applied Math.* J. Wiley, Interscience, New York (1978).

[*147] Gromov M.— *Structures métriques pour les variétés Riemanniennes.* Cedric/ Fernand Nathan, Paris (1981).

[*148] Günther M.— Conformal normal coordinates. *Ann. Global Anal. Geom.* **11** (1993) 173–184.

[*149] Hamilton R.— Harmonic Maps of Manifolds with Boundary. *Lectures Notes in Math.* **471**. Springer, Berlin (1975).

[*150] Hamilton R.— The inverse function theorem of Nash and Moser. *Bull. Am. Math. Soc.* **7** (1982) 65–222.

[*151] Hamilton R.— Three manifolds with positive Ricci curvature. *J. Differ. Geom.* **17** (1982) 255–306.

[*152] Hamilton R.— Four-manifolds with positive curvature operator. *J. Differ. Geom.* **24** (1986) 153–179.

[*153] Hamilton R.— The Ricci flow for surfaces. *Contemp. Math.* **71** (1988) 237–262.

[*154] Hamilton R.— The Harnack estimate for the Ricci flow. *J. Differ. Geom.* **37** (1993) 225–243.

[*155] Hamza H.— Sur les transformations conformes des variétés riemanniennes à bord. *J. Funct. Anal.* **92** (1990) 403–447.

[*156] Han Z.C.— Prescribing gaussian curvature on S^2. *Duke Math. J.* **61** (1990) 679–703.

[*157] Han Z-C.— Asymptotic approach to singular solutions for nonlinear elliptic equations involving critical Sobolev exponent. *Ann. Inst. Henri Poincaré. Analyse non lineaire* **8** (1991) 159–174.

[*158] Han Z-C. and Li Y-Y.— A note on the Kazdan-Warner type condition. *Ann. Inst. Henri Poincaré. Analyse non lineaire* **13** (1996) 283–292.

[*159] Hanani A.— Equations de type Monge-Ampère sur les variétés hermitiennes compactes. *J. Funct. Anal.* **137** (1996) 49–75.

[*160A] Hardt R.— Singularities of harmonic maps. *Bull. Am. Math. Soc.* **34** (1997) 15–34.

[*160B] Hardt R. and Lin F.H.— A remark on H^1 mappings. *Manuscr. Math.* **56** (1986) 1–10.

[*161] Hartman P.— On homothopic harmonic maps. *Can. J. Math.* **19** (1967) 673–687.

[*162] Hebey E.— Changements de métriques conformes sur la sphère. Le probème de Nirenberg. *Bull. Sci. Math.* **114** (1990) 215–242.

[*163] Hebey E.— La méthode d'isométrie-concentration dans le cas d'un problème non linéaire sur les variétés compactes à bord. *Bull. Sci. Math.* **116** (1992) 35–51.

[*164A] Hebey E.— A propos d'un théorème non linéaire de T. Aubin pour la transformation conforme de la courbure scalaire sur S_n. *C.R. Acad. Sci.* **314** (1992) 183–185.

[*164B] Hebey E.— Asymptotics for some quasilinear elliptic equations. *Differ. Integral Equations* **9** (1996) 89–106.

[*165A] Hebey E.— Optimal Sobolev inequalities on complete Riemannian manifolds with Ricci curvature bounded below and positive injectivity radius. *Am. J. Math.* **117** (1996) 1–10.

[*165B] Hebey E.— Sobolev spaces on Riemannian manifolds. *Lectures Notes in Math.* **1635**. Springer (1996).

[*166] Hebey E. et Vaugon M.— Remarque sur le problème de Yamabe. *Funct. Anal.* **96** (1991) 31–37.

[*167] Hebey E. et Vaugon M.— Meilleures constantes dans le théorème d'inclusion de Sobolev et multiplicité pour les problèmes de Nirenberg et Yamabe. *Indiana Univ. Math. J.* **41** (1992) 377–407.

[*168] Hebey E. et Vaugon M.— Le problème de Yamabe équivariant. *Bull. Sci. Math.* **117** (1993) 241–286.

[*169] Hebey E. et Vaugon M.— Courbure scalaire prescrite pour des variétés non conformément difféomorphes à la sphère. *C.R. Acad. Sci.* **316** (1993) 281.

[*170A] Hebey E. and Vaugon M.— Un théorème de pincement intégral sur la courbure concirculaire en géométrie conforme. *C. R. Acad. Sci.* **316** (1993) 483.

[*170B] Hebey E. and Vaugon M.— Effective L_p pinching for the concircular curvature. *J. Geom. Anal.* (to appear)

[*171] Hebey E. et Vaugon M.— Meilleures constantes dans le théorème d'inclusion de Sobolev. *Ann. Inst. Henri Poincaré, Anal. Non Linéaire* **123** (1996) 57–93.

[*172] Hebey E. and Vaugon M.— The best constant problem in the Sobolev embedding theorem for complete Riemannian manifolds. *Duke Math. J.* **79** (1995) 235–279.

[*173] Helein F.— Régularité des applications faiblement harmoniques entre une surface et une sphère. *C.R. Acad. Sci. Paris* **311** (1990) 519–524.

[*174] Helein F.— Régularité des applications faiblement harmoniques entre une surface et une variété riemannienne. *C.R. Acad. Sci. Paris* **312** (1991) 591–596.

[*175] Helein F.— *Applications harmoniques, lois de conservation et repère mobiles.* Diderot editeur 1996.

[*176] Hempel J.— 3 manifolds. *Am. Math. Studies* **86**. Princeton University Press, New Jersey (1976).

[*177] Hijazi O.— Première valeur propre de l'opérateur de Dirac et nombre de Yamabe. *C.R. Acad. Sci. Paris* **313** (1991) 865.

[*178] Hörmander L.— *An introduction to complex analysis in several variables.* North-Holland, Amsterdam London (1973).

[*179] Huisken G.— Ricci deformation of the metric on a riemannian manifold. *J. Differ. Geom.* **21** (1985) 47–62.

[*180] Hulin D.— Deformation conforme des surfaces non compactes à courbure negative. *J. Funct. Anal.* **111** (1993) 449–472.

[*181] Inoue A.— On Yamabe's problem by a modified direct method. *Tôhoku Math. J.* **34** (1982) 499–507.

[*182] Ilias S.— Inégalités de Sobolev et résultats d'isolement pour les applications harmoniques. *J. Funct. Anal.* (to appear)

[*183] Ishiara T.— A mapping of riemannian manifolds which preserves harmonic functions. *J. Math. Kyoto Univ.* **19** (1979) 215–229.

[*184] Itô S.— The fundamental solutions of the parabolic equation in a differentiable manifold II. *Osaka J. Math.* **6** (1954) 167–185.

[*185] Itô S.— Fundamental solutions of parabolic differential equations and boundary value problems. *Japan J. Math.* **27** (1957) 55–102.

[*186] Jäger W. and Kaul H.— Rotationally symmetric harmonic maps from a ball into a sphere and the regularity problem for weak solutions of elliptic systems. *J. Reine Angew. Math.* **343** (1983) 146–161.

[*187] Jerison D. and Lee J.— The Yamabe problem on CR manifolds. *J. Differ. Geom.* **25** (1987) 167–197.

[*188] Jerison D. and Lee J.— Extremals for the Sobolev inequality on the Heisenberg group and the CR Yamabe problem. *J. Am. Math. Soc.* **1** (1988) 1–13.

[*189] Jeune A.— Prescription de la courbure de Ricci sur C^n, $n > 2$. *Bull. Sci. Math.* **116** (1992) 185–226.

[*190] Jin Z.— A counter example to the Yamabe problem for complete non compact manifolds (S.S. Chern ed.). *Lecture Notes in Math.* **1306**. Springer (1986).

[*191A] Jin Z.— A relation between the size of a domain and the solvability of the Dirichlet problems for semilinear elliptic equations on the domain. *Indiana Univ. Math. J.* **42** (1993) 259–282.

[*191B] Jin Z.— A truncation method for semilinear elliptic equations. *Commun. Partial Differ. Equations* **19** (1994) 605–616.

[*192] Kato S.— Conformal deformation to prescribed scalar curvature on complete noncompact riemannian manifolds with nonpositive curvature. *Tôhoku Math. J.* **45** (1993) 205–230.

[*193] Kazdan J.— Some applications of partial differential equations to problems in Geometry. Surveys in Geometry Series. Tokyo Univ. (1983).

[*194] Kazdan J.— *Prescribing the curvature of a Riemannian manifold. Am. Math. Soc. Reg. Conf. Ser. Math. Conference Board* **57**. Providence RI (1985).

[*195] Kazdan J. and Warner F.— Curvature functions for compact 2-manifolds. *Ann. Math.* **99** (1974) 14–47.

[*196] Kazdan J. and Warner F.— Curvature functions for open 2-manifolds. *Ann. Math.* **99** (1974) 203–219.

[*197] Kazdan J. and Warner F.— Existence and conformal deformation of metrics with prescribed Gaussian and scalar curvatures. *Ann. Math.* **101** (1975) 317–331.

[*198] Kazdan J. and Warner F.— Scalar Curvature and conformal deformation of Riemannian structure. *J. Differ. Geom.* **10** (1975) 113–134.

[*199] Kazdan J. and Warner F.— Prescribing curvatures. *Proc. Symp. Pure Math.* **27**. Am. Math. Soc. (1975).

[*200] Kazdan J. and Warner F.— A direct approach to the determination of Gaussian and scalar curvature functions. *Invent. Math.* **28** (1975) 227–230.

[*201] Kobayashi R.— Kähler-Einstein metric on an open algebraic manifold. *Osaka J. Math.* **21** (1984) 399–418.

[*202] Kobayashi S. and Nomizu K.— *Foundations of Differential Geometry I and II.* Interscience Publishers, New York (1963).

[*203] Kodaira K.— *Complex manifolds and deformation of complex structures.* Springer, New York (1986).

[*204] Krylov N.V.— *Nonlinear elliptic and parabolic equations of the second order.* M.I.A. Reidel Publishing Company (1987).

[*205] Kuiper N.H.— On conformally flat spaces in the large. *Ann. Math.* **50** (1949) 916–924.

[*206] Ladyzenskaya O. and Ural'ceva N.— *Linear and quasilinear elliptic partial differential equations.* Academic Press, New York (1968).

[*207] Ladyzenskaja O., Solonnikov V., and Ural'ceva N.— Linear and quasilinear equations of parabolic type. Translations of Mathematical Monographs, vol. **23**. *Am. Math. Soc.* Providence (1968).

[*208] Lee J.M. and Parker T.H.— The Yamabe problem. *Bull. Am. Math. Soc.* **17** (1987) 37–91.

[*209] Le Gluher J.M.— Thesis, Paris VI, 1993.

[*210] Levine M.— A remark on extremal Kähler metrics. *J. Differ. Geom.* **21** (1985) 73–77.

[*211] Li P. and Yau S-T.— On the parabolic kernel of the Schrödinger operator. *Acta Math.* **156** (1986) 153–200.

[*212] Li Y. and Ni W-M.— On conformal scalar curvature equations in $\mathbb{R}^n$. *Duke Math. J.* **57** (1988) 895–924.

[*213] Li Y-Y.— On $\Delta u = K(x)u^5$ in R^3. *Commun. Pure Appl. Math.* **46** (1993) 303–340.

[*214] Li Y-Y.— Prescribing scalar curvature on S^3, S^4 and related problems. *J. Funct. Anal.* **118** (1993) 43–118.

[*215] Li Y-Y.— Prescribing scalar curvature on S^n and related problems. *C.R. Acad. Sci. Paris* **317** (1993) 159–164. Part I: J. Differ. Equations **120** (1995) 319–410. Part II: Existence and compactness. *Commun. Pure Appl. Math.* **49** (1996) 541–597.

[*216] Li Y-Y. and Zhu M.— Uniqueness theorems through the method of moving spheres. *Duke Math. J.* (to appear)

[*217] Li Y-Y. and Zhu M.— Sharp Sobolev inequalities involving boundary terms. (to appear)

[*218] Li Yiren— Etude spectrale d'une classe conforme. *Bull. Sc. Math.* **119** (1995) 485–508.

[*219] Licois J-R. et Véron L.— Un théorème d'annulation pour des équations elliptiques non linéaires sur des variétés Riemanniennes compactes. *C.R. Acad. Sci. Paris* **320** (1995) 1337.

[*220] Lin F.H.— Une remarque sur l'application $x/|x|$. *C.R. Acad. Sci. Paris* **305** (1987) 529–531.

[*221] Lions P.L.— *Generalized solutions of Hamilton-Jacobi equations.* Pitman, Boston (1981).

[*222] Lions P.L.— The concentration-compactness principle in the calculus of variations. *Rev. Mat. Iberoam.* **1.1** (1985) 145–201 and **1.2** (1985) 45–121.

[*223] Lions P.L.— Existence globale de solutions pour les équations de Navier Stokes compressibles isentropiques. *C.R. Acad. Sci. Paris* **316** (1993) 1335–1340.

[*224] Lions P.L.— Compacité des solutions des équations de Navier Stokes compressibles isentropiques. *C.R. Acad. Sci. Paris* **317** (1993) 115–120.

[*225] Lions P.L.— Limites incompressible et acoustique pour des fluides visqueux, compressibles et isentropiques. *C.R. Acad. Sci. Paris* **317** (1993) 1197–1202.

[*226] Lohkamp J.— Metrics of negative Ricci curvature. *Ann. Math.* **140** (1994) 655–683.

[*227] Lohkamp J.— The space of negative scalar curvature metrics. *Invent. Math.* **110** (1992) 403–407.

[*228] Lübke M.— Stability of Einstein-Hermitian vector bundles. *Manuscripta Math.* **42** (1983) 245–257.

[*229] Ma Li— A result on the Kazdan-Warner problem on S^n. *Bull. Sc. Math.* **119** (1995) 409–418.

[*230] Mabuchi T.— K-energy maps integrating Futaki invariants. *Tôhoku Math. J.* **38** (1986) 575–593.

[*231] Malliavin P.— *Géométrie différentielle stochastique.* Les presses de l'Université de Montréal (1978).

[*232] Marcus M. and Veron L.— Uniqueness of solutions with blowup at the boundary for a class of nonlinear elliptic equations. *C.R. Acad. Sci. Paris* **317** (1993) 559.

[*233] Margerin C.— Pointwise pinched manifolds are space forms. *Proc. Symp. Pure Math.* **44** (1986) 307–328.

[*234] Margerin C.— Un théorème optimal pour le pincement faible en dimension 4. *C.R. Acad. Sci. Paris* **303** (1986) 877–880.

[*235] Milgram A. and Rosenbloom P.— Harmonic forms and heat conduction. *Proc. Natl. Acad. Sci. USA* **37** (1951) 180–184.

[*236] Milnor J.— *Morse Theory.* Princeton Univ. Press, Princeton, New Jersey (1963).

[*237] Milnor J.— *Lectures on the h-cobordism theorem.* Princeton Univ. Press, Princeton, New Jersey (1965).

[*238] Milnor J.— *Topology from the differentiable viewpoint.* Univ. of Virginia Press, Charlottes-ville, Virginia (1965).

[*239] Mok N-M.— An embedding theorem of complete Kähler manifolds of positive bisectional curvature onto affine algebraic varieties. *Bull. Soc. Math. Fr.* **112** (1984) 197–258.

[*240] Mok N-M.— Compactification of complete Kähler surfaces of finite volume satisfying certain conditions. *Ann. Math.* **128** (1989) 383–425.

[*241] Mok N-M. and Zhong J-Q.— Compactifying complete Kähler manifolds of finite topological type and bounded curvature. *Ann. Math.* **129** (1989) 417–470.

[*242] Morrey C. Jr.— The problem of Plateau on a riemannian manifold. *Ann. Math.* **49** (1948) 807–851.

[*243] Morrey C. Jr.— *Multiple integrals in the Calculus of Variations.* Springer, New York (1966).

[*244] Moser J.— On Harnack's theorem for elliptic differential equations. *Commun. Pure Appl. Math.* **14** (1961) 577–591.

[*245] Moser J.— *On a nonlinear problem in differential geometry. In Dynamical Systems* (M. Peixoto ed.). Academic Press, New York (1973).

[*246] Müller S.— Det = det $-A$ remark on the distributional determinant. *C.R. Acad. Sc. Paris* **311** (1990) 13–17.

[*247] Myers S.— Riemannian manifolds with positive mean curvature. *Duke Math. J.* **8** (1941) 401–404.

[*248] Nadel A.— Multiplier ideal sheaves and Kähler-Einstein metrics of positive scalar curvature. *Ann. Math.* **132** (1990) 549–596.

[*249] Nadel A.— On complex manifolds which can be compactified by adding finitely many points. *Invent. Math.* **101** (1990) 173–189.

[*250] Naito H.— Asymptotic behavior of solutions to Eells-Sampson equations near stable harmonic maps. *Math. Z.* **202** (1989) 383–389.

[*251] Nash J.— C^1 isometric imbeddings. *Ann. Math.* **60** (1954) 383–396.

[*252] Nash J.— The imbedding problem for Riemannian manifolds. *Ann. Math.* **63** (1956) 20–63.

[*253] Nash J.— Continuity of solutions of parabolic and elliptic differential equations. *Am. J. Math.* **80** (1958) 931–953.

[*254] Nash J.— Positive Ricci curvature on fiber bundles. *J. Differ. Geom.* **14** (1979) 251–254.

[*255] Ni W.M.— Some minimax principles and their applications in nonlinear elliptic equations. *J. Anal. Math.* **37** (1980) 248–275.

[*256] Ni W.M.— On the elliptic equation $\Delta u + k(x)u^{(n+2)/(n-2)} = 0$, its generalizations and applications in Geometry. *Indiana Univ. Math. J.* **31** (1982) 493–529.

[*257] Nirenberg L.— Variational and topological methods in nonlinear problems. *Bull. Am. Math. Soc.* **4** (1981) 267–302.

[*258] Nirenberg L.— Variational methods in nonlinear problems. *Lectures Notes in Math.* **1365**. Springer, Berlin (1989).

[*259] Nishikawa S.— Deformation of Riemannian metrics and manifolds with bounded curvature ratios. *Proc. Symp. Pure Math.* **44** (1986) 343–352.

[*260] Obata M.— Certain conditions for a Riemannian manifold to be isometric with a sphere. *J. Math. Soc. Japan* **14** (1962) 333–340.

[*261] Onofri E.— On the positivity of the effective action in a theory of random surface. *Commun. Math. Phys.* **86** (1982) 321–326.

[*262] Ouyang T.— On the positive solutions of semilinear equations $\Delta u + \lambda u + hu^p = 0$ on compact manifolds. Part II. *Indiana Univ. Math. J.* **40** (1991) 1083–1141.

[*263] Peter S.— Convergence of Riemannian manifolds. *Compositio Math.* **62** (1987) 3–16.

[*264] Pluzhnikov I.— Harmonic mappings of riemannian surfaces and foliated manifolds. *Math. USSR* **Sb 41** (1982) 281–287.

[*265] Pogorzelski W.— Etude de la solution fondamentale de l'équation parabolique. *Ric. Mat.* **5** (1956) 25–56.

[*266] Pogorzelski W.— Propriétés des intégrales de l'équation parabolique normale. *Ann. Pol. Mat.* **4** (1957) 61–92.

[*267] Pohozaev S.I.— Eigenfunctions of the equation $\Delta u + \lambda f(u) = 0$. *Doklady* **165** (1965) 1408–1410.

[*268] Pucci C.— Limitazioni per soluzioni di equazioni ellittiche. *Ann. Mat. Pura Appl.* **74** (1966) 15–30.

[*269] Pucci P. and Serrin J.— The structure of the critical set in the mountain pass theorem. *Trans. Am. Math. Soc.* **299** (1987) 115–132.

[*270] Rabinowitz P.— Théorie du degré topologique et applications à des problèmes aux limites nonlinéaires. Lectures Notes, Anal. Num. Fonct. Univ. Paris VI (1975).

[*271] Rabinowitz P.— Minimax methods in critical point theory with applications to differential equations. *Conf. Board of Math. Sci. Reg. Conf. Ser. Math.* **65** A.M.S. (1986).

[*272] Ratto A., Rigoli M., and Veron L.— Scalar curvature and conformal deformation of hyperbolic space. *J. Funct. Anal.* **121** (1994) 15–77.

[*273] Rauzy A.— Courbure scalaire des variétés d'invariant conforme négatif. *C.R. Acad. Sci. Paris* **316** (1993) 273–276; *Trans. Am. Math. Soc.* **347** (1995) 4729–4745.

[*274] Rauzy A.— Multiplicité pour un problème de courbure scalaire prescrite. *Bull. Sc. Math.* **120** (1996) 153–194.

[*275] Real C.— Métriques d'Einstein-Kähler sur les variétés à première classe de Chern positive. *C.R. Acad. Sc. Paris* **322** (1996) 461.

[*276] Rey O.— Proof of two conjectures of H. Brezis and L.A. Peletier. *Manuscripta Math.* **65** (1989) 19–37.

[*277] Rey O.— The role of the Green's function in a nonlinear elliptic equation involving the critical Sobolev exponent. *J. Funct. Anal.* **89** (1990) 1–52.

[*278] Rivière T.— Applications harmoniques de B^3 dans S^2 partout discontinues. *C.R. Acad. Sci. Paris* **314** (1992) 719–723.

[*279] Sard A.— The measure of the critical values of differentiable maps. *Bull. Am. Math. Soc.* **48** (1942) 883–890.

[*280] Schoen R.— Conformal deformation of a Riemannian metric to constant scalar curvature. *J. Differ. Geom.* **20** (1984) 479–495.

[*281] Schoen R.— Variational theory for the total salar curvature functional for Riemannian metrics and related topics. Topics in Calculus of Variations. *Lectures Notes in Math.* **1365**. Springer, Berlin (1989).

[*282] Schoen R.— *On the number of constant scalar curvature metrics in a conformal class.* Differential Geometry: A symposium in honor of Manfredo Do Carmo (H.B. Lawson and K. Tenenblat, eds). Wiley (1991) 311–320.

[*283] Schoen R.— A report on some recent progress on nonlinear problems in Geometry. *J. Differ. Geom. Suppl.* **1** (1991) 201–241.

[*284] Schoen R. and Uhlenbeck K.— A regularity theory for harmonic maps. *J. Differ. Geom.* **17** (1982) 307–335.

[*285] Schoen R. et Uhlenbeck K.— Regularity of minimizing harmonic maps into the sphere. *Invent. Math.* **78** (1984) 89–100.

[*286] Schoen R. and Uhlenbeck K.— Boundary regularity and the Dirichlet problem for harmonic maps. *J. Differ. Geom.* **18** (1983) 253–268.

[*287] Schoen R. and Yau S.T.— Harmonic maps and the topology of stable hypersurfaces and manifolds of non-negative Ricci curvature. *Comment. Math. Helv.* **51** (1976) 333–341.

[*288] Schoen R. and Yau S-T.— On the proof of the positive mass conjecture in general relativity. *Comment. Math. Phys.* **65** (1979) 45–76.

[*289] Schoen R. and Yau S-T.— Conformally flat manifolds, Kleinian groups and scalar curvature. *Invent. Math.* **92** (1988) 47–71.

[*290] Simon L.— Asymptotics for a class of non-linear evolution equations, with applications to geometric problems. *Ann. Math.* **118** (1983) 525–571.

[*291] Siu Y-T.— The existence of Kähler-Einstein metrics on manifolds with positive anticanonical line bundle and a suitable finite symmetry group. *Ann. Math.* **127** (1988) 585–627.

[*292] Skoda H.— Sous-ensembles analytiques d'ordre fini ou infini dans $\mathbb{C}^n$. *Bull. Soc. Math. Fr.* **180** (1972) 353–408.

[*293] Smoller J.— *Schock waves and reaction diffusion equations.* Grundlehren **258**. Springer, New York (1982).

[*294] Sternberg S.— *Lectures on Differential Geometry.* Prentice-Hall, Englewood Cliffs, N.J. (1964).

[*295] Struwe M.— On the evolution of harmonic mappings of riemanian surfaces. *Comment. Math. Helv.* **60** (1985) 558–581.

[*296] Struwe M.— *The evolution of harmonic maps.* International congress of Mathematicians, Kyoto/Japan 1990. Springer (1991).

[*297] Tang J.— Prescribing curvature on manifolds with nonpositive Yamabe invariant and singularities. *Commun. Partial Differ. Equations* **19** (1994) 701–718.

[*298] Tang J.— Solvability of the equation $\Delta_g u + \tilde{S}u^\sigma = Su$ on manifolds. *Proc. Am. Math. Soc.* **121** (1994) 83–92.

[*299] Temam R. and Wang S.— Inertial forms of Navier Stokes equations on the sphere. *J. Funct. Anal.* **117** (1993) 215–242.

[*300] Tian G.— On Kähler-Einstein metrics on certain Kähler manifolds with $C_1(M) > 0$. *Invent. Math.* **89** (1987) 225–246.

[*301] Tian G.— A Harnack type inequality for certain complex Monge-Ampère equations. *J. Differ. Geom.* **29** (1989) 481–488.

[*302] Tian G.— On Calabi's conjecture for complex surfaces with positive first Chern class. *Invent. Math.* **101** (1990) 101–172.

[*303] Tian G. and Yau S-T.— Kähler–Einstein metrics on complex surfaces with $C_1 > 0$. *Comment. Math. Phys.* **112** (1987) 175–203.

[*304] Tian G. and Yau S-T.— Complete Kähler manifolds with zero Ricci curvature II. *Invent. Math.* **106** (1991) 27–60.

[*305] Tian G. and Yau S-T.— Existence of Kähler–Einstein metrics on complete Kähler manifolds and their applications to algebraic geometry. *Proc. Conf. San Diego. Adv. Ser. Math. Phys.* **1** (1987) 574–629.

[*306] Tian G. and Yau S-T.— Complete Kähler manifolds with zero Ricci curvature II. *Invent. Math.* **106** (1991) 27–60.

[*307] Uhlenbeck K.— Harmonic maps; a direct method in the calculus of variations. *Bull. Am. Math. Soc.* **76** (1970) 1082–1087.

[*308] Uhlenbeck K.— Morse theory by perturbation methods with applications to harmonic maps. *Trans. Am. Math. Soc.* **267** (1981) 569–583.

[*309] Vaugon M.— Equations différentielles nonlinéaires sur les variétés riemanniennes compactes I et II. *Bull. Sci. Math.* **106** (1982) 351–367; **107** (1983) 371–391.

[*310] Vaugon M.— Transformation conforme de la courbure scalaire sur la sphère. *Ann. Inst. Henri Poincaré* **3** (1986) 55–65.

[*311] Vaugon M.— Transformation conforme de la courbure scalaire sur une variété riemannienne compacte. *J. Funct. Anal.* **71** (1987) 182–194.

[*312] Vazquez J.L. et Véron L.— Solutions positives d'équations elliptiques semi-linéaires sur des variétés Riemanniennes compactes. *C.R. Acad. Sci. Paris* **312** (1991) 811–815.

[*313] Véron L.— Semilinear elliptic equations with uniform blow-up on the boundary. *J. Anal. Math.* **59** (1992) 231–250.

[*314] Véron L.— Singularities of solutions of second order quasilinear equations. (to appear)

[*315] Webster S.— Pseudohermitian structures on a real hypersurface. *J. Differ. Geom.* **13** (1978) 25–41.

[*316] White B.— Homothopy classes in Sobolev spaces and energy minimizing maps. *Bull. Am. Math. Soc.* **13** (1985) 166–168.

[*317] White B.— Nonunique tangent maps at isolated singularities of harmonic maps. *Bull. Am. Math. Soc.* **26** (1992) 125–129.

[*318] Witten E.— A new proof of the positive energy theorem. *Commun. Math. Phys.* **80** (1981) 381–402.

[*319] Xu X. and Yang P.— Remarks on prescribing Gauss curvature. *Trans. Am. Math. Soc.* **336** (1993) 831–840.

[*320] Ye R.— Global existence and convergence of Yamabe flow. *J. Differ. Geom.* **39** (1994) 35–50.

[*321] Yeung S.— Compactification of Kähler manifolds with negative Ricci curvature. *Invent. Math.* **106** (1991) 13–25.

Subject Index

absolutely continuous 77
adjoint operator 27, 75, 126
admissible function 253
Alexandroff–Bakelman–Pucci theorem 139
almost everywhere 76
a priori estimate 257, 263, 266, 270, 273, 289
arc length 5
Ascoli's theorem 74
asymptotically flat 164
atlas
 definition 1
 classe C^k 1
 equivalent 1

Banach
 space 70
 –Steinhaus theorem 73
 's theorem 74
Berger's problem 191
best constants 43, 58, 60, 65, 68
 definition 99
 in Sobolev spaces 41, 100, 139, 140, 153, 236
Betti numbers 29, 30, 255
Bianchi's identities 6
bifurcation theory 136
blow-up 142, 224
boundary 25
bracket 2

Calabi's conjecture 255, 263
Calderon–Zygmund inequality 90
Cauchy's theorem 72
chart 1
Cheeger's theorem 124
Chern class 252
Cherrier's problem 180
Christoffel symbols 3
Clarkson's inequalities 89
closed (see differentiable)
co-differential 27
cohomology class
 definition 252
 positive definite 255
compact mapping 74
complex manifold 251
concircular curvature tensor 335, 346
conformal normal coordinates 158, 159
conjugate point 17, 18
connection 3
 Riemannian 6
continuity method 257, 258, 269, 270, 290
convex
 ball 11, 19
 function 291, 306
  inequality for 309
 set 289
convolution product 89
coordinates 1
 geodesic 9
 normal 7
covariant derivative 4
covering manifold 24
critical point 40
CR structure 182
curvature
 definition 3
 Ricci 7, 252, 346
 scalar 7, 145, 179, 194, 196
 sectional 7, 321
 tensor 4, 6, 252
cut locus 14, 18

degenerate 40
degree (Leray–Schauder) 135
density problem 33 → 35
diffeomorphism 72
differentiable 71
 form 2
  closed 28

coclosed 28
co-exact 28
exact 28
harmonic 28
homologous 28
manifold 1
mapping 1
differential operator 83, 126
elliptic 83, 84, 125
leading part of 83
linear 83, 125
strongly elliptic 126
uniformly elliptic 83
Dirichlet problem 289, 290, 314
distance 5
divergence form 85
dual space 70

eigenfunction 101, 102
eigenvalue
definition 75, 102
of the Laplacian 31, 101, 102
eigenvector 75
Einstein–Kähler metric 254, 256, 259
Einstein metric 7
elliptic (see differential operator and linear)
elliptic equation (definition) 125
energy at P 168
of a functional 247
of a map 348
equicontinuous 74
equivariant Yamabe problem 187
Euler equation 42, 102, 103, 148, 192, 193
Euler–Poincaré characteristic χ 29, 191
exact (see differentiable) 28
exceptional case 63, 68
exponential mapping 9
exterior
differential p-form 2
differentiation 3

Fatou's theorem 77
finiteness theorem (Cheeger) 124
first Chern class 252, 255
first eigenvalue 115, 116, 190
first fundamental form 251
fixed point method 74, 134
form (see differentiable)
formal adjoint 84
Fredholm
alternative 126
mapping 132
nonlinear theorem 236, 237
theorem 75
Fubini's theorem 78
functional 105, 146, 150, 191, 306, 315

Gauss–Bonnet theorem 191
geodesic 8, 11
coordinates 9
global scalar product 28
Green's formula 107
Green's function 120, 122, 123, 162, 163, 167, 168
definition 107
existence properties 108, 112
Gromov's compactness theorem 125
Gromov's isoperimetric inequality 119

Hahn–Banach theorem 73
harmonic (see differentiable and spherical)
harmonic map 349
weakly harmonic map 350
Harnack inequality 127, 275
Hausdorff dimension 357
heat operator 129
solution 130
Hermitian metric 251
Hilbert space 70
Hodge decomposition theorem 29
Hölder's inequality 88
homeomorphism 72
homologous 28
Hopf–Rinow theorem 13

imbedding 1
immersion 1
implicite funtion theorem 72
improvement of the best constants 57, 68
index
form 18
inequality 19
inequality
Clarkson's 89
Hölder's 88
index 19
interpolation 93
isoperimetric 40
Minkowski 309
optimal 50
plurisubharmonic function 316, 317

inequality (Alexandrov–Bakelman–Pucci) 128
injectivity radius 15, 124, 125
inner product 70
integrable 76
integration over Riemannian manifolds 23, 24, 29
interpolation inequalities 93
inverse function theorem 72
isometry – concentration 134, 214
isoperimetric function 119
isoperimetric inequality 40, 119

Jacobi field 17
Jacobian matrix 71

Kähler
  Einstein metric (see Einstein–Kähler) 254, 256, 259
  manifold 251, 252
  metric 251, 253
Kondrakov theorem 53, 55
Kronecker's symbol 4
Korn–Lichtenstein theorem 90

Laplacian 27, 106
leading part 83
Lebesgue
  integral 75, 77
  measure 77
  theorems 76, 77
length
  arc 5
  minimizing 11
  second variation 15
Leray–Schauder degree 135
Leray–Schauder theorem 74
Lichnerowicz's theorem 115
linear
  elliptic equation 101, 113
    existence of solution for 104, 105, 113
  mapping 70
  tangent mapping 2
locally conformally flat manifold 117, 160
lower solution (see subsolution) 97
  method of 198, 200, 202, 267
L_p spaces 78

manifold
  complete 13, 14, 16, 45
  complex 251
  definition 1
  differentiable 1
  Riemannian 4
  with boundary 25
mapping
  differentiable 1
  rank 1
mass 165
maximum principle 96, 97, 138
  for parabolic equations 130
  generalized 98, 330
  in narrow domains 138
  second part 139
  when $u \leq 0$ 139
mean value theorem 71
measurable 76
measure 76
methods 134 → 138
metric
  Einstein 7
  hermitian 251
  Kähler 251
  Riemannian 4
  space 5
  tensor, estimates on 20
minimax methods 134
minimizing curve 11
Monge–Ampère equation 257, 269, 314
Moser's theorem 65, 209, 232
moutain pass lemma 134
moving planes method 136, 137, 138
Myer's theorem 16

Nash imbedding theorem 123
Neumann problem 88
Nirenberg's problem 230
norm 70
normal coordinates 7
normed space 70

Obata's theorem 176
open mapping theorem 73
optimal inequalities 50
orientable 23
orientation 23

Palais–Smale condition 134
parabolic equation 132
  linear (solution) 131
  local existence for nonlinear 133
  solution of the Eells–Sampson equation 132

parallel
  displacement 8
  vector field 8
parametrix for the Laplacian 106
plurisubharmonic function 314
  inequality 316, 317
Pohozaev identity 229
point of concentration 134, 141, 218, 219
  definition 215
positive mass 166
precompact 74
principal symbol 125
projective space
  complex 260, 263
  real 23, 209

Rademacker's theorem 77
Radon measure 75, 301, 315
Radon's theorem 81
rank 71
reflexive 70, 81
regularity
  interior 85, 86
  up to the boundary 87
regularity theorem 127
regularization 80
reflexive 70
relaxed energy 360
Ricci
  curvature 7, 322, 323, 346
  form 252
  tensor 7
Riemannian
  connection 6
  manifold 4
  metric 4

Sard theorem 123
scalar curvature 7, 145, 179
Schauder
  fixed point theorem 74
  interior estimates 88
Schouten's tensor 117
second variation of the length integral 15
sectional curvature 7, 321
semi-norm 73
sense of distribution 84
sliding method 138
Sobolev
  imbedding theorem 35, 44, 45, 50
    proof 37 → 39, 46, 47, 49, 51
  lemma 37
  space 32
spectrum 31
spherical harmonic 233, 235, 236
steepest descent 134
subsolution (see lower solution) 97
supersolution (see upper solution and lower solution) 97

tangent
  space 2
  vector 2
Taylor's formula 72
tension field 348, 349
tensor field 2
test functions Aubin's 155, 159
  Escobar's 181
  Hebey and Vaugon's 171
  Schoen's 161
topological methods 172, 213
torsion 3
totally geodesic 349
trace 69
tubular neighborhood 15

uniformly
  convex 81
  locally finite cover 48
upper solution (see supersolution) 97

variational problem 42, 101, 105, 146, 150, 154, 204, 209, 311, 317
vector field 2
volume element
  oriented 26
  Riemannian 30

weak
  convergence 74
  derivative 81, 84
  solution 84, 85
weakly sequentially compact 74
Weyl's tensor 117
Whitney's theorem 4

Yamabe's
  equation 146
  problem 145
  theorem 147, 150
Yamabe problem 145
  functional $J(\varphi)$ 150
Yang–Mills field IX

Notation

Basic Notation

We use the Einstein summation convention.
Compact manifold means compact manifold without boundary unless we say otherwise.
$\mathbb{N}$ is the set of positive integers, $n \in \mathbb{N}$.
$\mathbb{R}^n$: Euclidean n-space $n \geq 2$ with points $x = (x^1, x^2, \ldots, x^n)$ $x^i \in \mathbb{R}$ real numbers.
$\mathbb{C}^m$: Complex space with real dimension $2m$. z^λ $(\lambda = 1, 2, \ldots, m)$ are the complex coordinates $\bar{z}^\lambda = z^{\bar{\lambda}}$.
We often write ∂_i for $\partial/\partial x^i$, ∂_λ for $\partial/\partial z^\lambda$.
$\mathbb{H}_n$: hyperbolic space.

Notation Index

$b_p(M)$ 29
B_r ball of radius r in $\mathbb{R}^n$ generally with center at the origin
$B = B_1$
$B_p(\rho)$ Riemannian ball with center P and radius ρ
$B(P, \delta)$ ball of center P with radius δ
C the circle (or a constant)
$C(M, g)$ Set of conformal transformations 188
$C(S_n) = C(S_n, g_0)$
C^k, C^∞, C^ω differentiable manifold 1
$C^r\ (\overline{W})$, C^r_B 35
$C^{r+\alpha}$ or $C^{r,\alpha}$ 35, 36
$C_0(\mathscr{H})$ 75
$C^p(\Omega, G)$ 71
$C(K)$ or $C^0(K)$ 74
$\mathsf{C}_1(M)$ First Chern Class 252
$D^p f$ 71
$\widehat{dx_j}$ 26
dV 30
d', d'', d^c 251
d 3
$\mathcal{D}(M)$ space of C^∞ functions with compact support in M
$\mathscr{D}(M)$ 32
$D = B \cap \bar{E}$ 35
$d(P, Q)$ distance from P to Q
$E = \{x \in \mathbb{R}^n / x^1 < 0\}$ 35
$\mathcal{E}$ euclidean metric
$E_{ij} = R_{ij} - \frac{R}{n} g_{ij}$ 336, 346
$e(f)$ 348
$E(f)$ 348
$\exp_p(X)$ 9
f'_t. Suppose f is a function of two variables (x, t), then f'_t is the first partial derivative of f with respect to t
g: Riemannian metric 4
g_{ij} the components of g, g^{ij} 5
$|g|$ in real coordinates 26
$|g|$ in complex coordinates 252
$[g]$ conformal class of g
$G(x, y)$ Green function of the Laplacian
G_L Green function of L 156, 161
$\mathsf{G}(P, Q)$ Green's function 108
$GL(\mathbb{R}^n)^+$ 23

$H_k^p(M_n)$ or H_k^p when there is no ambiguity 32
$\overset{\circ}{H}{}_k^p(M_n)$ 32
H_k 33
$\mathsf{H}(P, Q)$ 106
$\mathscr{I}(\varphi)$ 306
$\mathfrak{I}(\varphi)$ 315
$I(M, g)$ group of isometries 187
$J(\varphi)$ the Yamabe functional (often) 150
$\mathsf{K}(n, q)$ 40
$K(n, 2)$ the best constant in the Sobolev inequality $H_1^2 \subset L_N$ 140, 153, 236
$\tilde{K}(n, 2)$ 181
L conformal Laplacian (often) 156, 161
$L_p(\mathscr{H})$ or L_p when there is no ambiguity 78
$\mathscr{L}(F, G)$ 70
M_n (or M) manifold of dimension n 1
$\mathfrak{M}(\varphi)$ 315
$\mathscr{M}(\varphi)$ 301
$\mathsf{M}(\varphi)$ 253
$M(\varphi)$ 290
(M_n, g) Riemannian manifold 4
MTM 357
N is generally equal to $2n/(n-2)$
$0(P)$ orbit of P
$0_G(P)$ orbit of P under the group G 188
$\mathbb{P}_n(\mathbb{R})$ real projective space
$\mathsf{P}_n(\mathbb{C})$ complex projective space
R_{ikl}^j 4
R_{ijkl} 6
R_{ij}, R 7
(S_n, g_0) the sphere of dimension n of radius 1 endowed with the standard metric g_0
S_f 357
$\mathbb{S}_n(\rho)$ sphere of dimension n and radius ρ
supp φ means support of φ
$T_P(M)$, $T(M)$, $T^*(M)$, $T_s^r(M)$ 2
V generally denotes the volume of the manifold when it is compact
$W_{ijk\ell}$ components of the Weyl tensor 117
$Z_{ijk\ell}$ components of the concircular curvature tensor 335
$|Z|^2 = Z_{ijk\ell}Z^{ijk\ell}$
$\alpha(M)$ 280
$\alpha_G(M)$ 281
$\Gamma(P, Q)$, $\Gamma_k(P, Q)$ 109
$\Gamma_1(M)$ 3
Γ_{ik}^j Christoffel symbols 3
Δ Laplacian Operator 27, 28
Δ laplacian (in chapter 9: $\Delta = -\nabla^i\nabla_i$ is the rought Laplacian)
$\Lambda^p(M)$ 3
Φ_*, Φ^* 2
Ψ 252
(Ω, φ) a local chart 1
δ codifferential 27
δ', δ'' 253
δ_j^k Kronecker's symbol 4
∂_n, ∂_ν normal derivative oriented to the outside (often)
μ_0 239
μ inf $J(\varphi)$ (often) 150
(or in chapter 6: $\mu = \lim \mu_{q_j}$) 222
$\overline{\mu}$ 189
$\mu(A)$ 76
μ_S 222
η 26
λ_1 first nonzero eigenvalue λ_i 31
$\chi(M)$ (or χ if there is no ambiguity) Euler–Poincaré characteristic 29
χ_E characteristic function of E 41
ω_n volume of $\mathbb{S}_n(1)$
ω 251
ω^m 254
$\tau(f)$ 348

Other symbols

$\| \ \|_0$ norm in C or C°
$\| \ \|_p$ with $p \geq 1$ norm in L_p 78
$\| \ \|_{H_k^p}$ 32
$\nabla_i Y$ 3
$\nabla_{\alpha_1\alpha_2\cdots\alpha\ell}$ 4
$|\nabla^k\varphi|$, $\nabla^k\varphi$ 32
$\partial\Omega$ the boundary of Ω
$\oslash$ empty set
$\left(\dfrac{\partial}{\partial x^i}\right)_P$ tangent vector at P 2
$[X, Y]$ bracket 2
$*$ adjoint operator 27
$\dfrac{\partial u}{\partial \nu}(x_0)$ 96

Convention

positive	means strictly positive
negative	means strictly negative
non-positive	means negative or zero
non-negative	means positive or zero
Compact manifold	means compact manifold without boundary unless we say otherwise
$a_{ij} \geq 0$	(resp. $a_{ij} > 0$) for a bilinear form means $a_{ij}\xi^i\xi^j \geq 0$ for any vector ξ (resp. $a_{ij}\xi^i\xi^j > 0$ for any vector $\xi \neq 0$).

Printing: Mercedesdruck, Berlin
Binding: Buchbinderei Lüderitz & Bauer, Berlin